Applied Optimal
Control Theory of
Distributed Systems

MATHEMATICAL CONCEPTS AND METHODS IN SCIENCE AND ENGINEERING

Series Editor: **Angelo Miele**
Mechanical Engineering and Mathematical Sciences
Rice University

Recent volumes in this series:

A Continuation Order Plan is available for this series. A continuation order will bring delivery of each new volume immediately upon publication. Volumes are billed only upon actual shipment. For further information please contact the publisher.

Applied Optimal Control Theory of Distributed Systems

K. A. Lurie

Worcester Polytechnic Institute
Worcester, Massachusetts

Plenum Press • New York and London

Library of Congress Cataloging-in-Publication Data

Lurie, K. A.
    Applied optimal control theory of distributed systems / K.A.
Lurie.
        p.   cm. -- (Mathematical concepts and methods in science and
engineering ; 43)
    Includes bibliographical references and index.
    ISBN 0-306-43993-X
    1. Control theory. 2. Mathematical optimization. 3. Mathematical
physics.   I. Title.   II. Series.
QA402.3.L87   1993
003'.78--dc20   92-39594
                                                                    CIP

ISBN 0-306-43993-X

© 1993 Plenum Press, New York
A Division of Plenum Publishing Corporation
233 Spring Street, New York, N.Y. 10013

Printed in the United States of America

Contents

1

Introduction

Many classical problems of mathematical physics can be formulated as extremal problems for certain functionals. Such a formulation allows a natural definition of solutions of the corresponding boundary-value problem (generalized solutions); in addition, the formulation is well-suited to the use of direct methods of the calculus of variations.

Isoperimetric problems are the traditional extremal problems of the calculus of variations. However, the calculus of variations also deals with the minimization of functionals with constraints of a more complicated type, e.g., those expressed by means of inequalities, differential equations, integral equations, and other functional equations. Although many such problems (Lagrange, Mayer, Bolza) were formulated long ago, their systematic investigation has begun only recently, stimulated by the increasing demands of practical applications.

These problems have become known as optimal control problems. Every problem of optimal control is characterized by three main features.

1. First, one is given state equations and boundary conditions that describe the behavior of a system. The nature of the state equations and of the boundary conditions is determined by the mathematical model adopted. In all cases, however, one can identify a group of dependent variables (state variables) and a group of control functions (controls), belonging to specified function classes and taking on values in definite sets.

2. The specification of these sets, the sets of admissible controls, is the second necessary element of every optimization problem.

3. Finally, a functional of the state variables and the controls is specified. The optimization problem is to find, from the set of admissible controls, those that minimize the functional. Controls that have this property are called *optimal controls*. Thus, the optimal control problem is the problem of finding the maximal working possibilities of a controlled system.

Most investigations in optimal control are devoted to the case where the behavior of the controlled system is represented by a model with lumped parameters; mathematically, such a model is described by ordinary differential equations. This scheme encompasses many optimization problems in different fields. The mathematical problems of the optimal control of systems

1

with lumped parameters have been fully developed. Important results have been obtained in this field*; among them, Pontryagin's maximum principle,[133] which is the basic necessary condition for a strong relative minimum. The maximum principle can be formulated with special simplicity if state differential equations are expressed in the Cauchy normal form; this is almost always possible and is generally done in investigations of optimal control with one independent variable.

In many applications, models with lumped parameters do not describe adequately a particular phenomenon. Frequently, it is found that a system which is optimal in the case of a simplified model is nonoptimal if the process is described more accurately; this means that one is not fully exploiting the possibilities of control inherent in the system. A better and more adequate description might be achieved within the framework of distributed parameter models; mathematically, these are described by partial differential equations. These then give rise to some corresponding optimization problems.

The systematic investigation of optimization problems of mathematical physics began about 25 years ago. In most of the published investigations, the state equations and the boundary conditions have been such that the control appears in a fairly simple manner (on the right-hand sides of linear and quasi-linear equations of higher order, as the coefficients of the lower derivatives of linear differential operators, as terms in linear boundary conditions, etc.). Regardless of the type of equation obtained for such problems, the necessary conditions for optimality (such as Pontryagin's maximum principle) appear as a direct generalization of the corresponding conditions for optimization problems with one independent variable. The results achieved in this direction are indisputable; but it must be pointed out that they are due to the comparative simplicity of the dependence of the system operators on the control functions. Moreover, they do not reveal important features that are inherent in optimization problems for distributed parameter systems and which have no analogues in one-dimensional optimization problems. The study of a number of such features is undertaken in this book.

In their most pronounced form, these new features appear in problems that contain the control in the principal part of the differential operator characterizing the state equation. In physical applications, we are dealing with problems of the optimal control of the properties of continuous media. The derivation of necessary conditions (such as those embodied in Pontry-

---

* The reviews of Krasovskii[82] and Gabasov and Kirillova[41] give a detailed account of the work done in this field in the Soviet Union.

agin's maximum principle) is considerably difficult for such problems. The following discussion concerns these difficulties.

Pontryagin's maximum principle is a necessary condition for a minimum with respect to strong local variations of the control. In other words, in the derivation of the maximum principle, the optimal control is compared with admissible controls that differ from the optimal control on sets of small Lebesgue measure. Within these sets, the increment of the control is by no means small and is determined solely by the restrictions imposed on the control. For problems with one independent variable (ordinary differential equations), a small interval of variation of the independent variable (or a finite number of such intervals of small measure) represents such a set. Essentially, the maximum principle is the condition that the principal part of the increment of the functional due to a strong local variation of the control be nonnegative.

In attempting to extend the maximum principle to problems involving partial differential equations, we must deal with the following question: What is to be the analogue of the small interval for such problems? In the case of two independent variables, for example, there exists a nondenumerable set of small, simply connected regions that differ in shape. Therefore, it is necessary to consider whether or not the necessary conditions of the type of the maximum principle depend on the shape of the small region in which the control is varied. The answer to this question depends on the way in which the control functions occur in the state equations. It turns out that, for the comparatively simple cases mentioned above (control of the type of a source on the right-hand sides of the equations, control which is a coefficient of lower derivatives, etc.), the necessary conditions *do not* depend on the shape of the region of variation. In more complicated cases, *there is* such a dependence. These include problems containing the control in the principal part of the linear differential operators occurring in the state equation. In these problems, regions of variation with different shapes can lead to necessary conditions having differing degrees of generality.

In each particular problem, it is important to know whether the necessary conditions for a strong relative minimum depend on the shape of the region of local variation or not. One can give a very general procedure for determining the answer to this question. This procedure consists of writing the basic equations of the problem in a standard form, which is analogous to the Cauchy normal form. Any system of partial differential equations can be written in the form

$$\frac{\partial z^j}{\partial x_i} = X_i^j(x, z, \zeta, u) \qquad (i = 1, \ldots, m, j = 1, \ldots, n) \qquad (1.1)$$

Here, $z^1, \ldots, z^n$ are the dependent variables and $x_1, \ldots, x_m$ the independent variables; u denotes the control functions. The system (1.1) is solved for the derivatives of all of the dependent variables with respect to all of the independent variables. It is important to note that, when $m > 1$, the system (1.1) contains not only x, z, and u, but also the so-called *parametric* variables ζ; it is these variables which allow one to satisfy integrability conditions for the system. In contrast to the variables u, the variables ζ are not specified, but are determined in the process of solving optimization problems. The dependent variables z are taken to be continuous,* whereas the parametric variables ζ and the controls u are generally discontinuous functions of the independent variables.

The structure of the right-hand sides of the equations in their standard form shows whether the necessary conditions for a minimum depend on the shape of the region of variation or not. At the same time, the standard form of the equations enables one to give a unified derivation of the necessary conditions of optimality for problems of various types; in the general case, the local variation is performed in such a way that one obtains the most accurate expression for the principal part of the increment of the functional. This then leads to the strongest necessary conditions which may be generated by local variations.

Let us elucidate the meaning of this assertion with the following example: to determine the smallest electrical resistance of a simply connected planar region G filled with a substance of resistivity $\rho(x, y)$ in the presence of distributed electromotive forces. The boundary of the region is assumed to be insulated everywhere except for two sections, the electrodes, which are connected through a load R (Fig. 1.1). The distribution of the current $\mathbf{j} = -\mathrm{curl}\,(\mathbf{i}, z^2)$ and the potential z^1 of the electric field are described by the equations

$$\mathrm{div}\,\mathbf{j} = 0, \qquad \rho\mathbf{j} = -\mathrm{grad}\,z^1 - \mathbf{E}_0 \qquad\qquad (1.2)$$

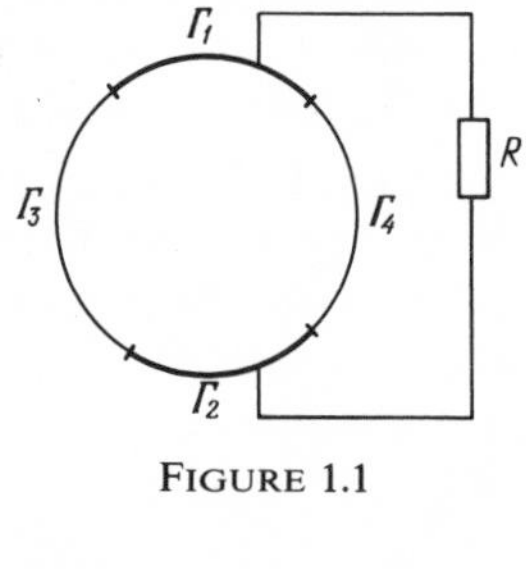

FIGURE 1.1

---

* This assumption remains valid for a vast class of problems; regarding more general cases, see Section 6.1.

or, in standard form,

$$z_x^1 = -\rho\zeta^1 - E_1, \qquad z_y^1 = -\rho\zeta^2 - E_2$$

$$z_x^2 = \zeta^2, \qquad\qquad z_y^2 = -\zeta^1 \tag{1.3}$$

Here, $\mathbf{E}_0(E_1, E_2)$ is a given vector function of position (an external field). The problem is to choose, from the bounded measurable functions $\rho(x, y)$ satisfying the inequalities

$$0 < \rho_{min} \leq \|\rho(x, y)\|_{L_\infty(G)} \leq \rho_{max} \leq \infty \tag{1.4}$$

the optimal control $\rho(x, y)$ that maximizes the functional (the power released across the load)

$$N = \left[\int_{\Gamma_1} (\mathbf{j}, \mathbf{n}) \, dt\right]^2 R = I^2 R \tag{1.5}$$

where the integration is along the electrode Γ_1 with normal $\mathbf{n}$.

In this problem, it is readily seen that the control ρ occurs in the principal part of the basic differential operator. The increment ΔN of the functional N from its optimal value, generated by the admissible increment $\Delta\rho$ of the control, is given by (see Section 2.3)

$$\Delta N = -\iint_G \Delta\rho(\mathbf{J}, \mathbf{k}) \, dx \, dy \tag{1.6}$$

where $\mathbf{J}(Z^1, Z^2) = (\zeta^1 + \Delta\zeta^1, \zeta^2 + \Delta\zeta^2)$ is the vector of the admissible current density corresponding to the control $P = \rho + \Delta\rho$, and $\mathbf{k}(-\omega_{1y}, \omega_{1x})$ is a vector determined from the boundary-value problem

$$\rho\omega_{1y} - \omega_{2x} = 0, \qquad \rho\omega_{1x} + \omega_{2y} = 0$$

$$\omega_2\big|_{\Gamma_1, \Gamma_2} = \omega_{2\pm} \qquad \text{at electrodes } \Gamma_1, \Gamma_2$$

$\omega_1 = \omega_{1+}$ on the right-hand insulating wall Γ_4 (see Fig. 1.1), $\omega_1 = \omega_{1-}$ on the left-hand insulating wall Γ_3; $\omega_{2+} - \omega_{2-} + 2IR = (\omega_{1+} - \omega_{1-})R$.

It can be assumed that the boundary-value problem determines the potential ω_2 of a fictitious electric field and a fictitious current of density $\mathbf{k}$ for the optimal function $\rho(x, y)$. The source of this fictitious potential and current is the electromotive force $2IR$ in the external circuit.

The expression (1.6) contains not only the increment $\Delta\rho$ of the control, but also the increments $\Delta\zeta^1$ and $\Delta\zeta^2$ of the parametric variables. This circumstance distinguishes a problem of this type from the simpler problems mentioned above. In the simpler cases, the increment of the functional depends only on the increment of the control, and not on increments of the parametric variables. In contrast to $\Delta\rho$, the increments $\Delta\zeta^1$ and $\Delta\zeta^2$ are not determined directly from outside; therefore, to deduce the sign of the increment ΔN from Eq. (1.6), it is necessary to eliminate $\Delta\zeta^1$ and $\Delta\zeta^2$, by expressing them in terms of $\Delta\rho$ and in terms of quantities defined for the optimal regime. The increments $\Delta\zeta^1$ and $\Delta\zeta^2$ cannot be ignored, if one is dealing with strong (not small) variations $\Delta\rho$: where $\Delta\rho \neq 0$, the increments $\Delta\zeta^1$ and $\Delta\zeta^2$ are of order $\Delta\rho$ and can therefore have a strong influence on the sign of (1.6).

If $\Delta\rho \neq 0$ everywhere in G, then such an elimination cannot be carried out in the general case. Therefore, we then resort to local variations made in a region G_ε of a small area ε and attempt to calculate the principal part of the vector $\mathbf{J}$ over ε. The increments of the parametric variables that determine this vector depend on the shape of the region of local variation; this leads naturally to the problem of determining the region which is best in this context; namely, that region and corresponding variations which lead to the strongest local necessary conditions.

It can be shown that the best small region is *a narrow strip* that contracts toward its center preserving its shape. The vector $\mathbf{J}$ within the strip can readily be calculated in terms of quantities of order ε; it depends explicitly on the inclination of the strip relative to the vectors $\mathbf{j}$ and $\mathbf{k}$ at its center. The increment ΔN determined by (1.6) must be nonpositive for all orientations of the strip; this requirement yields the most general criterion for a strong relative maximum in the class of local variations of the control with constant value of the admissible control within the region of variation:

$$\rho = \rho_{\min}, \quad \text{when } (\mathbf{j}, \mathbf{k}) \geq 0,\ 0 \leq \chi \leq \arccos p$$

$$\rho = \rho_{\max}, \quad \text{when } (\mathbf{j}, \mathbf{k}) \leq 0,\ \pi \geq \chi \geq \pi - \arccos p$$

Here, χ is the angle between the vectors $\mathbf{j}$ and $\mathbf{k}$, and

$$p = \frac{\rho_{\max} - \rho_{\min}}{\rho_{\max} + \rho_{\min}}$$

Evidently, Weierstrass's condition imposes a restriction on the mutual orientation of the true vector and the fictitious vector of current density.

Weaker results are obtained if the variation is performed within an ellipse; and weakest results follow from variations within a circle.

It is natural to call this phenomenon *anisotropy of the variation*. Physically, it is due to the fact that the dipole moment of the polarization charges which arise on the boundary of the region of variation depends on the shape of the boundary and its orientation with respect to the external field, whose role in the present case is played by the field of the optimal vector **j**. The dependence is most pronounced for inclusions that have the shape of thin bands.

In three dimensions, an analogous role would be played by *an oblate spheroid* (almost a disk), etc.

This result is a generalization of a theorem in geometrical theory of functions, known as *Grötzsch's principle,*[51,264] although the basic idea is actually due to Rayleigh and Maxwell. Namely, we assume that, in the ring $r < |z| < R$, $r \neq 0$, there is a finite number of nonoverlapping, simply connected regions S_k, $k = 1, 2, \ldots, n$, bounded by Jordan curves which have arcs on $|z| = r$ and $|z| = R$, and these arcs do not degenerate into points (Fig. 1.2). The regions S_k are regarded as strips extending from $|z| = r$ to $|z| = R$. If these regions are mapped conformally onto rectangles of the w plane with edges equal to a_k and b_k, respectively, in such a way that the arcs go over into sides of length a_k, then

$$\sum_{k=1}^{n} \frac{a_k}{b_k} \le \frac{2\pi}{\ln \frac{R}{r}}$$

and equality holds only when the S_k are sectorial regions of the form $r < |z| < R$, $\varphi_h < \arg z < \varphi_{h+1}$, which completely fill the ring.

This theorem admits the following physical interpretation. Suppose that the circle $|z| = R$ is at zero potential, and the circle $z = r$ is at potential 1. Suppose further that the regions S_k are filled with a homogeneous isotropic conductor with conductivity $\sigma = 1$, and that the remaining parts of the ring $r < |z| < R$ are filled with an insulating material ($\sigma = 0$). The current which flows through S_k is numerically equal to a_k/b_k; it is natural to call this ratio the conductance of the region S_k. If the entire ring $r < |z| < R$ were

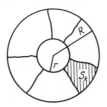

FIGURE 1.2

filled with a conductor with $\sigma = 1$, the current flowing through it would be $2\pi/\ln (R/r)$ (the conductance of the ring).

Grötzsch's theorem states that the introduction of insulating layers does not increase the conductance of a region. This assertion can be generalized in various directions. Above all, it is valid when the conducting sections are replaced by insulators of any shape; in addition, it is not necessary to require that the insulators be perfect; it is sufficient to reduce the conductivity of a section compared with its original value. Finally, it is not necessary to assume that the conductor is homogeneous and isotropic.

These generalizations are physically obvious; the corresponding mathematical proofs may be based on the methods of optimal control theory. The general principle given above is not restricted to two-dimensional current distributions; it holds equally well for any finite number of dimensions of the region. The same result may also be deduced by using the general method.

Grötzsch's theorem ceases to hold if an inhomogeneous distribution of external electromotive forces is admitted within the region. The current density $\mathbf{j}$ and the potential z^1 then satisfy Eqs. (1.3), where $\mathbf{E}_0 = \mathbf{E}_0(x, y)$ is the inhomogeneous external field. Under such conditions, the introduction of insulating inclusions in definite parts of the main region may increase the total current flowing through the electrodes. This is the generalization of Grötzsch's theorem mentioned above. Its derivation makes essential use of the concept of a variation within a strip, which plays the principal role in problems of this kind because of the anisotropy of the variation. On the other hand, there are various problems in which the "anisotropy" of the variation is not manifested (in the problem described above, there is no anisotropy if the variation is performed within a small disk, something which is obvious from a physical point of view).

Of this kind are the "comparatively simple" optimization problems mentioned above; they contain the control on the right-hand sides of the equations, in the coefficients of the lower derivatives of linear operators, etc.

A simple example is given by a problem based on Poisson's equation in the form

$$z_{xx} + z_{yy} = u(x, y)$$

with the right-hand side regarded as a control. If the control has an increment $\Delta u(x, y)$ in a region G_ε of sufficiently small diameter and small measure ε, then the principal part of the increment of the functional of the solution of this problem obviously depends on the increment of the "total charge" $\iint_{G_\varepsilon} \Delta u(x, y)\, dx\, dy$ of the region of variation and is independent of the other characteristics of the region (in particular, its shape in the limit $\varepsilon \to 0$ and its orientation on the plane). An analogous conclusion holds for any finite dimensional space.

The introduction of variations in a strip in order to take into account the anisotropy of the variation makes it possible to give a natural generalization of Pontryagin's maximum principle to a large class of optimization problems involving partial differential equations. The main features of this more general formulation are the need to eliminate increments of the parametric variables from the expression for the increment of the functional and the requirement that the sign of this increment be unchanged for all admissible orientations of the strip (or oblate spheroid).

The elimination procedure can be avoided if the basic equations are represented in local Cartesian coordinates $(\mathbf{n}, \mathbf{t})$, where $\mathbf{n}$ is normal and $\mathbf{t}$ tangent to the strip. Under these conditions, the right-hand sides of the equations contain combinations of the parametric variables which are continuous on the boundary of the strip, and the elimination process becomes unnecessary. However, it is still necessary to consider different orientations of the vectors $(\mathbf{n}, \mathbf{t})$, i.e., rotations of the strip.

In elliptic and parabolic problems, strips of any orientation are in general admitted for comparison. In this respect, hyperbolic problems are an exception; if the control function is the phase velocity of propagation of perturbations, the admissible slope of the variation strip is restricted by certain limits, which are determined by the phase velocities inside and outside the strip. Under this restriction, the corresponding solutions (the dependent variables) are continuous functions with piecewise continuous derivatives. Variation of the control in strips with nonadmissible slopes immediately carries the solution out of this class; strong discontinuities (discontinuities of the first kind) appear *in the solution.* In a number of cases this phenomenon gives rise to new optimization possibilities, shielding entire regions in state space from the penetration of perturbations produced by the initial state. The formation of discontinuities in the solution is due to the "collision of characteristics" that arrive at the line of the discontinuity from different sides and in this sense is analogous to the occurrence of discontinuities in the solutions of nonlinear hyperbolic equations. The reasons for the collision of the characteristics are different in the two cases: in the nonlinear case, it is due to the dependence of the phase velocity on the solution; in the optimization problem, it is due to the influence of the control.

Anisotropy of the variation reflects the physical features of optimization problems. In cases where an optimal control exists, allowance for this phenomenon leads to the strongest local necessary conditions of a strong relative minimum.

The problem of existence of optimal controls is of primary importance for a vast scope of problems concerning optimal distributions of material characteristics of continuous media.

One encounters here the specific features which have no analogues not only for one-dimensional controlled systems, but also for many simpler multidimensional problems of control. These features are connected first of all with the problem of chattering regimes.

Such regimes have long been known[43,447,523] in one-dimensional problems, and corresponding results can certainly be transferred to the multidimensional case. But besides that, in distributed parameter systems, there also arise *multidimensional* chattering regimes; these may disappear if the problem degenerates into a one-dimensional problem, e.g., into a problem possessing axial symmetry.

This idea will again be illustrated by the problem of the smallest electrical resistance of a planar region in the presence of distributed electromotive forces. Assume that we have only two conducting media at our disposal, both media being characterized by constant specific resistances ρ_1 and ρ_2 (with corresponding specific resistance tensors given by $\mathbb{D}_1 = \rho_1 \mathbb{E}$, $\mathbb{D}_2 = \rho_2 \mathbb{E}$, respectively, where $\mathbb{E}$ is the unit tensor). These two conductors now are to be used to assemble the conducting medium in such a way as to maximize the functional (1.5). It is clear that to obtain the required distribution, the flow of the electric current in *some definite favorable direction* should be facilitated as much as possible, whereas the current flow in the perpendicular direction should be blocked; this should be done at *any* point of the region filled in with currents. Now it becomes obvious that the specific resistance sought for at each point *should be direction dependent*; in other words, it should be a tensor function of the coordinates. But how to build the required anisotropic medium from the conducting materials we have at our disposal? In a broad sense, the difficulty which confronts us is that we do not possess conductors with suitable degree of anisotropy; particularly, the given materials may well be isotropic. The way out of this difficulty is to imitate the required medium by constructing a composite from the given isotropic compounds, resulting in a more or less complicated internal microstructure. The simplest one will be a layered microstructure; its effective characteristics (specific resistances) obviously differ along the layers and across them. One may also deal with more complicated microstructures. In many cases, however, such complications will be shown to be unnecessary (see Chapters 3 and 4). Clearly, there is no parameter in the problem which would restrict the dimensions of the microinclusions of given components from below. Therefore, one should expect (and this will be confirmed in what follows) that the optimal value of the functional is achieved under some *infinitely frequent* subdivision of the initially given domain into parts occupied by different components. In the following, references to chattering regimes and media with microstructure will be understood in this sense.

At this point, it is useful to review the techniques applied to the resolution of chattering regimes when they arise in one-dimensional control problems. Consider the controlled system

$$\frac{dx}{dt} = X(t, x, u), \qquad x = (x^1, \ldots, x^n), \qquad X = (X^1, \ldots, X^n)$$

and let the set U of admissible values of the (scalar) control u consist of precisely two points: $u = u_1$ and $u = u_2$. Then, according to Filippov's lemma,[447,523] one introduces an "averaged" (or "relaxed") problem with

$$\frac{dx}{dt} = \lambda X(t, x, u_1) + (1 - \lambda)X(t, x, u_2)$$

where the "concentration" $\lambda = \lambda(t)$, $0 \leq \lambda(t) \leq 1$, is to be determined along with $x(t)$. In the general case of an arbitrary set U, the right-hand side of the relaxed system is restructured to yield the least convex hull of the set $X(t, x, u)$, $u \in U$, for each fixed pair of (t, x).

This result actually provides a solution to the problem of averaging (relaxation) for the one-dimensional case. For the multidimensional case, there is no general procedure of relaxation, and different rules must be derived for the various types of problems. In Chapters 3 and 4, this derivation is performed for a number of cases where the material characteristics of continuous media play the role of controls. For those examples, one has to deal with the effective description of the equivalent characteristics of composites.

The techniques implied by the analogy with "chattering" regimes in the one-dimensional case do not immediately transfer to the relaxation in the multidimensional case. The new element which now arises is the dependence of effective properties of a composite not only upon the concentration, but also upon the form of inclusions of the initially given components. Continuity conditions show that this form substantially influences the jumps of parametric variables across the boundaries of the inclusions.

The effective properties of any composite are characterized by the tensor $\mathbb{D}^0$ connecting the values of the various state variables averaged over some elementary volume of the composite medium, a so-called physically small volume. This volume is taken to be sufficiently small in comparison to some length scale provided by the problem "in the large," i.e., by the distribution of the sources, the form of the basic domain, etc. At the same time, the volume is large enough to include a great number of parts occupied by different compounds belonging to the admissible set U. The state variables mentioned here are the current density and the potential gradient for the current problem discussed before; for the problems of elasticity, these variables represent the stress and strain tensors, respectively, and so on.

The problem of evaluation of $\mathbb{D}^0$ for composites of given microstructure is widely discussed both by mechanicians[422,423,425–427,441] and by mathematicians[403,405,406,421,482,525,526]; the cited papers include vast bibliographies. In these papers an asymptotic procedure is outlined which enables one to calculate the effective properties of a composite possessing some given microstructure. This procedure (called homogenization) requires the solution of certain accessory boundary-value problems which turn out to be fairly complicated and may be solved only numerically for more or less nontrivial microstructures. For our purposes, however, we require an alternative approach.

We possess a certain set U of initially given components at our disposal, these components generally being anisotropic materials themselves. We are interested in determination of effective material properties of the *whole set* of composites which may be obtained from the given ingredients with the aid of the process of mixing. Speaking about the "material properties" we, of course, bear in mind a description of the mentioned set in terms of the invariants of the corresponding tensors $\mathbb{D}^0$.

The necessity of obtaining an invariant description of the whole set of possible composites arises from the fact that we do not know in advance which element of this set will actually participate in the optimal distribution of materials at any point of the region. Therefore, if we know how to describe all possible composites, and, moreover, if we are able to point out at least one (possibly simple) microstructure which would actually represent any admissible composite, then we obtain a set of compound media which is in a sense complete. Having obtained this set at our disposal, we become able to solve the problem of existence of optimal controls for a vast class of functionals.

In order to illustrate this approach mathematically, we introduce some definitions.

Let $\{\mathbb{D}_k\}$, $\mathbb{D}_k \in U$, be a sequence of tensors of material properties related to the initially given set U of materials, and let $\{z_k = z_k(\mathbb{D}_k)\}$ be the corresponding sequence of solutions (e.g., flow functions, displacements, and so on). The sequence $\{z_k\}$ may be treated as bounded in the corresponding functional space, and consequently, as weakly convergent to the element z_0 of that space, i.e.,

$$z_k \rightarrow z_0$$

The element z_0, however, need not be generated (in the sense of the basic boundary-value problem) by some material of the initial set U. It thus becomes reasonable to introduce some *extended* set of materials, a set including (beside U) the elements which generate the weak limits z_0 of *all sequences* $\{z_k\}$. The set of materials obtained in this way from the initial set U will be designated GU and called the *G-closure* of U. The tensor $\mathbb{D}^0$ which generates the solution z_0 (the weak limit of $\{z_k\}$) will be called the G-limit of the sequence $\{D_k\}$. Evidently, $U \in GU$.

Now, so long as the materials from GU are admissible, we are free to prove the existence of the elements $\mathbb{D}^0 \in GU$ generating extremal points z of any *weakly continuous* functional $I[z]$. These functionals are continuous relative to the weak convergence of their argument, and therefore they clearly achieve their bounds at the points z_0 generated by some elements $\mathbb{D}^0$ *belonging to GU, and not to U, in general.*

A set which coincides with its G-closure will be called G-closed. Most of the commonly used initial sets U are not G-closed, and the main problem which arises is that of building their G-closures. In other words, it is required to describe all of the composites which may be assembled from the given components with the aid of arbitrary mixing.

Having G-closed set of admissible controls at our disposal, we can also guarantee that the investigation of the necessary conditions of optimality with the aid of variations within narrow strips (oblate spheroids) will provide the strongest conditions of extremum within the class of local variations.

If, however, the initially given set U is not G-closed and if we perform a variation within this set, then the procedure of variation in strips (spheroids) usually provides necessary conditions which generally are intrinsically contradictory. These contradictions arise because the initially formulated optimization problem is ill-posed; to relax it, it is necessary to extend the initial set U of controls to its G-closure GU. For it is this set, not U, which generally contains the desired optimal control and, as a consequence, the necessary conditions derived for the relaxed problem turn out to be compatible.

The problem of relaxation also arises naturally in connection with problems concerning the optimal shape of bodies in nonhomogeneous external fields. The requirement of optimality of the shape of a body leads to an additional boundary condition on its surface, which (together with the other data of the problem) serves to determine the most advantageous shape of the body.*

As a rule it is difficult, if not impossible, to prove the existence of a body satisfying all of the imposed conditions. In particular, this applies to

---

* Here we are speaking of the simplest problem or the isoperimetric problem of the calculus of variations. Such problems have been treated exhaustively by Pólya and Szegö,[132] and also in subsequent investigations of other authors, most of which are described in Payne's review.[334]

problems of the optimal shape of a region when there are differential constraints. Such problems arise, for example, when the body to be optimized is in a given inhomogeneous field. Here too, the concept of variation in a strip is frequently a natural way out of a difficult situation.

The point is that, in many cases, one imposes boundary conditions on the unknown surface of the body which express ideal properties of the surface or (which is the same) the medium surrounding the body (perfect conductor, insulator, absolutely rigid body, vacuum, etc.), while the material of the body itself is not ideal. A change in the shape of the body can then be interpreted as the introduction of isotropic inclusions of the surrounding medium within the body or of inclusions of the body material within the surrounding medium. Therefore, the problem of the influence of the deformation of the boundary of the body on the criterion functional is equivalent to the problem concerning the influence of inclusions of ideal material. The solution of the problem for inclusions of arbitrary shape and size is unknown; however, if the diameter of the inclusion is small in comparison with some characteristic dimension (e.g., the radius of curvature of a trajectory of the original vector field at the given point), one can specify a shape of the inclusion and an orientation of it that guarantee an extremum of the functional $\delta I / V$, where δI is the increment of the functional I produced by the inclusion of the volume V. For the current flow problem (1.3) in two independent variables, the optimal inclusion has the shape of an ellipse with an infinitesimally small diameter and unit eccentricity; in this case, the inclusion may be viewed as an infinitely thin strip (or a "needle").

What we have said is also true for inclusions of material with nonideal properties; moreover (and this is important), a needle-shaped inclusion is the best local inclusion, regardless of the properties of the functional I. To different functionals I, there correspond different orientations of the strip at each point yielding an extremum of the ratio $\delta I / V$. These orientations are determined by the directions of the basic vector field and the auxiliary field (the field of the Lagrangian multipliers), which contains information about the functional being minimized.

It is clear, then, that the problem of optimal inclusions at internal points of the region can be solved by the method that we have described. Furthermore, it is easy to see that these arguments also hold when the inclusion adjoins the boundary of the region. Here, we must recall the natural boundary condition which must be satisfied on the unknown optimal boundary. This condition is derived under the assumption of sufficient smoothness of the boundary; frequently, however, it cannot be satisfied by smooth boundaries, since these may not exist. Such a situation is typical for problems with inhomogeneous external fields, when an improvement in the properties of the boundary of the body is achieved by the introduction of thin needles,* emanating on the boundary and directed into the body, the functional being minimized decreasing with an increasing number of needles. This behavior of the control can also be regarded as the analogue of a chattering regime, as encountered in the one-dimensional case when an optimal control does not exist.

The region adjoining the boundary filled with needles can be essentially regarded as anisotropic, since the effective specific characteristics of the material in this region are different in directions along the needles and across them. Guided by these implications, it is natural to attempt to seek the best shape of the body by treating its material as anisotropic. After such relaxation, the original problem of the optimal shape of an isotropic body may be replaced by the new problem: the simultaneous determination of the optimal body and the optimal distribution of the anisotropic material within the body. Examples show that differential constraints can be avoided if the problem is posed in this way; one then simply has an isoperimetric problem, and the optimal solution frequently consists of an optimal body bounded by a smooth surface. Thus, in this class of problems, the relaxation eliminates the difficulties associated with the influence of inhomogeneous external fields.

The applications modeled by optimization problems with control in the principal part of the linear operators are not restricted to problems related to the choice of the distributions of the characteristics of continuous media. The practically important problems of the optimal choice of the geometrical characteristics of thin-walled elastic structures (the thicknesses of the membranes or plates, and the curvatures and thicknesses of shells) fall within the same category. Usually, it is necessary to minimize the volume (or weight) of a structure under additional restrictions of isoperimetric type or differential type. And here, too, alongside sufficiently smooth optimal control, one can have chattering regimes, i.e., infinitely frequent corrugation of the surfaces. The correct formulation of such problems for plates and shells is based on asymptotic equations derived with the allowance of fast oscillations of elastic moduli and thickness as functions of position.[536] Optimization problems of this type are of great interest because of their substantial practical significance.†

---

* Or "tongues" in the three-dimensional case.

† Problems relating to the optimization of engineering structures have been extensively studied and have been presented in a number of monographs. In this connection, see the reviews of Reitman and Shapiro,[142,501] Prager and Taylor,[339] Zavelani-Rossi,[63] Berke and Venkayya,[428] Felton,[445] Haftka and Prasad,[452] Niordson and Olhoff,[488] Tao,[500] Venkayya,[519] among others.

The mechanics of continuous media provides a wealth of optimal control problems. In the present book, we hardly touch on the problems that reduce to the determination of the optimal intensities of sources and sinks, irrespective of their mechanical content. One of the most interesting problems of this kind concerns the most advantageous extraction of oil deposits, studied by Meerov and Litvak and published in a monograph.[105] In this problem, the control possibilities are of necessity restricted to the control of sources and sinks (oil wells). In gas dynamics, problems of the optimal choice of the shape of profiles around which a flow occurs (wings, nozzles, etc.) are classical. This class of problems is characterized by well-developed methods of investigation and a large number of different problems solved. It should be noted that many investigations in this field contain results that are also of significance for the general theory of the optimal control of distributed parameter systems.* Some of these results are also of use within the present context.

Magnetohydrodynamics is another source of optimization problems, which arise principally in connection with the rational design of magnetohydrodynamic generators, installations for converting thermal energy into electrical energy, and magnetohydrodynamic engines. The problem is to eliminate internal losses of the current and thus maximize useful power in such constructions. Essentially, one requires that the streamlines in the channel of a magnetohydrodynamic generator (motor) have a shape dictated by the following requirement: the optimal operation of the system as a whole. Equations (1.2), together with appropriate boundary conditions, provide the mathematical description of this problem which is examined in detail in the main text of this book.

Recently, the methods of the theory of optimal control have been applied to the problem of plasma containment by a magnetic field.

In various practical situations, one encounters problems of controlling temperature fields in solids; these problems have long attracted the interest of specialists on optimal control. Usually, it is necessary to determine nonstationary controls that occur in the boundary conditions or on the right-hand side of the heat-conduction equation (the density of distributed heat sources). Similar problems are also studied in connection with other evolution equations.

These questions have been considered by many authors, and some of the results were included by Butkovskii in his monograph.[18] It has been found that the effective solution of problems of this class requires the use of special methods. For example, Butkovskii's investigations have confirmed

---

* See Ref. 160 and the Appendix to the Russian translation thereof written by Gonor and Kraiko.

the fruitfulness of using the method of moments to solve a large class of problems of optimization of nonstationary distributed systems. In Ref. 18, he applied the method of moments to the problem of the fastest heating of a massive body and to the problem of damping a vibrating distributed system in the shortest possible time. Later, Butkovskii developed the concept of a finite control, which has proved very helpful for solving the problem of the controllability of linear nonstationary systems.

In this book, we consider only selected questions of the optimal control of systems described by hyperbolic equations and parabolic equations. We have already referred to some of these problems in Chapter 6; we consider a number of general facts relating to the existence and uniqueness of optimal controls and the use of Pontryagin's maximum principle. We also consider some new results obtained in the problem of the optimal heating of bodies and optimization problems with moving boundaries.

The last chapter of the book is devoted to the application of Bellman's method of dynamic programming to optimization problems for systems described by partial differential equations. This method has not yet been used widely. The reason is that the functional equations of dynamic programming, in the case of problems involving partial differential equations, are fairly complicated even for numerical solution. In addition, Bellman's optimality principle is valid in the general case only for evolution systems. Nevertheless, the possibilities of applying dynamic programming to problems involving partial differential equations have by no means been exhausted, and in this field new results are to be expected.

In general, although many problems have been solved, there remain many open questions in the theory of the optimal control of distributed systems, and these include serious problems of considerable practical importance. Some of them have already been mentioned above. We may also mention the problem of the optimal shape of bodies in aerodynamics, whose investigation has been based hitherto almost exclusively on study of the first variation.* Very little can be said about the optimal shape of nozzles under flow conditions involving transition through the velocity of sound. Problems of the optimization of structures under conditions of elastic–plastic behavior need further development. Numerous examples of problems of optimal control arise in the theory of metallurgical and chemical–technical processes and also in the theory of nuclear reactors. Among the mathematical problems awaiting their solution, we mention optimization problems under restrictions on the state variables. Such restrictions are frequently encountered in design problems, and it is an urgent practical requirement to take

---

* In this connection, see Section 6.5.

them into account. Great importance must be attached to the development of effective methods of the numerical and approximate solution of the optimal control problems for distributed parameter systems, and to the simulation of such problems.

Sometimes, the criticism is voiced that the optimal solutions are seldom accompanied by a large gain in the value of the functional as compared with the value that one can find by heuristic arguments on a rule of thumb basis. This is not a failure of the theory; indeed, an engineer must be considered a poor engineer if his intuition and experience do not suggest solutions that are close to the optimum. It is also clear that the investigation of theoretical questions of optimal control of distributed parameter systems will not be fruitful without the necessary intuition and experience. The mathematical difficulties associated with the absence of an optimal solution, its nonuniqueness, etc., are frequently a consequence of bad formulation of the problem resulting from an inadequate understanding of its physical features. The advantage of the optimal solution is precisely this: it gives a qualitative characteristic of the most advantageous behavior of the system; it elucidates the dependence of this behavior on the parameters and external factors and thus it extends our initial intuitive ideas. It is my hope that I have succeeded, at least partly, in expressing this idea on the pages of this book.

2

The Mayer–Bolza Problem for Several Independent Variables: Necessary Conditions for a Minimum

2.1. The Normal Form of the Basic Equations

Partial differential equations describing the behavior of distributed parameter systems can be represented in different forms. For optimization problems, a special form of expression of the equations is particularly convenient, and by analogy with the case of a single independent variable we shall call it the *normal form.*[*]

Let $(x_1, \ldots, x_m)$ be the independent, and $(z^1, \ldots, z^n)$ the dependent variables; in an open bounded region G of the space $(x_1, \ldots, x_m)$ we consider the definite system

$$f_\alpha\left(x_1, \ldots, x_m; z^1, \ldots, z^n; \frac{\partial z^1}{\partial x_1}, \ldots, \frac{\partial z^n}{\partial x_m}; u_1, \ldots, u_p\right) = 0$$

$$(\alpha = 1, \ldots, n) \qquad (2.1)$$

The left-hand sides of Eqs. (2.1) are continuously differentiable functions of their arguments; they include, in particular, p functions $(u_1, \ldots, u_p)$ of the variables $(x_1, \ldots, x_m)$, which we call *volume controls*. We assume that the system (2.1) does not lead to finite dependences between x, z, and u; suppose further that (2.1) can be solved for some n of the mn derivatives $\partial z^j/\partial x_i$ of the function $z^j (j = 1, \ldots, n)$ with respect to certain of the independent variables x_i. Having done this, we take the remaining $n(m - 1)$ derivatives as new (*parametric*) variables ζ^s [$s = 1, \ldots, n(m - 1)$]; as a result, we obtain[†]

$$\Xi_{ji} \equiv \frac{\partial z^j}{\partial x_i} - X_i^j(x, z, \zeta, u) = 0 \qquad (i = 1, \ldots, m; j = 1, \ldots, n) \quad (2.2)$$

---

[*] This form is a special case of what is usually termed *Pfaffian* form (Ref. 141, p. 324).
[†] Parametric variables can also be introduced in other ways, some of which will be demonstrated below.

In these equations, the X_i^j are continuously differentiable functions of their arguments, and $n(m-1)$ of them are equal to the corresponding variables $\zeta^s[s = 1, \ldots, n(m-1)]$.

The determination of the integrals $z(x)$ of the system (2.2) requires fulfillment of the integrability conditions

$$\frac{\partial X_i^j}{\partial x_k} = \frac{\partial X_k^j}{\partial x_i} \qquad (j = 1, \ldots, n; i, k = 1, \ldots, m) \tag{2.3}$$

The differentiation in these equations is carried out with respect to all arguments of the functions X_i^j; for example, one thus has

$$\frac{\partial X_i^j}{\partial x_k} = \frac{\partial' X_i^j}{\partial x_k} + \frac{\partial X_i^j}{\partial z^r} \frac{\partial z^r}{\partial x_k} + \frac{\partial X_i^j}{\partial \zeta^s} \frac{\partial \zeta^s}{\partial x_k} + \frac{\partial X_i^j}{\partial u_t} \frac{\partial u_t}{\partial x_k}$$

$$= \frac{\partial' X_i^j}{\partial x_k} + \frac{\partial X_i^j}{\partial z^r} X_k^r + \frac{\partial X_i^j}{\partial \zeta^s} \frac{\partial \zeta^s}{\partial x_k} + \frac{\partial X_i^j}{\partial u_t} \frac{\partial u_t}{\partial x_k}$$

Here, the prime denotes differentiation with respect to the explicitly occurring variable and we have adopted the usual summation convention for repeated indices.*

A feature of Eqs. (2.2) is that they are solved for the derivatives of *all* of the dependent variables z with respect to *all* of the independent variables x; for $m > 1$, this is achieved at the price of introducing the parametric variables ζ. This form of expression is the normal form mentioned earlier. Any system of partial differential equations can be reduced to such a form; the representation is not unique, however.

The parametric variables are the main feature of the equations in their normal form. In the general form of (2.2), these variables and the controls u occur formally on an equal footing, but in every concrete optimization problem it is necessary to make a clear distinction between them since these variables play fundamentally different roles: Whereas the controls u can, by definition, be specified more or less arbitrarily from without, the values of the variables ζ may not be so specified but are found from given u by solving the problem. This is of great importance; it would be wrong, for example, to formulate an optimal problem for the equations†

$$z_x = u_1(x, y), \qquad z_y = u_2(x, y) \tag{2.4}$$

---

* This condition will be assumed throughout unless otherwise stated.

† Here and in what follows, a subscript denoting an independent variable will mean partial differentiation with respect to that variable.

since simultaneous specification of the controls $u_1(x, y)$ and $u_2(x, y)$ is in general impossible without violation of the integrability condition. In other words, the system (2.4) is overdetermined; it may be transformed to a determined system by reducing the number of controls to one by setting, for example,

$$z_x = u(x, y), \qquad z_y = \zeta$$

or

$$z_x = \zeta, \qquad z_y = u(x, y)$$

etc. Note that, in general, different optimization problems result from the two systems. The last example shows, among other things, that once the parametric variables have been introduced there is no need to take (2.3) into account explicitly; for, if a solution of the system (2.2) exists, (2.3) are certainly satisfied. This assertion is true if classical solutions are considered; for generalized solutions, it need not be true. One can, however, also avoid adding (2.3) to the system (2.2) when the functions X_i^j are piecewise continuous and the function z^j are continuous. It is easy to show that the requirement of continuity of z^j is then a natural generalization of the integrability condition (2.3).

As an illustration of the previous statements, consider the example $m = 2$, $n = 1$, $p = 1$. Suppose that the functions X_1 and X_2 vary rapidly in the direction normal to a layer of thickness ε (Fig. 2.1); within this layer, and also in its neighborhood, we shall assume that the functions X_1 and X_2 are continuously differentiable, so the condition of integrability

$$\frac{\partial X_1}{\partial x_2} = \frac{\partial X_2}{\partial x_1}$$

is satisfied everywhere in the region under consideration.

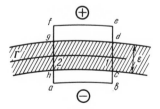

FIGURE 2.1

We form the linear integral of the vector (X_1, X_2) around the contour *abcdefgha* and apply Stokes's theorem; using the condition of integrability and collapsing the layer ε to the line Γ (see Fig. 2.1), and the lengths of the segments *bc, de, fg, ha* to zero, we obtain

$$z_t^+ - z_t^- = 0 \qquad (2.5)$$

where the subscript t denotes differentiation in the tangential direction to Γ, while the symbols z_t^+ and z_t^- denote the limiting values of z_t on opposite sides of Γ. After the passage to the limit, Γ is a line of discontinuity of the functions X_1 and X_2.

In the derivation of this condition, it is assumed that the behavior of the functions X_1 and X_2 in the layer ε is such that the linear integrals over the segments *cd* and *gh* tend to zero together with ε (physically, in the limit, this is equivalent to there being no double-layer-type singularity on the line Γ). It follows from this that the function z is continuous at the points 1 and 2 of Γ (the limit points for *c, d* and *g, h*, respectively); in conjunction with (2.5), this shows that z is also continuous along all of the line Γ.

In what follows, unless stated otherwise, we shall seek continuous solutions z^j of the system (2.2).* The parametric variables ζ, and also the controls u, may be discontinuous functions of x.

We now give some examples of the transformation of equations to their normal form.

The first-order equation

$$z_x^1 + z_y^1 = 0$$

is equivalent to the system

$$z_x^1 = \zeta, \qquad z_y^1 = -\zeta$$

The Laplace equation

$$z_{xx}^1 + z_{yy}^1 = 0$$

is equivalent to the system (Cauchy–Riemann)

$$z_x^1 = \zeta^1, \qquad z_y^1 = -\zeta^2, \qquad z_x^2 = \zeta^2, \qquad z_y^2 = \zeta^1$$

---

* Even if the functions X_i^j are not piecewise continuous.

The Helmholtz equation

$$z^1_{xx} + z^1_{yy} + uz^1 = 0$$

corresponds to the system

$$z^1_x = z^2, \quad z^1_y = z^3, \quad z^2_x = -\zeta^2 - uz^1, \quad z^2_y = \zeta^1, \quad z^3_x = \zeta^1, \quad z^3_y = \zeta^2$$

The wave equation

$$z^1_{yy} - (uz^1_x)_x = 0$$

corresponds to the system

$$z^1_x = -\zeta^1/u, \quad z^1_y = \zeta^2, \quad z^2_x = \zeta^2, \quad z^2_y = -\zeta^1$$

The last example shows that solutions of the system in the normal form can be regarded as generalized solutions of the corresponding equations of higher order. An analogous example is given by the system

$$z^1_x = \zeta^1, \quad z^1_y = -\zeta^2 + u, \quad z^2_x = \zeta^2, \quad z^2_y = \zeta^1$$

which is equivalent to the equations $\Delta z^1 = u_y$, $\Delta z^2 = u_x$ only for a continuously differentiable function $u(x, y)$.

Problems of mathematical physics are frequently formulated for equations (systems) of higher order; as we have already said, such equations (systems) can be reduced to normal form in many ways.

The condition of continuity of the new dependent variables z which are then introduced imposes certain restrictions on the choice of the procedures for reduction to normal form. To clarify this, let us consider an example. Suppose a system is described by the Poisson equation

$$z^1_{xx} + z^1_{yy} = u \tag{2.6}$$

where $u(x, y)$ is a control of bounded modulus which can have discontinuities of the first kind. At such discontinuities, the solution z^1 and its derivatives are continuous and, in general, the second derivatives are discontinuous; it is therefore natural to set

$$z^1_x = z^2, \quad z^1_y = z^3$$

(recall that the notation z is applied to *continuous* dependent variables). The Poisson equation is equivalent to the system

$$z_x^1 = z^2, \qquad z_y^1 = z^3, \qquad z_x^2 = \zeta^1$$

$$z_y^2 = \zeta^2, \qquad z_x^3 = \zeta^2, \qquad z_y^3 = u - \zeta^1 \tag{2.7}$$

Another procedure equivalent to the one given here consists of introducing an auxiliary continuous dependent variable z^2 by

$$z_x^2 = u, \qquad z_y^2 = \zeta^1$$

Eq. (2.6) is then equivalent to the system

$$z_x^1 - z^2 = z_y^3, \qquad z_y^1 = -z_x^3$$

which can be readily reduced to normal form.

We now consider a controlled system described by the equation

$$(uz_x^1)_x + (uz_y^1)_y = 0 \tag{2.8}$$

where $u(x, y)$ is a control of bounded modulus which may have discontinuities of the first kind. The first derivatives of the function z^1 have simultaneous discontinuities but an expression of the form

$$uz_x^1 y_t - uz_y^1 x_t \tag{2.9}$$

where x_t and y_t are the direction cosines of the tangent to the discontinuity line, is continuous. With the transition to the first-order system, it is natural to define the dependent variable z^2 in such a way that (2.9) coincides with the directional derivative of z^2 along the line of discontinuity. For this, it is sufficient to set

$$uz_x^1 = z_y^2, \qquad uz_y^1 = -z_x^2$$

With $z_x^2 = \zeta^2$, $z_y^2 = -\zeta^1$ (the variables ζ are separately discontinuous), we obtain the system in normal form

$$z_x^1 = -\zeta^1/u, \qquad z_y^1 = -\zeta^2/u, \qquad z_x^2 = \zeta^2, \qquad z_y^2 = -\zeta^1 \tag{2.10}$$

As we see, derivatives of u do not occur here, and the solutions of the system (2.10) can be regarded as the generalized solutions of (2.8).

Alternative representations of the basic first-order systems equivalent to (2.8) violate the requirement of continuity of the state variables z. For example, the system

$$z_x^1 = z^2/u, \qquad z_x^2 = \zeta^1, \qquad z_x^3 = \zeta^3$$

$$z_y^1 = z^3/u, \qquad z_y^2 = \zeta^2, \qquad z_y^3 = -\zeta^1$$

includes the variables z^2, z^3 which are themselves discontinuous across the discontinuity lines of the control u.

It should be mentioned that not every system of partial differential equations can be reduced to the form (2.2) with the additional requirement of continuity of the state variables z across the discontinuity lines of the control u. An important example of a system which does not allow such a reduction is provided by Kirchhoff's equations of bending of thin elastic plates. The system includes the equations of equilibrium

$$\frac{\partial Q_x}{\partial x} + \frac{\partial Q_y}{\partial y} + q = 0$$

$$\frac{\partial M_x}{\partial x} + \frac{\partial M_{xy}}{\partial y} - Q_x = 0$$

$$\frac{\partial M_{xy}}{\partial x} + \frac{\partial M_y}{\partial y} - Q_y = 0$$

which connect the components (Q_x, Q_y) of the shear force, the components (M_x, M_{xy}, M_y) of the moment tensor, and the density q of the transverse external load. This system is complemented by the equations representing Hooke's law

$$M_x = D\left(\frac{\partial \phi_1}{\partial x} + \nu \frac{\partial \phi_2}{\partial y}\right)$$

$$M_y = D\left(\frac{\partial \phi_2}{\partial y} + \nu \frac{\partial \phi_1}{\partial x}\right)$$

$$M_{xy} = \frac{D(1 - \nu)}{2}\left(\frac{\partial \phi_1}{\partial y} + \frac{\partial \phi_2}{\partial x}\right), \qquad D = Eh^3/12(1 - \nu^2)$$

and by the equations expressing the Kirchhoff hypothesis

$$\phi_1 = -\frac{\partial w}{\partial x}, \qquad \phi_2 = -\frac{\partial w}{\partial y}$$

Here, D denotes cylindrical rigidity of a plate (proportional to h^3, where h is a thickness), and w denotes the transverse displacement of a plate.

The whole system is readily seen to be equivalent to

$$z_x^1 = \zeta^1, \qquad z_x^2 = q, \qquad z_x^3 = \zeta^4, \qquad z_x^4 = z^1 + z^3 + \frac{D(1-\nu)}{2}(\zeta^6 + \zeta^7)$$

$$z_y^1 = -\zeta^2 - z^2, \qquad z_y^2 = \zeta^3, \qquad z_y^3 = z^2, \qquad z_y^4 = -D(\zeta^5 + \nu\zeta^8)$$

$$z_x^5 = -D(\zeta^8 + \nu\zeta^5), \qquad z_x^6 = \zeta^5, \qquad z_x^7 = \zeta^7, \qquad z_x^8 = -z^6$$

$$z_y^5 = -z^1 + \frac{D(1-\nu)}{2}(\zeta^6 + \zeta^7), \qquad z_y^6 = \zeta^6, \qquad z_y^7 = \zeta^8, \qquad z_y^8 = -z^7$$

Suppose that D represents the control function; continuity of tangential derivatives of the state variables z^1, z^4, and z^5 across the discontinuity line of D is equivalent to continuity of the shear force Q_n and of the moments M_n, M_{nt}; the state variables z^2 and z^3 may always be considered as continuous without any restrictions of generality. As to the state variables z^6, z^7, and z^8, they also are continuous because they represent components of the displacement vector. Altogether, for eight admissible values $Z^1, \ldots, Z^8$ of the parametric variables $\zeta^1, \ldots, \zeta^8$, we obtain seven conditions expressing the continuity of the tangential derivatives of the state variables $z^1, \ldots, z^7$ (note that the last pair of equations does not include any ζ-variables in its right-hand sides, and the tangential derivative of z^8 automatically becomes continuous). It is clear that this discrepancy arises because of too many requirements imposed along the line of discontinuity. Indeed, in Kirchhoff's theory, instead of three conditions of continuity for forces and moments, only two are considered, namely, the moment component M_n and $Q_n - \partial M_{nt}/\partial t$ of the *generalized shear force* are the only state variables assumed continuous. The latter requirement is of approximate character; it does not follow from the basic equations, and it is for this reason that the Kirchhoff system cannot be reduced to normal form.

These circumstances arise from the Kirchhoff hypothesis; if instead one introduces the (more exact) Reissner hypothesis, then it is possible to derive a standard system satisfying all of the additional requirements. This system differs from the previous one in the last two equations which now have the form

$$z_x^8 = -z^6 - (12/5)(1 + \nu)\zeta^2/Eh, \qquad z_y^8 = -z^7 + (12/5)(1 + \nu)\zeta^1/Eh$$

These equations express the Reissner hypothesis. Reissner's theory is known to preserve three conditions of continuity for the shear force and moments.

The impossibility of a standard representation of Kirchhoff's equations only means that corresponding optimization problems should be based on equations in traditional form.

The examples so far considered have involved two independent variables. It is clear that all of our assertions remain in force for a larger number of independent variables. For $m = 3$, the equation

$$\text{div } u \text{ grad } z^1 = 0$$

is satisfied by the substitution

$$u \text{ grad } z^1 = \text{curl } \mathbf{A} = \text{curl } (\mathbf{A}^* + \text{grad } \varphi)$$

Let $x_3 = 0$ be the equation of a discontinuity plane of the function u; on this plane, the derivatives $z_{x_1}^1$ and $z_{x_2}^1$ and the expression $uz_{x_3}^1$ given by

$$\frac{\partial A_2}{\partial x_1} - \frac{\partial A_1}{\partial x_2}$$

are continuous.

By the choice of the function $\varphi(x_1, x_2, x_3)$, one can make all three components of the vector $\mathbf{A}$ continuous on the discontinuity plane. Let $[A_i^*]_-^+$ be the magnitude of the discontinuity of component i of $\mathbf{A}^*$; we set div grad $\varphi = -\text{div } \mathbf{A}^*$. Then the conditions

$$[A_1^*]_-^+ = -[\varphi_{x_1}]_-^+, \qquad [A_2^*]_-^+ = -[\varphi_{x_2}]_-^+, \qquad [A_3^*]_-^+ = -[\varphi_{x_3}]_-^+$$

guarantee continuity of the difference $(A_2)_{x_1} - (A_1)_{x_2}$ and simultaneously express the continuity of the vector $\mathbf{A}$ on the discontinuity plane. On the other hand, these conditions suffice to relate the values of the function φ on the two sides of the discontinuity plane u.

We set $A_1 = z^2$, $A_2 = z^3$, $A_3 = z^4$; with this notation, the original equations take the form

$$uz_{x_1}^1 = z_{x_2}^4 - z_{x_3}^3, \qquad uz_{x_2}^1 = z_{x_3}^2 - z_{x_1}^4$$

$$uz_{x_3}^1 = z_{x_1}^3 - z_{x_2}^2, \qquad z_{x_1}^2 + z_{x_2}^3 + z_{x_3}^4 = 0$$

This system can be readily reduced to normal form. In the examples we have given, there has been no more than one volume control. It is easy to see that the outlined procedure also encompasses the case of several controls. For example, the equation

$$\operatorname{div} u_1 \operatorname{grad} z^1 = u_2$$

for $m = 2$ is reduced to normal form by setting

$$z_x^2 = u_2, \qquad z_y^2 = \zeta^1, \qquad u_1 z_x^1 - z^2 = z_y^3, \qquad u_1 z_y^1 = -z_x^3$$

etc.

We give one further form of expression of the basic equations which we shall find useful for what follows. Regarding $x_1, \ldots, x_m$ as Cartesian coordinates, we construct at the point $(x_1, \ldots, x_m)$ a local rectangular coordinate system $q_1, \ldots, q_m$ determined by the unit vectors $\mathbf{e}_1, \ldots, \mathbf{e}_m$ and the Lamé coefficients $H_1, \ldots, H_m$. The orientation of the new coordinate axes with respect to the axes $x_1, \ldots, x_m$ is given by the (symmetric) matrix $\{c_{\alpha\beta} = (\mathbf{e}_\alpha, \mathbf{i}_\beta)\}$ of direction cosines; in the general case, no restrictions are imposed on this orientation. Using the matrix $\{c_{\alpha\beta}\}$ and (2.2), we form the expressions (no summation over i)

$$\frac{1}{H_i} \frac{\partial z^j}{\partial q_i} = \frac{\partial z^j}{\partial x_k} c_{ik} = X_k^j c_{ik} \qquad (i = 1, \ldots, m; j = 1, \ldots, n)$$

From the resulting system of equations we eliminate the parametric variables; this can always be done by suitably changing the matrix $\{c_{\alpha\beta}\}$ if necessary. As a result, we obtain a system of n equations of the form (no summation over i)

$$F_s\left(\frac{1}{H_i} \frac{\partial z^j}{\partial q_i}, z, u, x, c_{\alpha\beta}\right) = 0 \qquad (s = 1, \ldots, n)$$

which, by assumption, is a consistent set. We now separate one of the coordinates, e.g., q_m, and solve this last system for the derivatives $\partial z^j / \partial q_m (j = 1, \ldots, n)$. We have

$$\frac{1}{H_m} \frac{\partial z^j}{\partial q_m} = Q^j\left(\frac{1}{H_1} \frac{\partial z}{\partial q_1}, \ldots, \frac{1}{H_{m-1}} \frac{\partial z}{\partial q_{m-1}}, z, u, x, c_{\alpha\beta}\right) \qquad (j = 1, \ldots, n)$$

$$(2.11)$$

   This system is solved for the derivatives of the dependent variable z with respect to one of the distinguished coordinates (q_m); we emphasize that its right-hand sides depend on the elements of the matrix $\{c_{\alpha\beta}\}$, i.e., on the mutual orientation of the old (fixed) and new (variable) axes at every point of the region G; this dependence also occurs indirectly through the derivatives $\partial z/\partial q_i$ $(i = 1, \ldots, m - 1)$.

   Essentially, it is this circumstance which determines the particular significance of the system (2.11): It can be shown that the set of controls u and elements $c_{\alpha\beta}$, in a certain sense, can be conveniently interpreted as a unified system of control parameters. This question will be considered in more detail in the example in Section 2.3.

   If we introduce the differential $dt_k = H_k dq_k$ of an arc of the coordinate line q_k, then (2.11) can be written in the final form

$$\frac{\partial z^j}{\partial t_m} = Q^j\left(\frac{\partial z}{\partial t_1}, \ldots, \frac{\partial z}{\partial t_{m-1}}, z, u, x, c_{\alpha\beta}\right) \qquad (j = 1, \ldots, n) \qquad (2.12)$$

   The system (2.11)–(2.12), expressed in curvilinear coordinates, is inconvenient for practical use; therefore, in what follows we shall work with equations in the form (2.2), using (2.11) and (2.12) to illustrate the various questions that arise when necessary conditions for optimality are derived for equations of various types.

   In equations containing the time, it is sometimes convenient to distinguish this variable. The basic equations are then written in the form of a system in a Banach space:

$$\frac{dz}{dt} = f(z, u, t) \qquad (2.13)$$

where f stands for some operator.

   Returning to (2.2) and (2.3), we fix the control functions u and define the concept of a generalized solution for these equations. We take the set $D(G)$ of infinitely differentiable vectors $\varphi(x) = (\varphi_1, \ldots, \varphi_m)$ and scalars $\psi(x)$ with support in the region G [supp $(\varphi, \psi) \subset G$] of Euclidean space R_m and define a generalized solution of the system (1.2) and (1.3) as a set of generalized functions $z = (z^1, \ldots, z^n)$, $\zeta = (\zeta^1, \ldots, \zeta^{n(m-1)})$ which satisfy the system of integral identities

$$\int_G \{z^j \nabla \cdot \varphi + (\varphi, X^j)\} \, dx = 0 \qquad (j = 1, \ldots, n) \qquad (2.14)$$

$$\int_G [\nabla \psi, X^j] \, dx = 0 \qquad (j = 1, \ldots, n) \qquad (2.15)$$

for all $(\varphi, \psi) \in D(G)$. In these expressions, X^j stands for the vector $(X^j_1, \ldots, X^j_m)$; the symbols $(\ ,\ )$ and $[\ ,\ ]$ denote the scalar and vector product, respectively.

If the functions X, ζ, z, and u are continuously differentiable with respect to all of their arguments, the generalized solution defined in this manner is a classical solution, i.e., (2.2) and (2.3) are satisfied in the ordinary sense. Obviously, the converse is also true. Thus, the concept of a generalized solution is obtained by giving up the requirement of continuous differentiability of the functions X, ζ, z, and u.

The identities (2.15) correspond to (2.3); we see that, in general, these identities do not follow from (2.14), so that, as we have already said, (2.14) alone are insufficient when one is considering generalized solutions. Such solutions must be determined by the complete system of identities (2.13) and (2.14).*

In what follows, the definition of a generalized solution will be modified in accordance with the requirements of a particular problem.

With regard to the boundary $\partial G = \Gamma$ of the region G, we shall assume that it is composed of a finite number of $(m - 1)$-dimensional surfaces, each of which may be mapped onto an $(m - 1)$-dimensional hyperplane, the mapping and its derivatives being continuous up to some specified order.

With regard to boundary conditions, those that specify the values of the functions z^j on definite parts of the boundary Γ (conditions of the first type) are of particular interest. Suppose that for every j $(j = 1, \ldots, n)$ we are given a connected open part γ_j of the boundary Γ; we note that one may have cases when the set $\Gamma_j = \mathrm{cl}\ \gamma_j$ coincides with Γ and also cases when Γ_j is the empty set; the sets Γ_j and Γ_k for $j \neq k$ can have nonempty intersection. We set

$$z^j(x) = z^j_1(x), \qquad x \in \gamma_j \qquad (j = 1, \ldots, n) \qquad (2.16)$$

where the $z^j_1(x)$ are given functions belonging to a known function space (the space depending on the particular problem). If for some j the set γ_j is empty, then the corresponding condition (2.16) must be eliminated.

There exist problems for which some of the functions $z^j_1(x)$ are not specified but are treated as *boundary controls*; for such controls, we adopt the notation $v^j(x)$ (conditions of the second type).

Thus, suppose $z^j_1(x) = v^j(x)$ for $j = j_1, \ldots, j_{n_1}$ $(n_1 \leq n)$.

---

* Previously, (2.2) were obtained under the assumption of continuous differentiability of the functions X; now, allowing generalized solutions, we take (2.2) and (2.3) as basic, regarding them in the generalized sense (2.14)–(2.15); for such equations, the requirement of continuous differentiability generally is not satisfied.

In a number of cases, boundary controls are introduced in an even more complicated way. Namely, instead of conditions of the type (2.16), one considers differential equations on $(m - 1)$-dimensional boundary manifolds; these equations contain the boundary controls. We shall say that these are conditions of the third type.

Suppose $\delta_j = \text{int } \Delta_j \subset \Gamma$ is a boundary manifold, and that the intersection $\delta_j \cap \gamma_j$ is, for simplicity, empty; on δ_j we introduce a system of curvilinear coordinates q_i, $i = 1, \ldots, m - 1$. The values of the functions $z(x)$ for $x \in \delta_j$ are related by the system of differential equations*

$$\Theta_{ji} \equiv \frac{\partial z^j}{\partial q_i} - T_i^j(q, z, \kappa^j, v^j) = 0 \qquad (i = 1, \ldots, m - 1) \qquad (2.17)$$

whose right-hand sides are continuously differentiable functions containing the boundary controls $v_j = (v_1^j, \ldots, v_\pi^j)$ and the parametric variables $\kappa^j = (\kappa_1^j, \ldots, \kappa_{m-2}^j)$; the functions T_i^j satisfy integrability conditions

$$\frac{\partial T_i^j}{\partial q_k} = \frac{\partial T_k^j}{\partial q_i} \qquad (i, k = 1, \ldots, m - 1) \qquad (2.18)$$

which are analogous to (2.3). For a certain system of indices $j = \{j_s\}$ on the boundaries $\partial\delta_j$ of the regions δ_j, the functions z^j are assumed given:

$$z^j(x) = z_2^j(x), \qquad x \in \partial\delta_j, \qquad j = \{j_s\} \qquad (2.19)$$

As before, we could assume here that some of the functions $z_2^j(x)$, $x \in \partial\delta_j$, are controls, either by introducing differential equations on $(m - 2)$-dimensional manifolds $\partial\delta_j$, etc.; the problem would then contain controls of ever higher rank. Since such problems occur seldom in practice, we shall restrict ourselves to the case when the boundary controls either replace some of the functions $z_1^j(x)$, $x \in \gamma_j$, in (2.16) or occur in the boundary equations (2.17).

Generalized solutions $z^j(q)$ of (2.17) and (2.18) are defined by analogy with such solutions for (2.2) and (2.3); the functions $z^k(q)$, $k = 1, \ldots, (j - 1), (j + 1), \ldots, n$, and the controls v are fixed, and the relations (2.18) play the role of (2.3).

Finally, in a number of cases the region G is not specified in advance; to be specific, we assume that an (open) part σ of its boundary Γ is not known *a priori* but must be determined together with the control functions.

---

* If δ_j is the empty set, then the conditions (2.17) are not specified.

The set σ may have a nonempty intersection with the sets $\gamma_j(\delta_j)$ on which the conditions of the above types are specified; the boundary $\partial\sigma$ of the set σ is assumed known.

Before we turn to the description of the sets of values of the control functions u and v, and also of the possible positions of the surface σ, we fix the controls u and v and the surface σ and define for the boundary-value problem (2.2), (2.3), (2.16)–(2.19) obtained in this manner the concept of a generalized solution. To this end, we take a set $D(R_m)$ of finite, infinitely differentiable vectors $\varphi(x) = (\varphi_1, \ldots, \varphi_m)$ and scalars $\psi(x)$; obviously, our previously introduced set $D(G) \subset D(R_m)$. In addition, in the $(m-1)$-dimensional space R_{m-1} we introduce a set $D(R_{m-1})$ of finite infinitely differentiable vectors $x(q) = (x_1, \ldots, x_{m-1})$ and scalars $\omega(q)$; the assumption made previously about the nature of the boundary Γ enables us to assert that $\delta_j \subset \mathrm{supp}\,(x, w)$, where δ_j is any open component of Γ. A generalized solution of the problem (2.2), (2.3), (2.16)–(2.19) is now defined as the set of generalized functions $z = (z^1, \ldots, z^n)$, $\zeta = (\zeta^1, \ldots, \zeta^{n(m-1)})$, $\kappa = (\kappa_1^j, \ldots, \kappa_{m-2}^j)$, $j = 1, \ldots, n$ which for all $(\varphi, \psi) \subset D(R_m)$, $(x, \omega) \subset D(R_{m-1})$ satisfy the system of integral identities* (no summation over j)

$$\int_{\gamma_{j_s}} z_1^j(\varphi, \mathbf{N}^j)\, dx - \int_G \{z^j \nabla \cdot \varphi + (\varphi, X^j)\}\, dx + \int_{\partial\delta_j} z_2^{j_s}(\chi, \mathbf{n}^{j_s}) J\, dq$$

$$- \int_{\delta_j} \left\{ z^j \frac{1}{J} \nabla_q \cdot J\chi + (\chi, T^j) \right\} J\, dq = 0 \qquad (j = 1, \ldots, n) \qquad (2.20)$$

$$\int_G [\nabla\psi, X^j]\, dx = 0, \qquad \int_{\delta_j} [\nabla_q J\omega, T^j]\, dq = 0 \qquad (j = 1, \ldots, n) \quad (2.21)$$

Here, $\mathbf{N}^j = (N_1^j, \ldots, N_m^j)$ is the unit vector of the external normal to the surface γ_j, and $\mathbf{n}^j = (n_1^j, \ldots, n_{m-1}^j)$ is the unit vector of the outer normal to the boundary $\partial\delta_j$ of the region δ_j lying in the manifold δ_j; the vector χ vanishes at the points of $\partial\delta_j$; $j \neq \{j_s\}$.

The identities (2.21) [of which the first coincides with (2.15)] correspond to (2.3) and (2.18); with regard to the role which they play, we may repeat the remarks made earlier in connection with the identities (2.15).

---

* The symbol $J\,dq = H_1 \cdots H_{m-1}\, dq_1 \cdots dq_{m-1}$ denotes the element of area of the surface δ_j in the orthogonal curvilinear coordinates $q_1, \ldots, q_{m-1}$.

The optimum problem is as follows: to determine the controls u and v, the surface σ, and the solutions corresponding to them of the problem (2.20)–(2.21) in such a way that the functional

$$I[u, v] = \int_{\Gamma/\cup\Gamma_j} C^j(q) z^j(q) J \, dq + \int_{\partial\delta} c^j(q) z^j(q) J \, dq \qquad (2.22)$$

takes the smallest possible value.

In (2.22) we have adopted the notation $\partial\delta = \cup\partial\delta_j$. Note that the form of (2.22), in conjunction with the constraint equations, encompasses integral functionals of the form

$$I[u, v] = \int_G f_0(z, \zeta, u, x) \, dx + \int_\Gamma f_1(z, \kappa, v, q) J \, dq \qquad (2.23)$$

To prove this, it is sufficient to introduce additional dependent variables—the vector $z_0 = (z^{n+1}, \ldots, z^{n+m})$ and the vector $z_1 = (z^{n+m+1}, \ldots, z^{n+2m-1})$—by means of the expressions*

$$\nabla \cdot z_0 = f_0, \qquad z_{x_k}^{n+i} - z_{x_i}^{n+k} = 0 \qquad (i, k = 1, \ldots, m) \qquad (2.24)$$

$$\nabla \cdot z_1 = f_1, \qquad z_{x_k}^{n+m+i} - z_{x_i}^{n+m+k} = 0 \qquad (i, k = 1, \ldots, m - 1) \qquad (2.25)$$

These relations can be included in the systems (2.2) and (2.16) by introducing the additional parametric variables

$$\zeta^{n(m-1)+i} \qquad \left(i = 1, \ldots, \frac{m(m + 1)}{2} - 1 \right)$$

$$\zeta^{n(m-2)+k} \qquad \left(k = 1, \ldots, \frac{m(m - 1)}{2} - 1 \right)$$

Let us consider, for example, the case $m = 2$; then $z_0 = (z^{n+1}, z^{n+2})$, and (2.24) take the form

$$z_{x_1}^{n+1} + z_{x_2}^{n+2} = f_0, \qquad z_{x_2}^{n+1} - z_{x_1}^{n+2} = 0$$

---

* For simplicity, we assume that x_i are Cartesian coordinates and Γ are the planes $x_m = \text{const.}$

We need to introduce two parametric variables ζ^{n+1} and ζ^{n+2}; we do this in accordance with

$$z_{x_1}^{n+1} = -\zeta^{n+2} + f_0, \qquad z_{x_2}^{n+2} = \zeta^{n+1}$$

$$z_{x_1}^{n+2} = \zeta^{n+1}, \qquad z_{x_2}^{n+2} = \zeta^{n+2} \tag{2.26}$$

The resulting system has the form (2.2). The functional

$$\iint_G f_0 \, dx_1 \, dx_2$$

is equal to

$$\int_\Gamma z^{n+1} \, dx_2 - z^{n+2} \, dx_1$$

i.e., it is reduced to the form (2.22). Equations (2.25) may be similarly transformed.

With regard to the boundary conditions for the new variables z_0 and z_1, it is a matter of preference whether or not they are added to the systems (2.24) and (2.25) (in the latter case, the missing relations are obtained from the necessary conditions for a minimum).

With regard to the permissible controls u and v, we shall assume that they belong to sets A and B (either open or closed) of certain function spaces. In particular, we shall consider restrictions which have the form of equalities

$$a_k(u; x) = 0, \qquad x \in G \qquad (k = 1, \ldots, r_1) \tag{2.27}$$

and inequalities

$$a_k(u; x) \geq 0, \qquad x \in G \qquad (k = r_1 + 1, \ldots, r \leq p) \tag{2.28}$$

along with (for $x \in \gamma_j$ or δ_j)

$$b_k(v; q) = 0 \qquad (k = 1, \ldots, \rho_1) \tag{2.29}$$

and

$$b_k(v; q) \geq 0 \qquad (k = \rho_1 + 1, \ldots, \rho \leq \pi) \tag{2.30}$$

The left-hand sides of these relations for any fixed x are single-valued functions of u and v.

It is convenient to express the restrictions (2.28) and (2.30) in the form of equivalent equalities. To this end, we introduce a system of (real) additional control functions $u_{k*}(x)$ and $v_{k*}(x)$ by means of

$$a_k(u; x) - u_{k*}^2 = 0 \qquad (k = r_1 + 1, \ldots, r \le p) \qquad (2.28)_*$$

$$b_k(v; q) - v_{k*}^2 = 0 \qquad (k = \rho_1 + 1, \ldots, \rho \le \pi) \qquad (2.30)_*$$

The additional controls u_{k*} and v_{k*} are not subject to any restrictions apart from $(2.28)_*$ and $(2.30)_*$. The relations $(2.28)_*$ and $(2.30)_*$ have the form (2.27) and (2.29); for the set of controls $u = (u_k, u_{k*})$, $v = (v_k, v_{k*})$, the restrictions can be regarded as given by (2.27) and (2.29). Henceforth, this will be taken to be the case. It is natural to require the sets A (2.27) and B (2.29) to satisfy the following conditions:

a. For every choice of the control $u \in A$, $v \in B$, there exists a (not necessarily unique) generalized solution of the problem (2.2), (2.3), (2.16)–(2.19).

b. A solution $z(x)$ of the problem depends continuously on the control in the following sense: if the sequence u_m, v_m of controls converges in the norm (of the space of controls) to the controls u, v, then the corresponding sequence z_m of solutions converges (at least weakly) in the space of solutions to the solution z.

c. In the sets A and B there exist optimal controls u and v, i.e., controls such that

$$I[u, v] \le I[U, V], \qquad \forall U \subset A, \qquad \forall V \subset B$$

The verification of these conditions, especially the last, is of very great difficulty in every optimum problem. We call the sets A, B of controls satisfying the first two conditions the set of admissible controls. With regard to the unknown part σ of the boundary Γ, certain smoothness requirements can be specified specially on this part of the boundary.

Such are the conditions which occur in the formulation of a typical optimum problem for partial differential equations; possible modifications of these conditions can be readily taken into account.

2.2. General Scheme for Obtaining Necessary Conditions of Stationarity and Realization of the Scheme for a Number of Special Cases

In this section, we shall give a formal derivation of the relations which determine the boundary-value problem that is the adjoint with the problem (2.2), (2.3), (2.16)–(2.19). The normal form (2.2)–(2.3) of the basic equations corresponds to the standard expression of the equations of the adjoint problem. These can be written down directly using the known functions X_i^j and T_i^j and the boundary conditions (2.16) and (2.19). We assume that all of the necessary derivatives exist and are continuous; cases of discontinuity will be considered separately.

We take the restrictions of the controls to have the form of (2.27) and (2.29). We make the following assumptions about the structure of the boundary: (a) We assume that the boundary Γ is composed of n nonintersecting closed sets $\Gamma_1, \ldots, \Gamma_n$ and a closed set Σ; (b) at the points of each of the sets $\gamma_j = \text{int } \Gamma_j$ we specify the values of the corresponding function z^j [(2.16)], and at the points of $\sigma = \text{int } \Sigma$ we specify (2.17) for a certain system of indices $j = \{j_k\}$; (c) on the boundary $\partial\sigma$ we specify the values of z^j for a subset $\{j_s\}$ of the system of indices $\{j_k\}$ [(2.19)]. We shall assume that the set σ is known.

The formal procedure involves using the ordinary method of Lagrangian multipliers. We take the functions*

$$\xi_{ij}(x) \qquad (i = 1, \ldots, m; j = 1, \ldots, n)$$

$$\alpha_k(x) \qquad (k = 1, \ldots, r_1)$$

$$\theta_{ij}(q) \qquad (j = \{j_k\}; i = 1, \ldots, m - 1)$$

$$\beta_k(q) \qquad (k = 1, \ldots, \rho_1)$$

which correspond to the restrictions (2.2), (2.27), (2.17), (2.30), and we form the functional

$$\Pi = I + \int_G (\xi_{ij}\Xi_{ji} + \alpha_k a_k)\, dx + \int_\sigma (\theta_{ij}\Theta_{ji} + \beta_k b_k) J\, dq \qquad (2.31)$$

The rectangular matrices $\{\xi_{ij}\}$, $\{\Xi_{ij}\}$, $\{\theta_{ij}\}$, $\{\Theta_{ij}\}$ are denoted by ξ, Ξ, θ, Θ and the vectors (α_k), (a_k), (β_k), (b_k), (C^j), (c^{j_k}) by α, a, β, b, C, c. Furthermore, we introduce the Lagrangians

$$L = \text{tr}\,(\xi\Xi) + (\alpha, a), \qquad l = \text{tr}\,(\theta\Theta) + (\beta, b) + (C, z) \qquad (2.32)$$

The expression for the first variation of Π is made up of integrals over the region G and integrals over the boundary Γ and the boundary $\partial\sigma$. Together, terms of the first type in the expression for the first variation

---

* The elements θ_{ij} can be assumed equal to zero for the j that do not occur in the system of indices $\{j_k\}$. Integral restrictions can also be taken into account by means of corresponding multipliers (see examples 1 and 3).

result in

$$\int_G \left(\frac{\partial L}{\partial z^j} - \frac{\partial \xi_{ij}}{\partial x_i} \right) \delta z^j \, dx + \int_G \frac{\partial L}{\partial \zeta^k} \delta \zeta^k \, dx + \int_G \frac{\partial L}{\partial u_s} \delta u_s \, dx \qquad (2.33)$$

In order to write down the terms that represent the integral over Γ, we consider the variation of the functional $\int_\Gamma fJ \, dq$, where f is the value on Γ of a function defined in the region bounded by the closed surface Γ, and it is continuously differentiable everywhere in this region, including on Γ. The variation of the functional is formed in accordance with the rule

$$\delta \int_\Gamma fJ \, dq = \int_\Gamma \delta fJ \, dq + \int_\Gamma \left(kf + \frac{\partial f}{\partial N} \right) \delta NJ \, dq \qquad (2.34)$$

Here, k is the mean curvature of the surface Γ, δN is the variation of the exterior normal to Γ, and $\partial/\partial N$ denotes the total derivative along the normal, i.e.,

$$\frac{\partial}{\partial N} f(z, \zeta, u, x) = \frac{\partial' f}{\partial N} + \frac{\partial f}{\partial z} \frac{\partial z}{\partial N} + \frac{\partial f}{\partial \zeta} \frac{\partial \zeta}{\partial N} + \frac{\partial f}{\partial u} \frac{\partial u}{\partial N}$$

where $\partial'/\partial N$ is the derivative with respect to the independent variables which occur explicitly in f. Equation (2.34) is of course true only if the function f is continuously differentiable and the surface Γ admits continuous second derivatives. The integrals over the manifold Γ in the expression for the first variation result in

$$\int_\Gamma (\xi_j, \mathbf{N}) \, \delta z^j J \, dq + \int_\sigma \left(\frac{\partial l}{\partial z^j} - \frac{1}{J} \frac{\partial J\theta_{ij}}{\partial q_i} \right) \delta z^j J \, dq$$

$$+ \int_\sigma \left(\frac{\partial l}{\partial \kappa_i^j} \delta \kappa_i^j + \frac{\partial l}{\partial v_k} \delta v_k \right) J \, dq + \int_{\partial \sigma} (\theta_j, \mathbf{n}) \, \delta z^j \, dx \qquad (2.35)$$

Here, $\xi_j = (\xi_{1j}, \ldots, \xi_{mj})$, $\theta_j = (\theta_{1j}, \ldots, \theta_{(m-1)j})$, $J = H_1 H_2 \ldots H_{m-1}$.

We assume further that a smooth $(m-1)$-dimensional manifold Γ_0 of discontinuity of the control u divides the region G into two parts G_+ and G_-. Here, the surface Γ_0 is to be determined together with the control functions.

The total variation Δf of f on the surface Γ_0 is made up of the variation δf of the function f itself and the variation $(\partial f/\partial N) \, \delta N$ associated with the displacement of the surface. The contribution of the corresponding terms to the expression for the first variation of the functional Π is

$$\int_{\Gamma_0} \left\{ [(\xi_j, \mathbf{N}) \, \Delta z^j]_-^+ + \left[L - (\xi_j, \mathbf{N}) \frac{\partial z^j}{\partial N} \right] \delta N \right\} dx \qquad (2.36)$$

The symbol $[\ ]_-^+$ stands for the difference of the limiting values of the quantity in the brackets on the two sides of the surface Γ_0. In particular, if the functions z^j remain continuous on the transition through Γ_0, then $[\Delta z^j]_-^+ = 0$, and (2.36) is replaced by

$$\int_{\Gamma_0} \left\{ [(\xi_j, \mathbf{N})]_-^+ \, \Delta z^j + \left[L - (\xi_j, \mathbf{N}) \frac{\partial z^j}{\partial N} \right]_-^+ \delta N \right\} dx \qquad (2.37)$$

Note that the parametric variables ζ are discontinuous on Γ_0 together with the controls u.

Finally, we assume that the region σ is split into two parts σ_+ and σ_- by a smooth $(m-2)$-dimensional manifold $\partial\sigma_0$ of discontinuity of the controls v, the surface $\partial\sigma_0$ being determined together with the controls.

The total variation Δf of the function f on the surface $\partial\sigma_0$ is made up of the variation δf of the function f itself and the variation $(\text{grad } f, \delta \mathbf{r})$ due to displacement of the surface $\partial\sigma_0$, which, in general, is a discontinuity surface of the manifold σ.

The corresponding term in the expression for the first variation is

$$\int_{\partial\sigma_0} \left\{ [(\theta_j, \mathbf{n}) \, \Delta z^j]_-^+ - ([(\theta_j, \mathbf{n}) \, \text{grad } z^j]_-^+, \delta \mathbf{r}) \right\} J \, dq \qquad (2.38)$$

Here, $[\ ]_-^+$ denotes the difference of the limiting values on $\partial\sigma_0$ of the quantity in the brackets taken on the surface σ on the two sides of the manifold $\partial\sigma_0$.

We are now in a position, using the usual argumentation of the calculus of variation, to write down the necessary conditions of stationarity of the functional I. These conditions have the form:

in the region G

$$\left. \begin{array}{ll} \dfrac{\partial \xi_{ij}}{\partial x_i} - \dfrac{\partial L}{\partial z^j} = 0 & (j = 1, \ldots, n) \\[4mm] \dfrac{\partial L}{\partial \zeta^k} = 0 & [k = 1, \ldots, n(m-1)] \\[4mm] \dfrac{\partial L}{\partial u_s} = 0 & (s = 1, \ldots, p) \end{array} \right\} \qquad (2.39)$$

on the boundary Γ

$$
\left.
\begin{aligned}
&(\xi_j, \mathbf{N}) = 0 \qquad (j = 1, \ldots, k-1, k+1, \ldots, n), x \in \Gamma_k \\[2mm]
&C^j + (\xi_j, \mathbf{N}) = 0 \qquad (j = 1, \ldots, n), x \in \Gamma/(\cup\Gamma_j) \cup \sigma \\[2mm]
&C^j + (\xi_j, \mathbf{N}) + \frac{\partial l}{\partial z^j} - \frac{1}{J}\frac{\partial J\theta_{ij}}{\partial q_i} = 0 \qquad (j = \{j_k\}), x \in \sigma \\[2mm]
&\frac{\partial l}{\partial \kappa_i^j} = 0 \qquad (i = 1, \ldots, m-2; j = \{j_k\}), x \in \sigma \\[2mm]
&\frac{\partial l}{\partial v_k} = 0 \qquad (k = 1, \ldots, \pi), x \in \sigma
\end{aligned}
\right\} \quad (2.40)
$$

and on the manifold $\partial\sigma$

$$
c^{j_k} + (\theta_{j_k}, \mathbf{n}) = 0 \qquad \text{for all } j_k \neq j_s \tag{2.41}
$$

On the manifold Γ_0 of discontinuity of the controls u we have the Weierstrass-Erdmann conditions

$$
\left.
\begin{aligned}
&(1) \; [(\xi_j, \mathbf{N})]_-^+ = 0 && \text{if } [\Delta z^j]_-^+ = 0 \\[2mm]
&(2) \; (\xi_j, \mathbf{N})_+ = (\xi_j, \mathbf{N})_- = 0 && \text{if the variations } (\Delta z^j)_+ \\
& && \text{and } (\Delta z^j)_- \text{ are arbitrary} \\[2mm]
&(3) \; \left[L - (\xi_j, \mathbf{N})\frac{\partial z^j}{\partial N} \right]_-^+ = 0 \quad (j = 1, \ldots, n)
\end{aligned}
\right\} \quad (2.42)
$$

On the manifold $\partial\sigma_0$ of discontinuity of the controls v

$$
\left.
\begin{aligned}
&(1) \; [(\theta_j, \mathbf{n})]_-^+ = 0 && \text{if } [\Delta z^j]_-^+ = 0 \\[2mm]
&(2) \; (\theta_j, \mathbf{n})_+ = (\theta_j, \mathbf{n})_- = 0 && \text{if the variations } (\Delta z^j)_+ \\
& && \text{and } (\Delta z^j)_- \text{ are arbitrary} \\[2mm]
&(3) \; [(\theta_j, \mathbf{n}) \,\text{grad}\, z^j]_-^+ = 0 && (j = \{j_k\})
\end{aligned}
\right\} \quad (2.43)
$$

Equations (2.39)–(2.43) together with (2.16) and (2.19) show that the first variation of the functional Π—the sum of the expressions (2.33), (2.35), (2.36), and (2.38)—vanishes.

Equations (2.39)–(2.43) can be written in a different form. To this end, we consider the "moments" $\partial L/\partial z^j_{x_i}$ and $\partial l/\partial z^j_{q_i}$ and show that they are equal to the Lagrangian multipliers ξ_{ij} and θ_{ij}. We introduce the "Hamilton function"

$$H(x, z, \zeta, u, \xi) = (z^j_{x_i} L_{z^j_{x_i}} - L)_{z^j_{x_i} = X^j_i} = \xi_{ij} X^j_i - (\alpha, a)$$
$$h(q, z, \kappa, v, \theta) = (z^j_{q_i} l_{z^j_{q_i}} - l)_{z^j_{q_i} = T^j_i} = \theta_{ij} T^j_i - (\beta, b) - (C, z) \tag{2.44}$$

We have the obvious equations

$$H_{x_i} = -L_{x_i}, \qquad H_{z^j} = -L_{z^j}, \qquad H_{\zeta^j} = -L_{\zeta^j}$$
$$H_{u_k} = -L_{u_k}, \qquad H_{\xi_{ij}} = X^j_i \tag{2.45}$$

and

$$h_{q_i} = -l_{q_i}, \qquad h_{z^j} = -l_{z^j}, \qquad h_{\kappa^j_i} = -l_{\kappa^j_i}$$
$$h_{v_k} = -l_{v_k}, \qquad h_{\theta_{ij}} = T^j_i \tag{2.46}$$

Using (2.45), we replace (2.2) and the first n equations of (2.39) by

$$\frac{\partial z^j}{\partial x_i} = \frac{\partial H}{\partial \xi_{ij}}, \qquad \frac{\partial \xi_{ij}}{\partial x_i} = -\frac{\partial H}{\partial z^j} \tag{2.47}$$

These equations have the form of canonical Volterra equations.[377] However, this result is of a formal nature since the function H cannot be regarded as a true Hamilton function until the variables ζ and u have been eliminated from it by means of the remaining equations (2.39). These last equations can be written in the form

$$\frac{\partial H}{\partial \zeta^k} = 0 \qquad [k = 1, \ldots, n(m-1)]$$
$$\frac{\partial H}{\partial u_s} = 0 \qquad (s = 1, \ldots, p) \tag{2.48}$$

Similarly, (2.16) and the last three groups of (2.40) can be represented in the equivalent form

$$\frac{\partial z^j}{\partial q_j} = \frac{\partial h}{\partial \theta_{ij}}, \qquad \frac{1}{J}\frac{\partial J\theta_{ij}}{\partial q_i} = -\frac{\partial h}{\partial z^j} + (\xi_j, \mathbf{N}) \tag{2.49}$$

$$\left.\begin{array}{ll}
\dfrac{\partial h}{\partial \kappa_i^j} = 0 & (i = 1, \ldots, m - 2; j = \{j_k\}), x \in \sigma \\[4mm]
\dfrac{\partial h}{\partial v_k} = 0 & (k = 1, \ldots, \pi), x \in \sigma
\end{array}\right\} \tag{2.50}$$

We assume that the variations Δz^j are continuous on the manifolds Γ_0 and $\partial\sigma_0$ of discontinuity of the controls u and v. Then, using the Hadamard–Hugoniot theorem[55] and equations* $(2.42)_1$ and $(2.43)_1$, we can write $(2.42)_3$ and $(2.43)_3$ in the form

$$(\xi_{ij})_+ [X_i^j]_-^+ = 0 \tag{$2.42)_3$}$$

$$(\theta_{ij})_+ \left[\frac{1}{H_i} T_i^j\right]_-^+ = 0 \tag{$2.43)_3$}$$

In deriving the relations of this section, we assumed that all necessary derivatives exist and are continuous, and that the corresponding solutions are classical. Going over to generalized solutions, we must, as mentioned above, take into account (2.3) and (2.18) or, more precisely, the integral identities (2.21) corresponding to them. It is, however, easily seen that, when (2.3) and (2.18) are taken into account, the Lagrangian multipliers

$$\Phi_{ik}^j = \Phi_{ki}^j \qquad (j = 1, \ldots, n; i, k = 1, \ldots, m)$$

$$\varphi_{ik}^j = \varphi_{ki}^j \qquad (j = 1, \ldots, n; i, k = 1, \ldots, m - 1)$$

are introduced, the functions z^j are assumed to be continuous, and L and l are assumed to be continuously differentiable with respect to z, ζ, u, κ, v, then we arrive at the same conditions of stationarity (2.39)–(2.43), in which it is only necessary to replace ξ_{ij} and θ_{ij} by the corresponding combinations $\xi_{ij}(\theta_{ij})$ and the derivatives $\partial\Phi_{ik}^j/\partial x_s(\partial\varphi_{ik}^j/J\,dq_p)$. For example, if $m = 2$, then ξ_{1j} is replaced by $\xi_{1j} - \partial\Phi_{12}^j/\partial x_1$, ξ_{2j} by $\xi_{2j} + \partial\Phi_{12}^j/\partial x_2$, etc. Without detriment to the result, we can assume $\Phi = \varphi = 0$.

Let us give some examples.

---

* Here and in what follows, a subscript appended to the number of an equation denotes the serial number of the equation in the group of equations having the given equation number.

Example 1. In a simply connected bounded planar region G with boundary Γ, we consider the optimum problem

$$\left.\begin{array}{c} \Delta z^1 = u(x, y), \qquad \dfrac{\partial z^1}{\partial N}\bigg|_\Gamma = f(t), \qquad I = \displaystyle\int_\Gamma z^1\, dt = \min \\[3mm] u_{\min} \le \text{vrai max } u(x, y) \le u_{\max} \\[3mm] u_{\min} \le \dfrac{1}{S}\displaystyle\int_\Gamma f(t)\, dt \le u_{\max} \end{array}\right\} \tag{2.51}$$

Here, t is the arc length of the boundary Γ of G measured from some origin; S is the area of G.

We set $x_1 = x$, $x_2 = y$; the Poisson equation is equivalent to the system [cf. (2.7)]

$$z_x^1 = z^2, \qquad z_y^1 = z^3, \qquad z_x^2 = \zeta^1, \qquad z_y^2 = \zeta^2$$

$$z_x^3 = \zeta^2, \qquad z_y^3 = u - \zeta^1 \tag{2.52}$$

We set $\xi_{1j} = \xi_j$, $\xi_{2j} = \eta_j$; we have

$$H = \xi_1 z^2 + \eta_1 z^3 + \xi_2 \zeta^1 + \eta_2 \zeta^2 + \xi_3 \zeta^2 + \eta_3(u - \zeta^1) - \gamma u$$

(γ is the Lagrangian multiplier corresponding to the solvability condition of the Neumann problem), $z_N^1 = z^2 y_t - z^3 x_t$, where x_t and y_t are the direction cosines of the tangent to Γ; we can set

$$l = \rho(z^2 y_t - z^3 x_t - f)$$

where $\rho = \rho(t)$ is a Lagrangian multiplier. The adjoint system

$$\xi_{1x} + \eta_{1y} = 0, \qquad \xi_{2x} + \eta_{2y} = -\xi_1, \qquad \xi_{3x} + \eta_{3y} = -\eta_1$$

$$\xi_2 - \eta_3 = 0, \qquad \eta_2 + \xi_3 = 0 \tag{2.53}$$

is integrated with the boundary conditions

$$\xi_1 y_t - \eta_1 x_t + 1 = 0, \qquad \xi_2 y_t - \eta_2 x_t + \rho y_t = 0$$

$$\xi_3 y_t - \eta_3 x_t - \rho x_t = 0 \tag{2.54}$$

We assume that the functions z^j are continuous; on the line Γ_0 of possible discontinuity of the control u, the following conditions must be satisfied:

$$[\xi_j y_t - \eta_j x_t]_-^+ = 0 \qquad (j = 1, 2, 3)$$

$$[H]_-^+ = (z_x^j)_-[\xi_j]_-^+ + (z_y^j)_-[\eta_j]_-^+ \tag{2.55}$$

Example 2. (Problem of the minimum of the electrical resistance of a planar region)   The bounded planar region G and the boundary Γ are the same as in Example 1; on Γ (see Fig. 1.1) we distinguish two arcs Γ_1 and Γ_2 without common points (electrodes); the remaining two arcs, Γ_3 and Γ_4, are insulators. The electrodes are connected through the load R.

The distribution of the current $\mathbf{j}$ and the potential z^1 is described by the equations

$$\operatorname{div} \mathbf{j} = 0, \qquad u\mathbf{j} = -\operatorname{grad} z^1 - \mathbf{E}_0 \tag{2.56}$$

where $\mathbf{E}_0 = (E_1, E_2)$ is a given vector function of position (external field), and $u(x, y)$ is the resistivity of the medium. The boundary conditions are $[\mathbf{j} = (-z_y^2, z_x^2)]$

$$z^1|_{\Gamma_1} = \text{const}, \qquad z^1|_{\Gamma_2} = \text{const}, \qquad z^2|_{\Gamma_3} = 0, \qquad z^2|_{\Gamma_4} = I$$

$$z^1|_{\Gamma_1} - z^1|_{\Gamma_2} = IR \tag{2.57}$$

The problem is to maximize the function ($\mathbf{N}$ is the outer normal to Γ_1)

$$I = \int_{\Gamma_1} (\mathbf{j}, \mathbf{N})\, dt$$

by choosing a bounded measurable function $u(x, y)$ which satisfies the inequalities

$$0 < u_{\min} \leq \text{vrai max } u(x, y) \leq u_{\max} \tag{2.58}$$

An equivalent expression of the conditions (2.56) and (2.57) is given by

$$z_x^1 = -u\zeta^1 - E_1, \qquad z_y^1 = -u\zeta^2 - E_2$$

$$z_x^2 = \zeta^2, \qquad z_y^2 = -\zeta^1$$

$$z^1|_{\Gamma_1} = z_+^1, \qquad z^1|_{\Gamma_2} = z_-^1, \qquad z^2|_{\Gamma_3} = 0, \qquad z^2|_{\Gamma_4} = I$$

$$z_+^1 - z_-^1 = IR$$

We introduce the factors ξ_1, η_1, ξ_2, η_2; we then have

$$H = -\xi_1(u\zeta^1 + E_1) - \eta_1(u\zeta^2 + E_2) + \xi_2\zeta^2 - \eta_2\zeta^1 \qquad (2.59)$$

The adjoint system

$$\xi_{1x} + \eta_{1y} = 0, \qquad \xi_{2x} + \eta_{2y} = 0$$

$$u\xi_1 + \eta_2 = 0, \qquad u\eta_1 - \xi_2 = 0 \qquad (2.60)$$

can be integrated subject to the boundary conditions

$$\left.\begin{array}{ll}
\xi_1 y_t - \eta_1 x_t = 0 & \text{on } \Gamma_3, \Gamma_4 \\[2mm]
\xi_2 y_t - \eta_2 x_t = 0 & \text{on } \Gamma_1, \Gamma_2 \\[2mm]
\displaystyle\int_{\Gamma_4} (\xi_2 y_t - \eta_2 x_t)\, dt - 1 + R \int_{\Gamma_1} (\xi_1 y_t - \eta_1 x_t)\, dt = 0
\end{array}\right\} \qquad (2.61)$$

(t is the arc length of Γ). The functions z^j are continuous; on the discontinuity line Γ_0 of the control, the following conditions are satisfied:

$$[\xi_j y_t - \eta_j x_t]_-^+ = 0 \qquad (j = 1, 2)$$

$$[H]_-^+ = (z_x^j)_-[\xi_j]_-^+ + (z_y^j)_-[\eta_j]_-^+ \qquad (2.62)$$

Example 3. (Extremum problem for the torsional rigidity of an inhomogeneous planar region) The stressed state in the case of torsion of a simply connected, isotropic, inhomogeneous planar region G with boundary Γ is described by the equations

$$z_x^1 = -u\zeta^1 + y, \qquad z_y^1 = -u\zeta^2 - x$$

$$z_x^2 = \zeta^2, \qquad\qquad z_y^2 = -\zeta^1 \qquad (2.63)$$

with the boundary condition

$$z^2 = 0 \qquad \text{on } \Gamma \qquad (2.64)$$

Here, $u = u(x, y) = \mu^{-1}(x, y)$ is the reciprocal of the shear modulus (the compliance): the bounded measurable function $u(x, y)$ satisfies the restrictions

$$0 < u_{min} \leq \text{vrai max } u(x, y) \leq u_{max}, \qquad \frac{1}{S}\int\int_G u\, dx\, dy = u_* < u_{max} \quad (2.65)$$

where u is a given constant and S is the area of G. The warping of the rod section and the Prandtl function are denoted by $z^1(x, y)$ and $z^2(x, y)$, respectively. The problem consists of determining the function $u(x, y)$, which maximizes (minimizes) the functional

$$I = \int\int_G z^2\, dx\, dy \qquad (2.66)$$

which differs from the torsional rigidity functional by a factor 2. We introduce the multipliers ξ_1, η_1, ξ_2, η_2, γ; we then have

$$H = \pm z^2 + \xi_1(-u\zeta^1 + y) - \eta_1(u\zeta^2 + x) + \xi_2\zeta^2 - \eta_2\zeta^1 - \gamma u \quad (2.67)$$

(the upper sign in the first term corresponds to a maximum and the lower to a minimum of the functional I). The adjoint system

$$\xi_{1x} + \eta_{1y} = 0, \qquad \xi_{2x} + \eta_{2y} = \mp 1$$

$$\xi_1 u + \eta_2 = 0, \qquad \eta_1 u - \xi_2 = 0 \qquad (2.68)$$

is integrated subject to the boundary condition

$$\xi_1 y_t - \eta_1 x_t = 0 \qquad \text{on } \Gamma \qquad (2.69)$$

On the discontinuity line Γ_0 of the control, (2.62) are satisfied.

Example 4. (Extremum problem for the first eigenvalue of an elliptic operator) In the region $\bar{G} = G \cup \Gamma$, consider the operator

$$Lz^1 = -\sum_{i,k=1}^{m} \frac{\partial}{\partial x_i}\left(u_{ik}\frac{\partial z^1}{\partial x_k}\right) \qquad (2.70)$$

whose coefficients $u_{ik}(x)$ satisfy the condition

$$\mu_1 \sum_{k=1}^{m} t_k^2 \geq \sum_{i,k=1}^{m} u_{ik}t_it_k \geq \mu_2 \sum_{k=1}^{m} t_k^2, \qquad \mu_1 > \mu_2 > 0 \qquad (2.71)$$

in $\bar{G}$, for all real $t_1, \ldots, t_m$ (uniform ellipticity). Suppose that on the boundary Γ

$$z^1 = 0 \tag{2.72}$$

then the operator L is positive definite.[110]

The problem is to determine the controls $u_{ik}(x)$ ($i, k = 1, \ldots, m$) under the restrictions (2.71) and (2.72) in such a way that the solution of the boundary-value problem

$$Lz^1 = \Lambda z^1, \qquad \Lambda = \text{const} \tag{2.73}$$

$$\int_G (z^1)^2 \, dx = 1, \qquad z^1(x) > 0, \qquad \forall x \in G$$

(which obviously exists only for some $\Lambda = \Lambda_1$) minimizes (maximizes) the functional

$$I = \Lambda_1 \tag{2.74}$$

Equation $(2.73)_1$ may be converted to normal form in a straightforward manner. We illustrate the procedure for $m = 4$. We set

$$z^1_{x_1} = \zeta^1, \qquad z^1_{x_2} = \zeta^2, \qquad z^1_{x_3} = \zeta^3, \qquad z^1_{x_4} = \zeta^4$$

$$z^2_{x_1} = z^1, \qquad z^2_{x_2} = \zeta^5, \qquad z^2_{x_3} = \zeta^6, \qquad z^2_{x_4} = \zeta^7 \tag{2.75}$$

Equation (2.73) can be satisfied by introducing functions z^3, z^4, z^5, z^6, z^7, z^8 by the relations

$$z^3_{x_2} + z^4_{x_3} + z^5_{x_4} = u_{1k}\zeta^k + \Lambda z^2$$

$$-z^3_{x_1} + z^6_{x_3} + z^7_{x_4} = u_{2k}\zeta^k$$

$$-z^4_{x_1} - z^6_{x_2} + z^8_{x_4} = u_{3k}\zeta^k$$

$$-z^5_{x_1} - z^7_{x_2} - z^8_{x_3} = u_{4k}\zeta^k$$

on the right-hand sides of these equations, summation over k from 1 to 4 is understood. These equations can be readily reduced to normal form, after which the necessary conditions of stationarity are written down.

Example 5. (Extremal problems for harmonic functions; control of the shape of a surface)   We consider a closed surface Σ bounding a body B; we denote the complement of the set $B \cup \Sigma$ in the complete space by O. At points of O we consider the harmonic function

$$\Delta z^1 = 0 \tag{2.76}$$

which vanishes at infinity: $z^1|_\infty = 0$.

We shall consider three types of boundary condition on Σ and three variants of behavior at infinity, including the corresponding physical interpretation of the function z^1:

$$\text{(I)} \qquad z^1|_\Sigma = 1, \qquad z^1|_\infty = \frac{C}{r} + o\!\left(\frac{1}{r}\right) \tag{2.77}$$

(z^1 is the electric potential produced in O by a perfectly conducting surface Σ charged to potential 1 with respect to the infinitely distant point, and $o(\,\cdot\,)$ means the order of magnitude higher than $(\,\cdot\,)$);

$$\text{(II)} \qquad \left.\begin{array}{c} z^1|_\Sigma = (\mathbf{h}, \mathbf{r}_0) + \text{const} \\[2mm] z^1|_\infty = \dfrac{\text{const}}{r^2} + o\!\left(\dfrac{1}{r^2}\right) \\[2mm] \mathbf{h} = \text{const} \end{array}\right\} \tag{2.78}$$

$|\mathbf{h}| = 1$, and $\mathbf{r}_0$ is the radius vector of a point on Σ (z^1 is the electric potential produced at O by the perfectly conducting surface Σ placed in a homogeneous electric field $\mathbf{h}$);

$$\text{(III)} \qquad \left.\begin{array}{c} -\dfrac{\partial z^1}{\partial N}\bigg|_\Sigma = (\mathbf{h}, \mathbf{N}) \\[2mm] z^1|_\infty = \dfrac{\text{const}}{r^2} + o\!\left(\dfrac{1}{r^2}\right) \\[2mm] \mathbf{h} = \text{const}, \qquad |\mathbf{h}| = 1 \end{array}\right\} \tag{2.79}$$

and $\mathbf{N}$ is the exterior normal to Σ (z^1 is the velocity potential of an incompressible perfect fluid produced by an obstacle in the form of a solid body B placed in a uniform flow that moves with velocity $\mathbf{h}$).

In all three cases, we wish to extremize the functional

$$I = \iiint_O |\text{grad } z^1|^2 \, dx_1 \, dx_2 \, dx_3 \tag{2.80}$$

under different restrictions on the choice of Σ.

By means of Green's theorem, we can readily show that in Case (I),

$$I = 4\pi C \tag{2.81}$$

where C is the constant in the second condition (2.77) (the capacitance of the body B). Similarly, in Case (II),

$$I = P = \sum_{i,k=1}^{3} P_{ik}h_i h_k \tag{2.82}$$

is the polarization in the direction of h, and in Case (III),

$$I = W = \sum_{i,k=1}^{3} W_{ik}h_i h_k \tag{2.83}$$

is the apparent mass in the same direction.

Usually, one considers the following restrictions on Σ: (a) the area S of Σ is given; (b) the volume V of the body B is given; (c) the diameter D of the body B is given.

The quantities C, W, P, like S and V, represent geometrical characteristics of the body B; therefore, a restriction can be formulated by specifying any of these quantities.

In particular, we can pose the following optimization problems: (d) Find an extremum of C for a given W [and (or) P]; (e) find an extremum of W for a given P [and (or) C]; (f) find an extremum of P for a given C [and (or) W].

The restrictions (a)–(c) are classical restrictions; the corresponding results are given in Ref. 132. The restrictions (d)–(f) are not classical.

The boundary Σ of the body B may be partly known. For convenience, we consider the case of two independent variables (x, y), and we shall assume that the boundary Σ has a rectilinear part γ of known length l, and that we wish to find the remaining part $\Sigma - \gamma$ in such a way that under the restriction (a) (the area

$$V = \tfrac{1}{2} \oint_{\Sigma} x\,dy - y\,dx \tag{2.84}$$

of the planar figure B is given) the apparent mass of the body B be minimal. We have

$$z_x^1 = \zeta^1, \qquad z_y^1 = \zeta^2, \qquad z_x^2 = -\zeta^2, \qquad z_y^2 = \zeta^1,$$

$$H = -(\zeta^1)^2 - (\zeta^2)^2 + \xi_1\zeta^1 + \eta_1\zeta^2 - \xi_2\zeta^2 + \eta_2\zeta^1 \tag{2.85}$$

$$\left.\begin{array}{ll} \xi_{1x} - \eta_{1y} = 0, & \xi_{2x} + \eta_{2y} = 0 \\[2mm] \xi_1 + \eta_2 - 2\zeta^1 = 0, & \eta_1 - \xi_2 - 2\zeta^2 = 0 \end{array}\right\} \tag{2.86}$$

$$\left.\begin{array}{ll} \xi_1 y_t - \eta_1 x_t = 0 & \text{along } \Sigma \\[2mm] (\zeta^1 + h_x)^2 + (\zeta^2 + h_y)^2 = \rho & \text{along } \Sigma - \gamma \end{array}\right\} \tag{2.87}$$

The last condition is an additional relation which serves to determine the unknown part $\Sigma - \gamma$ of the boundary Σ; the unknown constant ρ is determined in accordance with condition (2.84). Essentially, this result was obtained by Riabouschinsky.[349]

If part of the contour consists of rectilinear segments, the problem can be completely solved for we then have the known flow pattern of a flow around a polygonal arc with formation of a cavity of given area V.

In a similar manner one can pose the planar problem of minimizing the capacity of a system of two figures B and B' such that $B' \subset B$, the figure B being completely known whereas only the area of the figure B' is known.

Example 6. (Optimization problems for hyperbolic equations; control of the phase velocity of propagating perturbations) In the strip ($0 \le t \le T$, $-\infty < x < \infty$) of the t, x plane, we are given a quasi-linear first-order equation

$$z_t = f_0(t, x, z; u) + f_1(t, x, z; u)z_x \tag{2.88}$$

with free term f_0, and containing the control $u(t, x)$ in the coefficient f_1. To this equation are added the initial conditions

$$z(0, x) = z_0(x) \tag{2.89}$$

no boundary conditions are specified.

The piecewise continuous function $u(t, x)$ is not known in advance; it is to be chosen in such a way as to minimize the functional

$$I = \int_{-\infty}^{\infty} \rho(x)z(T, x)\, dx \tag{2.90}$$

where $\rho(x)$ is a function with compact support. The functions f_0, f_1, and z_0 are continuously differentiable with respect to all of their arguments. Equation (2.88) is equivalent to the system

$$z_t = f_0 + f_1\zeta, \qquad z_x = \zeta$$

The function H has the form

$$H = \xi(f_0 + f_1\zeta) + \eta\zeta$$

The adjoint system is

$$\xi_t + \eta_x = -\xi(f_{0z} + f_{1z}\zeta), \qquad \xi f_1 + \eta = 0$$

and it is equivalent to the single equation

$$\xi_t - (\xi f_1)_x = -\xi(f_{0z} + f_{1z}\zeta) \tag{2.91}$$

which is to be solved subject to the boundary condition

$$\xi(T, x) = -\rho(x) \tag{2.92}$$

Suppose we consider the one-dimensional wave equation

$$z_{tt}^1 - (uz_x^1)_x = 0 \tag{2.93}$$

where the phase velocity $u^{1/2}$ depends on the two variables t and x. Certain initial and boundary conditions are specified on the part Γ_1 of the boundary Γ of the principal region G; the problem is to choose a function $u(t, x)$ satisfying the restrictions

$$0 < u_{\min} \le u(t, x) \le u_{\max} < \infty \tag{2.94}$$

and minimizing the functional

$$I = \int_{\Gamma/\Gamma_1} F(z^1, \tau) \, d\tau \qquad (\tau = \text{length of boundary curve}) \tag{2.95}$$

The function F is continuously differentiable with respect to all of its arguments; no part of Γ is a segment of a characteristic. Equation (2.92) is equivalent to the system

$$z_t^1 = \zeta^1, \qquad z_x^1 = \zeta^2, \qquad z_t^2 = u\zeta^2, \qquad z_x^2 = \zeta^1$$

The Hamiltonian

$$H = \xi_1\zeta^1 + \eta_1\zeta^2 + \xi_2 u\zeta^2 + \eta_2\zeta^1$$

leads to the adjoint system

$$\xi_{1t} + \eta_{1x} = 0, \qquad \xi_{2t} + \eta_{2x} = 0$$

$$\xi_1 + \eta_2 = 0, \qquad \eta_1 + \xi_2 u = 0 \tag{2.96}$$

2.3. Weierstrass's Necessary Condition (Example)

The derivation of this condition involves a special formulation of the increment ΔI of the functional I due to the fact that the optimal control u is replaced by the admissible control $U = u + \Delta u$. There are several variants of this formulation, and the first one was described in Rozonoér's well-known paper.[144] We first indicate the construction for a special example; we then follow it with the general case.

We consider the problem of minimizing the electrical resistance of a planar region (Example 2 in Section 2.2). Denoting quantities that correspond to the optimal regime by lowercase letters, and the quantities for an admissible regime by the corresponding uppercase letters, we may write down the equations

$$z_x^1 - Z_x^1 = -u\zeta^1 + UZ^1, \qquad z_y^1 - Z_y^1 = -u\zeta^2 + UZ^2$$

$$z_x^2 - Z_x^2 = \zeta^2 - Z^2, \qquad z_y^2 - Z_y^2 = -\zeta^1 + Z^1 \tag{2.97}$$

which follow from (2.56).

The multipliers ξ_1, η_1, ξ_2, η_2 are used to derive an integral identity corresponding to the system (2.97) ($\Delta z_x^1 = Z_x^1 - z_x^1$, etc.):

$$\iint_G [\xi_1(\Delta z_x^1 - u\zeta^1 + UZ^1) + \eta_1(\Delta z_y^1 - u\zeta^2 + UZ^2)$$

$$+ \xi_2(\Delta z_x^2 + \zeta^2 - Z^2) + \eta_2(\Delta z_y^2 - \zeta^1 + Z^1)] \, dx \, dy = 0$$

An integration by parts of the first terms in the parentheses and the use of the first pair of equations (2.60) yield

$$\int_\Gamma (\xi_1 y_t - \eta_1 x_t) \, \Delta z^1 \, dt + \int_\Gamma (\xi_2 y_t - \eta_2 x_t) \Delta z^2 \, dt$$

$$+ \iint_G [\xi_1(UZ^1 - u\zeta^1) + \eta_1(UZ^2 - u\zeta^2)$$

$$+ \xi_2(\zeta^2 - Z^2) + \eta_2(Z^1 - \zeta^1)] \, dx \, dy = 0$$

To transform the sum of the curvilinear integrals, we use the conditions (2.57) and (2.61); this sum is equal to ΔI. We transform the integral over G by means of the second pair of equations (2.60); collectively, these results imply [see (2.59)]

$$\Delta I = -\iint_G [H(\xi, \eta, \zeta, u, x, y) - H(\xi, \eta, Z, U, x, y)]\, dx\, dy$$

$$= \iint_G \frac{\Delta u}{u}(-\xi_2 Z^2 + \eta_2 Z^1)\, dx\, dy \tag{2.98}$$

The first two equations of (2.60) are satisfied by the substitution

$$\xi_i = -\omega_{iy}, \qquad \eta_i = \omega_{ix} \qquad (i = 1, 2) \tag{2.99}$$

and the second pair of equations (2.60) then results in

$$u\omega_{1y} - \omega_{2x} = 0, \qquad u\omega_{1x} + \omega_{2y} = 0 \tag{2.100}$$

with boundary conditions (2.61) of the form

$$\left.\begin{array}{ll} \omega_2|_{\Gamma_1} = \omega_{2+} = \text{const}, & \omega_2|_{\Gamma_2} = \omega_{2-} = \text{const} \\[2mm] \omega_1|_{\Gamma_3} = \omega_{1-} = \text{const}, & \omega_1|_{\Gamma_4} = \omega_{1+} = \text{const} \\[2mm] \omega_{2+} - \omega_{2-} + 1 = R(\omega_{1+} - \omega_1) \end{array}\right\} \tag{2.101}$$

Let $J(Z^1, Z^2)$ denote an admissible vector of the current density (i.e., a vector corresponding to an admissible control U); then the expression for ΔI becomes

$$\Delta I = \iint_G \frac{\Delta u}{u}(J, \text{grad } \omega_2)\, dx\, dy \tag{2.102}$$

This expression is exact; as we see, the increment ΔI of the functional I depends not only on Δu and the characteristics of the optimal regime (u and grad ω_2), but also on the admissible vector $J(Z^1, Z^2)$ of the current density determined by the admissible values Z^1 and Z^2 of the parametric variables. Whereas the increment Δu of the control may be independently specified, the vector J is determined by solving the boundary-value problem

(2.56)–(2.57), where the admissible control U is to be taken instead of u. It is impossible to solve this problem effectively for arbitrary admissible control, and one must therefore introduce additional special assumptions about the method of variation of the control u; every such assumption generates a definite necessary condition of an extremum. Before we turn to the description of these special variations, let us consider an auxiliary problem, which is important for what follows. We take an ellipse D with center at the point P of the region G and semiaxes a and b oriented along the unit vectors $\boldsymbol{\alpha}$ and $\boldsymbol{\beta}$, respectively; the ellipse is made of a homogeneous conductor of resistivity U. We assume that the ellipse D is placed in a homogeneous field of currents $\mathbf{j}$ that form an angle φ with the axis $\boldsymbol{\beta}$ (Fig. 2.2) and flow in an infinite conducting medium with constant resistivity u. The components of the current density $\mathbf{J}$ within the ellipse are determined by the expressions[155]

$$J_\alpha = -|\mathbf{j}|u\frac{a+b}{ub+Ua}\sin\varphi, \qquad J_\beta = |\mathbf{j}|u\frac{a+b}{ua+Ub}\cos\varphi \qquad (2.103)$$

The field of $\mathbf{J}$ is homogeneous within the ellipse; its orientation with respect to the field $\mathbf{j}$ depends on the ratio $m = b/a$ of the semiaxes of the ellipse and on the orientation of the ellipse with respect to $\mathbf{j}$ (i.e., on the angle φ). If $m = b/a = 1$ (the case of a circular disk), then $\mathbf{J} = [2u/(u+U)]\mathbf{j}$; i.e., the vectors $\mathbf{J}$ and $\mathbf{j}$ are parallel for any φ, and there is no dependence of the mutual orientation on φ (this is obvious on physical grounds). If $b/a \sim 0$ (narrow strip along the $\boldsymbol{\alpha}$ axis), then to terms of $O(b/a)$ one has

$$J_\alpha = -|\mathbf{j}|\frac{u}{U}\sin\varphi = \frac{u}{U}j_\alpha, \qquad J_\beta = |\mathbf{j}|\cos\varphi = j_\beta \qquad (2.104)$$

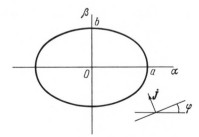

FIGURE 2.2

Analogous expressions hold for $b/a \sim \infty$, a narrow strip along the β axis. It is obvious that narrow strips correspond to the most pronounced dependence of the mutual orientation of the vectors $\mathbf{J}$ and $\mathbf{j}$ on the angle φ, i.e., on the orientation of the ellipse D with respect to the $\mathbf{j}$ field.

With this in mind, we return to the problem of variation of the control in the minimum resistance problem. We shall assume that the function U is piecewise smooth; i.e., continuously differentiable everywhere except at a finite number of smooth lines Γ_0, along which it may have discontinuities of the first kind. We also assume that the functions E_1 and E_2 [see (2.56)] have Hölder-continuous derivatives. It is known[91] that under these conditions, the solution of the problem (2.56)–(2.57) is classical if it satisfies the continuity conditions

$$[z^1]_-^+ = 0, \qquad [z^2]_-^+ = 0 \qquad\qquad (2.105)$$

on the lines Γ_0. We shall assume that these are satisfied.

We define an admissible control U by

$$U = \begin{cases} u + \Delta u, & x \in D_\varepsilon \\ u, & x \notin D_\varepsilon \end{cases} \qquad\qquad (2.106)$$

where D_ε is an ellipse with center at the point P of the region G and with semiaxes $a\varepsilon$ and $b\varepsilon$ oriented along the unit vectors $\boldsymbol{\alpha}$ and $\boldsymbol{\beta}$; the axis $\boldsymbol{\beta}$ makes an angle φ with the optimal vector $\mathbf{j}$ of the current density at the point P. We shall assume that ε is a small parameter ($\varepsilon \ll 1$); suppose that the point P is not a singular point of the field $\mathbf{j}$; then the vector $\mathbf{J}$ within the ellipse differs from the vector (2.103) by an amount of order $o(1)$ if $\mathbf{j}$ is regarded as the optimal current density vector at the point P in these last equations; in addition, the increment Δu should then be regarded as constant in D_ε and equal to the value of Δu at the same point.* What we have said enables us to present (2.102) in the form

$$\Delta I = -\iint_{D_\varepsilon} \Delta u \frac{1 + m}{mu + U}\left[\Delta u \frac{m - 1}{u + mU} j_\beta \omega_{2\beta} - (\mathbf{j}, \operatorname{grad} \omega_2) \right] dx\, dy + o(\varepsilon)$$

$$(2.107)$$

where j_β and $\omega_{2\beta}$ are the components of the vectors $\mathbf{j}$ and of $\operatorname{grad} \omega_2$ along the $\boldsymbol{\beta}$ axis.

---

* This assertion will be proved in Section 2.4; here, we shall assume that the point P is not a point of discontinuity of the optimal control u since it would then be a singular point of the $\mathbf{j}$ field.

The point P, being a continuity point of the integrand in (2.107), is a Lebesque point of this function. Denoting by χ_ε the characteristic function of the set D_ε, we obtain*

$$\lim_{\varepsilon \to 0} \frac{\Delta I}{\displaystyle\iint_{D_\varepsilon} \chi_\varepsilon \, dx \, dy} = -\Delta u \frac{1 + m}{mu + U} \left[\Delta u \frac{m - 1}{u + mU} j_\beta \omega_{2\beta} - (\mathbf{j}, \operatorname{grad} \omega_2) \right]$$

(2.108)

If the functional I is to attain a maximum, it is necessary that $\Delta I \leq 0$; the same inequality must obviously be satisfied by the left-hand side of (2.108) and, therefore, by its right-hand side as well. We thus arrive at the inequality

$$\Delta u \frac{1 + m}{mu + U} \left[\Delta u \frac{m - 1}{u + mU} j_\beta \omega_{2\beta} - (\mathbf{j}, \operatorname{grad} \omega_2) \right] \geq 0 \qquad (2.109)$$

which expresses Weierstrass's necessary condition of a strong relative minimum. The inequality (2.109) must be satisfied at all continuity points of the optimal control u; continuity arguments show that it must also be satisfied up to (smooth) discontinuity lines of this function. The left-hand side of (2.109) contains three parameters characterizing the variation of the control: Δu, m, and φ (see Fig. 2.2). Without loss of generality, we may assume $0 < m \leq 1$ (the case $\infty > m \geq 1$ reduces to this case if the roles of the α and β coordinate axes are interchanged). The inequality (2.109) must be satisfied for all possible values of the parameters. We note first that only the following two cases are possible:

Case 1: $\quad \Delta u \leq 0$, $\quad A^{(1)} = \Delta u \dfrac{m - 1}{u + mU} j_\beta \omega_{2\beta} - (\mathbf{j}, \operatorname{grad} \omega_2) \leq 0 \qquad (2.110)$

Case 2: $\quad \Delta u \geq 0$, $\quad A^{(2)} = \Delta u \dfrac{m - 1}{u + mU} j_\beta \omega_{2\beta} - (\mathbf{j}, \operatorname{grad} \omega_2) \geq 0 \qquad (2.111)$

In the first of these cases, $u = u_{\max}$ [see (2.58)] and $(\mathbf{j}, \operatorname{grad} \omega_2) \geq 0$ [this last follows from the second inequality in (2.110) if the β axis is chosen such that $j_\beta = 0$]; in the second case, $u = u_{\min}$ and $(\mathbf{j}, \operatorname{grad} \omega_2) \leq 0$. The case $(\mathbf{j}, \operatorname{grad} \omega_2) = 0$ must be rejected (if one ignores the trivial possibility $\mathbf{j} = 0$ or $\operatorname{grad} \omega_2 = 0$ at the point P) since the vectors $\mathbf{j}$ and $\operatorname{grad} \omega_2$ then form

---

* The limit equation (2.108) is valid if the passage to the limit is made with respect to a system of sets that *contract regularly* (see Ref. 114, pp. 397–398) to the point P. For $0 < m < \infty$, the sets D_ε contract regularly to P as $\varepsilon \to 0$.

a right angle, and the β axis can be chosen in such a way that the expression

$$(\Delta u)^2 \frac{m^2 - 1}{(mu + U)(u + mU)} j_\beta \omega_{2\beta}$$

is either positive or negative.*

Let us investigate Case 1 in more detail. Let χ be the angle formed by the vectors $\mathbf{j}$ and grad ω_2 at the point P (obviously, $\pi/2 \geq \gamma \geq 0$ in Case 1). If the direction of β lies within the hatched sectors (Fig. 2.3) bounded by the straight lines aa and bb perpendicular to the vectors $\mathbf{j}$ and grad ω_2, respectively, then $j_\beta \omega_{2\beta} < 0$, and the inequality (2.110) is satisfied; if the vector β lies outside these sectors, then a maximum of the function

$$f(\varphi, \psi) = (1 - m) \frac{u_{max} - u_{min}}{u_{max} + mu_{min}} j_\beta \omega_{2\beta}$$

$$= (1 - m) \frac{u_{max} - u_{min}}{u_{max} + mu_{min}} |\mathbf{j}| \, |\text{grad } \omega_2| \cos \varphi \cos \psi$$

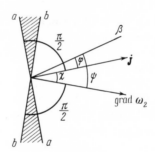

FIGURE 2.3

---

* The following heuristic considerations demonstrate that only the limiting values u_{min} and u_{max} of the control u should be admitted in the optimal region. Consider the "lumped parameter" approximation of the conducting medium occupying the region G (Example 2 in Section 2.2). This approximation is provided by some arbitrarily complicated linear network containing lumped electromotive forces and resistances; the load R may also be included in this network. It is easily seen that the current $I(u)$ flowing through any of these resistances (in particular, through R) depends on some other arbitrary resistance u of the network according to the law

$$I(u) = \frac{au + b}{cu + d}$$

where the coefficients a, b, c, d may be at most linearly dependent on the remaining resistances, the first two including a possible dependence on the electromotive forces. The function $I(u)$ is monotonic and thus attains extreme values at the endpoints of the segment $[u_{min}, u_{max}]$. We rule out the trivial case $ad - bc = 0$ when $I(u)$ is independent of u.

must be found under the additional conditions $\psi = \chi + \varphi$, $\chi = $ const, with the requirement that the corresponding value of $A^{(1)}$ be nonpositive. It is easy to verify that the function $f(\varphi, \chi + \varphi)$ attains a maximum at $\varphi = -\chi/2$, i.e., for the direction of β that bisects the acute angle χ (we shall say that the corresponding position of the ellipse is critical). For the critical position,

$$f = f_{max} = (1 - m) \frac{u_{max} - u_{min}}{u_{max} + mu_{min}} |\mathbf{j}| |\text{grad } \omega_2| \cos^2 (\chi/2)$$

The corresponding value of $A^{(1)}$ is

$$A^{(1)}_{max} = |\mathbf{j}| |\text{grad } \omega_2| \left[(1 - m) \frac{u_{max} - u_{min}}{u_{max} + mu_{min}} \cos^2 (\chi/2) - \cos \chi \right]$$

The requirement $A^{(1)}_{max} \leq 0$ now leads to the inequality

$$0 \leq \chi \leq \arccos \frac{\lambda(u_{max} - u_{min})}{2u_{max} - \lambda(u_{max} - u_{min})} \qquad (2.112)$$

where

$$\lambda = \frac{1 - m}{1 + mu_{min}/u_{max}}, \qquad 0 \leq \lambda \leq 1$$

With increasing λ (i.e., with decreasing m), the right-hand side of the inequality (2.112) decreases monotonically; for $m = 1$ (the case of a disk), $\lambda = 0$, and (2.112) yields

$$0 \leq \chi \leq \pi/2 \qquad (2.113)$$

If $m = 0$ (narrow strip), then $\lambda = 1$, so that $u = u_{max}$ if $(\mathbf{j}, \text{grad } \omega_2) \geq 0$ and

$$0 \leq \chi \leq \arccos p, \qquad p = \frac{u_{max} - u_{min}}{u_{max} + u_{min}} \qquad (2.114)$$

Obviously, the last inequality is the strongest condition; if it is satisfied, then so are (2.112) and (2.113) for all intermediate values of λ (or m).

In the second case [see (2.111)] the analogue of (2.113) is the condition

$$\pi \geq \chi \geq \pi/2 \qquad (2.115)$$

and the condition

$$u = u_{\min}, \qquad \text{if } (\mathbf{j}, \text{grad } \omega_2) \le 0 \quad \text{and} \quad \pi \ge \chi \ge \pi - \arccos p \quad (2.116)$$

the analogue to (2.114) may be derived in a completely analogous manner.

Our argument shows, first, that the strongest conditions for the chosen method of variation (within an ellipse) are obtained when the control is varied in a narrow strip; second, the critical positions of the strip are those when its normal (the direction of $\boldsymbol{\beta}$) bisects the angle χ made by the vectors $\mathbf{j}$ and grad ω_2 at the given point (obviously, the sign of the normal vector is left undefined).

The transition to a strip can be made right at the start by setting $m = 0$ in (2.109). The inequality

$$\frac{\Delta u}{U}\left[\frac{\Delta u}{u}j_\beta \omega_{2\beta} + (\mathbf{j}, \text{grad } \omega_2)\right] \le 0 \qquad (2.117)$$

then can be obtained from the condition

$$\frac{\Delta u}{u}(\mathbf{J}, \text{grad } \omega_2) \le 0 \qquad (2.118)$$

which, in turn, can be derived from (2.102) by means of arguments similar to those that lead to (2.109).

In order to make the transition from (2.118) to (2.117), we assume that $\Delta u \ne 0$ within a narrow strip (of width ε) with normal $\mathbf{n}$; the values of the current density $\mathbf{J}$ on the boundary of the strip, obtained from the values outside the strip in the limit as $\varepsilon \to 0$, are the same as those which would be obtained in the absence of the strip, i.e., they are equal to the corresponding values of the components of the optimal vector $\mathbf{j}$ at the given position.* On the other hand, if $\mathbf{t}$ is the unit vector directed along the strip, then to $o(1)$, the following relations hold between the limiting values of the vectors $\mathbf{j}$ and $\mathbf{J}$ on the rectilinear boundary of the strip.[53],†

$$j_n = J_n, \qquad uj_t = UJ_t \qquad (2.119)$$

---

* This assertion is false only in the neighborhood of the ends of the strip; the final result (2.119), however, remains unaffected.
† See preceding footnote.

Equations (2.119) enable us to eliminate the components of the vector **J** from (2.118); the result will hold to $o(1)$, since this is the order of the difference between the components of the vector **J** inside the strip and on its boundary. We obtain

$$\frac{\Delta u}{u}\left(j_n \omega_{2n} + \frac{u}{U} j_t \omega_{2t} \right) \leq 0$$

i.e., the inequality (2.117). Equations (2.119) express nothing more than the continuity of the limiting values of the derivatives of the functions z^1 and z^2 in the direction along the strip on the boundary of the strip. These conditions follow from (2.105), which express the continuity of the functions z^1 and z^2 on the boundary of the strip.

These remarks make it possible to present the Weierstrass condition in a form corresponding to the expression of the basic equations of the problem in the form (2.12).

Suppose $s_1 = n$, $s_2 = t$, $ds_1 = dn = H_1\,dq_1$, $ds_2 = dt = H_2\,dq_2$ are local orthogonal curvilinear coordinates; in these coordinates, the basic equations (2.56) take the form

$$z_n^1 = uz_t^2 - E_{0n}, \qquad z_n^2 = -\frac{1}{u}z_t^1 - \frac{1}{u}E_{0t} \qquad (2.120)$$

The Hamiltonian is

$$H = -\omega_{1t}(uz_t^2 - E_{0n}) + \omega_{2t}\left(\frac{1}{u}z_t^1 + \frac{1}{u}E_{0t} \right) \qquad (2.121)$$

and the functions ω_1 and ω_2 satisfy the adjoint equations

$$u\omega_{1t} - \omega_{2n} = 0, \qquad u\omega_{1n} + \omega_{2t} = 0 \qquad (2.122)$$

which are the same as those of the system (2.100) if $\mathbf{n} = (y_t, -x_t)$, $\mathbf{t} = (x_t, y_t)$; the functions also satisfy the boundary conditions (2.101).

It now is a simple matter to obtain the inequality (2.117). We assume that $\Delta u \neq 0$ in a strip with normal **n** and tangent **t**, and form the difference

$$\Delta H = H(\omega_1, \omega_2, z_t^1, z_t^2, U, x, y) - H(\omega_1, \omega_2, z_t^1, z_t^2, u, x, y)$$

$$= -\omega_{1t}(U - u)z_t^2 - \omega_{2t}\left(\frac{1}{u} - \frac{1}{U} \right)(z_t^1 + E_{0t}) \qquad (2.123)$$

The requirement $\Delta H \leq 0$ then leads to (2.117) if the derivative ω_{1t} is eliminated by means of the first of (2.122) and the derivative z_t^1 is eliminated by means of the second equation in (2.120).

The difference between the function H determined in accordance with (2.59) and the function (2.121) is that the former depends on the parametric variables ζ^1 and ζ^2, which, in general, are discontinuous on the boundaries of the strip [their combinations $\zeta^1 y_t - \zeta^2 x_t = j_n = -z_t^2$ and $u(\zeta^1 x_t + \zeta^2 y_t) = uj_t = -z_t^1 - E_{0t}$, are continuous]; this circumstance is reflected in (2.98). The function (2.121), in contrast, is constructed in such a way that the parametric variables ζ^1 and ζ^2 enter it through the continuous combinations z_t^1 and z_t^2, so that only the control u has a discontinuity on the boundary of the strip. This is achieved by virtue of the fact that $\mathbf{n}$ and $\mathbf{t}$ are related to the orientation of the strip in the special manner specified above; a change in the orientation of the strip is accompanied by a new choice of coordinates. The final inequalities (2.114) and (2.116) are obtained after all possible orientations of the vector pairs $(\mathbf{n}, \mathbf{t})$ have been compared, as was done above.

It is noteworthy that it is precisely the inequality $\Delta H \leq 0$, written in the form (2.123), which is the complete formal analogue of Pontryagin's maximum principle; by contrast, the inequality $\Delta H \leq 0$ containing the function H defined by (2.59) is not such an analogue since the inequality then contains admissible values of the parametric variables.

What we have said clarifies the remark made in Section 2.1 with regard to the role of (2.11) and (2.12). The elements of the matrix $\{c_{kl}\} = \{(\mathbf{e}_k, \mathbf{i}_l)\}$, $k, l = 1, 2$, and the control u form a unified system of control parameters in the sense that the corresponding Weierstrass inequality (2.117) must be satisfied for all possible admissible values of the variables c_{kl} and U. The use of variations in a strip makes it possible to speak with a certain amount of justification of the "anisotropy" of the variation of the control u (see Introduction).

Our results make it possible to give a simple proof of Grötzsch's theorem (see Introduction). Suppose that the distribution of the current $\mathbf{j}$ and the potential z^1 in the region G is described by (2.26), where $u = u_{\min}$ and $\mathbf{E}_0 = 0$, and by the boundary conditions

$$z^1|_{\Gamma_1} - z^1|_{\Gamma_2} + 1 = IR, \qquad j_N|_{\Gamma_3} = j_N|_{\Gamma_4} = 0 \qquad (2.124)$$

where I again denotes the functional ($\mathbf{N}$ is the outer normal to Γ)

$$I = \int_{\Gamma_1} j_N \, dt \qquad (2.125)$$

The first condition (2.124) expresses the presence of a unit electromotive force in the external circuit. Grötzsch's theorem asserts that under the imposed conditions, the introduction of sections on which $u(x, y) > u_{min}$ into G decreases the value of the functional I.

We shall assume that the function $u(x, y)$ is chosen from among the piecewise continuous functions* satisfying the condition (2.28). It is easy to show that under the imposed conditions the equation

$$\mathbf{j} = -\frac{1}{u} \operatorname{grad} \omega_2 \qquad (2.126)$$

holds for any function $u(x, y)$ [to see this, it suffices to compare (2.56) and the condition (2.124) with the adjoint system (2.100) and the boundary conditions (2.101)].

It follows from (2.126), in particular, that $\chi = \pi$ and Weierstrass's inequality (2.116) shows that the functional is maximized by the control $u = u_{min}$ for every point of the region. The fact that $\chi = \pi$ assures that (2.116) is not violated for any possible value of p [see (2.114)] and this, in turn, means that the choice $u = u_{min}$ is optimal for any value of u_{max}, including $u_{max} = \infty$ (when perfect insulators are permitted). In other words, the restriction $u_{min} \leq u(x, y) \leq u_{max}$ can be replaced by the restriction $u(x, y) \geq u_{min}$ without affecting the validity of Grötzsch's theorem.†

It is also apparent that to prove the theorem it suffices to use Weierstrass's inequalities in the weakened form (2.113) and (2.115); the true value of the stronger inequalities (2.114) and (2.116) becomes apparent only in the proof of more general propositions that reduce to the case $\mathbf{E}_0 \neq 0$ when the assertion of Grötzsch's theorem generally is not true. This problem will be considered in detail in Chapter 3.

Finally, one can show (see Appendix A) that the assertion of Grötzsch's theorem is also true with respect to strong global variations of the resistivity, as mentioned in the Introduction.

2.4. Weierstrass's Necessary Condition (Continued): An Example

The arguments of the preceding section require some explanation. Within the framework of the assumptions made above—piecewise smooth

---

* In what follows, it will be shown that the result of the present argument also remains true for the larger class of measurable functions whose $L_\infty(G)$ norm satisfies the condition (2.58).
† Grötzsch himself stated and proved the theorem for the case when the sections introduced into the region have infinite resistivity ($U = u_{max} = \infty$).

admissible controls—we must show that (2.103) represent the principal part of the vector $\mathbf{J}$ inside the ellipse D_ε when $\varepsilon \to 0$, or equivalently that

$$\Delta z^2(x, y) = -|\mathbf{j}| \, \Delta u\left(\frac{b \cos \varphi}{ua + Ub}x + \frac{a \sin \varphi}{ub + Ua}y\right) \qquad (x, y) \in D_\varepsilon \quad (2.127)$$

yields the principal part of the increment Δz^2 for $(x, y) \in D_\varepsilon$ as $\varepsilon \to 0$.

We shall find it convenient to give this proof first for piecewise constant control and then to extend the result to the general case of measurable bounded controls in Section 2.6.

Thus, suppose that the control $u(x, y)$ is piecewise constant; i.e., constant everywhere except at a finite number of smooth lines Γ_0, along which it may have discontinuities of the first kind. Under fairly general assumptions about the boundary Γ of the domain G, there exists a generalized Green's function of the boundary-value problem (2.56)–(2.57). Stated differently, there exists a fundamental solution of the equation

$$M_u g \equiv \text{div } u \text{ grad } g = \delta(Q) \qquad (2.128)$$

satisfying the boundary conditions (see Fig. 1.1, $\mathbf{N}$ is the outer normal to Γ)

$$g|_{\Gamma_3} = 0, \qquad Rg|_{\Gamma_4} + \int_{\Gamma_4} u\frac{\partial g}{\partial N}\,dt = 0$$

$$\left.\frac{\partial g}{\partial N}\right|_{\Gamma_1} = \left.\frac{\partial g}{\partial N}\right|_{\Gamma_2} = 0 \qquad (2.129)$$

and the conditions

$$[g]_-^+ = 0, \qquad \left[u\frac{\partial g}{\partial N}\right]_-^+ = 0 \qquad (2.130)$$

on the discontinuity lines Γ_0 of the control $u(x, y)$.

If $f \in L_p(G)$, $1 < p < \infty$, then

$$z(P) = \int_G f(Q)g(P, Q)\,dS_Q$$

$$dS_Q = dx_Q\,dy_Q \qquad (2.131)$$

defines a generalized solution of the inhomogeneous equation $M_u z = f$ satisfying the conditions (2.129) and (2.130).

Let $u_1(x, y)$ and $u_2(x, y)$ be two different piecewise constant controls, let Q_1 and Q_2 be two arbitrary points in G, and let $g^1(P, Q_1)$ and $g^2(P, Q_2)$ be the Green's functions of the operators M_{u_1} and M_{u_2} with poles Q_1 and Q_2, respectively. In the integral identity

$$\int_G \varphi \nabla \cdot \mathbf{A} \, dS = \int_\Gamma \varphi A_N \, dt - \int_G (\mathbf{A}, \nabla \varphi) \, dS \qquad (2.132)$$

we first set $\varphi = g^2$, $\mathbf{A} = u_1 \nabla g^1$, and then $\varphi = g^1$, $\mathbf{A} = u_2 \nabla g^2$; subtracting the results term by term and using (2.128) and the conditions (2.120) and (2.130), we obtain

$$g^2(Q_1, Q_2) = g^1(Q_2, Q_1) = \int_G (u_2 - u_1)(\nabla_P g^1(P, Q_1), \nabla_P g^2(P, Q_2)) \, dS_P \qquad (2.133)$$

In particular, if $u_1 = u_2$, then (2.133) leads to

$$g(Q_1, Q_2) = g(Q_2, Q_1) \qquad (2.134)$$

expressing the symmetry property of the Green's function.

Equation (2.133) can be differentiated term by term with respect to the coordinates of the point Q_1, for example; the result of such a differentiation must be understood in the sense of generalized functions. We have

$$\nabla_{Q_1} g^2(Q_1, Q_2) - \int_G (u_2 - u_1)\left(\frac{d\nabla_P g^1(P, Q_1)}{d\mathbf{r}_{Q_1}}, \nabla_P g^2(P, Q_2)\right) dS_P$$

$$= \nabla_{Q_1} g^1(Q_2, Q_1) \qquad (2.135)$$

In order to express the components of the tensor $d\nabla_P g^1(P, Q_1)/d\mathbf{r}_{Q_1}$, we note that the Green's function $g^1(P, Q_1)$ admits the representation

$$g^1(P, Q_1) = \frac{1}{2\pi u_1(Q_1)} \ln r_{Q_1 P} + g^1_*(P, Q_1) \qquad (2.136)$$

where $g^1(P, Q_1)$ is twice continuously differentiable with respect to the points P and Q_1 everywhere except the point $P = Q_1$, where $\lim_{P \to Q_1} g^1_*(P, Q_1) = O(1)$, and also the lines Γ_0; we have also introduced the notation

$$r_{Q_1 P} = [(x_{Q_1} - x_P)^2 + (y_{Q_1} - y_P)^2]^{1/2}$$

The components of the vector $u_1(Q_1)\nabla_P g^1(P, Q_1)$ are equal to

$$u_1(Q_1)g^1_{x_P} = \frac{1}{2\pi}\frac{\cos\varphi}{r_{Q_1 P}} + u_1(Q_1)\frac{\partial g^1_*(P, Q_1)}{\partial x_P}$$

$$u_1(Q_1)g^1_{y_P} = \frac{1}{2\pi}\frac{\sin\varphi}{r_{Q_1 P}} + u_1(Q_1)\frac{\partial g^1_*(P, Q_1)}{\partial y_P}$$

(2.137)

here, φ denotes the polar angle of the radius vector $\mathbf{r}_{Q_1 P}$ of the point P with respect to the origin Q_1.

The expressions for the components of the tensor $d\nabla_P g^1(P, Q_1)/d\mathbf{r}_{Q_1}$ contain generalized functions;[46] for example,

$$u_1(Q_1)g^1_{x_{Q_1} x_P} = \left\{\frac{1}{2\pi}\frac{\cos\varphi}{r_{Q_1 P}}\right\}_{x_{Q_1}} - \delta(P, Q_1)\frac{1}{2\pi}\int_0^{2\pi}\cos^2\varphi\, d\varphi$$

$$+ u_1(Q_1)\frac{\partial^2 g^1_*(P, Q_1)}{\partial x_{Q_1}\partial x_P}$$

$$\left\{\frac{1}{2\pi}\frac{\cos\varphi}{r_{Q_1 P}}\right\}_{x_{Q_1}} = -\frac{1}{2\pi r^2_{Q_1 P}} + \frac{1}{2\pi}\frac{2(x_P - x_Q)^2}{r^4_{Q_1 P}} = \frac{1}{2\pi}\frac{\cos 2\varphi}{r^2_{Q_1 P}}$$

The expressions for the remaining components are similar. We obtain

$$u_1(Q_1)\frac{d\nabla_P g^1(P, Q_1)}{d\mathbf{r}_{Q_1}}$$

$$= \frac{1}{2\pi}\left\{\begin{matrix} r^{-2}_{Q_1 P}\cos 2\varphi & r^{-2}_{Q_1 P}\sin 2\varphi \\ r^{-2}_{Q_1 P}\sin 2\varphi & -r^{-2}_{Q_1 P}\cos 2\varphi \end{matrix}\right\}$$

$$-\frac{1}{2}\left\{\begin{matrix} \delta(P, Q_1) & 0 \\ 0 & \delta(P, Q_1) \end{matrix}\right\} + u_1(Q_1)\frac{d\nabla_P g^1_*(P, Q_1)}{d\mathbf{r}_{Q_1}}$$

(2.138)

$$u_1(Q_1)\left(\frac{d\nabla_P g^1(P, Q_1)}{d\mathbf{r}_{Q_1}}, \nabla_P g^2(P, Q_2)\right) = \mathbf{i}_1 A_1 + \mathbf{i}_2 A_2$$

(2.139)

where

$$
\left.
\begin{aligned}
A_1 &= g_{x_P}^2 \left[\frac{1}{2\pi} \frac{\cos 2\varphi}{r_{Q_1 P}^2} - \frac{1}{2} \delta(P, Q_1) \right] + g_{y_P}^2 \frac{1}{\pi} \frac{\sin 2\varphi}{r_{Q_1 P}^2} + u_1(Q_1) A_{1*} \\[2mm]
A_2 &= g_{x_P}^2 \frac{1}{2\pi} \frac{\sin 2\varphi}{r_{Q_1 P}^2} - g_{y_P}^2 \left[\frac{1}{2\pi} \frac{\cos 2\varphi}{r_{Q_1 P}^2} + \frac{1}{2} \delta(P, Q_1) \right] + u_1(Q_1) A_{2*} \\[2mm]
A_{1*} &= g_{x_P}^2 \frac{\partial^2 g_*^1}{\partial x_{Q_1} \partial x_P} + g_{y_P}^2 \frac{\partial^2 g_*^1}{\partial y_{Q_1} \partial y_P} \\[2mm]
A_{2*} &= g_{x_P}^2 \frac{\partial^2 g_*^1}{\partial x_{Q_1} \partial y_P} + g_{y_P}^2 \frac{\partial^2 g_*^1}{\partial y_{Q_1} \partial y_P}
\end{aligned}
\right\} \quad (2.140)
$$

We now suppose that the function $g^1(Q_2, Q_1)$ is known; then (2.135) is equivalent to a system of two singular integral equations for the unknown functions $g_{x_{Q_1}}^2(Q_1, Q_2), g_{y_{Q_1}}^2(Q_1, Q_2)$; the compatibility of this system follows directly from its construction.

The system has the form (we assume that $Q_1 \in G$)

$$
\left.
\begin{aligned}
&\frac{u_2(Q_1) + u_1(Q_1)}{2u_1(Q_1)} g_{x_{Q_1}}^2(Q_1, Q_2) - \frac{1}{2\pi} \int_G \frac{u_2(P) - u_1(P)}{u_1(Q_1)} \\[2mm]
&\qquad \times \left[g_{x_P}^2(P, Q_2) \frac{\cos 2\varphi}{r_{Q_1 P}^2} + g_{y_P}^2(P, Q_2) \frac{\sin 2\varphi}{r_{Q_1 P}^2} + u_1(Q_1) A_{1*} \right] dS_P \\[2mm]
&\qquad\qquad = g_{x_{Q_1}}^1(Q_1, Q_2) \\[3mm]
&\frac{u_2(Q_1) + u_1(Q_1)}{2u_1(Q_1)} g_{y_{Q_1}}^2(Q_1, Q_2) - \frac{1}{2\pi} \int_G \frac{u_2(P) - u_1(P)}{u_1(Q_1)} \\[2mm]
&\qquad \times \left[g_{x_P}^2(P, Q_2) \frac{\sin 2\varphi}{r_{Q_1 P}^2} - g_{y_P}^2(P, Q_2) \frac{\cos 2\varphi}{r_{Q_1 P}^2} + u_1(Q_1) A_{2*} \right] dS_P \\[2mm]
&\qquad\qquad = g_{y_{Q_1}}^1(Q_1, Q_2)
\end{aligned}
\right\} \quad (2.141)
$$

and to write down the expressions on the right-hand sides we use the symmetry of the Green's function $g^1(Q_1, Q_2)$.

We now suppose that

$$u_2(P) - u_1(P) \neq 0, \qquad P \in G_\varepsilon$$

$$u_2(P) - u_1(P) = 0, \qquad P \in G_\varepsilon \tag{2.142}$$

we shall assume that the set G_ε belongs to a system of sets which can be regularly contracted to a point O common to all of these sets.* Seeking a solution of the system (2.141) that is continuously differentiable in G_ε and taking into account what we have said above, we may write (2.141) in the form $(O = Q_1 \in G_\varepsilon)$

$$\left.\begin{aligned}
&\frac{u_1(Q_1) + u_2(Q_1)}{2u_1(Q_1)} g^2_{x_{Q_1}}(Q_1, Q_2) - \frac{u_2(Q_1) - u_1(Q_1)}{2u_1(Q_1)} \\
&\quad \times [Cg^2_{x_{Q_1}}(Q_1, Q_2) + Sg^2_{y_{Q_1}}(Q_1, Q_2)] = g^1_{x_{Q_1}}(Q_1, Q_2) + o(1) \\
&\frac{u_1(Q_1) + u_2(Q_1)}{2u_1(Q_1)} g^2_{y_{Q_1}}(Q_1, Q_2) - \frac{u_2(Q_1) - u_1(Q_1)}{2u_1(Q_1)} \\
&\quad \times [Sg^2_{x_{Q_1}}(Q_1, Q_2) - Cg^2_{y_{Q_1}}(Q_1, Q_2)] = g^1_{y_{Q_1}}(Q_1, Q_2) + o(1)
\end{aligned}\right\} \tag{2.143}$$

where

$$C = \frac{1}{\pi} \int_{G_\varepsilon} \frac{\cos 2\varphi}{r^2_{Q_1 P}} \, dS_P, \qquad S = \frac{1}{\pi} \int_{G_\varepsilon} \frac{\sin 2\varphi}{r^2_{Q_1 P}} \, dS_P \tag{2.144}$$

and $o(1)$ denotes a quantity that tends to zero as ε tends to zero. A solution of the system (2.143) is given by

$$g^2_{x_{Q_1}}(Q_1, Q_2) = \frac{\Delta_x}{\Delta} + o(1)$$

$$g^2_{y_{Q_1}}(Q_1, Q_2) = \frac{\Delta_y}{\Delta} + o(1) \tag{2.145}$$

---

* If the sets G_ε include sets of arbitrarily small diameter and if, for every set G_ε of this system, there exists a square containing it with center at the point 0 and edge h such that

$$h^2 \leq \alpha\varepsilon, \qquad \varepsilon = \text{mes } G_\varepsilon$$

where α is a number which does not depend on the choice of the set G_ε, then the system of sets G_ε is said to be *regularly contractible* to the point 0 as $\varepsilon \to 0$. In particular, a system of concentric disks and a system of similar ellipses are regularly contractible to their common center. For more details about regularly contractible sets, see Ref. 114.

where

$$
\begin{aligned}
\Delta_x &= \frac{u_1(Q_1)(1-C) + u_2(Q_1)(1+C)}{2u_1(Q_1)} g^1_{x_{Q_1}}(Q_1, Q_2) \\
&\quad + \frac{u_2(Q_1) - u_1(Q_1)}{2u_1(Q_1)} Sg^1_{y_{Q_1}}(Q_1, Q_2) \\
\Delta_y &= \frac{u_2(Q_1) - u_1(Q_1)}{2u_1(Q_1)} Sg^1_{x_{Q_1}}(Q_1, Q_2) \\
&\quad + \frac{u_1(Q_1)(1+C) + u_2(Q_1)(1-C)}{2u_1(Q_1)} g^1_{y_{Q_1}}(Q_1, Q_2) \\
\Delta &= \frac{1}{4u_1^2(Q_1)} \{[u_1(Q_1) + u_2(Q_1)]^2 - (C^2 + S^2)[u_1(Q_1) - u_2(Q_1)]^2\}
\end{aligned}
\right\} \quad (2.146)
$$

Let G_ε be the interior of the ellipse D_ε with center Q_1; the polar equation of the ellipse has the form $r^2 = \varepsilon^2(a^2 \cos^2 \varphi + b^2 \sin^2 \varphi)$. We have

$$
\begin{aligned}
C &= \frac{1}{2\pi} \int_0^{2\pi} \ln(a^2 \cos^2 \varphi + b^2 \sin^2 \varphi) \cos 2\varphi \, d\varphi = \frac{a-b}{a+b} \\
S &= \frac{1}{2\pi} \int_0^{2\pi} \ln(a^2 \cos^2 \varphi + b^2 \sin^2 \varphi) \sin 2\varphi \, d\varphi = 0 \\
g^2_{x_{Q_1}}(Q_1, Q_2) &= \frac{u_1(Q_1)(a+b)}{au_1(Q_1) + bu_2(Q_1)} g^1_{x_{Q_1}}(Q_1, Q_2) + o(1) \\
g^2_{y_{Q_1}}(Q_1, Q_2) &= \frac{u_1(Q_1)(a+b)}{au_2(Q_1) + bu_1(Q_1)} g^1_{y_{Q_1}}(Q_1, Q_2) + o(1)
\end{aligned}
\right\} \quad (2.147)
$$

The argument Q_2 in these equations indicates that the potentials g^1 and g^2 are produced by a point source placed at Q^2. It is clear that the same equations describe the potentials produced by any distribution of the sources within the region.* We now set $u_1 = u$, $u_2 = U$, $g^2(Q_1, Q_2) = Z^2(Q_1, Q_2)$, $g^1(Q_1, Q_2) = z^2(Q_1, Q_2)$, $g^2_{x_{Q_1}} = J_y$, $g^2_{y_{Q_1}} = -J_x$, $g^1_{x_{Q_1}} = j_y$,

---

* The point Q_1 is placed outside the region G_ε; (2.135) may be written in the form

$$
g^2_{x_{Q_1}}(Q_1, Q_2) = g^1_{x_{Q_1}}(Q_1, Q_2) + O(\text{mes } G_\varepsilon)
$$

$$
g^2_{y_{Q_1}}(Q_1, Q_2) = g^1_{y_{Q_1}}(Q_1, Q_2) + O(\text{mes } G_\varepsilon)
$$

$g^1_{y_{Q_1}} = -j_x$; we then arrive at the expressions

$$J_x = \frac{u(a+b)}{ub + Ua}j_x + o(1), \qquad J_y = \frac{u(a+b)}{ua + Ub}j_y + o(1) \qquad (2.148)$$

which coincide with (2.103) [to within terms of $o(1)$]. Equations (2.144)–(2.146) may be used to obtain analogues of (2.103) for small regions of arbitrary shape. However, it must be borne in mind that when the terms of $o(1)$ are omitted, Eqs. (2.148) are an exact solution of the system (2.141), where

$$u_1 = u = \text{const}, \qquad u_2 = \begin{cases} U = \text{const}, & x \in D_\varepsilon \\ u = \text{const}, & x \notin D_\varepsilon \end{cases} \qquad (2.149)$$

(G_ε is an ellipse D_ε), and where we have taken a flow function of the homogeneous field of currents **j** instead of the function $g^1(Q_1, Q_2)$. The exact nature of this solution (the homogeneity of the **J** field within the ellipse) is due to the fact that the first two equations in (2.147) remain true if any point within the ellipse is taken as the pole Q_1.* If the form of the region G_ε is not elliptic, with the other conditions remaining as before, the **J** field within the region G_ε is no longer strictly homogeneous (it will be more nearly so, the more nearly G_ε is an ellipse).

On the other hand, if the external field **j** is produced by a point source placed at Q_2 (i.e., at a finite distance from D_ε), then the **J** field in the elliptic region D_ε will be only approximately homogeneous (the deviations from homogeneity decrease with increasing distance of the source or, which is the same thing, with decreasing size of the ellipse compared with the radius of curvature R of the streamline of the original field at the point Q_1 to which the ellipse is contracted).

If, finally we place a region G_ε which is not elliptic into an inhomogeneous external **j** field (with source at Q_2), the inhomogeneity of the **J** field in this region will be due to both factors—inhomogeneity of the external field and the nonellipticity of the region. The influence of the former gets weaker as the region G_ε is contracted to the pole Q_1 irrespective of the shape of this region. For a sufficiently small (compared with R) region G_ε, the region can be assumed to be situated in a homogeneous field

---

* It is sufficient to note that the derivatives with respect to x and y of the projections of the force of attraction by the homogeneous ellipse of a material point outside the ellipse are proportional to the integrals that occur on the left-hand sides of (2.147). The projections of this force onto the x and y axes of the ellipse are proportional to the corresponding coordinates (see Ref. 156, p. 102, for the analogous expressions in the case of a homogeneous ellipsoid).

of currents equal to $\mathbf{j}(Q_1)$. The $\mathbf{J}$ field within G_ε will nevertheless remain inhomogeneous; although the size of the region is immaterial, the shape of the region still remains important. We therefore encounter the problem of determining the shape of the region of variation which leads to the strongest necessary conditions for an extremum. In the following section, we shall show that in this respect strips and narrow layers which derive from elliptic (ellipsoidal) regions play a distinguished role.

2.5. An Extremal Property of Strip-Shaped Regions

The sign of expression (2.102) for ΔI is of particular interest; under the assumptions made, we may take $\Delta u = \text{const}$, $u = \text{const}$, grad $\omega_2 = \text{const} = \mathbf{a}$ in G_ε; it is required that $\Delta I \leq 0$ for all admissible G_ε. It is convenient to specify the area S of the region G_ε and consider the functional

$$\frac{1}{S}\Delta I = \frac{1}{S}\int\int_{G_\varepsilon} \frac{\Delta u}{u}(\mathbf{J}, \mathbf{a})\, dx\, dy$$

Thus, we arrive at the following auxiliary problem. A homogeneous field of currents $\mathbf{j}$ is created in an infinite homogeneous medium with resistivity $u = \text{const}$. A cylindrical body of resistivity $U = \text{const} > u$ is introduced into the medium; the area of the cross section G_ε of the body is given and equal to S. Let $\mathbf{a} = \mathbf{i}_1 a_1 + \mathbf{i}_2 a_2$ be a constant vector such that $(\mathbf{j}, \mathbf{a}) \leq 0$. We wish to determine the shape of the section G_ε that maximizes the following functional

$$W_a = \frac{1}{S}\int\int_{G_\varepsilon}(\mathbf{J}, \mathbf{a})\, dx\, dy \qquad (2.150)$$

A comparison with (2.120) shows that $W_a = (\Delta I / S\, \Delta u)u$; since $u > 0$, $\Delta u > 0$, it follows that max $W_a = c$ max ΔI, $c > 0$. If the increment ΔI is to be nonpositive, we must have max $\Delta I \leq 0$, and hence max $W_a \leq 0$ as well. Therefore, the conditions under which the functional (2.150) attains a maximum are of primary interest.

Similarly, if $U = \text{const} < u$, then one chooses a constant vector $\mathbf{b}$ such that $(\mathbf{j}, \mathbf{b}) \geq 0$ and seeks a shape of the cross section G_ε of area S which minimizes the functional

$$W_b = \frac{1}{S}\int\int_{G_\varepsilon}(\mathbf{J}, \mathbf{b})\, dx\, dy \qquad (2.151)$$

We consider the case $W_a = \max$, $U > u$ (the case $W_b = \min$, $U < u$ is treated similarly). Among the stationarity conditions corresponding to this problem, we distinguish the requirement

$$(\mathbf{J}, \mathbf{a})_+ + \omega_{1t}[z_n^1]_-^+ + \omega_{2t}[z_n^2]_-^+ = \lambda = \text{const} \qquad \text{along } \Sigma_\varepsilon \qquad (2.152)$$

which follows from the last equation in (2.42); the constant Lagrangian multiplier λ is associated with the constraint

$$\iint_{G_\varepsilon} dx\, dy = S$$

and is determined from this equation. The symbol $[\ \ ]_-^+$ denotes the difference between the limiting values along the boundary Σ_ε of the corresponding quantities taken from within $(+)$ and from without $(-)$ the inclusion G_ε. The potential z^1, the flow function z^2, and the Lagrangian multipliers ω_1, ω_2 satisfy equations analogous to (2.56) and (2.100), respectively, and corresponding boundary conditions.

In order to satisfy (2.152), let us first replace it by

$$(\mathbf{J}, \mathbf{a})_+ = \text{const} \qquad \text{along } \Sigma_\varepsilon \qquad (2.153)$$

No Lagrangian multipliers are needed to determine the region satisfying this condition. One just observes that the flow function $Z^2(x, y)$ is harmonic in G_ε; this also applies to the function $(\mathbf{J}, \mathbf{a}) = -Z_y^2 a_1 + Z_x^2 a_2$. By virtue of (2.153), this last function takes on a constant value on the boundary of the region; hence, it is constant everywhere in G_ε. Under these conditions, this is possible only in the case where an ellipse is the boundary of the region.

Indeed, wishing to write down the system of singular integral equations (2.135) corresponding to this problem, we take $g^1(P, Q_1)$ to be the function

$$g^1(P, Q_1) = \frac{1}{2\pi u} \ln r_{Q_1 P}$$

We multiply the (vector) equation (2.135) by $\text{const} \cdot \mathbf{r}_{Q_2}$, after which we go to the limit $Q_2 \to \infty$. The right-hand side of the resulting equation is a constant vector (does not depend on Q_1); the vector $\lim_{Q_2 \to \infty} r_{Q_2} \nabla_{Q_1} g^2(Q_1, Q_2)$ for $Q_1 \in G_\varepsilon$ does not depend on Q_1, in accordance with what we have proved above. The limiting equation then shows that for $Q_1 \in G_\varepsilon$ the expression

$$\int_{G_\varepsilon} \nabla_P g^1(P, Q_1) \, dS_P$$

is a linear vector function of the coordinates x_{Q_1} and y_{Q_1}. But

$$\int_{G_\varepsilon} \nabla_P g^1(P, Q_1) \, dS_P = -\nabla_{Q_1} \int_{G_\varepsilon} g^1(P, Q_1) \, dS_P$$

from which it follows that at the points of G_ε the expression

$$\int_{G_\varepsilon} g^1(P, Q_1) \, dS_P \tag{2.154}$$

is a polynomial of second degree in x_{Q_1} and y_{Q_1}. By a suitable orthogonal transformation of the coordinates, this polynomial can be reduced to a form that does not contain the first powers and products of coordinates; we shall assume that this reduction has already been made. The coefficients of the resulting polynomial must be positive since, in accordance with (2.154), this polynomial is the internal potential of the homogeneous body G_ε of positive density. As a consequence of Dive's theorem[156,227] (or rather of its two-dimensional analogue proved by Holder[268]), these conditions imply that the boundary of the body can only be an ellipse, which is what we wanted to prove.

Note that specification of the area S of the ellipse G_ε [see (2.153)] does not yet determine this ellipse uniquely. At this point we have only proved that the region G_ε which satisfies (2.153) must be sought among ellipses of the given area. Using (2.103), which are valid for any ellipse, we calculate the functional (2.150); the value of this functional depends only on the ratio b/a of the semiaxes of the ellipse. The extremal values of this ratio are 0 or ∞, depending on how the axes α and β (see Fig. 2.3) are oriented with respect to the vectors $\mathbf{j}$ and $\mathbf{a}$; therefore, fixing $S \neq 0$, we obtain an "infinitesimally thin strip of infinite length" as the limiting ellipse. However, for such a strip, the terms removed from the left-hand side of (2.152) under the transition to (2.153), are obviously constant. We now see that the strip is the desired extremal region. In order to avoid the problems associated with the infinite length of the strip, we must reject the requirement of specifying S. In that case, the extremal value of the functional (2.150) is attained for "an ellipse of infinitesimally small diameter with unit eccentricity"; however, nothing is changed if we speak of an infinitesimally thin strip of finite length ("needle")[132].

Similar results are also valid in the case of three independent variables; let us consider this case briefly. Let x_1, x_2, x_3 be a system of Cartesian coordinates with origin at the center Q_1 of the ellipsoid D_ε with semiaxes a, b, c ($a \geq b \leq c$) oriented along x_1, x_2, x_3; the resistivity of the material of the ellipsoid is $U = $ const, and that of the surrounding medium, $u = $ const. Equation (2.102) for the increment of the functional remains in force; if $\mathbf{j}$ is a homogeneous current density in the medium with resistivity u, then the current density $\mathbf{J}$ within the ellipsoid is also homogeneous and is given by

$$\mathbf{J} = \frac{u}{U}\left[\frac{j_1\mathbf{i}_1}{1 + \dfrac{abc}{2U}(u - U)A_1} + \frac{j_2\mathbf{i}_2}{1 + \dfrac{abc}{2U}(u - U)A_2} + \frac{j_3\mathbf{i}_3}{1 + \dfrac{abc}{2U}(u - U)A_3}\right]$$

where

$$A_1 = \int_0^\infty \frac{ds}{(s + a^2)R_s}, \qquad A_2 = \int_0^\infty \frac{ds}{(s + b^2)R_s}, \qquad A_3 = \int_0^\infty \frac{ds}{(s + c^2)R_s}$$

$$R_s = \sqrt{(s + a^2)(s + b^2)(s + c^2)}$$

We consider two special cases.

Case 1. $a \geq b = c$ (prolate spheroid). Then

$$A_1 = \frac{1}{a^3 e^3}\left(-2e + \ln\frac{1 + e}{1 - e}\right)$$

$$A_2 = A_3 = \frac{1}{a^3 e^3}\left(\frac{e}{1 - e^2} + \frac{1}{2}\ln\frac{1 - e}{1 + e}\right)$$

$$e = \sqrt{1 - \frac{b^2}{a^2}}$$

If $b = c \to 0$, then $ab^2 A_1 \to 0$, $ab^2 A_2 = ab^2 A_3 \to 1$, and $\mathbf{J}$ is given by the limiting expression

$$\mathbf{J} = \frac{u}{U}\left[j_1\mathbf{i}_1 + \frac{2U}{u + U}(j_2\mathbf{i}_2 + j_3\mathbf{i}_3)\right]$$

For the integrand of (2.102), we obtain

$$\frac{\Delta u}{U}\left[\frac{2U}{u+U}(\mathbf{j},\text{grad }\omega_2)+\frac{u-U}{u+U}j_1\omega_{2x_1}\right]=\frac{2\Delta u}{u+U}\left[(\mathbf{j},\text{grad }\omega_2)+\frac{u-U}{2U}j_1\omega_{2x_1}\right]$$

This last expression is less than or equal to zero if

$$\Delta u\leq 0,\qquad (\mathbf{j},\text{grad }\omega_2)+\frac{u-U}{2U}j_1\omega_{2x_1}\geq 0$$

or if

$$\Delta u\geq 0,\qquad (\mathbf{j},\text{grad }\omega_2)+\frac{u-U}{2U}j_1\omega_{2x_1}\leq 0$$

Let φ be the angle between the vectors $\mathbf{j}$ and grad ω_2; then in the first of these two cases

$$u=u_{\max},\qquad \cos\varphi-\frac{u_{\max}-u_{\min}}{2u_{\min}}\sin^2\frac{\varphi}{2}\geq 0$$

or

$$0\leq\varphi\leq\arccos\frac{u_{\max}-u_{\min}}{u_{\max}+3u_{\min}}\qquad(2.155)$$

while in the second

$$u=u_{\min},\qquad \cos\varphi-\frac{u_{\min}-u_{\max}}{2u_{\max}}\sin^2\frac{\varphi}{2}\leq 0$$

or

$$\pi-\arccos\frac{u_{\max}-u_{\min}}{u_{\max}+3u_{\min}}\leq\varphi\leq\pi\qquad(2.156)$$

Case 2. $a=b\geq c$ (oblate spheroid). In this case

$$A_1=A_2=\frac{\pi}{2a^3g^3}-\frac{c}{a^4g^2}-\frac{1}{a^3g^3}\arctan\frac{c}{ag}$$

$$A_2=\frac{2}{a^3g^3}\left(\frac{ag}{c}-\arctan\frac{ag}{c}\right),\qquad g=\sqrt{1-\frac{c^2}{a^2}}$$

If $c \to 0$, then $a^2 c A_1 = a^2 c A_2 \to 0$, $a^2 c A_3 \to 2$, and $\mathbf{J}$ is given by the limiting expression

$$\mathbf{J} = \frac{u}{U}\left(j_1\mathbf{i}_1 + j_2\mathbf{i}_2 + \frac{U}{u}j_3\mathbf{i}_3\right)$$

The integrand of (2.102) is

$$\frac{\Delta u}{U}\left[(\mathbf{j}, \text{grad } \omega_2) + \frac{U - u}{u}j_3\omega_{2x_3}\right]$$

This last expression is less than or equal to zero if

$$\Delta u \leq 0, \qquad (\mathbf{j}, \text{grad } \omega_2) + \frac{U - u}{u}j_3\omega_{2x_3} \geq 0$$

or if

$$\Delta u \geq 0, \qquad (\mathbf{j}, \text{grad } \omega_2) + \frac{U - u}{u}j_3\omega_{2x_3} \leq 0$$

In the first case (φ again is the angle between $\mathbf{j}$ and grad ω_2)

$$u = u_{\max}, \qquad 0 \leq \varphi \leq \arccos\frac{u_{\max} - u_{\min}}{u_{\max} + u_{\min}} \qquad (2.157)$$

and in the second

$$u = u_{\min}, \qquad \pi - \arccos\frac{u_{\max} - u_{\min}}{u_{\max} + u_{\min}} \leq \varphi \leq \pi \qquad (2.158)$$

Direct comparison shows that the conditions (2.157) and (2.158) are stronger than the conditions (2.155) and (2.156), i.e., variation in an oblate spheroid is more effective than in a prolate spheroid. The critical position of the oblate spheroid is one in which the symmetry axis (the x_3 axis) and the vectors $\mathbf{j}$ and grad ω_2 lie in one (the meridian) plane, the axis x_3 bisecting the angle between the vectors $\mathbf{j}$ and grad ω_2.

2.6. The Case of Bounded Measurable Controls

In this section, we shall dispense with the assumption of piecewise constancy of the controls, replacing it by the requirement that the set of controls belong to $L_\infty(G)$ as defined by the condition

$$u_{min} \leq \text{vrai max } u(x, y) \leq u_{max} \quad (2.159)$$

To this end, we define a generalized solution $Z^2(x, y)$ of the boundary-value problem (2.56)-(2.57) as a function $Z^2 \in W_2^1(G)$ which is zero on Γ_3, equal to $I = $ const on Γ_4, and satisfies the integral identity

$$\iint_G U(\nabla Z^2, \nabla\varphi)\, dx\, dy + \varphi_+ IR = \iint_G (E_1\varphi_y - E_2\varphi_x)\, dx\, dy \quad (2.160)$$

for any $\varphi \in W_2^1(G)$ equal to $\varphi_+ = $ const on Γ_4 and zero on Γ_3. Note that in this definition Z^2 is an admissible function corresponding to any admissible control U; in particular, Z^2 and U can also be optimal functions z^2 and u.

The function $\omega_1(x, y)$ is defined as a generalized solution of the adjoint boundary-value problem (2.100)-(2.101); i.e., as an element of the space $W_2^1(G)$ which satisfies $\omega_{1+} = $ const on Γ_4, ω_1 equal to zero on Γ_3, and the integral identity

$$\iint_G u(\nabla\omega_1, \nabla\psi)\, dx\, dy - \psi_+(1 - R\omega_{1+}) = 0 \quad (2.161)$$

for any $\psi \in W_2^1(G)$, ψ equal to zero on Γ_3, and $\psi_+ = $ const on Γ_4. In this definition, the function u plays the role of an optimal control.

If the function $U(x, y)$ is sufficiently smooth, the first of these definitions is equivalent to the conditions

$$\text{div } U \nabla Z^2 = E_{1y} - E_{2x}$$

$$Z^2|_{\Gamma_3} = 0, \qquad Z^2|_{\Gamma_4} = I$$

$$U\frac{\partial Z^2}{\partial N}\bigg|_{\Gamma_1} = -E_{0t}|_{\Gamma_1}, \qquad U\frac{\partial Z^2}{\partial N}\bigg|_{\Gamma_2} = -E_{0t}|_{\Gamma_2}$$

$$IR = -\int_{\Gamma_4} U\frac{\partial Z^2}{\partial N}\, dt - \int_{\Gamma_4} E_{0t}\, dt, \qquad E_{0t} = E_1 x_t + E_2 y_t$$

which determine $Z^2(x, y)$ in accordance with (2.56) and (2.57).

Similarly, for sufficiently smooth $u(x, y)$ the definition (2.161) is equivalent to the relations

$$\text{div } u\nabla\omega_1 = 0$$

$$\omega_1|_{\Gamma_3} = 0, \quad \omega_1|_{\Gamma_4} = \omega_{1+}, \quad \frac{\partial\omega_1}{\partial N}\bigg|_{\Gamma_1} = \frac{\partial\omega_1}{\partial N}\bigg|_{\Gamma_2} = 0$$

$$\int_{\Gamma_4} u\frac{\partial\omega_1}{\partial N} \, dt = 1 - R\omega_{1+}$$

which determine $\omega_1(x, y)$ in accordance with (2.100)-(2.101).

Equation (2.102) for smooth $u(x, y)$ is equivalent to the equation [see (2.100)]

$$\Delta I = -\iint_G \Delta u(\nabla Z^2, \nabla\omega_1) \, dx \, dy \tag{2.162}$$

This last expression also remains true for the more general definition of the functions Z^2 and ω_1 as implied by the identities (2.160) and (2.161). To prove this, we set $\varphi = \omega_1$ in (2.160) and write down this identity first for the admissible control $U = u + \Delta u$ and its corresponding function $Z^2 = z^2 + \Delta z^2$, then for the optimal control u and the corresponding function z^2, and subtract the results term by term, obtaining

$$\iint_G u(\nabla(\Delta z^2), \nabla\omega_1) \, dx \, dy + \iint_G \Delta u(\nabla Z^2, \nabla\omega_1) \, dx \, dy = -\omega_{1+}R \, \Delta I$$

Similarly, in (2.161) we first set $\psi = Z^2 = z^2 + \Delta z^2$, then $\psi = z^2$, and subtract the results; we obtain

$$\iint_G u(\nabla\omega_1, \nabla(\Delta z^2)) \, dx \, dy = (1 - R\omega_{1+}) \, \Delta I$$

Comparison of the last two equations leads to (2.162), which is what we wanted to prove.

On the other hand, since $Z^2, \omega_1 \in W_2^1(G)$, we have $\nabla Z^2, \nabla\omega_1 \in L_2(G)$, from which it follows that $(\nabla Z^2, \nabla\omega_1) \in L_1(G)$. The function Δu is bounded and measurable, so that $\Delta u(\nabla Z^2, \nabla\omega_1) \in L_1(G)$.

We now take an arbitrary internal point $Q(x, y)$ of the region G and a region G_ε of small measure mes $G_\varepsilon = \varepsilon > 0$ lying entirely within G and containing Q as an interior point. We specify the admissible control $U(x, y)$ by

$$U(x, y) = \begin{cases} u + \Delta u, & (x, y) \in G_\varepsilon \\ u, & (x, y) \notin G_\varepsilon \end{cases} \tag{2.163}$$

(the admissible control $u + \Delta u$ is assumed constant at the points of G_ε).

We show that the principal part of the vector ∇Z^2 in the parameter ε can be found by solving a certain auxiliary problem. Let $u(Q)$ be the value of the optimal control at the point Q; we then define the piecewise constant control U_0 by

$$U_0(x, y) = \begin{cases} u + \Delta u, & (x, y) \in G_\varepsilon \\ u(Q), & (x, y) \notin G_\varepsilon \end{cases}$$

The controls U and U_0 differ only at points which do not belong to G_ε.

We introduce a function $Z_0^2(x, y) \in W_2^1(G)$, equal to $z_{0+}^2 = $ const on Γ_4, zero on Γ_3, and satisfying the integral identity

$$u(Q) \iint_G (\nabla z_0^2, \nabla \varphi) \, dx \, dy + \varphi_+ z_{0+}^2 R = \iint_G (E_1^* \varphi_y - E_2^* \varphi_x) \, dx \, dy$$

for all $\varphi \in W_2^1(G)$ which are equal to φ_+ on Γ_4 and zero on Γ_3. We shall assume that the functions E_1^* and E_2^* are continuous and satisfy the requirement

$$\nabla z_0^2(Q) = \nabla z^2(Q) \tag{2.164}$$

which is imposed at the single point Q.

Similarly, we define the function w by means of the identity

$$\iint_G U_0[(\nabla z_0^2, \nabla \varphi) + (\nabla w, \nabla \varphi)] \, dx \, dy + \varphi_+ (z_{0+}^2 + w_+) R$$

$$= \iint_G (E_1^* \varphi_y - E_2^* \varphi_x) \, dx \, dy$$

which together with the foregoing shows that

$$\iint_{G_\varepsilon} [u + \Delta u - u(Q)](\nabla z_0^2, \nabla \varphi) \, dx \, dy$$

$$+ \iint_G U_0(\nabla w, \nabla \varphi) \, dx \, dy + \varphi_+ w_+ R = 0 \tag{2.165}$$

We set

$$Z^2 = z^2 + w + \delta z^2 \tag{2.166}$$

Our aim is to show that the function w represents the principal part of the difference $Z^2 - z^2$ with respect to the parameter ε. Using (2.160), we can readily verify that δz^2 satisfies the identity

$$\iint_G U(\nabla \delta z^2, \nabla \varphi) \, dx \, dy + \varphi_+ \delta z_+^2 R$$

$$= -\iint_G U(\nabla w, \nabla \varphi) \, dx \, dy - \iint_{G_\varepsilon} \Delta u(\nabla z^2, \nabla \varphi) \, dx \, dy - \varphi_+ w_+ R \tag{2.167}$$

By virtue of (2.165), the right-hand side can be represented in the form

$$-\iint_G (U - U_0)(\nabla w, \nabla \varphi) \, dx \, dy$$

$$+ \iint_{G_\varepsilon} [u + \Delta u - u(Q)](\nabla z_0^2, \nabla \varphi) \, dx \, dy - \iint_{G_\varepsilon} \Delta u(\nabla z^2, \nabla \varphi) \, dx \, dy \tag{2.168}$$

In (2.167), we now set $\varphi = \delta z^2$ and estimate the different terms in this identity. The left-hand side is estimated as

$$u_{\min} \| \nabla \delta z^2 \|_G^2$$

where the symbol $\| \cdot \|_G$ denotes the $L_2(G)$ norm. The terms in (2.168) admit upper bounds of the form

$$\left| \iint_G (U - U_0)(\nabla w, \nabla \delta z^2) \, dx \, dy \right| \leq u_{\max} \| \nabla \delta z^2 \|_G \| \nabla w \|_{G/G_\varepsilon}$$

$$\left| \iint_{G_\varepsilon} [u - u(Q)](\nabla z_0^2, \nabla \delta z^2) \, dx \, dy + \iint_{G_\varepsilon} \Delta u(\nabla (z_0^2 - z^2), \nabla \delta z^2) \, dx \, dy \right|$$

$$\leq \| (u - u(Q)) \nabla z_0^2 \|_{G_\varepsilon} \| \nabla \delta z^2 \|_G + u_{\max} \| \nabla (z_0^2 - z^2) \|_{G_\varepsilon} \| \nabla \delta z^2 \|_G$$

Collecting our results, we arrive at the following estimate of the norm $\|\delta z^2\|_G$:

$$\|\nabla \delta z^2\|_G \leq c_1 \|\nabla w\|_{G/G_\varepsilon} + c_2 \|(u - u(Q))\nabla z_0^2\|_{G_\varepsilon} + c_3 \|\nabla(z^2 - z_0^2)\|_{G_\varepsilon}$$

$$(2.169)$$

In accordance with what we have proved earlier, the function ∇w is of order ε at the points G/G_ε; the first term on the right-hand side of (2.169) has the same order. We now estimate the second term by assuming that the set G_ε belongs to a system of sets that are regularly contractible* at the point Q; for such sets, it is well known that

$$\lim_{\varepsilon \to 0} \frac{\|u - u(Q)\|_{G_\varepsilon}}{\varepsilon^{1/2}} = 0$$

and then one therefore has

$$\|u - u(Q)\|_{G_\varepsilon} = o(\varepsilon^{1/2})$$

for almost all points Q.

The third term on the right-hand side in (2.169) may be similarly estimated: One need only write $\nabla z^2 - \nabla z_0^2 = \nabla z^2 - \nabla z^2(Q) + \nabla z^2(Q) - \nabla z_0^2$ and note that in accordance with what we have proved in Section 2.4 the values of ∇z_0^2 at the point of G_ε differ by terms $o(1)$ from $\nabla z_0^2(Q) = \nabla z^2(Q)$. Thus,

$$\|\nabla \delta z^2\|_G = o(\varepsilon^{1/2}) \qquad (2.170)$$

If G_ε is an ellipse with semiaxes $a\varepsilon^{1/2}$ and $b\varepsilon^{1/2}$ oriented along the directions of the axes $\boldsymbol{\alpha}$ and $\boldsymbol{\beta}$, respectively, then, as is shown in Section 2.4, the derivatives $z_\alpha^2 + w_\alpha$ and $z_\beta^2 + w_\beta$ differ from J_β and $-J_\alpha$ by terms of $o(1)$ at points of G_ε. This remark in conjunction with (2.170) shows that in (2.162) or, equivalently, in (2.102), we may assume the vector $\mathbf{J}$ to have components defined by (2.103), which is what we wanted to prove.

Note that a definite system of sets G_ε (similar ellipses) was introduced only in the final stage of the argument. The proof remains true if this system

---

* See the footnote following (2.142).

is replaced by any other system of regularly contractible sets and defines the components of the vector $\mathbf{J}$ by expressions analogous to (2.148). (See also the discussion in Section 2.4.)

However, in view of what was proven in Section 2.5, the system of ellipses leads to a strip which provides the strongest necessary conditions in the case of piecewise constant controls; furthermore, this conclusion is valid for almost all points of G for bounded and measurable controls.

Equations (2.103) arose as a consequence of an argument based on premises of a physical nature that were intuitively clear in the case of piecewise constant controls. As we have seen, these expressions remained valid when the controls belonged to the much larger class of bounded and measurable functions although the original intuitive arguments could not be extended to this class of controls.

2.7. Formula for the Increment of a Functional: Weierstrass's Necessary Condition (the General Case)

In the general case, the derivation of Weierstrass's condition entails the derivation of an expression for the increment ΔI of the minimized functional for the transition from the optimal control u to an admissible control $U = u + \Delta u$. The measure ε of the set G_ε on which the variation is performed $(G_\varepsilon : \Delta u \neq 0)$ is the main small parameter in the expression for ΔI. The problem is to separate the principal part in ε in this expression; the requirement of nonnegativity of this principal part then provides Weierstrass's necessary condition for a strong relative minimum.

We take the basic equations and boundary conditions of the optimization problem in the form of (2.2), (2.27), (2.17), (2.19), (2.29); we first write these equations down for the admissible controls U and V, then for the optimal controls u and v, and subtract the results term by term. By definition, the corresponding differences $(\Delta\Xi_{ij}, \Delta a_k, \Delta\Theta_{ij}, \Delta b_k)$ are equal to zero. If the minimized functional is given by (2.22), then the functional [see (2.31)]

$$\Delta I + \int_G (\xi_{ij}\Delta\Xi_{ji} + \alpha_k\Delta a_k)\, dx + \int_{\cup\delta_j} (\theta_{ij}\Delta\Theta_{ji} + \beta_k\Delta b_k)\, dx$$

differs from ΔI by zero terms.

We shall assume that the boundary Γ of G is fixed, and that the functions X_i^j $(i = 1, 2, \ldots, m; j = 1, 2, \ldots, n)$ and T_i^j $(i = 1, 2, \ldots, m - 1;$

$j = 1, 2, \ldots, n$) have continuous second-order partial derivatives with respect to the arguments z, ζ, and κ.

Using the notation (2.44) and taking the stationarity conditions (2.39)–(2.43) into account, we can readily establish that

$$
\Delta I = - \int_G [H(x, z + \Delta z, \zeta + \Delta\zeta, u + \Delta u, \xi)
$$

$$
- H(x, z, \zeta, u, \xi)]\, dx + \int_G \frac{\partial H(x, z, \zeta, u, \xi)}{\partial z^j} \Delta z^j\, dx
$$

$$
- \int_{\cup \delta_j} [h(x, z + \Delta z, \kappa + \Delta\kappa, v + \Delta v, \theta) - h(x, z, \kappa, v, \theta)]\, dx
$$

$$
+ \int_{\cup \delta_j} \frac{\partial h(x, z, \kappa, v, \theta)}{\partial z^j} \Delta z^j\, dx \tag{2.171}
$$

On the right-hand side of this equation, we combine the first two and the last two integrals, transforming the integrands in accordance with the equations

$$
H(x, z + \Delta z, \zeta + \Delta\zeta, u + \Delta u, \xi) - H(x, z, \zeta, u, \xi) - \frac{\partial H(x, z, \zeta, u, \xi)}{\partial z^j} \Delta z^j
$$

$$
= H(x, z, \zeta + \Delta\zeta, u + \Delta u, \xi) - H(x, z, \zeta, u, \xi)
$$

$$
+ \left[\frac{\partial H(x, z, \zeta + \Delta\zeta, u + \Delta u, \xi)}{\partial z^j} - \frac{\partial H(x, z, \zeta, u, \xi)}{\partial z^j} \right] \Delta z^j
$$

$$
+ \frac{1}{2} \frac{\partial^2 H(x, z + \vartheta_1 \Delta z, \zeta + \Delta\zeta, u + \Delta u, \xi)}{\partial z^j\, \partial z^k} \Delta z^j \Delta z^k \qquad 0 < \vartheta_1 < 1
$$

A similar approach is used for h. The result is

$$
\Delta I = - \int_G [H(x, z, \zeta + \Delta\zeta, u + \Delta u, \xi) - H(x, z, \zeta, u, \xi)]\, dx
$$

$$
- \int_{\cup \delta_j} [h(x, z, \kappa + \Delta\kappa, v + \Delta v, \theta) - h(x, z, \kappa, v, \theta)]\, dx + \eta_3
$$

$$
\tag{2.172}
$$

$$\eta_3 = -\int_G \left[\frac{\partial H(x, z, \zeta + \Delta\zeta, u + \Delta u, \xi)}{\partial z^j} - \frac{\partial H(x, z, \zeta, u, \xi)}{\partial z^j} \right] \Delta z^j \, dx$$

$$-\frac{1}{2} \int_G \frac{\partial^2 H(x, z + \vartheta_1 \Delta z, \zeta + \Delta\zeta, u + \Delta u, \xi)}{\partial z^j \, \partial z^k} \Delta z^j \, \Delta z^k \, dx$$

$$-\int_{\cup \delta_j} \left[\frac{\partial h(x, z, \kappa + \Delta\kappa, v + \Delta v, \theta)}{\partial z^j} - \frac{\partial h(x, z, \kappa, v, \theta)}{\partial z^j} \right] \Delta z^j \, dx$$

$$-\frac{1}{2} \int_{\cup \delta_j} \frac{\partial^2 h(x, z + \vartheta_2 \Delta z, \kappa + \Delta\kappa, v + \Delta v, \theta)}{\partial z^j \, \partial z^k} \Delta z^j \, \Delta z^k \, dx$$

$$0 < \vartheta_2 < 1 \quad (2.173)$$

One says that the functions $z^j(x)$, $j = 1, 2, \ldots, n$, satisfy Weierstrass's necessary conditions if the inequality

$$H(x, z, \zeta, u, \xi) - H(x, z, Z, U, \xi) \geq 0 \quad (2.174)$$

is satisfied for almost all $x \in G$, for all admissible U and Z, and if, in addition,

$$h(x, z, \kappa, v, \theta) - h(x, z, K, V, \theta) \geq 0 \quad (2.175)$$

for almost all $x \in \delta = \cup \delta_j$ and all admissible V and K.

As a rule, the derivation of the necessary conditions entails the use of variations $\Delta u = U - u$ ($\Delta v = V - v$) of control functions concentrated on sets of small m-dimensional $[(m - 1)$-dimensional] measure. Let G_ε and Γ_ε be such sets of measure ε_1 and ε_2; we have*

$$U = \begin{cases} u + \Delta u, & x \in G_{\varepsilon_1}, \\ u, & x \in G/G_{\varepsilon_1}; \end{cases} \qquad V = \begin{cases} v + \Delta v, & x \in \Gamma_{\varepsilon_2} \\ v, & x \in \delta/\Gamma_{\varepsilon_2} \end{cases} \quad (2.176)$$

The expression (2.172) for ΔI can be represented in the form

$$\Delta I = -\int_{G_{\varepsilon_1}} [H(x, z, \zeta + \Delta\zeta, u + \Delta u, \xi) - H(x, z, \zeta, u, \xi)] \, dx$$

$$-\int_{\Gamma_{\varepsilon_2}} [h(x, z, \kappa + \Delta\kappa, v + \Delta v, \xi) - h(x, z, \kappa, v, \xi)] \, dx + \eta_1 + \eta_2 + \eta_3$$

$$(2.177)$$

---

* In a number of cases, the imposed restrictions can be satisfied by varying the controls in a finite number of sets $G_\varepsilon^{(i)}$ and $\Gamma_\varepsilon^{(j)}$. In this connection, see Refs. 128 and 129 along with Section 3.2.

where $(0 < \vartheta_3 < 1)$,

$$\eta_1 = -\int_{G/G_{\varepsilon_1}} [H(x, z, \zeta + \Delta\zeta, u, \xi) - H(x, z, \zeta, u, \xi)] \, dx$$

$$= -\frac{1}{2} \int_{G/G_{\varepsilon_1}} \frac{\partial^2 H(x, z, \zeta + \vartheta_3 \Delta\zeta, u, \xi)}{\partial \zeta^j \, \partial \zeta^k} \Delta\zeta^j \, \Delta\zeta^k \, dx \qquad (2.178)$$

$$\eta_2 = -\int_{\delta/\Gamma_{\varepsilon_2}} [h(x, z, \kappa + \Delta\kappa, v, \theta) - h(x, z, \kappa, v, \theta)] \, dx$$

$$= -\frac{1}{2} \int_{\delta/\Gamma_{\varepsilon_2}} \frac{\partial^2 h(x, z, \kappa + \vartheta_4 \Delta\kappa, v, \theta)}{\partial \kappa^j \, \partial \kappa^k} \Delta\kappa^j \, \Delta\kappa^k \, dx \qquad 0 < \vartheta_4 < 1$$

$$(2.179)$$

and where η_3 is given by (2.173).

In many cases, one can show that for $\varepsilon_1, \varepsilon_2 \to 0$ the terms η_1, η_2, η_3 are small, of higher order in ε_1 and ε_2 than the integrals on the right-hand side of (2.177); these integrals are small to first order in ε_1 and ε_2. Usually, it can be shown that nonpositivity of these integrals implies the fulfillment of Weierstrass's conditions (2.174) and (2.175), which thus are necessary for a minimum.

We shall assume that the controls u and v are measurable functions and that to each optimal solution $z \in W_p^1(G)$, $z \in W_{p*}^1(\Gamma)$ there corresponds a solution $\xi \in W_q^1(G)$, $\theta \in W_{q*}^1(\delta)$ of the adjoint system, and that $p^{-1} + q^{-1} \le 1$, $(p^*)^{-1} + (q^*)^{-1} \le 1$.

The validity of (2.174) and (2.175) as necessary conditions for a minimum can also be proved under a different set of assumptions. We now consider several such cases.

Case 1. If the functions X_i^j and T_i^j have the structure

$$X_i^j = A_{ir}^j(x)z^r + B_{is}^j(x)\zeta^s + C_i^j(x, u)$$

$$T_i^j = a_{ir}^j(x)z^r + b_{is}^j(x)\kappa^s + c_i^j(x, v)$$

$$(2.180)$$

where A, B, C, a, b, c are continuous functions, and the sets of admissible values of u and v are compact,* then Weierstrass's conditions are necessary and sufficient conditions for a minimum. The proof is based on the fact that the term η_3 in (2.172) is equal to zero by virtue of (2.180), and the integrands in (2.172) do not depend on ζ, $\Delta\zeta$, κ, $\Delta\kappa$ by virtue of (2.180) and of the stationarity conditions (2.48) and (2.50). If Weierstrass's conditions (2.174) and (2.175) are satisfied, (2.172) shows that $\Delta I \geq 0$, and the sufficiency part of the proof follows.

To prove necessity, we use the fact that $\eta_1 = \eta_2 = 0$ by virtue of (2.180); the remaining arguments are the same as those in Section 2.3 [see (2.109)].

As an illustration, we take Example 1 of Section 2.2.

Case 2. We assume that the measurable controls U and V, and the corresponding adjoint variables ξ and θ are bounded functions, i.e., $|U(x)| < \mu_1$, $|V(x)| < \mu_2$, $|\xi(x)| < \mu_3$, $|\theta(x)| < \mu_4$; in addition,

$$\|\Delta z\|_{L_p(G)} \leq \mu_5 \|\Delta u\|_{L(G)}$$

$$\|\Delta \zeta\|_{L_p(G/G_{\varepsilon_1})} \leq \mu_6 \|\Delta u\|_{L(G)}, \qquad \|\Delta \zeta\|_{L_\infty(G_{\varepsilon_1})} \leq \mu_7 \|\Delta u\|_{L_\infty(G_{\varepsilon_1})}$$

$$\|\Delta z\|_{L_{p*}(\delta)} \leq \mu_8 \|\Delta v\|_{L(\delta)}, \qquad \|\Delta \kappa\|_{L_{p*}(\delta/\Gamma_{\varepsilon_2})} \leq \mu_9 \|\Delta v\|_{L(\delta)}$$

$$\|\Delta \kappa\|_{L_\infty(\Gamma_{\varepsilon_2})} \leq \mu_{10} \|\Delta v\|_{L_\infty(\Gamma_{\varepsilon_2})}$$

Suppose further that $p = p^* = 2$. The specification of the controls U and V by means of (2.176) yields the estimates

$$|\eta_1| \leq K_1 \int_{G/G_{\varepsilon_1}} |\Delta \zeta|^2 \, dx \leq K_1 \mu_6^2 \|\Delta u\|_{L(G)}^2$$

$$|\eta_2| \leq K_2 \int_{\delta/\Gamma_{\varepsilon_2}} |\Delta \kappa|^2 \, dx \leq K_2 \mu_9^2 \|\Delta v\|_{L(\delta)}$$

$$|\eta_3| \leq K_3 \left(\int_G |\Delta \zeta| |\Delta z| \, dx + \int_G |\Delta u| |\Delta z| \, dx + \int_G |\Delta z|^2 \, dx \right)$$

$$+ K_4 \left(\int_\delta |\Delta \kappa| |\Delta z| \, dx + \int_\delta |\Delta v| |\Delta z| \, dx + \int_\delta |\Delta z|^2 \, dx \right)$$

---

* If the functions C_i^j and c_i^j are linear with respect to the controls, it is sufficient to assume weak compactness of the sets of admissible values of u and v.

$$\leq K_3\left(\mu_5\mu_6\|\Delta u\|^2_{L(G)} + \mu_7 \int_{G_{\varepsilon_1}} |\Delta u||\Delta z|\, dx\right.$$

$$\left. + \int_{G_{\varepsilon_1}} |\Delta u||\Delta z|\, dx + \mu_5\|\Delta u\|^2_{L(G)}\right) + K_4\left(\mu_8\mu_9\|\Delta v\|^2_{L(\delta)}\right.$$

$$\left. + \mu_{10} \int_{\Gamma_{\varepsilon_2}} |\Delta v||\Delta z|\, dx + \int_{\Gamma_{\varepsilon_2}} |\Delta v||\Delta z|\, dx + \mu_8\|\Delta v\|^2_{L(\delta)}\right)$$

The estimation of the integrals

$$\int_{G_{\varepsilon_1}} |\Delta u||\Delta z|\, dx, \qquad \int_{\Gamma_{\varepsilon_2}} |\Delta v||\Delta z|\, dx$$

by means of the Cauchy–Schwarz inequality and the use of (2.176) yield

$$|\eta_3| \leq \gamma_1\varepsilon_1^{3/2} + \gamma_2\varepsilon_1^2 + \gamma_3\varepsilon_2^{3/2} + \gamma_4\varepsilon_2^2$$

$$|\eta_1| \leq \gamma_5\varepsilon_1^2, \qquad |\eta_2| \leq \gamma_6\varepsilon_2^2$$

where $\gamma_1 > 0, \ldots, \gamma_6 > 0$ are constants.

Standard arguments (e.g., see ref. 144) may now be used to prove the necessity of conditions (2.174) and (2.175).

Examples 2 and 3 of Section 2.2 may be used to illustrate the approach for the present case. In these examples, we have seen that $\eta_1 = \eta_2 = \eta_3 = 0$ by virtue of the special structure of the constraint equations (2.56) and (2.63); in view of what has been shown, conditions (2.174) and (2.175) are necessary if a minimum is to be attained. The proof made essential use of variations of the type (2.176); conditions (2.174) and (2.175) of this derivation retain meaning only for such variations. On the other hand, for Examples 2 and 3 of Section 2.2 conditions (2.174) and (2.175) also retain their validity for arbitrary variations Δu and associated variations $\Delta\zeta$; when interpreted this way, these conditions will also be sufficient in these examples.* In the general case, however, conditions (2.174) and (2.175) acquire a constructive nature only as necessary conditions, i.e., as conditions derived for variations of the type (2.176). In order to obtain information

---

* Conditions (2.174) and (2.175) for the variations (2.176) are not in general sufficient. Indeed, because of the nonlinear nature of the problem, the fulfillment of these conditions for the variations (2.176) does not guarantee their fulfillment for variations of general form. For the linear conditions (2.180), the necessary and the sufficient conditions are identical.

about the nature of the optimal controls from these conditions, it is necessary to eliminate the increments $\Delta \zeta$ and $\Delta \kappa$ of the parametric variables from (2.174) and (2.175).† This may be done by expressing these increments in terms of the optimal functions z, ζ, γ, u, v and the increments Δu and Δv of the controls. However, the methods of Section 2.3 may be used for this elimination only for variations of the type (2.176), i.e., for variations that lead to Weierstrass's necessary conditions. In more general problems than those of Examples 2 and 3 of Section 2.2, conditions (2.174) and (2.175) based on arbitrary variations Δu and Δv cease to be sufficient conditions, although they remain as necessary conditions for the variations (2.176).

There exists a large class of optimization problems for which the expressions (2.174) and (2.175) do not depend on the admissible values Z and K of the parametric variables; under such conditions these quantities need not be eliminated, of course. This class includes, in particular, problems in which the control occurs on the right-hand sides of the basic equations as a source density, and also problems containing the control as a coefficient of lower derivatives in linear and quasi-linear differential constraint equations. What we have said is also true of equations of the form

$$Lz^j = f^j(x, z, z_x, u) \qquad (j = 1, \ldots, n)$$

where L is an elliptic (hyperbolic) operator that depends only on the second derivatives; here, the first derivatives z_x are assumed to be continuous on lines of possible discontinuity of the control $u(x)$. The same applies to the case when Lz^j is a parabolic operator of the form

$$\frac{\partial z^j}{\partial t} - a_{rs}^{jp}(x) \frac{\partial^2 z^p}{\partial x_r \, \partial x_s}$$

and the functions f^j do not depend on z_t. In the hyperbolic case, there are certain special features if the control is varied in strips (spheroids) oriented along the characteristic directions x_τ and t_τ, while in the parabolic case they arise for variation in strips perpendicular to the t axis.[68,69]

In the following chapters, some of these special problems will be considered in detail; here we merely point out that a characteristic feature unites them all: there is no dependence of the necessary conditions on the shape of the region of local variation.

---

† This is characteristic of problems in partial differential equations involving the specification of the constraint equations in the normal form (2.2) and (2.17) (see Section 2.3).

Finally, there may be cases when the dependence of the admissible controls (or only the differences $\Delta u = U - u$) on the independent variables is specified in advance.

In these cases, the problem of deriving the necessary conditions simplifies because the local variations are concentrated on sets of a definite unchanged structure (e.g., in narrow strips perpendicular to the t axis[68]).

To conclude this section, we note that for systems of a very general form a method of deriving Weierstrass's necessary conditions was proposed by Plotnikov.[128,129]

2.8. Legendre's Necessary Condition

This condition can be obtained from Weierstrass's condition under the assumption that the variations Δu and Δv, and therefore $\Delta \zeta$ and $\Delta \kappa$ as well, are small compared with the optimal values of the corresponding quantities. We shall derive Legendre's condition first for an example, and then consider the general case.

We take Example 2 of Section 2.2.

The increment $\Delta(-I)$ of the functional resulting from variation Δu of the control is determined by (2.98):

$$\Delta(-I) = \iint_G [H(\xi, \eta, \zeta, u, x, y) - H(\xi, \eta, Z, U, x, y)] \, dx \, dy$$

where the function H has the form [see (2.28), (2.58), and (2.59)]

$$H(\xi, \eta, \zeta, u, u_*, x, y) = -\xi_1(u\zeta^1 + E_1) - \eta_1(u\zeta^2 + E_2)$$

$$+\xi_2\zeta^2 - \eta_2\zeta^1 - \alpha^*[(u_{max} - u)(u - u_{min}) - u_*^2]$$

Here, $\alpha^*(x, y)$ is the Lagrangian multiplier corresponding to the constraint

$$(u_{max} - u)(u - u_{min}) - u_*^2 = 0 \qquad (2.181)$$

[see $(2.28)_*$]. Taking into account the stationarity condition (2.60) and the conditions

$$-\xi_1\zeta^1 - \eta_1\zeta^2 + \alpha^*(2u - u_{max} - u_{min}) = 0$$

$$\qquad\qquad (2.182)$$

$$\alpha^* u_* = 0$$

which follow from (2.48) (second row), we obtain the following expression for the increment $\Delta H = H(\xi, \eta, Z, U, x, y) - H(\xi, \eta, \zeta, u, x, y)$ of the function H,

$$\Delta H = -\xi_1 \Delta u \, \Delta \zeta^1 - \eta_1 \Delta u \, \Delta \zeta^2 + \alpha^*(\Delta u)^2 + \alpha^*(\Delta u_*)^2 \quad (2.183)$$

where only the quadratic terms in the small variations Δu, Δu_*, and $\Delta \zeta$ have been retained. Note that in this example the function H is a second-degree polynomial in u, u_*, ζ^1, ζ^2, so that the expression for ΔH is exact. In the general case, an expression of the type (2.183) holds to terms of third order of smallness in Δu and $\Delta \zeta$.

The variations Δu and Δu_* are related by an equation obtained by varying (2.181); the result is

$$\Delta u = -\frac{2u_* \, \Delta u_*}{2u - u_{max} - u_{min}} \quad (2.184)$$

Equations $(2.182)_1$ and (2.184) are now used to eliminate α^* and Δu, and to write the inequality $\Delta H \leq 0$ [see (2.183)] in the form

$$\Delta H = \frac{1}{2u - u_{max} - u_{min}} \left\{ 2(\xi_1 \Delta \zeta^1 + \eta_1 \Delta \zeta^2) u_* \, \Delta u_* \right.$$

$$\left. + (\xi_1 \zeta^1 + \eta_1 \zeta^2) \left[\frac{4u_*^2}{(2u - u_{max} - u_{min})^2} + 1 \right] (\Delta u_*)^2 \right\} \leq 0 \quad (2.185)$$

This inequality expresses Legendre's condition, a necessary condition for a weak relative minimum.

Let us note some of the consequences of (2.185). From $(2.182)_2$ we conclude that three cases are possible: (1) $\alpha^* \neq 0$, $u_* = 0$; (2) $\alpha^* = 0$, $u_* \neq 0$; (3) $\alpha^* = 0$, $u_* = 0$. In the first of these cases, it follows from (2.181) that either $u = u_{max}$ or $u = u_{min}$. Condition (2.185) also shows that [see also (2.50) and (2.100)]

$$u = u_{max}, \qquad \text{when } (\mathbf{j}, \text{grad } \omega_2) \geq 0$$

$$\tag{2.186}$$

$$u = u_{min}, \qquad \text{when } (\mathbf{j}, \text{grad } \omega_2) \leq 0$$

Comparing these conditions with (2.114) and (2.116), we see that the latter are more stringent; this is not surprising if one bears in mind that Legendre's condition is based on small variations.*

In the second of the cases mentioned, $u_{min} < u < u_{max}$; however, $(\mathbf{j}, \text{grad } \omega_2) = 0$ [see (2.182)$_1$ and (2.60), (2.100)], and we thus obtain

$$\Delta H = -(\xi_1 \, \Delta\zeta^1 + \eta_1 \, \Delta\zeta^2) \, \Delta u \qquad (2.187)$$

We introduce a "variation in a strip." The use of (2.119) (**t** is a vector tangent to the boundary of the strip) yields

$$\Delta\zeta^1 = Z^1 - \zeta^1 = -\frac{\Delta u}{u} j_t x_t, \qquad \Delta\zeta^2 = Z^2 - \zeta^2 = -\frac{\Delta u}{u} j_t y_t$$

The substitution of the expressions from (2.187) and the use of (2.60) and (2.100) result in

$$\Delta H = -j_2 \omega_{2t} \frac{(\Delta u)^2}{u^2} \qquad (2.188)$$

If $(\mathbf{j}, \text{grad } \omega_2) = 0$ and $|\mathbf{j}| \neq 0$, $|\text{grad } \omega_2| \neq 0$, then (2.188) could be made either positive or negative by the choice of **t** (an analogous result was obtained in Section 2.3); therefore, this case must be rejected.

The third case $\alpha^* = 0$, $u = 0$ requires the simultaneous satisfaction of the conditions $(\mathbf{j}, \text{grad } \omega_2) = 0$ and $u = u_{max}$ (or $u = u_{min}$), which is a contradiction.

It is now easy to write down Legendre's condition for the general case. With δ denoting small variations, we obtain the equations

$$\frac{\partial \delta z^j}{\partial x_i} = \frac{\partial X_i^j}{\partial \zeta^t} \delta\zeta^t + \frac{\partial X_i^j}{\partial u_r} \delta u_r \qquad (2.189)$$

---

* In this example the smallness of the variations Δu and Δu_* essentially was used only to replace (2.181) by (2.184). As we have pointed out, the representation of ΔH in the form (2.183) does not require the variations to be small. It is readily seen that, with the retention of (2.181) instead of (2.184), we can reduce the expression for ΔH to the form

$$\Delta H = -\frac{\Delta u}{u}(\mathbf{J}, \text{grad } \omega_2), \qquad \mathbf{J} = \mathbf{j} + \Delta\mathbf{j}$$

[see (2.102)], from which (2.114) and (2.116) are directly obtained. We may add that in problems of the type (2.180), linear in the controls (see Example 1 of Section 2.2), the Legendre and Weierstrass necessary conditions are the same.

$$\frac{\partial a_k}{\partial u_r} \delta u_r = 0 \tag{2.190}$$

$$\frac{\partial \delta z^j}{\partial q_i} = \frac{\partial T_i^j}{\partial \kappa^t} \delta \kappa^t + \frac{\partial T_i^j}{\partial v_r} \delta v_r \tag{2.191}$$

$$\frac{\partial b_k}{\partial v_r} \delta v_r = 0 \tag{2.192}$$

They are deduced by varying (2.2), (2.17), (2.27), and (2.29) with respect to z_x^j, z_q^j, ζ, κ, u, v, and by expanding the left-hand sides of (2.174) and (2.175) in series in powers of $\delta\zeta$, $\delta\kappa$, δu, δv, retaining only the first quadratic terms. The result is

$$\frac{\partial^2 H}{\partial \zeta^t \partial \zeta^{t'}} \delta\zeta^t \delta\zeta^{t'} + 2\frac{\partial^2 H}{\partial \zeta^t \partial u_r} \delta\zeta^t \delta u_r + \frac{\partial^2 H}{\partial u_r \partial u_{r'}} \delta u_r \delta u_{r'} \leq 0 \tag{2.193}$$

$$\frac{\partial^2 h}{\partial \kappa^t \partial \kappa^{t'}} \delta\kappa^t \delta\kappa^{t'} + 2\frac{\partial^2 H}{\partial \kappa^t \partial v_r} \delta\kappa^t \delta v_r + \frac{\partial^2 H}{\partial v_r \partial v_{r'}} \delta v_r \delta v_{r'} \leq 0 \tag{2.194}$$

These inequalities in conjunction with (2.189)–(2.192) and the corresponding boundary conditions* are the Legendre conditions. The role of (2.189) and (2.191) is to specify the variations $\delta\zeta$ and $\delta\kappa$; in the case of one independent variable, when the parametric variables vanish, (2.189) and (2.191) are eliminated from the system of relations that form the Legendre conditions.

The inequalities (2.193) and (2.194) are necessary conditions for a maximum of the functions $H(x, z, \zeta, u, \xi)$ and $h(x, z, \kappa, v, \theta)$ with respect to the variables ζ, u and κ, v.

2.9. Jacobi's Necessary Condition

This condition is associated with the investigation of the second variation of the functional I. Suppose δu and δv are small but otherwise arbitrary variations of the controls, and δz, $\delta\xi$, $\delta\kappa$ are the corresponding variations of the basic and of the parametric variables. We introduce the following variational equations:

$$\frac{\partial \delta z^j}{\partial x_i} = \frac{\partial X_i^j}{\partial z^k} \delta z^k + \frac{\partial X_i^j}{\partial \zeta^t} \delta\zeta^t + \frac{\partial X_i^j}{\partial u_r} \delta u_r \tag{2.195}$$

---

* Obtained by varying the boundary conditions (2.16) and (2.19).

$$\frac{\partial a_k}{\partial u_r} \delta u_r = 0 \tag{2.196}$$

$$\frac{\partial \, \delta z^j}{\partial q_i} = \frac{\partial T_i^j}{\partial z^k} \delta z^k + \frac{\partial T_i^j}{\partial \kappa^t} \delta \kappa^t + \frac{\partial T_i^j}{\partial v_r} \delta v_r \tag{2.197}$$

$$\frac{\partial b_k}{\partial v_r} \delta v_r = 0 \tag{2.198}$$

with corresponding boundary conditions

$$\delta z^j(x) = 0, \qquad x \in \gamma_j \tag{2.199}$$

$$\delta z^j(x) = 0, \qquad x \in \partial \delta_j \tag{2.200}$$

Assuming that the region G is fixed, and using (2.171), we readily obtain the following expression for the second variation of the functional I:

$$\delta^2 I = \int_G 2\omega \, dx + \int_\delta 2\varphi \, dx \tag{2.201}$$

where we have denoted

$$
\left.
\begin{aligned}
-2\omega &= \frac{\partial^2 H}{\partial z^i \, \partial z^{i'}} \delta z^i \, \delta z^{i'} + \frac{\partial^2 H}{\partial \zeta^t \, \partial \zeta^{t'}} \delta \zeta^t \, \delta \zeta^{t'} \\[2mm]
&\quad + \frac{\partial^2 H}{\partial u_r \, \partial u_{r'}} \delta u_r \, \delta u_{r'} + 2 \frac{\partial^2 H}{\partial z^i \, \partial \zeta^t} \delta z^i \, \delta \zeta^t \\[2mm]
&\quad + 2 \frac{\partial^2 H}{\partial z^i \, \partial u_r} \delta z^i \, \delta u_r + 2 \frac{\partial^2 H}{\partial \zeta^t \, \partial u_r} \delta \zeta^t \, \delta u_r \\[2mm]
-2\varphi &= \frac{\partial^2 h}{\partial z^i \, \partial z^{i'}} \delta z^i \, \delta z^{i'} + \frac{\partial^2 h}{\partial \kappa^t \, \partial \kappa^{t'}} \delta \kappa^t \, \delta \kappa^{t'} \\[2mm]
&\quad + \frac{\partial^2 h}{\partial v_r \, \partial v_r'} \delta v_r \, \delta v_{r'} + 2 \frac{\partial^2 h}{\partial z^i \, \partial \kappa^t} \delta z^i \, \delta \kappa^t \\[2mm]
&\quad + 2 \frac{\partial^2 h}{\partial z^i \, \partial v_r} \delta z^i \, \delta v_r + 2 \frac{\partial^2 h}{\partial \kappa^t \, \partial v_r} \delta \kappa^t \, \delta v_r
\end{aligned}
\right\} \tag{2.202}
$$

The requirement of nonnegativity of the second variation is Jacobi's necessary condition. One considers the auxiliary problem of minimizing the functional (2.201) subject to the constraints (2.195)–(2.200). This is a problem of the Mayer–Bolza type with quadratic functional (2.201) and linear constraint equations. All of the necessary conditions given earlier are used in its solution. The set of variations

$$\delta z = \delta \zeta = \delta \kappa = \delta u = \delta v = 0 \qquad (2.203)$$

has a corresponding zero value of the functional $\delta^2 I$; if the conditions of stationarity of the second variation (i.e., of the auxiliary problem) are satisfied by only this set, then zero is an extremal value of the functional (2.201). If the necessary condition of Weierstrass (Legendre) is also satisfied for the original problem, then the extremum of $\delta^2 I$ is a minimum, and Jacobi's condition is therefore satisfied.

If there exists a set of variations that differs from (2.203) and satisfies the stationarity conditions of the auxiliary problem, then the sign of the second variation must also be investigated. The stationarity conditions of the functional (2.201) subject to constraints—variational equations—are identical with the equations obtained by varying the stationarity conditions of the original problem. The set of these conditions and (2.195)–(2.200) constitutes a homogeneous boundary-value problem, from which one must determine the variations δz, $\delta \zeta$, $\delta \kappa$, δu, δv and the corresponding conjugate variables. The trivial solution of this problem may not be unique; i.e., there exists another, nonzero solution. To this second solution there also corresponds a zero value of the functional $\delta^2 I$, and under these conditions one can usually show that this zero value is not a minimum.

Thus, the minimal value of $\delta^2 I$ is negative, and Jacobi's condition is violated. A detailed examination of the possibilities which then arise will be given in an example in Chapter 3.

3

Optimal Distribution of the Resistivity of the Working Medium in the Channel of a Magnetohydrodynamic Generator

In this chapter, we apply the results obtained earlier to the problem of the optimal distribution of the resistivity of a conducting medium in the channel of a magnetohydrodynamic (MHD) generator. This problem has independent applied significance and simultaneously enables us to give a detailed illustration of the necessary conditions for optimality. Indeed, in a large number of cases the example illustrates many important facts that enable one to draw conclusions about the classical or generalized nature of an optimal solution, if one exists. It may happen that the originally chosen class of admissible controls does not contain an optimal control; in these cases, the necessary conditions indicate a natural way to relax the problem, i.e., a way of extending the class of admissible controls to guarantee the existence of an optimal control.

From the point of view of practical applications, it is not always required that an optimal control exists; in practice, the existence of only a minimizing sequence often suffices. Nevertheless, the investigation of relaxed problems is of great interest because it casts light on the actual nature of the optimization, clarifying the factors that prevent optimal behavior of the system. In addition, the optimal control in the regularized problem frequently possesses a physical transparency absent in the minimizing sequences corresponding to the original formulation of the problem.

The exposition of the basic material of the chapter is preceded by a brief description of longitudinal end effects in MHD channels and of the methods used to suppress them.

3.1. Longitudinal End Effects in MHD Channels

The technical tasks that arise in the construction of MHD installations (generators, accelerators, pumps) give rise to a large number of both theoretical and applied problems. The difficulties of creating strong magnetic fields in fairly large volumes, the maintaining of a high electrical conductivity of the gas, and the reduction of heat losses on the walls—these and other problems require maximal use of the existing technical possibilities. Therefore, the problem of the optimal choice of the factors that determine the characteristics of various MHD devices is an important problem of modern magnetohydrodynamics.

The problem of the flow of a conducting fluid in channels is the basis of the theoretical investigation of the ranges of operation of MHD devices. The corresponding calculations are frequently made by means of the so-called one-dimensional approximation, and enable one with a certain accuracy to determine the basic parameters of the system, whose behavior is described by ordinary differential equations in this case. Although the one-dimensional approximation does not take into account effects associated with the curvature of the lines of the electric current in the channel, or other spatial phenomena, it explains many of the factors that hinder effective operation of the system.

For example, to eliminate the decrease in conductivity when there is sufficiently strong cooling of the working material in the channel of an MHD generator, one must intervene from outside, either by introducing readily ionizing additives or by additional heating. The distribution of the intensity with which the additives are introduced or the density of the heat sources along the channel can be found by solving a suitable optimization problem.

Important questions which can readily be analyzed within the framework of the one-dimensional approximation are the determination of the optimal choice (within definite limits) of the magnetic field strength along the channel, the transverse dimensions and the length of the channel, and of the initial data. These may be based on various optimality criteria. Some problems of this kind were considered in Refs. 101 and 367.

A more detailed study of the most advantageous regimes of operation of MHD devices requires the formulation and solution of optimization problems for systems described by partial differential equations.

Even the simplest "two-dimensional" model of an MHD generator with finite electrodes[29,97] reveals a serious loss of current and useful power in such a device; this is due to the end effects associated with the reverse flow of currents in regions in which the external magnetic field changes sufficiently rapidly. The elimination of losses of this kind is an important technical problem, whose solution requires all of the available control possibilities. The most important of these factors are the distribution of the external magnetic field and the distribution of the conductivity of the working medium. By an appropriate choice of the corresponding functions, the end effects can be reduced to the smallest possible amounts.

Equally, one can pose the problem of minimizing the Joule losses in such a channel; one can also consider more complicated optimization problems: maximization of the generator efficiency,* optimization problems with allowance for the Hall effect, etc.

Two-dimensional effects in channels of MHD generators with finite electrodes have been variously investigated.[28-34,193,368] The investigations were made within the framework of the approximation of small magnetic Reynolds number and interaction parameter, which are well-known in magnetohydrodynamics; in this approximation, one can ignore the induced fields and assume that the hydrodynamic parameters of the flow (velocity, pressure, and gas temperature) are known based on this approximation† and on the condition of rectilinearity of the flow. The basic differential equations of the problem are obtained from the basic equations of magnetohydrodynamics.[30] They form an elliptic system for two unknown functions—the electric field potential z^1 and the flow function z^2 corresponding to the current density vector $\mathbf{j} = -\text{curl } \mathbf{i}_3 z^2$:

$$z_x^1 = \rho z_y^2, \quad z_y^1 = -\rho z_x^2 + c^{-1} VB \tag{3.1}$$

Here, $\rho = \rho(x, y)$ is the resistivity of the working medium (gas), $\mathbf{v} = V(y)\mathbf{i}_1$ is the gas velocity, and $\mathbf{B} = -B(x)\mathbf{i}_3$ is the external magnetic field.‡ The velocity distribution of the fluid is assumed known and taken from the solutions of the corresponding hydrodynamic problems.§

In Refs. 28–34, 193, and 368, solutions of the system (3.1) were obtained and investigated under different boundary conditions on the channel walls for different types of functions $\rho(x)$, $V(y)$, and $B(x)$. In all cases, these three functions were assumed given, and the main attention was devoted to studying the corresponding solutions and integrated characteristics of the device.

The optimization problems corresponding to the system (3.1) are characterized by the fact that the functions $\rho(x, y)$, $V(y)$, and $B(x)$ are regarded as controls that must be determined along with the functions z^1 and z^2. In practice, the velocity can be controlled by various hydraulic

---

* "Efficiency" here is taken to be the ratio of the useful power to the sum of the useful power and the magnitude of the Joule losses.

† It is also assumed that the Hall effect is negligibly small.

‡ In what follows, unless otherwise stated, we shall assume that $B(x)$ is a nonnegative even function.

§ A detailed derivation of (3.1) and discussion of the quantities contained in this equation can be found in Ref. 33, pp. 362–368.

measures: variation of the cross-sectional area of the channel, creation of artificial obstacles, etc. The magnetic field is controlled by an appropriate choice of the currents in the exciting electromagnets, and also by an appropriate choice of the geometry of the magnetic circuits. Control of the conductivity can be achieved by subjecting the working medium to different kinds of radiation, to heating, or to ionization by electric fields. Of greatest practical value is the method of introducing small amounts of conductivity-increasing additives into the flow (usually compounds of alkali metals) or water vapor, which lowers the conductivity; important control factors may also be nonconducting baffles set up in the channel parallel to the flow,[34,319,367] modulation of the temperature along the channel,[249] etc.

Ultimately, the important thing is to realize not just any distribution of the conductivity but the optimal one. Therefore, in theoretical investigations it is not sensible to restrict the class of admissible controls to those that are physically realizable distributions of the conductivity: the controls may be sought in the class of piecewise continuous differentiable functions of two independent variables, or even in the class of measurable bounded functions. The optimal conductivity distributions then obtained can be approximated with sufficient accuracy in practice by available means of control.

It is important to emphasize that the controls ρ, V, and B are almost always subject to known restrictions, which characterize the actually existing possibilities of control. These constraints are frequently specified by ordinary or isoperimetric inequalities; as a result, the range of variation of the control may be closed. This fact is particularly important in optimization problems whose constraint equations are linear in the control; in these cases, absence of restrictions on the control may be equivalent to nonexistence of an optimal control, and therefore of solutions to the optimization problem.

It is important to consider a substantial difference between optimization problems for the system (3.1), in which the resistivity $\rho(x, y)$ is fixed and the functions $V(y)$ or $B(x)$ are controls, and problems in which the resistivity is a control and $V(y)$ and $B(x)$ are fixed.

Problems of the first type contain the controls in the free term of the basic equations (3.1), and they can therefore be reduced, in principle, to the simple (containing no constraints) variational problem (e.g., by means of a Green's function). Only when the construction of the corresponding Green's function leads to difficulties, the reduction of the problem to one of the simple kind may be impossible to carry out in practice.

In contrast, problems of the second type cannot be reduced to those of the simple kind at all, since the control $\rho(x, y)$ then occurs in the principal part of the basic differential operator of the system (3.1), and a Green's function cannot be constructed until the optimal control is known.

The optimization problem is even more complicated when all three functions $\rho(x, y)$, $V(y)$, and $B(x)$ are chosen as controls.

Various investigations have been made concerning the most advantageous conditions of operation of MHD generators when allowance is made for two-dimensional end effects. Some indication of the effect of these conditions can be deduced from the solutions obtained by Hurwitz, Kilb, and Sutton,[269] Sutton and Carlson,[368] Vatazhin,[29,32,34] and other authors for various special forms of the functions $\rho(x, y)$, $V(y)$, and $B(x)$.

We shall present the main results of these investigations, basically following Refs. 29 and 32.

If $\rho(x, y) = \rho = $ const, $B = B_0 = $ const, $V = $ const, then the total current I_∞ across the load R and the Joule losses Q_∞ in an infinitely long channel are determined by the expressions

$$I_\infty = \frac{\varepsilon \alpha_\infty^*}{\rho + R\alpha_\infty^*}, \qquad Q_\infty = \frac{\varepsilon^2 \alpha_\infty^* \rho}{(\rho + R\alpha_\infty^*)^2} \qquad (3.2)$$

Here, $\varepsilon = 2\delta c^{-1} VB_0$, $\alpha_\infty^* = K(\kappa_\infty')/K(\kappa_\infty)$, $\kappa_\infty^{-1} = \cosh(\pi\lambda/2\delta)$; $\kappa_\infty^2 + \kappa_\infty'^2 = 1$; 2λ is the length of the electrodes, 2δ is the width of the channel, c is the velocity of light, and $K(\kappa)$ is the complete elliptic integral of the first kind.

If $\rho(x, y) = \rho = $ const, $V = $ const, and the function $B(x)$ is zero for $|x| > \lambda$ and arbitrary in the electrode zone $|x| < \lambda$, then the analogous expressions have the form

$$I_\lambda = \frac{G_1}{\rho + R\alpha_\infty^*}, \qquad Q_\lambda = \frac{2V^2\delta}{\rho c^2}\left[G_2 - \frac{R(2\rho + R\alpha_\infty^*)}{2\delta(\rho + R\alpha_\infty^*)^2} G_1^2\right] \qquad (3.3)$$

Here

$$G_1 = \int_{-\lambda}^{\lambda} B(x)\, dx, \qquad G_2 = \int_{-\lambda}^{\lambda} B^2(x)\, dx$$

The function $\alpha_\infty^*(\lambda/\delta)$ increases monotonically from zero to infinity as the argument increases from zero to infinity. This function is plotted in Fig. 3.1 which is taken from Ref. 29. For small λ/δ,

$$\alpha_\infty^* = \frac{\pi}{2}\ln^{-1}\frac{8\delta}{\pi\lambda} - \frac{\pi}{24}\left(\frac{\pi\lambda}{2\delta}\right)^2 \ln^{-2}\frac{8\delta}{\pi\lambda} + O\left[\left(\frac{\pi\lambda}{2\delta}\right)^4 \ln^{-2}\frac{8\delta}{\pi\lambda}\right]$$

while for large λ/δ,

$$\alpha_\infty^* = \frac{2\ln 2}{\pi} + \frac{\lambda}{\delta} - \frac{1}{4\pi}\exp\left(-\frac{2\pi\lambda}{\delta}\right) + O\left[\exp\left(-\frac{3\pi\lambda}{\delta}\right)\right]$$

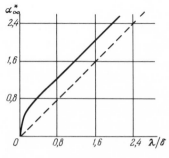

FIGURE 3.1

Equations (3.2) express the integral characteristics of the generator in the absence of the end effect associated with the vortex currents in the region in which the magnetic field $B(x)$ drops to zero; (3.3) gives the characteristics of the generator in the presence of the end effect.

To compare these results, we set $B(x) = B_0 = \text{const}$ for $|x| < \lambda$ in (3.3). For all values of the ratio λ/δ, $I_\infty > I_\lambda$; with regard to the Joule losses, in the limit $\lambda/\delta \to 0$ both Q_∞ and Q_λ tend to zero; as $\lambda/\delta \to \infty$, $Q_\infty \to 0$, and

$$Q_\lambda \to \frac{8 \ln 2}{\pi^2} \frac{V^2 B^2 \delta^2}{\rho c^2}$$

The difference between the limiting values of Q is explained by the fact that Q_∞ is calculated from the solution of the problem for a magnetic field that extends infinitely far from the region of the electrodes under the condition of separation of the charges at infinity, i.e., without allowance for the end effect, while Q_λ can be calculated from the current distribution in an infinite channel whose walls are electrodes for $x < 0$ where the potential difference λ is applied, while they are insulators for $x > 0$; the corresponding value of the dissipation is $0.5 Q_\lambda$.

In Ref. 31 it is shown that the end effect still has a similar influence on the magnitude of the Joule losses if the region of decrease of the magnetic field $B(x)$ is taken beyond the electrodes, i.e., if one considers fields of the form

$$B(x) = \begin{cases} B_0 = \text{const}, & 0 \le |x| < x_1, \quad x_1 > \lambda \\ 0, & |x| > x_1 \end{cases}$$

Taking the magnetic field beyond the electrode zone by an amount of order λ for $\lambda/\delta > 1$ leads to the establishment of a limiting current distribution in the channel, to which there correspond the characteristics $I = I_\infty$, $Q = Q_\infty + 2Q_c$, where

$$Q_c = \frac{16\delta^2 V^2 B_0^2}{\pi^3 \rho c^2} \Sigma, \qquad \Sigma = \sum_{\nu=1}^{\infty} \frac{1}{(2\nu - 1)^3} = 1.052$$

The dissipation Q_c can be calculated by considering currents in a channel with dielectric walls and a magnetic field equal to $B_0 = \text{const}$ in half of the channel and zero in the other. It is easy to verify that $2Q_c/Q_\lambda = 0.616$; this value characterizes the limiting relative decrease of the Joule losses that can be achieved by taking the field out to a distance of order λ from both sides of the electrode zone. The corresponding limiting value of the current is I_∞. The results we have given are based on the assumption that we have at our disposal a magnetic system capable of producing a homogeneous field of appreciable strength in a fairly large volume. Actual magnetic systems are characterized above all by the magnetic flux they produce, i.e., by the quantity $2\Phi = 2\int_0^\infty B(x)\, dx$.

The maximal value of the induction also is usually given; thus, we can assume that the function $B(x)$ satisfies conditions of the form

$$\int_0^\infty B(x)\, dx = \Phi, \qquad 0 \le B(x) \le B_{\max} \tag{3.4}$$

If the resistivity of the medium satisfies $\rho(x, y) = \rho = \text{const}$, and, in addition, $V = \text{const}$, then the current which can be taken from the generator is[29]

$$I = \frac{2V\lambda}{c} \frac{1}{\rho + R\alpha_\infty^*} \int_0^\infty B(x) T\left(\frac{x}{\lambda}\right) d\left(\frac{x}{\lambda}\right) \tag{3.5}$$

where the function $T(x/\lambda)$ is continuous and given by

$$T\left(\frac{x}{\lambda}\right) = \begin{cases} 1, & 0 \le x/\lambda \le 1 \\ F\left(\dfrac{x}{\lambda}\right) = \dfrac{2}{\pi} r \sqrt{\dfrac{r^2 - \kappa_\infty^2}{r^2 - 1}} \left[\Pi\left(\dfrac{\pi}{2}, \dfrac{\kappa_\infty'^2}{r^2 - 1}, \kappa_\infty'\right) \right. \\ \left. - \alpha_\infty^* \dfrac{r^2 - 1}{r^2} \Pi\left(\dfrac{\pi}{2}, -\dfrac{\kappa_\infty^2}{r^2}, \kappa_\infty\right) \right], & 1 \le x/\lambda < \infty \end{cases} \tag{3.6}$$

Here, we have introduced the notation

$$r = \kappa_\infty \cosh \frac{\pi x}{2\delta}$$

and

$$\Pi\left(\frac{\pi}{2}, h, k\right) = \int_0^{\pi/2} \frac{dt}{(1 + h \sin^2 t)\sqrt{1 - k^2 \sin^2 t}}$$

is the complete elliptic integral of the third kind. The function $T(x/\lambda)$ has a graph of the type shown in Fig. 3.2; for large values of the argument, $T(x) \approx \text{const} \cdot \exp(-\pi x/2\delta)$.

Let us consider the maximization of the functional I [see (3.5)] based on a choice of the function $B(x)$ under the restrictions (3.4).[101] The necessary conditions for a maximum are the Euler equations

$$-T\left(\frac{x}{\lambda}\right) + \Lambda + \mu(B_{max} - 2B) = 0 \qquad \mu b = 0 \qquad (3.7)$$

where $\Lambda = \text{const}$ and $\mu = \mu(x)$ are the Lagrangian multipliers corresponding to the first constraint (3.4) and the second constraint expressed in the form of the equivalent equation

$$(B_{max} - B)B - b^2 = 0 \qquad (3.8)$$

by means of the auxiliary real control $b(x)$. The Legendre condition $\mu < 0$ together with the Euler equations (3.7) leads to the following classification of extremal control:

1. $B = 0$ if $S(x/\lambda) \equiv -T(x/\lambda) + \Lambda \geq 0$
2. $B = B_{max}$ if $S(x/\lambda) \leq 0$
3. B is not defined if $S(x/\lambda) = 0$

In this region, $B(x)$ is an arbitrary function satisfying the constraints (3.4); in particular, it may take the values B_{max} and 0. The condition $S(x/\lambda) = 0$ must also be satisfied at points of switching of the control regions.

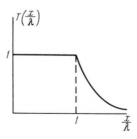

FIGURE 3.2

The positive function $T(x/\lambda)$ does not increase, so that only the following switching sequences can be realized with increasing x:

a. Switch from $2 \to 3$
b. Switch from $3 \to 1$
c. Switch from $2 \to 3$ with the subsequent switch $3 \to 1$
d. Switch from $2 \to 1$

Case (a) can be ruled out since the function $T(x/\lambda)$ does not remain constant for sufficiently large x/λ, contradicting the condition of existence of the control range (3).

In case (b), the range (3) can be realized only in the electrode zone $x/\lambda \leq 1$; the necessary condition for this is $B_{max} \geq \Phi(\lambda)$. If this condition is satisfied, any distribution $B(x)$ is optimal if it is such that $0 \leq B(x) \leq B_{max}$, $\int_\Delta B(x)\,dx = \Phi$ where Δ denotes the set of points of the interval $[0, \lambda]$ for which $B(x) \neq 0$.

Case (c) is possible only under the condition $B_{max} \geq \Phi(\lambda)$; an optimal control is such that $B = B_{max}$ for $x/\lambda \leq \beta$, $0 \leq \beta \leq 1$, $\int_\beta^\gamma B(x)d(x/\lambda) + B_{max}\beta = \Phi/\lambda$, $\beta \leq \gamma \leq 1$, and $B(x) = 0$ for $x/\lambda > \gamma$. Here, β and γ are any numbers satisfying $0 \leq \beta \leq \gamma \leq 1$.

Case (d) is possible both under the condition $B_{max} \geq \Phi/\lambda$ and under the condition $B_{max} < \Phi/\lambda$. If $B_{max} \geq \Phi/\lambda$, then $B = B_{max}$ for $x/\lambda \leq \delta$, where $\delta = \Phi/\lambda B_{max} \leq 1$, and $B = 0$ for $x/\lambda > \delta$. If $B_{max} < \Phi/\lambda$, then $B = B_{max}$ for $x/\lambda \leq \varepsilon$, $\varepsilon = \Phi/\lambda B_{max} > 1$, and $B = 0$ for $x/\lambda > \varepsilon$.

Thus, for $B_{max} \geq \Phi/\lambda$ the variants (b), (c), and (d) are possible; for $B_{max} < \Phi/\lambda$, only variant (d).

If $B_{max} \geq \Phi/\lambda$, then the maximal value of the functional (3.5) is

$$I_{max} = \frac{2V\Phi}{c}\frac{1}{\rho + R\alpha_\infty^*} \tag{3.9}$$

If $B_{max} < \Phi/\lambda$, then the maximal value of the same functional is given by

$$I_{max} = \frac{2V}{c}\frac{1}{\rho + R\alpha_\infty^*}B_{max}\lambda\left[1 + \int_1^{\Phi/\lambda B_{max}} F\left(\frac{x}{\lambda}\right)d\left(\frac{x}{\lambda}\right)\right] \tag{3.10}$$

We fix B_{max} and Φ and consider the variation of the maximal current I_{max} as the parameter λ is reduced. If the value of λ is sufficiently large (long electrodes), then the maximum of the current is given by (3.9). Since the function $\alpha_\infty^*(\lambda/\delta)$ decreases with decreasing λ, it is advantageous to reduce the value of this parameter; the functional I_{max} of (3.9) takes the largest value with respect to λ for $\lambda = \Phi/B_{max}$; we obtain

$$\max_{\lambda} I_{max} = \frac{2V\Phi}{c} \frac{1}{\rho + R\alpha_{\infty}^{*}(\Phi/B_{max}\delta)} \tag{3.11}$$

At even smaller λ the values of the functional I_{max} are determined by (3.10). These values decrease monotonically from (3.11) at $\lambda = \Phi/B_{max}$ to zero at $\lambda = 0$ [the latter follows readily from the properties of the function $F(x/\lambda)$ (see Ref. 29)].

Thus, under the restrictions (3.4) imposed on the function $B(x)$ there exists a critical length $2\lambda = 2\Phi/B_{max}$ of the electrodes yielding maximal current from the electrodes. The magnitude of this current is determined by (3.11), and the corresponding field distribution is: $B = B_{max}$ for $x \leq \lambda$, and $B = 0$ for $x > \lambda$.

In the derivation of this result, the resistivity of the working medium in the channel was assumed constant. Under these conditions, there are current losses in the regions in which the magnetic field changes strongly; this effect increases with decreasing length of the electrodes. Taking the field outside the electrode zone in the case $B_{max} < \Phi/\lambda$ partly compensates for the current loss. However, the mechanism of this compensation (extension of the field) begins to operate only for sufficiently small λ ($\lambda < \Phi/B_{max}$), whereas the losses due to the boundary effect are important at larger λ (such that $\lambda \geq \Phi/B_{max}$), when the optimal field $B(x)$ is concentrated only in the electrode zone.

From this we conclude that it would be desirable to introduce a new external controlling factor capable of changing the current distribution in the channel in such a way that the losses due to the boundary effect are as small as possible. The natural thing to do is to control the distribution of the resistivity of the working medium.

A number of authors[34,88,89,113,319] have studied the possibility of suppressing the boundary effect by introducing infinitesimally thin insulating baffles into the flow. In all investigations, the baffles were assumed to be rectilinear and they were placed parallel to the velocity vector of the working medium (the change in the velocity distribution due to the influence of viscosity was ignored). This is a very simple arrangement of the baffles; however, it is the only arrangement for which one can determine the characteristics of the system by analytic means.

It was shown that the introduction of baffles into a channel filled with a homogeneous conductor reduces the current taken from the electrodes if the external field is homogeneous; but if the field is nonzero only within the electrode zone, the current may be increased by placing baffles outside the electrode zone and therefore preventing the formation of closed end currents. If the baffles are arranged parallel to the walls of the channel and extend from $x = \lambda$ to $x = \infty$, then one can approximate the characteristics of a channel with $\rho = \infty$ arbitrarily closely for $|x| > \lambda$ if a sufficiently large number of baffles is chosen.

The main aim of all of the above investigations was to establish the most favorable (in the sense of reducing the current loss) distribution of baffles in the channel. To achieve this aim, an attempt was made to obtain solutions for admissible positions of the baffles. Because of the great analytic difficulties, such solutions can be constructed only in a very limited number of cases (the problem for a medium whose electrical conductivity varies with respect to two coordinates x and y is hopeless in this respect). However, if one is interested in constructing optimal solutions, there is no need to obtain admissible solutions. The methods of Chapter 2 provide possible rules for the direct construction of optimal solutions without using admissible solutions, which are thus made redundant.

This circumstance greatly extends the class of possible dependences of the resistivity on the coordinates; here, we choose bounded measurable scalar functions of the two independent variables (x, y). The requirements of boundedness and measurability have a mathematical nature. On the one hand, they guarantee the existence of an optimal control in the given class under certain additional conditions; simultaneously, this class contains a fairly extensive reserve of functions which permit practical realization. On the other hand, the class of admissible controls includes functions whose practical realization entails considerable difficulties; if the optimal function turns out to be such a function, it is necessary to approximate it as accurately as possible by means of readily realizable functions. Such a situation is generally characteristic of optimization problems concerning the maximal possibilities of systems; once an optimal solution has been obtained, one can establish to what extent it can be approximated using simpler means of control (e.g., horizontal baffles).

In the following sections of this chapter, we solve the general problem of maximizing the current obtained from the electrodes of an MHD generator by the optimal choice of the resistivity of the medium in the class of bounded measurable functions of two independent variables ($0 < \rho_{\min} \leq$ vrai max $\rho(x, y) \leq \rho_{\max} \leq \infty$).

3.2. Statement of the Problem (the Case of Scalar Resistivity)

We consider a planar channel (Fig. 3.3) of width 2δ whose walls are dielectric everywhere except at the two sections BB' of equal length 2λ along the walls of the channel opposite one another and made of a perfectly conducting material. The conducting sections are connected through the load R.

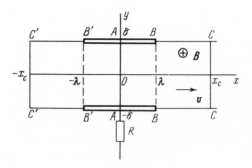

FIGURE 3.3

The working medium, whose point-to-point variation in resistivity is given by a scalar function $\rho(x, y)$, is taken to move through the channel with velocity $\mathbf{v}(V(y), 0)$. If a magnetic field $\mathbf{B}(0, 0, -B(x))$, $B(x) = B(-x) \geq 0$, is applied, an electric current $\mathbf{j}(\zeta^1, \zeta^2)$ flows in the channel, and a current I equal to

$$I = \int_{-\lambda}^{\lambda} \zeta^2(x, \pm\delta)\, dx \tag{3.12}$$

flows through the external resistance R. The distribution of the electric field potential z^1 and the current density in the channel are described by the equations [see (3.1)]

$$\operatorname{div} \mathbf{j} = 0, \qquad \rho\mathbf{j} = -\operatorname{grad} z^1 + \frac{1}{c}[\mathbf{v}, \mathbf{B}] \tag{3.13}$$

In the longitudinal direction, the channel can be assumed infinitely long; the components ζ^1 and ζ^2 of the vector $\mathbf{j}$ vanish at infinity.

Instead of this, one can also assume that the channel is bounded by a pair of vertical insulating walls CC and $C'C'$ (Fig. 3.3), at the distances $x = \pm x_c$, $x_c > \lambda$ from the coordinate origin through which the fluid can penetrate; on these walls,

$$\zeta^1\big|_{|x|=x_c} = 0 \tag{3.14}$$

By introducing the flow function $z^2(x, y)$ with the relation $\mathbf{j} = -\operatorname{curl} \mathbf{i}_3 z^2$, we can write the system (3.12) in the equivalent form

$$z_x^1 = -\rho\zeta^1, \qquad z_y^1 = -\rho\zeta^2 + \frac{1}{c}\, VB$$

$$z_x^2 = \zeta^2, \qquad z_y^2 = -\zeta^1 \tag{3.15}$$

These equations are augmented by the boundary conditions

$$
\left.
\begin{aligned}
z^1(x, \pm\delta) &= z^1_\pm = \text{const}, \qquad |x| < \lambda \\[2mm]
z^2(x, \pm\delta)\big|_{x>\lambda} &= z^2_+ = \text{const} \\[2mm]
z^2(x, \pm\delta)\big|_{x<-\lambda} &= z^2_- = \text{const} \\[2mm]
z^1_+ - z^1_- &= (z^2_+ - z^2_-)R = IR
\end{aligned}
\right\}
\tag{3.16}
$$

the last condition reflects Ohm's law for the external circuit of the generator.

The values of the function $\rho(x, y)$ at any point of the channel are determined by the possibilities available to the designer for controlling the resistivity of the medium. As a rule, these possibilities are restricted, and at best one can achieve a certain (nonzero) minimal value of the resistivity $\rho_{min} = \text{const}$; the maximal possible value $\rho_{max} = \text{const}$ may be infinitely large. Thus, it can be assumed that the function $\rho(x, y)$ satisfies the inequalities

$$
0 < \rho_{min} \le \rho(x, y) \le \rho_{max} \le \infty
\tag{3.17}
$$

in all cases.

The problem is to choose, among all piecewise continuous functions $\rho(x, y)$ satisfying (3.17), the optimal control $\rho(x, y)$ that maximizes the function I.

It is expedient to immediately extend the class of admissible controls $\rho(x, y)$ to the set of measurable functions $\rho(x, y)$ satisfying

$$
0 < \rho_{min} \le \text{vrai max } \rho(x, y) \le \rho_{max} \le \infty
\tag{3.18}
$$

With this modification, the posed problem differs from the problem in Example 2 in Section 2.2 only in notation (in particular, it is necessary to write ρ instead of u and set $E_1 = 0$, $E_2 = -c^{-1}VB$); if an optimal solution exists, then the previously obtained results may be used for its determination.

3.3. Optimal Distributions of Scalar Resistivity

Weierstrass's necessary condition [see (2.114) and (2.116)] shows that the optimal function $\rho(x, y)$ can take on only two values, namely

$$\left.\begin{array}{ll} \rho = \rho_{max}, & \text{for } 0 \le \chi(x, y) \le \arccos p \\[2mm] \rho = \rho_{min}, & \text{for } \pi - \arccos p \le \chi(x, y) \le \pi \\[2mm] \quad p = \dfrac{\rho_{max} - \rho_{min}}{\rho_{max} + \rho_{min}} & \end{array}\right\} \qquad (3.19)$$

The symbol $\chi(x, y)$ denotes the angle between the vectors $\mathbf{j}$ and grad ω_2 at the point (x, y); the function $\omega_2(x, y)$ is determined by the solution of the adjoint Eqs. (2.100)–(2.101) (in which u must be replaced by ρ).

The range G of variation of the independent variables is the union of the set G_1 in which $\rho = \rho_{min}$ and the set G_2 in which $\rho = \rho_{max}$, and the sets of points that do not belong to the union $G_1 \cup G_2$, from where it follows by Weierstrass's condition that mes $G = $ mes $G_1 \cup G_2$. The structure of the sets G_1 and G_2 is dictated by the fixed parameters of the problem, among which the principal role is played by the functions $B(x)$ and $V(y)$ and the constants ρ_{min}/R, ρ_{max}/R, λ/δ, x_c/δ.

Let us consider some special cases.

1. Suppose $B(x) = B_0 = \text{const} > 0$ (homogeneous field), $V = V/y \ge 0$. We set

$$w = z^1 - \frac{B_0}{c} \int_0^y V(y)\, dy \qquad (3.20)$$

Equations (3.15) can be written in the form

$$\begin{array}{ll} w_x = -\rho \zeta^1, & w_y = -\rho \zeta^2 \\[2mm] z_x^2 = \zeta^2, & z_y^2 = -\zeta^1 \end{array} \qquad (3.21)$$

and the vector $\mathbf{j}$ is equal to

$$\mathbf{j} = -\frac{1}{\rho}\, \text{grad } w \qquad (3.22)$$

The boundary conditions $(3.16)_{1-3}$ retain their form (see Fig. 3.3) if z^1 is replaced by w; condition $(3.16)_4$ then takes on the form

$$w_+ - w_- + \varepsilon = R(z_+^2 - z_-^2), \qquad \varepsilon = \frac{B_0}{c} \int_{-\delta}^{\delta} V\, dy \ge 0 \qquad (3.23)$$

A comparison of the boundary-value problem (3.21), (3.16)$_{1-3}$, and (3.23) with problem (2.100)-(2.101) indicates that for any $\rho(x, y)$ one has

$$z^2 = \varepsilon \omega_1, \qquad w = \varepsilon \omega_2 \tag{3.24}$$

Hence, taking into account (3.22) we obtain

$$\mathbf{j} = -\frac{\varepsilon}{\rho} \operatorname{grad} \omega_2 \tag{3.25}$$

which shows that the vectors $\mathbf{j}$ and grad ω_2 are everywhere antiparallel in the optimal range, i.e., $\chi = \pi$. Weierstrass's condition (3.19) now leads to the conclusion that the optimal control is equal to $\rho_{\min}$ in all of the region G.

Our result shows that a homogeneous external field is equivalent to an electromotive force ε in the external circuit. For $\rho = \rho_{\min} = $ const, it is easy to calculate the optimal value of the functional I; for $V = $ const, this value is equal to (see Appendix A.1)

$$\left. \begin{array}{c} I_{x_c}[B_0] = \dfrac{\varepsilon \alpha^*}{\rho_{\min} + R\alpha^*}, \qquad \varepsilon = \dfrac{2VB_0\delta}{c} \\[2ex] \alpha^* = K(\kappa')/K(\kappa), \qquad \kappa = \sqrt{1 - \kappa'^2} = \operatorname{dn}\left(\dfrac{\lambda}{\delta}K(k), k'\right) \\[2ex] k^2 + k'^2 = 1, \qquad K(k')/K(k) = x_c/\delta \end{array} \right\} \tag{3.26}$$

In the limit $x_c \to \infty$, the value of $I_{x_c}[B_0]$ increases monotonically and tends to the value I_∞ determined by the first of equations (3.2).

Figure 3.4 shows the vector lines of grad ω_2; the lines of $\mathbf{j}$ for $B = B_0 > 0$, $V \geq 0$ differ from these lines only in the direction of the arrows [see (3.25)].

2. We assume that $B(x) \neq$ const, $V(y) = $ const ≥ 0. We show that under these conditions the regime $\rho = $ const generally is not optimal. Consider, for example, the case when the continuous function $B(x) = B(-x)$

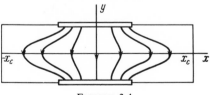

FIGURE 3.4

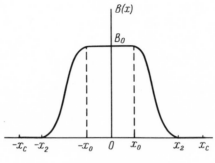

FIGURE 3.5

is positive for $|x| < x_2$, $x_2 < x_c$, and vanishes for $|x| > x_2$ (a possible graph of such a function is shown in Fig. 3.5*). For such a field, the distribution of the currents $\mathbf{j}$ in the case $\rho = \mathrm{const}$ is characterized by the vector lines shown in Fig. 3.6a for $x_2 < \lambda$ and in Fig. 3.6b for $x_2 > \lambda$ (the corresponding solutions for $x_c = \infty$ were constructed by Vatazhin in Refs. 28 and 29; solutions can also be readily constructed for the case $x_c < \infty$; these solutions do not provide anything qualitatively new).

Comparing Fig. 3.6a, b with Fig. 3.4, we see that Weierstrass's condition must be violated in some region since the lines of $\mathbf{j}$ and grad ω_2 intersect at both obtuse and acute angles, and the resistivity $\rho(x, y)$ is assumed to

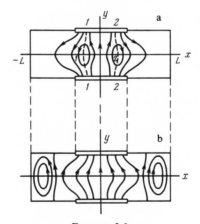

FIGURE 3.6

---

* It is assumed that x_c is fairly large.

be everywhere the same.* The same comparison suggests how the distribution of the resistivity should be altered in order to satisfy Weierstrass's condition. Namely, in the central region of the channel (to the right of the line 111 and to the left of the line 222 in Fig. 3.6a), where the vector lines of j and grad ω_2 form an obtuse angle, one should set $\rho = \rho_{min}$, and in the lateral regions (to the left of the line 111 and to the right of the line 222), where these lines form an acute angle, $\rho = \rho_{max}$ should be chosen. It is natural to attempt to determine the true shape and position of the lines 111 and 222 along which the control has a discontinuity.

In the preceding discussion, it must be borne in mind that the function $\chi(x, y)$ also has a discontinuity on the lines 111 and 222 of discontinuity of the resistivity. Along the lines 111 and 222, the vector lines of j and grad ω_2 are refracted: the former in accordance with the conditions

$$[j_n]_-^+ = 0, \qquad [\rho j_t]_-^+ = 0 \tag{3.27}$$

---

* This example gives a clear idea of the physical meaning of the critical positions of the strip of variation determined in Section 2.3. Suppose that in a well-conducting central region of the channel, where $\rho = \rho_{min}$ and the currents flow in a favorable direction (from the lower to the upper electrode in Fig. 3.6a, b), a poorly conducting ($\rho = \rho_{max}$) small strip is inserted. The influence of this variation strip on the functional is determined by its orientation relative to the vectors j and grad ω_2 at the given point (at the center of the strip). If the strip is in the critical position, then it decreases the functional I to the smallest extent compared with its value in the absence of the strip [at the same time, if $\pi/2 \leq \chi \leq \pi\text{-cos}^{-1} p$ (nonoptimal range), then the functional even increases at the critical position of the strip]. If the strip is oriented at a right angle to the critical direction, then the functional I is reduced the most. This is particularly apparent for sufficiently large values of ρ_{max}. In this case, $\cos^{-1} \rho \sim 0$ and the critical position of the strip is such that it is directed almost parallel to the lines of the current j; in this case the strip provides the least reduction of the functional. Conversely, if the strip is oriented at a right angle to the critical direction (across the current lines), it reduces the functional to the greatest possible extent.

For the lateral regions of the channel (see Fig. 3.6a, b), where $\rho = \rho_{max}$ and the currents flow in an unfavorable direction (they form current vortices and do not reach the electrodes), the situation is completely analogous. The introduction of a well-conducting strip ($\rho = \rho_{min}$) here, indicates (Section 2.3) that, in the critical position (perpendicular to the lines of j for $\rho_{min} \sim 0$), the strip reduces the functional the least. Conversely, if a strip with $\rho = \rho_{min} \sim 0$ here is placed along the j lines, it reduces the functional the most.

The difference between the critical positions of the strip with respect to the j lines in these two cases is explained by the fact that in the first case we are dealing with a poorly conducting strip ($\rho = \rho_{max} \sim \infty$), and in the second with a well-conducting strip ($\rho = \rho_{min} \sim 0$). For the poorly conducting strip, the direction of least influence on the functional (the critical direction) coincides with the direction of the vector j at the given point, while for a well-conducting strip the critical direction is perpendicular to the direction of j. At the same time, it is important to note that we are speaking of liquid strips, i.e., strip-shaped zones of changed resistivity of a fluid medium. The velocity $V(y)$ and the magnetic induction $B(x)$ are assumed to be continuous functions within the strips, right up to their boundaries.

and the latter in accordance with the equations

$$[\omega_{2t}]_-^+ = 0, \qquad \left[\frac{1}{\rho}\,\omega_{2n}\right]_-^+ = 0 \qquad (3.28)$$

where t and n are the unit tangent and the unit normal to the line of discontinuity, respectively. These last relations represent the Weierstrass–Erdmann conditions $(2.42)_1$ stated with allowance for (2.99) and (2.100). The refraction of the j and grad ω_2 lines must be reflected in figures such as those in Fig. 3.6a, b constructed for piecewise constant control $\rho(x, y)$.

In the optimal range, Weierstrass's condition eliminates the interval $(\cos^{-1} p, \pi - \cos^{-1} p)$ from the range of values of the function $\chi(x, y)$. We now prove that if there exists a continuously differentiable curve Γ_0 (111 or 222) separating the regions G_1 (where $\rho = \rho_{\min}$) and G_2 (where $\rho = \rho_{\max}$), then, in the optimal range, $\lim \chi(x, y) = \pi - \cos^{-1} p$ if the point (x, y) tends to Γ_0 from the side of G_1, and $\lim \chi(x, y) = \cos^{-1} p$ if the point (x, y) tends to Γ_0 from from the side of G_2. Simultaneously, we shall show that in the optimal range, the j and grad ω_2 vector lines are arranged on the line Γ_0 in such a way that the normal n to Γ_0 is a common bisector of the angles χ formed by the vectors j and grad ω_2 in their limiting positions on the different sides of Γ_0 (Fig. 3.7).

Indeed, the Weierstrass–Erdmann condition $(2.42)_3$, stated with allowance for (2.99) and (2.100), show that, for example, the limiting values of the vectors j and grad ω_2 at a point of the line Γ_0 on the side of $G_2(\rho = \rho_{\max})$ satisfy the condition

$$\frac{\rho_{\max} - \rho_{\min}}{\rho_{\max}}\, j_n \frac{\partial \omega_2}{\partial n} - (j, \text{grad } \omega_2) = 0 \qquad (3.29)$$

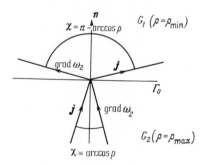

FIGURE 3.7

On the other hand, if we set $m = 0$ in Section 2.3 in the passage preceding the derivation of (2.112) (i.e., if we go over to variation in a strip), then it will follow from the arguments given there that for $(\mathbf{j}, \text{grad } \omega_2) > 0$ the expression

$$\frac{\rho_{max} - \rho_{min}}{\rho_{max}} j_\beta \frac{\partial \omega_2}{\partial \beta} - (\mathbf{j}, \text{grad } \omega_2) \qquad (3.30)$$

as a function of the direction $\boldsymbol{\beta}$, attains a maximum when $\boldsymbol{\beta}$ bisects the angle χ between the vectors $\mathbf{j}$ and grad ω_2 (Fig. 3.8). The corresponding maximum is

$$|\mathbf{j}| |\text{grad } \omega_2| \left(\frac{\rho_{max} - \rho_{min}}{\rho_{max}} \cos^2 \frac{\chi}{2} - \cos \chi \right)$$

For $0 \le \chi < \cos^{-1} p$ (i.e., at points of G_2) it is negative, and for $\chi = \cos^{-1} p$ it is zero. Continuity arguments now show that if the boundary Γ_0 is adjoined by a region G_2 at whose points Weierstrass's condition is satisfied, then the limiting value (on the side of G_2) of the angle χ on the curve Γ_0 is equal to $\cos^{-1} p$, and this value is the maximal value of χ in G_2. It follows from this, in particular, that the normal $\mathbf{n}$ to Γ_0 bisects this limiting angle, which is what we wanted to prove. The assertion relating to the limiting positions of the vectors $\mathbf{j}$ and grad ω_2 on the other side of Γ_0 may be proved in a similar fashion. Thus, assuming the existence of a continuously differentiable curve Γ_0 separating the regions G_1 and G_2, we arrive at the conclusion that there must be a consistency of the Weierstrass–Erdmann conditions and the Weierstrass condition in the optimal control range. The meaning of this consistency is that the lines of discontinuity split the basic region in such a way that in all of the resulting regions of constancy of the control the controls are used "to the limit," i.e., until the function χ reaches the corresponding limiting critical value on the line of discontinuity of the control.

Our conclusion that the Weierstrass–Erdmann and Weierstrass conditions are to be consistent makes essential use of the concept of variation

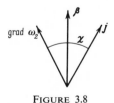

FIGURE 3.8

in a strip: variations in a disk ($m = 1$) would be insufficient to establish such consistency, while variations in an ellipse ($0 < m < 1$) would lead to weakened consistency conditions; in particular, the limiting value of the angle χ on the side of G_2 would be equal to [see (2.112)]

$$\arccos \frac{\lambda(\rho_{max} - \rho_{min})}{2\rho_{max} - \lambda(\rho_{max} - \rho_{min})}$$

We now turn to the problem of the optimal distribution of the scalar resistivity in the channel of an MHD generator when the function $B(x)$ is given by a graph of the type shown in Fig. 3.5 for $V(y) = \text{const} > 0$. If $p \to 0(\rho_{max} \to \rho_{min})$, the optimal function $\rho(x, y)$ tends to a constant equal to ρ_{min} everywhere in the channel. The discontinuity lines of the control disappear as such, but the points of these lines tend to certain limiting positions. These limiting positions can be readily specified: It is sufficient to superimpose Fig. 3.6a (or Fig. 3.6b) on Fig. 3.4 [constructing the vector lines in these figures for the case $\rho(x, y) = \text{const}$] and join the points at which $(\mathbf{j}, \text{grad } \omega_2) = 0$ by smooth curves; these curves then are the desired limiting curves. If we now construct the normals to these curves at every point, we find that, in general, they do not bisect the angle $\chi = \pi/2$ formed by the vectors $\mathbf{j}$ and grad ω_2 at the points of these curves. This assertion can be dirctly verified in the case $x_c = \infty$ by using the exact solution[28,29] available for this case. For $x_c < \infty$, one can also obtain an exact solution and verify the assertion.

Our argument has been based on tacit assumption that the limiting (as $p \to 0$) positions of the normals to the lines of discontinuity of the control coincide with the normals to the limiting position of the (vanishing) discontinuity lines at the corresponding points. In other words, it is assumed that the discontinuity lines tend to their limiting positions in a continuously differentiable manner (i.e., together with the directions of the normals). There is not a prior justification for such an assertion; moreover, we shall show that, in general, for $p \neq 0$ the problem of determining continuously differentiable curves along which the Weierstrass–Erdmann conditions and for which the Weierstrass conditions are satisfied within the regions which they bound, is overdetermined if one requires that these conditions be consistent in the above sense. This will show that the argument just given for the case $p \to 0$ does indeed lead to a contradiction.

Assigning the indices $+$ and $-$ to the limiting values on Γ_0 of the vectors $\mathbf{j}$ and grad ω_2 on the G_2 and G_1 side, respectively, we write the conditions (3.27) and (3.28) in the form (Fig. 3.9)

$$|\mathbf{j}_+| \cos \theta_+ = |\mathbf{j}_-| \cos \theta_-$$

$$\rho_{max}|\mathbf{j}_+| \sin \theta_+ = \rho_{min}|\mathbf{j}_-| \sin \theta_-$$

$$(3.31)$$

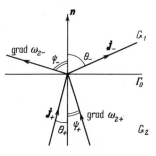

FIGURE 3.9

$$|\text{grad } \omega_{2+}| \sin \psi_+ = |\text{grad } \omega_{2-}| \sin \psi_-$$
$$\rho_{max}^{-1}|\text{grad } \omega_{2+}| \cos \psi_+ = \rho_{min}^{-1}|\text{grad } \omega_{2-}| \cos \psi_- \tag{3.32}$$

Condition (3.29) takes the form

$$\frac{\rho_{max} - \rho_{min}}{\rho_{max}} \cos \theta_+ \cos \psi_+ - \cos (\theta_+ + \psi_+) = 0 \tag{3.33}$$

From (3.31) and (3.32) we obtain the equations

$$\tan \theta_+ = \frac{\rho_{min}}{\rho_{max}} \tan \theta_-, \qquad \tan \psi_+ = \frac{\rho_{min}}{\rho_{max}} \tan \psi_- \tag{3.34}$$

which express the law of refraction of the **j** and grad ω_2 lines on Γ_0. It follows from this, in particular, that

$$\tan \theta_- \tan \psi_+ = \tan \theta_+ \tan \psi_- \tag{3.35}$$

Equation (3.33) can be written in the form

$$\frac{\rho_{max} - \rho_{min}}{\rho_{max}} = 1 - \tan \theta_+ \tan \psi_+$$

Hence, using (3.34), we find

$$1 - \frac{\rho_{max}}{\rho_{min}} = \frac{\tan \theta_+ \tan \psi_+ - 1}{\tan \theta_+ \tan \psi_+} = \frac{\dfrac{\rho_{min}^2}{\rho_{max}^2} \tan \theta_- \tan \psi_- - 1}{\dfrac{\rho_{min}^2}{\rho_{max}^2} \tan \theta_- \tan \psi_-}$$

and

$$1 - \tan \theta_- \tan \psi_- = \frac{\rho_{\min} - \rho_{\max}}{\rho_{\min}}$$

Finally,

$$\frac{\rho_{\min} - \rho_{\max}}{\rho_{\min}} \cos \theta_- \cos \psi_- - \cos (\theta_- + \psi_-) = 0 \qquad (3.36)$$

The last equation could have been written down directly by formulating the Weierstrass-Erdmann condition in terms of the limiting values on Γ_0. We obtained this equation as a consequence of (3.31)-(3.33). From (3.33) and (3.36), we now obtain

$$\frac{\rho_{\min} - \rho_{\max}}{\rho_{\min}} = \frac{\cos (\theta_- + \psi_-)}{\cos \theta_- \cos \psi_-} = \frac{\dfrac{\rho_{\max} - \rho_{\min}}{\rho_{\max}}}{\dfrac{\rho_{\max} - \rho_{\min}}{\rho_{\max}} - 1}$$

$$= -\frac{\dfrac{\cos (\theta_+ + \psi_+)}{\cos \theta_+ \cos \psi_+}}{\dfrac{\sin \theta_+ \sin \psi_+}{\cos \theta_+ \cos \psi_+}} = -\frac{\cos (\theta_+ + \psi_+)}{\sin \theta_+ \sin \psi_+} \qquad (3.37)$$

Similarly,

$$\frac{\rho_{\max} - \rho_{\min}}{\rho_{\max}} = \frac{\cos (\theta_+ + \psi_+)}{\cos \theta_+ \cos \psi_+} = \frac{\dfrac{\rho_{\min} - \rho_{\max}}{\rho_{\min}}}{\dfrac{\rho_{\min} - \rho_{\max}}{\rho_{\min}} - 1}$$

$$= -\frac{\dfrac{\cos (\theta_- + \psi_-)}{\cos \theta_- \cos \psi_-}}{\dfrac{\sin \theta_- \sin \psi_-}{\cos \theta_- \cos \psi_-}} = -\frac{\cos (\theta_- + \psi_-)}{\sin \theta_- \sin \psi_-} \qquad (3.38)$$

Equations (3.37) and (3.38) lead to the relations

$$\frac{\cos (\theta_- + \psi_-)}{\cos (\theta_+ + \psi_+)} = -\frac{\cos \theta_- \cos \psi_-}{\sin \theta_+ \sin \psi_+} = -\frac{\sin \theta_- \sin \psi_-}{\cos \theta_+ \cos \psi_+}$$

whence

$$\tan \theta_+ \tan \theta_- \tan \psi_+ \tan \psi_- = 1$$

Now, taking into account (3.35), we obtain

$$\tan \theta_- \tan \psi_+ = \tan \theta_+ \tan \psi_- = \pm 1 \tag{3.39}$$

Obviously, the lower sign must be rejected, and we then arrive at the expressions

$$\theta_- + \psi_+ = \theta_+ + \psi_- = \frac{\pi}{2} \tag{3.40}$$

from which, in particular,

$$\theta_+ + \psi_+ = \pi - (\theta_- + \psi_-) \tag{3.41}$$

Our derivation shows that condition (3.41) can be regarded as equivalent to the Weierstrass-Erdmann condition (3.33) [or (3.36)].

We now recall the need for consistency between the Weierstrass-Erdmann conditions and the Weierstrass condition*; the requirement of such consistency imposes one further condition on the angles $\theta_\pm$ and $\psi_\pm$, namely, for example

$$\theta_- = \psi_- \tag{3.42}$$

whence with allowance for (3.40) we immediately obtain $\theta_+ = \psi_+$.

---

* The conditions of such consistency can also be obtained directly. We have [see (3.34), (3.40)]

$$\tan \psi_- \cot \psi_+ = \tan \psi_- \tan \theta_- = \rho_{max}/\rho_{min}$$

whence

$$\tan \left(\frac{\pi}{2} + 0 \right) \le \tan (\theta_- + \psi_-) = \frac{\tan \theta_- + \tan \psi_-}{1 - \tan \theta_- \tan \psi_-}$$

$$= \frac{\dfrac{\rho_{max}}{\rho_{min}} \cot \psi_- + \tan \psi_-}{1 - \dfrac{\rho_{max}}{\rho_{min}}} \le \frac{2\sqrt{\rho_{max}/\rho_{min}}}{1 - \rho_{max}/\rho_{min}} = \tan (\pi - \arccos p)$$

thus, we arrive at the inequality $\pi/2 \le \theta_- + \psi_- \le \pi - \cos^{-1} p$, which in conjunction with the inequality (3.1)$_2$ (the Weierstrass condition) shows that $\theta_- + \psi_- = \pi - \cos^{-1} p$. Hence, using the condition $\tan \psi_- \tan \theta_- = \rho_{max}/\rho_{min}$, we find

$$\theta_- = \psi_- = \tfrac{1}{2}(\pi - \arccos p)$$

which is what we wanted to prove.

Altogether, for the determination of the vectors $\mathbf{j}$ and grad ω_2 in the complete region G, we have (2.34) and (2.100) with the corresponding conditions on the boundary of the region [conditions (2.35) and (2.101)], the conditions of consistency (3.27) and (3.28) on the discontinuity lines of the control, and the two conditions (3.41) and (3.42) for determining these lines.* The last circumstance (two conditions at each point of the discontinuity lines) renders the problem overdetermined if the function $B(x)$ is taken in some general form, which is what we wanted to prove.

Thus, if the posed problem has a solution, this is necessarily a generalized solution since there does not exist a sufficiently smooth curve separating the regions with different values of the control. Using variation in strips, one can construct a sequence of controls that corresponds to an ascending sequence of values of the functional I (see Appendix A.2). Because the functional is bounded, this sequence of its values tends to a limit, which is, in general, a local maximum of the total current. With regard to a limiting control function that realizes this maximum, we have not yet been able to prove existence.

In optimal control problems, there are many situations where optimal control does not exist within the class of admissible functions.† However, in practice, there always exists a sequence of admissible controls such that the sequence of values of the functional corresponding to it tends to the lower (upper) bound of the functional. The control functions that minimize the sequence are distinguished by considerable irregularity. These functions

---

* The additional conditions on the discontinuity lines of the control can be chosen in different forms equivalent to (3.41) and (3.42); it is here important that the number of these conditions is always equal to two. For example, one can require that

$$\theta_- + \psi_- = \pi - \arccos p \qquad (a)$$

and

$$\theta_+ + \psi_+ = \arccos p \qquad (b)$$

It is necessary that along the discontinuity line, both of these conditions are satisfied simultaneously; the fulfillment of one of them is insufficient. If, for example, only condition (b) is satisfied, then the left-hand side of (3.29) will, in general, be negative [see the discussion following (3.30)]; i.e., the $(+)$ limiting values will satisfy the Weierstrass condition in the strong sense (as an inequality). Using (3.34) [see the derivation of (3.36)], we can show that the Weierstrass condition will then be violated for the $(-)$ limiting values (the corresponding inequality will be satisfied in the opposite sense). Simultaneously, the Weierstrass–Erdmann condition (3.29) will be violated. The fulfillment of all conditions is guaranteed if the requirement (b) is augmented by the condition (a), or the condition (3.42), or the condition $\theta_+ = \psi_+$, etc.

† In this connection, see Section 4.7.

do not have a limit but fluctuate between certain definite values (*chattering ranges*). The more frequent are the fluctuations, the closer the values of the functional are to its bound.

A similar situation arises in the problem we are considering. To show this, let us recall the argument which applies to the limiting case $p \to 0 (\rho_{max} \to \rho_{min})$, for which we noted that Weierstrass's condition is not satisfied because the normal to the limiting position of the (disappearing) discontinuity line does not bisect the angle $\pi/2$ formed on this line by the vectors $\mathbf{j}$ and grad ω_2. One could avoid a contradiction by assuming that as the line tends to its limiting ($p = 0$) position the normals to this line do not tend to the normals of the limiting curve at the corresponding points. For sufficiently small p, the discontinuity line would then be a curve that executes frequent (in the limit $p = 0$, infinitely frequent) fluctuations about its limiting position (Fig. 3.10) with the frequency of these fluctuations increasing to infinity faster than the amplitude decreases to zero in the limit as $p \to 0$. An examination of Fig. 3.10 shows that layers of material with resistivity ρ_{max} and ρ_{min} alternate in the neighborhood of the curve Γ_0 ($p = \varepsilon \ll 1$). If the alternating layers are arranged sufficiently often, one can assume that the effective resistivity at the points of this layered structure is different in different directions: The effective resistivity in the direction β (see Fig. 3.10) differs from the resistivity in the direction α. A similar picture is obtained in the case when $\rho_{max}/\rho_{min} \gg 1$ ($p \to 1$). Weierstrass's conditions (2.113)–(2.116) show that the vectors $\mathbf{j}$ and grad ω_2 in the optimal range are almost parallel where $\rho = \rho_{max}$ and almost antiparallel where $\rho = \rho_{min}$. To be specific, we assume that $\rho_{max} \to \infty$ for fixed ρ_{min}. In the limit $\rho_{max} = \infty$, the currents are concentrated only in the region G_1 (where $\rho = \rho_{min}$), and on the boundary Γ_0 separating the regions (which is smooth by hypothesis) the limit equation $j_{n-} = 0$ is satisfied. The Weierstrass and Weierstrass–Erdmann conditions show that on Γ_0 the $(-)$ limiting value of the tangential component of the current density vanishes simultaneously: $j_{t-} = 0$; at the same time, on Γ_0 we have $\omega_{2n-} = 0$; however, $\omega_{2t-} \neq 0$.

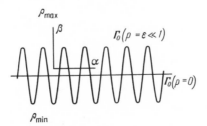

FIGURE 3.10

In addition, in $G_1(\rho = \rho_{\min})$ we must have

$$\mathbf{j} = F(x, y) \, \text{grad} \, \omega_2 \tag{3.43}$$

where $F(x, y) \le 0$ everywhere in G_1.

It is clear that the imposed conditions cannot be satisfied by any general function $B(x)$ if Γ_0 is assumed to be a smooth curve and the resistivity homogeneous and equal to $\rho = \rho_{\min}$ in G_1.

On the other hand, if we introduce infinitesimally thin insulating baffles ($\rho = \infty$) into this region, separated by thin strips of matter with $\rho = \rho_{\min}$, then current channels are formed within these strips in which (3.43) will be satisfied the more accurately the smaller the width of the channel (obviously, satisfaction of this equation in all of G_1 cannot be achieved by any finite number of baffles). One might expect, however, that by a suitable arrangement of such layers, it might be possible to satisfy all of the imposed conditions.

In this connection, we thus again arrive at the concept of infinitely frequently alternating, infinitesimally thin regions of different values of the resistivity. As before, our conclusion remains in force that at the points of these layers the effective resistivity depends on the direction; in particular, along a layer, the resistivity is different from the resistivity at right angles to the layer.

Our arguments are heuristic: they show how one should modify the formulation of the problem in order to guarantee the existence of an optimal control. Namely, we shall assume that the resistivity of the medium is described by a symmetric tensor $\wp(x, y)$, which varies from point to point.

The tensor $\wp(x, y)$ will be assumed to be generated by an infinite sequence of infinitesimal layers filled with conductors having the specific resistances $\rho_{\max}$ and $\rho_{\min}$; the principal values (eigenvalues) of such a tensor are connected by some relationship. In the next section, we will calculate the effective specific resistance of some more general layered composite assembled of two different anisotropic components.

3.4. Effective Resistivity of Laminated Composites

Consider a composite material such that any elemental volume Σ thereof contains an infinite number of alternating layers consisting of (anisotropic) conductors characterized by specific resistance tensors $\wp_+$ and $\wp_-$, respectively (Fig. 3.11). Let $\boldsymbol{\alpha}$ and $\boldsymbol{\beta}$ be unit vectors normal and tangent to the layers, respectively, and let m ($0 \le m \le 1$) be the relative concentration of the material $\wp_+$ within the elemental volume.

If we insert this volume into a locally homogeneous field $\mathbf{j}$ of electric currents flowing through some homogeneous (and generally anisotropic)

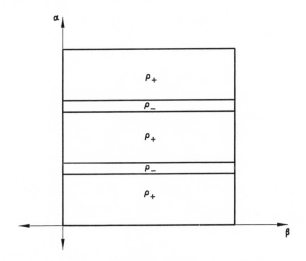

FIGURE 3.11

medium, then the field of currents will remain homogeneous within the layers, and the average values

$$\mathbf{j}_{av} = \frac{1}{\text{mes } \Sigma} \int\int_{\Sigma} \mathbf{j} \, dx \, dy, \qquad \text{grad } z_{av}^1 = \frac{1}{\text{mes } \Sigma} \int\int_{\Sigma} \text{grad } z^1 \, dx \, dy$$

of the vectors $\mathbf{j}$ and grad z^1, will be connected by the relationship

$$\mathfrak{p} \cdot \mathbf{j}_{av} = -\text{grad } z_{av}^1 + \frac{1}{c} [\mathbf{v}, \mathbf{B}] \tag{3.44}$$

where $\mathfrak{p}$ is the *effective specific resistance* tensor. The determination of this tensor is straightforward and requires no further comment.

$$\mathbf{j}_{av} = m\mathbf{j}_+ + (1 - m)\mathbf{j}_-, \qquad \text{grad } z_{av}^1 = m \text{ grad } z_+^1 + (1 - m) \text{ grad } z_-^1$$

$$\mathfrak{p}_+ \cdot \mathbf{j}_+ = -\text{grad } z_+^1 + \frac{1}{c} [\mathbf{v}, \mathbf{B}], \qquad \mathfrak{p}_- \cdot \mathbf{j}_- = -\text{grad } z_-^1 + \frac{1}{c} [\mathbf{v}, \mathbf{B}]$$

$$\mathbf{j}_+ \cdot \boldsymbol{\alpha} = \mathbf{j}_- \cdot \boldsymbol{\alpha}, \qquad \boldsymbol{\beta} \cdot \mathfrak{p}_+ \cdot \mathbf{j}_+ = \boldsymbol{\beta} \cdot \mathfrak{p}_- \cdot \mathbf{j}_-$$

$$\mathbf{j}_{av} \cdot \boldsymbol{\alpha} = m\mathbf{j}_+ \cdot \boldsymbol{\alpha} + (1 - m)\mathbf{j}_- \cdot \boldsymbol{\alpha} = \mathbf{j}_+ \cdot \boldsymbol{\alpha} = \mathbf{j}_- \cdot \boldsymbol{\alpha}$$

$$\mathbf{j}_+ = \mathbf{j}_{av} + a\boldsymbol{\beta}, \qquad \mathbf{j}_- = \mathbf{j}_{av} + b\boldsymbol{\beta}$$

$$\boldsymbol{\beta} \cdot \rho_+ \cdot \mathbf{j}_{av} + a\boldsymbol{\beta} \cdot \rho_+ \cdot \boldsymbol{\beta} - \boldsymbol{\beta} \cdot \rho_- \cdot \mathbf{j}_{av} - b\boldsymbol{\beta} \cdot \rho_- \cdot \boldsymbol{\beta} = 0$$

$$ma + (1 - m)b = 0$$

$$a = -\frac{1-m}{\boldsymbol{\beta} \cdot \tilde{\rho} \cdot \boldsymbol{\beta}} \boldsymbol{\beta} \cdot (\rho_+ - \rho_-) \cdot \mathbf{j}_{av} = -\frac{1-m}{m} b$$

$$\tilde{\rho} = \rho_+ + \rho_- - \rho^*, \qquad \rho^* = m\rho_+ + (1-m)\rho_-$$

$$m\rho_+ \cdot \mathbf{j}_+ + (1 - m)\rho_- \cdot \mathbf{j}_-$$

$$= \rho \cdot \mathbf{j}_{av} = m\rho_+ \cdot (\mathbf{j}_{av} + a\boldsymbol{\beta}) + (1 - m)\rho_- \cdot (\mathbf{j}_{av} + b\boldsymbol{\beta})$$

$$= \left[\rho^* - \frac{m(1-m)}{\boldsymbol{\beta} \cdot \tilde{\rho} \cdot \boldsymbol{\beta}} (\rho_+ - \rho_-) \cdot \boldsymbol{\beta}\boldsymbol{\beta} \cdot (\rho_+ - \rho_-) \right] \cdot \mathbf{j}_{av}$$

$$= -\text{grad}\, z_{av}^1 + \frac{1}{c} [\mathbf{v}, \mathbf{B}]$$

$$\rho = \rho^* - \frac{(\rho^* - \rho_-) \cdot \boldsymbol{\beta}\boldsymbol{\beta} \cdot (\rho_+ - \rho^*)}{\boldsymbol{\beta} \cdot (\rho_+ + \rho_- - \rho^*) \cdot \boldsymbol{\beta}} \tag{3.45}$$

The tensor $\sigma = \rho^{-1}$ of effective conductivity of the laminated composite consisting of conductors with specific conductivities $\sigma_+ = \rho_+^{-1}$, $\sigma_- = \rho_-^{-1}$ and respective concentrations m and $1 - m$, may be similarly expressed as

$$\sigma = \sigma^* - \frac{(\sigma^* - \sigma_-) \cdot \boldsymbol{\alpha}\boldsymbol{\alpha} \cdot (\sigma_+ - \sigma^*)}{\boldsymbol{\alpha} \cdot (\sigma_+ + \sigma_- - \sigma^*) \cdot \boldsymbol{\alpha}} \tag{3.46}$$

where, by definition,

$$\sigma^* = m\sigma_+ + (1 - m)\sigma_-$$

In particular, for isotropic components

$$\rho_+ = \rho_{max}\mathbb{E}, \qquad \rho_- = \rho_{min}\mathbb{E}$$

Equation (3.45) reduces to

$$\rho = [m\rho_{max} + (1 - m)\rho_{min}]\mathbb{E} - \frac{m(1-m)(\rho_{max} - \rho_{min})^2}{(1-m)\rho_{max} + m\rho_{min}} \boldsymbol{\beta}\boldsymbol{\beta}$$

or, since $\mathbb{E} = \boldsymbol{\alpha}\boldsymbol{\alpha} + \boldsymbol{\beta}\boldsymbol{\beta}$, to

$$\rho = \rho_1 \boldsymbol{\alpha}\boldsymbol{\alpha} + \rho_2 \boldsymbol{\beta}\boldsymbol{\beta}$$

where the principal values ρ_1 and ρ_2 are expressed by the formulas

$$\rho_1 = m\rho_{max} + (1 - m)\rho_{min}$$
$$\rho_2 = [m\rho_{max}^{-1} + (1 - m)\rho_{min}^{-1}]^{-1}$$

(3.47)

and α and β represent the unit vectors of the corresponding principal axes, coinciding for this case with the normal and tangent directions to the layers, respectively.

The elimination of the concentration m from (3.47) yields the relationship

$$\rho_2 = \frac{\rho_{max}\rho_{min}}{\rho_{max} + \rho_{min} - \rho_1}, \qquad \rho_1 \geq \rho_2$$

(3.48)

between the eigenvalues ρ_1, ρ_2 of ρ; this needs to be augmented by the inequality

$$\rho_{min} \leq \rho_1 \leq \rho_{max}$$

(3.49)

corresponding to $0 \leq m \leq 1$.

Note that the transition from the specific resistance ρ to the specific conductivity σ according to the formulas

$$\rho_1 = \sigma_1^{-1}, \qquad \rho_2 = \sigma_2^{-1}$$
$$\rho_{max} = \sigma_{min}^{-1}, \qquad \rho_{min} = \sigma_{max}^{-1}$$

transforms the relationships (3.47) into

$$\sigma_1 = [m\sigma_{min}^{-1} + (1 - m)\sigma_{max}^{-1}]^{-1}$$
$$\sigma_2 = m\sigma_{min} + (1 - m)\sigma_{max}$$

(3.50)

The eigenvalues σ_1, σ_2 of the effective tensor of conductivity

$$\sigma = \sigma_1\alpha\alpha + \sigma_2\beta\beta$$

then are connected by the formula

$$\sigma_1 = \frac{\sigma_{max}\sigma_{min}}{\sigma_{max} + \sigma_{min} - \sigma_2}, \qquad \sigma_{min} \leq \sigma_1 \leq \sigma_2 \leq \sigma_{max}$$

(3.51)

Assume now that $0 < \rho_{\min} \leq \rho_{\max} < \infty$; if ρ_1 coincides either with $\rho_{\max}$ or $\rho_{\min}$, the same will be true for ρ_2. The cases when either $\rho_{\min} = 0$ or $\rho_{\max} = \infty$ are exceptional; e.g., for $\rho_{\max} = 0$, $\rho_{\min} > 0$, the region $\rho_1 = \infty$, $\rho_2 = \rho_{\min}/(1 - m)$ will be admissible.

3.5. Basic Equations in the Case of Tensorial Resistivity

The distributions of the current $\mathbf{j}$ and the electric field potential z^1 in the channel now are described by the equations

$$\operatorname{div} \mathbf{j} = 0, \qquad \wp \cdot \mathbf{j} = -\operatorname{grad} z^1 + \frac{1}{c}[\mathbf{v}, \mathbf{B}] \qquad (3.52)$$

which differ from (3.13) only by the tensor factor $\wp(x, y)$ which now replaces the scalar factor $\rho(x, y)$. The tensor $\wp(x, y)$ will be associated with the layered composite generated by isotropic components with specific resistances $\rho_{\max}$ and $\rho_{\min}$; the symbols $\mathbf{j}$ and grad z^1 will designate the averaged values $\mathbf{j}_{av}$ and grad z^1_{av} of the current density and the field potential introduced in the previous section. Accordingly, the eigenvalues ρ_1 and ρ_2 of the tensor $\wp$ will be assumed to satisfy (3.48), (3.49); we use $\gamma(x, y)$ to denote the angle between the positive x axis and the unit vector $\boldsymbol{\alpha}$ normal to the layers.

The Cartesian components of the tensor $\wp$ are expressed by

$$\left.\begin{aligned}
\rho_{xx} &= (1/2)[\rho_1 + \rho_2 + (\rho_1 - \rho_2)\cos 2\gamma] \\[2mm]
\rho_{yy} &= (1/2)[\rho_1 + \rho_2 - (\rho_1 - \rho_2)\cos 2\gamma] \\[2mm]
\rho_{xy} &= \rho_{yx} = (1/2)(\rho_1 - \rho_2)\sin 2\gamma
\end{aligned}\right\} \qquad (3.53)$$

The equivalent system of the first-order equations has the form [see (3.15)]

$$z^1_x = -\rho_{xx}\zeta^1 - \rho_{xy}\zeta^2, \qquad z^1_y = -\rho_{yx}\zeta^1 - \rho_{yy}\zeta^2 + c^{-1}VB$$
$$z^2_x = \zeta^2, \qquad z^2_y = -\zeta^1 \qquad (3.54)$$

where we have introduced $z^2(\mathbf{j} = -\operatorname{curl} \mathbf{i}_3 z^2)$.

The boundary conditions still have the form (3.14) and (3.16), and the functional to be maximized is given by (3.12), as before.

We pose the problem of determining a pair of measurable functions $\rho_1(x, y)$, $\rho_2(x, y)$, satisfying the relationships (3.48), (3.49), and a measurable function $\gamma(x, y)$ such that the functional (3.12) takes the largest possible value, subject to the constraints (3.54), (3.14), and (3.16).

3.6. Necessary Conditions for Stationarity (Tensor Case)

We introduce Lagrange multipliers ξ_1, η_1, ξ_2, η_2 corresponding to the four equations (3.54), and form the function

$$H = -(1/2)\xi_1\{[\rho_1 + \rho_2 + (\rho_1 - \rho_2)\cos 2\gamma]\zeta^1 + (\rho_1 - \rho_2)$$

$$\cdot \sin 2\gamma \cdot \zeta^2\} - (1/2)\eta_1\{(\rho_1 - \rho_2)\sin 2\gamma \cdot \zeta^1$$

$$+ [\rho_1 + \rho_2 - (\rho_1 - \rho_2)\cos 2\gamma]\zeta^2 - c^{-1} VB\}$$

$$+ \xi_2\zeta^2 - \eta_2\zeta^1 - \mu_2[\rho_2 - (\rho_{max}\rho_{min})/(\rho_{max} + \rho_{min} - \rho_1)]$$

$$- \mu_1[(\rho_{max} - \rho_1)(\rho_1 - \rho_{min}) - \rho_{1*}^2] \tag{3.55}$$

where the Cartesian components of the tensor ρ have been expressed in terms of the controls ρ_1, ρ_2, and γ in accordance with (3.53). The multiplier μ_1 corresponds to the constraint (3.49), written in the equivalent form [see (2.28)]

$$(\rho_{max} - \rho_1)(\rho_1 - \rho_{min}) - \rho_{1*}^2 = 0$$

The stationarity conditions are given by

$$\xi_{1x} + \eta_{1y} = 0, \qquad \xi_{2x} + \eta_{2y} = 0 \tag{3.56}$$

$$\partial H/\partial \zeta^1 = 0, \qquad \partial H/\partial \zeta^2 = 0 \tag{3.57}$$

$$\partial H/\partial \rho_1 = 0, \qquad \partial H/\partial \rho_2 = 0, \qquad \partial H/\partial \gamma = 0 \tag{3.58}$$

$$\partial H/\partial \rho_{1*} = 0$$

We introduce the functions $\omega_1(x, y)$ and $\omega_2(x, y)$ defined by (2.99). Equations (3.56) are then satisfied identically. With the use of (3.57), (3.58) and some algebraic manipulation, they can be represented in the form (we omit easy calculations)

$$\rho_2\omega_{1\alpha} + \omega_{2\beta} = 0 \tag{3.59}$$

$$\rho_1 \omega_{1\beta} - \omega_{2\alpha} = 0 \tag{3.60}$$

$$\rho_1^{-1} j_\alpha \omega_{2\alpha} + \mu_1 (2\rho_1 - \rho_{max} - \rho_{min}) + \mu_2 \rho_2^2 / (\rho_{max} \rho_{min}) = 0 \tag{3.61}$$

$$\rho_2^{-1} j_\beta \omega_{2\beta} - \mu_2 = 0$$

$$(\rho_1 - \rho_2)(\rho_1 \omega_{2\beta} j_\alpha + \rho_2 \omega_{2\alpha} j_\beta) = 0 \tag{3.62}$$

$$\mu_1 \rho_{1*} = 0 \tag{3.63}$$

Here

$$\omega_{i\alpha} = \omega_{ix} \cos \gamma + \omega_{iy} \sin \gamma$$
$$(i = 1, 2) \tag{3.64}$$
$$\omega_{i\beta} = -\omega_{ix} \sin \gamma + \omega_{iy} \cos \gamma$$

$$\left. \begin{array}{l} j_\alpha = -z_\beta^2 = \zeta^1 \cos \gamma + \zeta^2 \sin \gamma \\ j_\beta = z_\alpha^2 = -\zeta^1 \sin \gamma + \zeta^2 \cos \gamma \end{array} \right\} \tag{3.65}$$

The functions $\omega_{i\alpha}$, $\omega_{i\beta}$ and j_α, j_β are the physical components of the vectors grad ω_i and **j** along the axes α and β.

The boundary conditions for the functions $\omega_i(x, y)$ retain the form (2.101).

Equations (3.59) and (3.60), in conjunction with (2.101) can be interpreted as the equations and boundary conditions which describe the distribution in a channel of fictitious currents with density $-(\rho^{-1}, \text{grad } \omega_2)$ due to the "potential difference" $\omega_{2+} - \omega_{2-} + 1$ on the electrodes in the absence of other external electromotive forces. The function ω_1 here plays the role of the corresponding "stream function."

Equation (3.63) may be satisfied either by setting $\rho_{1*} = 0$, $\mu_1 \neq 0$, or $\rho_{1*} \neq 0$, $\mu_1 = 0$, or finally, by assuming that $\rho_{1*} = \mu_1 = 0$. In the first case, we obtain boundary ranges of control ($\rho_1 = \rho_2 = \rho_{max}$ or $\rho_1 = \rho_2 = \rho_{min}$), in the second case, an anisotropic intermediate range of control ($\rho_1 \neq \rho_2$). The third case reduces to the first one.

The intermediate range is characterized by the relationship

$$\rho_1^{-1} j_\alpha \omega_{2\alpha} + \rho_2 (\rho_{max} \rho_{min})^{-1} j_\beta \omega_{2\beta} = 0 \tag{3.66}$$

which follows from (3.61) with $\mu_1 = 0$. Taking (3.59), (3.60), and (3.65) into account, (3.66) and (3.62) may be reduced to the respective forms

$$z_\alpha^2 \omega_{1\alpha} \rho_2^2 / (\rho_{\max} \rho_{\min}) + z_\beta^2 \omega_{1\beta} = 0$$
$$z_\alpha^2 \omega_{1\beta} + z_\beta^2 \omega_{1\alpha} = 0 \tag{3.67}$$

We introduce the angles φ and ψ (see Fig. 3.12) between the vectors grad z^2 and $\boldsymbol{\alpha}$, and $\boldsymbol{\alpha}$ and grad ω_1, respectively. Then the second of Eqs. (3.67) implies that either

$$\varphi = \psi \tag{3.68}$$

or

$$\varphi = \pi + \psi \tag{3.69}$$

the first Eq. (3.67) for both cases reduces to

$$\tan^2 \varphi = \rho_2^2 / (\rho_{\max} \rho_{\min}) \tag{3.70}$$

In particular, for $\rho_{\max} \to \infty$, $\rho_2 \neq \infty$, we obtain $\varphi \to 0(\pi)$.

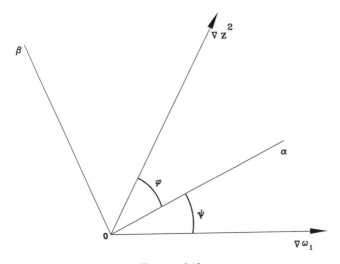

FIGURE 3.12

3.7. Weierstrass's Necessary Condition (Tensor Case)

To derive the Weierstrass condition, it is simplest to use the following expression for the increment ΔI of the functional I due to arbitrary variation $\Delta \wp$ of the tensor $\wp$:

$$\Delta I = -\int\int_B \left[\xi_1(\Delta \rho_{xx} Z^1 + \Delta \rho_{xy} Z^2) + \eta_1(\Delta \rho_{yx} Z^1 + \Delta \rho_{yy} Z^2) \right] dx \, dy$$

$$(3.71)$$

Here, Z^1 and Z^2 denote the Cartesian components of the vector $\mathbf{J}$ of the current density corresponding to the admissible resistivity tensor $\mathbb{P} = \wp + \Delta \wp$.

Equation (3.71) is exact; it can be readily derived from the basic equations when the stationarity conditions are taken into account [see the derivation of (2.98)].

The integrand in (3.71) can be written in the form

$$- ((\wp^{-1}, \text{grad } \omega_2), (\Delta \wp, \mathbf{J}))$$

when (2.97), (3.59), and (3.60) are used.

If we now introduce the dyad

$$\mathbf{J}(\wp^{-1}, \text{grad } \omega_2)$$

then the increment ΔI of the functional I may be written as

$$\Delta I = \int\int_G \text{tr}\,(\Delta \wp, \mathbf{J}(\wp^{-1}, \text{grad } \omega_2))\, dx \, dy \qquad (3.72)$$

If $\wp = \rho \mathbb{E}$, $\Delta \wp = \Delta \rho \mathbb{E}$, where $\mathbb{E}$ is the identity tensor, then (3.72) reduces to (2.102). If the functional I is to attain its maximum for the control $\wp$, it is necessary and sufficient to make ΔI nonpositive for all admissible $\Delta \wp$ and $\mathbf{J}$.

As before, we introduce variation in a strip; namely, we assume that the increment $\Delta \wp$ is nonzero only within a narrow strip of width $\varepsilon^2 a$ and length εa (ε is the small parameter of the problem). The admissible tensor within the strip satisfies the restrictions (3.48) and (3.49), but is otherwise arbitrary.

Equation (3.72) shows that for such a special variation the vector **J** needs only be calculated within the strip, and that it is sufficient to find the ε-independent part of this vector. If the strip is sufficiently narrow and if the field of the vector **j** within the strip does not have singularities, then the ε-independent part of **J** can be calculated under the assumption that the strip is in an external homogeneous field of the currents **j** (see Chapter 2). At the same time, the increment $\Delta\rho$ can be regarded as a constant tensor within the strip.

The solution of this latter problem is well known. We assume that the strip is oriented along the **t** axis and has outer normal **n** (Fig. 3.13), and that the resistivity of the strip is described by the tensor $\mathbb{P} = \rho + \Delta\rho$. If the strip is placed in the homogeneous field of currents **j** flowing in a medium with resistivity ρ, then the components of the current density inside the strip are given by

$$J_n = j_n + O(\varepsilon)$$

$$J_t = j_t + \frac{\rho_{tt} - P_{tt}}{P_{tt}} j_t + \frac{\rho_{tn} - P_{tn}}{P_{tt}} j_n + O(\varepsilon)$$

(3.73)

These expressions are then used to eliminate the vector **J** from the left-hand side of the inequality

$$\mathrm{tr}\,(\Delta\rho, \mathbf{J}(\rho^{-1}, \mathrm{grad}\,\omega_2)) \leq 0 \qquad (3.74)$$

which now follows from the inequality $\Delta I \leq 0$ when (3.72) is taken into account. The inequality must be satisfied almost everywhere as a necessary condition of a strong relative minimum.*

In terms of the principal axes α and β of the tensor ρ, this inequality may be written in the following equivalent form (E denotes the Weierstrass

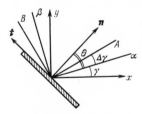

FIGURE 3.13

---

* We mean a minimum of the functional $-I$.

function):

$$E = -[(P_{\alpha\alpha} - \rho_1)J_\alpha + P_{\alpha\beta}J_\beta]\rho_1^{-1}\omega_{2\alpha} - [P_{\beta\alpha}J_\alpha + (P_{\beta\beta} - \rho_2)J_\beta]\rho_2^{-1}\omega_{2\beta} \geq 0$$

$$(3.75)$$

The components J_α and J_β of $\mathbf{J}$ must be expressed in terms of j_α and j_β, and the components $P_{\alpha\alpha}$, $P_{\alpha\beta}$, $P_{\beta\beta}$ of the tensor $\mathbb{P}$ in terms of the quantities that characterize this tensor with reference to its principal axes. As a result, the function E depends on the principal values P_1 and P_2 of the tensor $\mathbb{P}$, the angle $\Delta\gamma$ between the principal axes $\mathbf{A}$ and $\boldsymbol{\alpha}$ of the tensors $\mathbb{P}$ and $\mathfrak{p}$, and also on the angle θ between the direction of the normal $\mathbf{n}$ to the strip and the $\boldsymbol{\alpha}$ axis (Fig. 3.13). The remaining quantities in the expression for E characterize the optimal range and they are assumed fixed.

Making the necessary calculations (see Appendix A.3) we arrive at the following equivalent expression of the inequality (3.75):

$$M \equiv \frac{P_1 - \rho_1}{\rho_1} K_\alpha \omega_{2\alpha} + \frac{P_2 - \rho_2}{\rho_2} K_\beta \omega_{2\beta}$$

$$- (P_1 - P_2)(j_\alpha \omega_{2\alpha} - j_\beta \omega_{2\beta})(1 - \cos 2\Delta\gamma)$$

$$+ (P_1 - P_2)\frac{\rho_1 + \rho_2}{2\rho_1\rho_2}(\rho_1 j_\alpha \omega_{2\beta} + \rho_2 j_\beta \omega_{2\alpha}) \sin 2\Delta\gamma \leq 0$$

$$(3.76)$$

Here

$$K_\alpha = j_\alpha[P_2 + \rho_1 + (P_2 - \rho_1) \cos 2\theta] + j_\beta(P_2 - \rho_2) \sin 2\theta$$

$$(3.77)$$

$$K_\beta = j_\beta[P_1 + \rho_2 - (P_1 - \rho_2) \cos 2\theta] + j_\alpha(P_1 - \rho_1) \sin 2\theta$$

Consider inequality (3.76) and begin with the case of anisotropic intermediate regime of control; for this one, there holds (3.66) and the relationship

$$\rho_1 \omega_{2\beta} j_\alpha + \rho_2 \omega_{2\alpha} j_\beta = 0 \tag{3.78}$$

following from (3.62).

The expressions (3.77) may be represented in the form

$$K_\alpha = k_\alpha + \bar{K}_\alpha, \qquad K_\beta = k_\beta + \bar{K}_\beta$$

where

$$k_\alpha = j_\alpha[\,\rho_2 + \rho_1 + (\rho_2 - \rho_1)\cos 2\theta\,]$$

$$k_\beta = j_\beta[\,\rho_2 + \rho_1 + (\rho_2 - \rho_1)\cos 2\theta\,]$$

$$\bar{K}_\alpha = 2(P_2 - \rho_2)(j_\alpha \cos^2\theta + j_\beta \sin\theta \cos\theta)$$ $$(3.79)$$

$$\bar{K}_\beta = 2(P_1 - \rho_1)(j_\beta \sin^2\theta + j_\alpha \sin\theta \cos\theta)$$

Take $\Delta\gamma = 0$; the expression (3.76) for M will then be rewritten as

$$\frac{P_1 - \rho_1}{\rho_1} k_\alpha \omega_{2\alpha} + \frac{P_2 - \rho_2}{\rho_2} k_\beta \omega_{2\beta} + \frac{P_1 - \rho_1}{\rho_1} \bar{K}_\alpha \omega_{2\alpha} + \frac{P_2 - \rho_2}{\rho_2} \bar{K}_\beta \omega_{2\beta}$$

$$= \frac{P_1 - \rho_1}{\rho_1} k_\alpha \omega_{2\alpha} + \frac{P_1 - \rho_1}{\rho_{\max}\rho_{\min}} \rho_2 k_\beta \omega_{2\beta} - \frac{P_1 - \rho_1}{\rho_{\max}\rho_{\min}} \rho_2 k_\beta \omega_{2\beta}$$

$$+ 2(P_1 - \rho_1)(P_2 - \rho_2)[\,\rho_1^{-1} j_\alpha \omega_{2\alpha} \cos^2\theta + \rho_1^{-1} j_\beta \omega_{2\alpha} \sin\theta \cos\theta$$

$$+ \rho_2^{-1} j_\beta \omega_{2\beta} \sin^2\theta + \rho_2^{-1} j_\alpha \omega_{2\beta} \sin\theta \cos\theta\,]$$

The first two terms on the right-hand side vanish because of (3.79) and (3.66), and the terms proportional to $\sin\theta \cos\theta$ in the square brackets sum to zero because of (3.78). In view of (3.48), (3.66), and (3.79), the remaining terms are reduced to

$$2(P_1 - \rho_1)(P_2 - \rho_2)(\rho_1\rho_2)^{-1} j_\beta \omega_{2\beta}[\,\rho_1\rho_2(\rho_{\max}\rho_{\min})^{-1}(\rho_2 \cos^2\theta + \rho_1 \sin^2\theta)$$

$$- \rho_1\rho_2^2(\rho_{\max}\rho_{\min})^{-1}\cos^2\theta + \rho_1 \sin^2\theta\,]$$

this expression is nonpositive only if

$$j_\beta \omega_{2\beta} \leq 0 \qquad\qquad (3.80)$$

This inequality is obtained for the special case $\Delta\gamma = 0$; it is readily seen to be valid in the general case also, since the term

$$- (P_1 - P_2)(j_\alpha \omega_{2\alpha} - j_\beta \omega_{2\beta})(1 - \cos 2\gamma)$$

is nonpositive because of (3.66), (3.80), and (3.48) [note that $P_1 \geq P_2$ by definition (3.48)].

Consider now the boundary ranges of control and take the case $\rho_1 = \rho_2 = \rho_{\min}$. Equation (3.76) is not convenient for this case since it includes the directions α, β and the angles θ and $\Delta\gamma$ which now are unimportant. For this reason, we will return to the inequality (3.74) and express its left-hand side in terms of the components along the axes $\mathbf{n}$ and $\mathbf{t}$.

We have (see Fig. 3.13)

$$\rho = \rho_{\min}(\mathbf{nn} + \mathbf{tt})$$

$$\mathbb{P} = P_1\mathbf{AA} + P_2\mathbf{BB} = (P_1c^2 + P_2s^2)\mathbf{nn} + (P_2 - P_1)cs(\mathbf{nt} + \mathbf{tn})$$

$$+ (P_1s^2 + P_2c^2)\mathbf{tt}$$

$$c = \cos(\theta - \Delta\gamma), \qquad s = \sin(\theta - \Delta\gamma)$$

$$\Delta\rho = \mathbb{P} - \rho = (P_1c^2 + P_2s^2 - \rho_{\min})\mathbf{nn} + (P_2 - P_1)cs(\mathbf{nt} + \mathbf{tn})$$

$$+ (P_1s^2 + P_2c^2 - \rho_{\min})\mathbf{tt}$$

From (3.73) we obtain

$$J_n = j_n, \qquad J_t = \frac{1}{P_1s^2 + P_2c^2}[\rho_{\min}j_t - (P_2 - P_1)\, cs\, j_n]$$

The following calculations need no comments:

$$(\Delta\rho, \mathbf{J}) = N\mathbf{n} + T\mathbf{t}$$

$$N = (P_2 - P_1)csJ_t + (P_1c^2 + P_2s^2 - \rho_{\min})j_n$$

$$= \frac{1}{P_1s^2 + P_2c^2}[\rho_{\min}j_t - (P_2 - P_1)csj_n](P_2 - P_1)cs$$

$$+ (P_1c^2 + P_2s^2 - \rho_{\min})j_n$$

$$T = (P_1s^2 + P_2c^2 - \rho_{\min})J_t + (P_t - P_1)csj_n$$

$$= \frac{1}{P_1s^2 + P_2c^2}[\rho_{\min}j_t - (P_2 - P_1)csj_n]$$

$$\times (P_1s^2 + P_2c^2 - \rho_{\min}) + (P_2 - P_1)csj_n$$

$$-\rho_{\min}E = \rho_{\min}((\rho^{-1}, \operatorname{grad}\omega_2), (\Delta\rho, \mathbf{J})) = j_n\omega_{2n}p_{nn} + j_t\omega_{2n}p_{tn}$$

$$+ j_n\omega_{2n}p_{nt} + j_t\omega_{2t}p_{tt}$$

$$p_{nn} = P_1c^2 + P_2s^2 - \rho_{\min} - \frac{(P_2 - P_1)^2c^2s^2}{P_1s^2 + P_2c^2} = P_1 - \rho_{\min} + P_1\frac{(P_2 - P_1)s^2}{P_1s^2 + P_2c^2}$$

$$p_{tn} = (P_2 - P_1)\rho_{\min}\frac{cs}{P_1s^2 + P_2c^2} = p_{nt}$$

$$p_{tt} = \rho_{\min} \frac{P_1 s^2 + P_2 c^2 - \rho_{\min}}{P_1 s^2 + P_2 c^2} = \rho_{\min} \frac{P_2 - \rho_{\min}}{P_2}$$

$$+ \rho_{\min} \left(\frac{P_1 s^2 + P_2 c^2 - \rho_{\min}}{P_1 s^2 + P_2 c^2} - \frac{P_2 - \rho_{\min}}{P_2} \right)$$

$$= \rho_{\min} \frac{P_2 - \rho_{\min}}{P_2} - \frac{\rho_{\min}^2}{P_2} \frac{P_2 - P_1}{P_1 s^2 + P_2 c^2} s^2$$

The results may be summarized with

$$- \rho_{\min} E = (P_1 - \rho_{\min}) j_n \omega_{2n} + \frac{\rho_{\min}(P_2 - \rho_{\min})}{P_2} j_t \omega_{2t}$$

$$+ \frac{P_2 - P_1}{P_1 s^2 + P_2 c^2} \left[P_1 s^2 j_n \omega_{2n} + \rho_{\min} cs(j_t \omega_{2n} + j_n \omega_{2t}) - \frac{\rho_{\min}^2}{P_2} s^2 j_t \omega_{2t} \right]$$

$$(3.81)$$

The first two terms on the right-hand side depend on the direction **n** of the normal to the strip of variation and do not depend on the angle $\theta - \Delta\gamma$ between **n** and the principal direction **A** of the tensor $\mathbb{P}$; the other terms depend on both **n** and $\theta - \Delta\gamma$.

Consider first the case $\theta - \Delta\gamma = 0$ ($s = 0$, $c = 1$); then

$$- \rho_{\min} E = (P_1 - \rho_{\min}) j_n \omega_{2n} + \frac{\rho_{\min}}{P_2} (P_2 - \rho_{\min}) j_t \omega_{2t} \qquad (3.82)$$

This expression should be nonpositive for all pairs **(n, t)**, i.e., for all orientations of a strip.

Recall now that as a consequence of (3.48) we have

$$P_2 - \rho_{\min} = \frac{\rho_{\max} \rho_{\min}}{\rho_{\max} + \rho_{\min} - P_1} - \rho_{\min} = (P_1 - \rho_{\min}) \frac{P_2}{\rho_{\max}}$$

and hence we may write (3.82) in the form

$$- \rho_{\min} E = (P_1 - \rho_{\min}) \left(j_n \omega_{2n} + \frac{\rho_{\min}}{\rho_{\max}} j_t \omega_{2t} \right)$$

$$- \rho_{\min} E = (P_1 - \rho_{\min}) \left[\frac{\rho_{\min}}{\rho_{\max}} (\mathbf{j}, \operatorname{grad} \omega_2) + \frac{\rho_{\max} - \rho_{\min}}{\rho_{\max}} j_n \omega_{2n} \right]$$

The condition $-\rho_{\min} E \leq 0$ shows that $(\mathbf{j}, \text{grad } \omega_2) \leq 0$; this inequality may be strengthened if we take $\mathbf{n}$ along the bisector of the obtuse angle χ between the vectors $\mathbf{j}$ and grad ω_2. Finally, we then arrive at the inequalities

$$\pi \geq \chi \geq \pi - \arccos p$$

$$p = \frac{\rho_{\max} - \rho_{\min}}{\rho_{\max} + \rho_{\min}}$$

(3.83)

coinciding with those of (2.116).

The result will remain the same if we assume $\theta - \Delta\gamma = \pi/2$ ($s = 1$, $c = 0$). With this assumption,

$$-\rho_{\min} E = (P_1 - \rho_{\min}) j_n \omega_{2n} + \frac{\rho_{\min}(P_2 - \rho_{\min})}{P_2} j_t \omega_{2t}$$

$$+ \frac{P_2 - P_1}{P_1} \left(P_1 j_n \omega_{2n} - \frac{\rho_{\min}^2}{P_2} j_t \omega_{2t} \right)$$

$$= (P_2 - \rho_{\min}) j_n \omega_{2n} + \frac{\rho_{\min}}{P_1} (P_1 - \rho_{\min}) j_t \omega_{2t}$$

$$= \frac{(P_1 - \rho_{\min}) \rho_{\min}}{\rho_{\max} + \rho_{\min} - P_1} \left(j_n \omega_{2n} + \frac{\rho_{\max} + \rho_{\min} - P_1}{P_1} j_t \omega_{2t} \right)$$

$$= \frac{(P_1 - \rho_{\min}) \rho_{\min}}{P_1} \left[(\mathbf{j}, \text{grad } \omega_2) + \frac{2P_1 - \rho_{\max} - \rho_{\min}}{\rho_{\max} + \rho_{\min} - P_1} j_n \omega_{2n} \right]$$

The expression in the square brackets should be nonpositive which means that $(\mathbf{j}, \text{grad } \omega_2) \leq 0$; the second term in the square brackets is maximal for $P_1 = \rho_{\max}$ and with the unit vector $\mathbf{n}$ bisecting the obtuse angle between $\mathbf{j}$ and grad ω_2; together they result in

$$\cos \chi + \frac{\rho_{\max} - \rho_{\min}}{\rho_{\min}} \cos^2 \chi/2 \leq 0$$

which is the same as the inequality (3.83).

In general, (3.83) follows from the requirement $-\rho_{\min} E \leq 0$ [where the left-hand side is determined by (3.81)] imposed for any value of $\theta - \Delta\gamma$. To demonstrate, consider the extremal points of the function (3.81) based on the parameters $\mathbf{n}$ and $\theta - \Delta\gamma$. First we write (3.81) in the form

$$-\rho_{\min}E = \left(P_1 - \rho_{\min} + \frac{P_2 - P_1}{P_1 s^2 + P_2 c^2} P_1 s^2\right)(\mathbf{j}, \mathrm{grad}\ \omega_2)$$

$$+ \left[-P_1 + \rho_{\min} + \rho_{\min} - \frac{\rho_{\min}^2}{P_2} - \frac{P_2 - P_1}{P_1 s^2 + P_2 c^2} s^2\left(P_1 + \frac{\rho_{\min}^2}{P_2}\right)\right] j_t \omega_{2t}$$

$$+ \frac{P_2 - P_1}{P_1 s^2 + P_2 c^2} \rho_{\min} cs(j_t \omega_{2n} + j_n \omega_{2t})$$

or the form

$$-\rho_{\min}E = \left(\frac{P_1 P_2}{P_1 s^2 + P_2 c^2} - \rho_{\min}\right)(\mathbf{j}, \mathrm{grad}\ \omega_2) + \left(2\rho_{\min} - \frac{P_1 P_2 + \rho_{\min}^2}{P_1 s^2 + P_2 c^2}\right) j_t \omega_{2t}$$

$$+ \frac{P_2 - P_1}{P_1 s^2 + P_2 c^2} \rho_{\min} cs(j_t \omega_{2n} + j_n \omega_{2t}) \tag{3.84}$$

after similar forms have been canceled.

The requirement $-\rho_{\min}E \leq 0$ means that $(\mathbf{j}, \mathrm{grad}\ \omega_2) \leq 0$. At the extremal points of the function (3.84), its derivatives with respect to $\mathbf{n}$ and $\theta - \Delta\gamma$ vanish. This condition is satisfied for $\theta - \Delta\gamma = 0$, for $\theta - \Delta\gamma = \pi/2$, and for the unit vector $\mathbf{n}$ bisecting the obtuse angle between $\mathbf{j}$ and $\mathrm{grad}\ \omega_2$. To prove this, we observe that all of the terms in the expression for the partial derivative of $(\rho_{\min}E)$ with respect to $\theta - \Delta\gamma$ are proportional to cs, with the exception of the last one which includes the factor $j_t \omega_{2n} + j_n \omega_{2t}$, but this one vanishes for the accepted orientation of $\mathbf{n}$. The function $(-\rho_{\min}E)$ is extremal with respect to $\mathbf{n}$ for

$$\frac{\partial}{\partial n} j_t \omega_{2t} = 0$$

However, this condition is also satisfied by the same vector $\mathbf{n}$.

Isotropic control $\rho = \rho_{\min}$ thus is optimal subject to condition (3.83) or, equivalently, to inequality (2.116). Analogously, it follows that inequality (2.114) is necessary for optimality of the isotropic control $\rho = \rho_{\max}$. Thus, (2.114) and (2.116) remain valid for the tensor regime, too.

Now we may compare the necessary conditions of optimality for scalar and tensor regimes of control. For the first case, it was shown in Section 3.3 that the necessary conditions become controversial because of the absence of the intermediate regime of control. Namely, (2.114) and (2.116) exclude the interval $(\cos^{-1} p, \pi - \cos^{-1} p)$ from the set of values of an angle χ, admissible in optimal regimes, and for this reason the regions $\rho = \rho_{\min}$

and $\rho = \rho_{max}$ cannot be divided by some smooth boundary line. In the tensor case, however, besides isotropic regimes of control, there also arises an anisotropic intermediate regime, and this one provides a continuous transition of one isotropic regime into another. This transition will be realized within some region occupied by the control belonging to the intermediate regime. To illustrate this possibility, suppose that the region $\rho = \rho_{min}$ is bounded by some (smooth) line along which $\chi = \pi - \cos^{-1} p$. If we dispose at the other side of this line some layered microstructure so that along the line the normal $\mathbf{n}$ to the layers bisects the obtuse angle $\chi = \pi - \cos^{-1} p$ between the vectors $\mathbf{j}$ and grad ω_2, then inequality (3.80) will be satisfied. Continuity of the transition is achieved because along the boundary line the concentration m of the material with the specific resistance ρ_{max} is zero; this concentration increases when we go farther into the anisotropic zone occupied by the layered composite. Note that along the boundary line, the normal $\mathbf{n}$ to the layers need not coincide with that to the boundary as was required in Section 3.3 where we considered the strip of variation in a narrow neighborhood of the boundary line dividing the two zones of isotropic control. Now we have an extra parameter—the concentration—at our disposal, and it is this parameter that provides continuity of the transition.

Within the zone of the intermediate range, the angle χ may take on any value between the limits $\cos^{-1} p$ and $\pi - \cos^{-1} p$ [this follows directly from (3.62) and (3.66)], and the role of the anisotropic zone is to supply a continuous transition of χ from the value $\pi - \cos^{-1} p$ to that of $\cos^{-1} p$ as the concentration m increases from zero to unity. For $m = 1$, the value $\chi = \cos^{-1} p$ is achieved along some line within the intermediate control zone; this line is actually the boundary line of this zone since the isotropic range of control, $\rho = \rho_{max}$, where $0 \leq \chi \leq \cos^{-1} p$, may be in effect on its other side, and since the transition across the line is continuous as before. Here, also, the normal to the layers adjacent to the boundary line of the region of intermediate control, bisects the acute angle $\chi = \cos^{-1} p$ between the vectors $\mathbf{j}$ and grad ω_2, and this normal need not coincide with the normal to the boundary line itself.

We thus conclude that the layered microstructure of a composite is sufficient for the regularization of the necessary conditions of optimality. In the next chapter, such a microstructure will be shown to guarantee the existence of the optimal control, in general.

3.8. The Asymptotic Case $\rho_{max} \to \infty$

In this and in some of the subsequent sections, we shall consider the case when ρ_{max} tends to infinity. The optimal value of concentration m is determined by the set of necessary conditions; in Section 3.12 we will see that $m \to 0$ as $\rho_{max} \to \infty$. The influence of ideal insulator nevertheless remains substantial because we will assume that $\lim \rho_{max} \, m = \infty$. Bearing this in mind, we observe that the formulas (3.47) take on the form

$$\rho_1 = \infty, \qquad \rho_2 = \rho_{min} \qquad (3.85)$$

We shall construct asymptotic equations that describe an optimal range, and we shall characterize some of the main properties of the optimal control.

First, we consider the stationarity condition (3.62) for the control γ. Assuming $\rho_1 \neq \rho_2$, we may write

$$\rho_1 j_\alpha \omega_{2\beta} + \rho_2 j_\beta \omega_{2\alpha} = 0$$

A change from the principal axes **a** and **b** to Cartesian coordinates x, y [for this, we must invert (3.64) and (3.65)] yields the equivalent equation

$$(\rho_1 \zeta^2 \omega_{2x} + \rho_2 \zeta^1 \omega_{2y}) \tan^2 \gamma - (\rho_1 + \rho_2)(\omega_{2y}\zeta^2 - \omega_{2x}\zeta^1) \tan \gamma$$

$$- (\rho_1 \zeta^1 \omega_{2y} + \rho_2 \zeta^2 \omega_{2x}) = 0 \qquad (3.86)$$

The Weierstrass conditions (Section 3.7) enable us to give a rule for choosing the roots of this equation. Direct calculation (see Appendix A4) shows that to the root $(\tan \gamma)_1$ (the upper sign in the expression for the roots), there correspond the inequalities $j_\alpha \omega_{2\alpha} \geq 0$, $j_\beta \omega_{2\beta} \leq 0$, and to the root $(\tan \gamma)_2$ (lower sign in the expression for the roots), the inequalities $j_\alpha \omega_{2\alpha} \leq 0$, $j_\beta \omega_{2\beta} \geq 0$.

The Weierstrass condition is satisfied in the first case; as to the second case, one should rotate the layers of the microstructure by the angle $\pi/2$ which results in changing places of the parameters ρ_1 and ρ_2.

It can be checked directly that under these conditions, both roots of (3.86) determine the same pair of principal axes. This agrees with the fact that (3.86) is invariant under the substitution $\tan \gamma \to -\cot \gamma$, $\rho_1 \to \rho_2$, $\rho_2 \to \rho_1$.

We assume that $\rho_{max} \to \infty$, for fixed $\rho_{min} > 0$. The root $(\tan \gamma)_1$ of (3.86) then tends to*

$$(\tan \gamma)_0 = \begin{cases} (\omega_{2y}/\omega_{2x})_0, & \text{if } (\mathbf{j}, \text{grad } \omega_2)_0 > 0 \\ -(\zeta^1/\zeta^2)_0, & \text{if } (\mathbf{j}, \text{grad } \omega_2)_0 < 0 \end{cases} \qquad (3.87)$$

---

* We emphasize that we are speaking here of a passage to the limit with respect to the parameter $\mu = \rho_{min}/\rho_{max}$ which occurs explicitly in (3.86); the quantities ζ^1, ζ^2, ω_{2x}, ω_{2y} which occur in this equation [and in (3.87)] depend on μ. This is reflected in the notation $(\cdot)_0$.

It can be seen from this that the principal axes of the tensor ρ are oriented in the limit in such a way that either the vector grad ω_2 is parallel to the principal direction corresponding to the principal value $\rho_{max} = \infty$, or the vector $\mathbf{j}$ is parallel to the principal direction corresponding to ρ_{min}. It is natural to expect that for the first of these cases $(\mathbf{j}, \text{grad } \omega_2) = 0$ in the limit since the vector $\mathbf{j}$ cannot have a component in the principal direction which corresponds to infinitely large resistivity (this assertion will be proved later). Therefore, instead of the first of the inequalities (3.87) we must take $(\mathbf{j}, \text{grad } \omega_2) = 0$.

The second possibility is characterized by the fact that the scalar product $(\mathbf{j}, \text{grad } \omega_2)$ is, in general, nonzero and negative. This can be readily noted by recalling the interpretation of $-\text{grad } \omega_2$ as the intensity vector of a fictitious electric field in the problem for the conjugate variables (Section 3.6). Whereas the vector $\mathbf{j}$ (the current density) has a component only along the ρ_{min} principal direction in the limit, the vector $-\text{grad } \omega_2$ (the electric field intensity) in general has components along both principal directions.* Therefore, the second row of (3.87) generally includes an inequality.

3.9. Asymptotic Form of the Equations in the Optimal Range

One feature of these equations is that they have different forms in the parts of G in which the scalar product $(\mathbf{j}, \text{grad } \omega_2)$ has different signs. Processes in the channel are described by the system (3.54) or, equivalently, by the system

$$z_x^1 = \rho_{xx}z_y^2 - \rho_{xy}z_x^2, \qquad z_y^1 = \rho_{yx}z_y^2 - \rho_{yy}z_x^2 + c^{-1}VB \qquad (3.88)$$

together with the equations for the conjugate variables [see (3.59), (3.60), and (3.53)]

$$\rho_1\rho_2\omega_{1x} = \rho_{xy}\omega_{2x} - \rho_{xx}\omega_{2y}, \qquad \rho_1\rho_2\omega_{1y} = \rho_{yy}\omega_{2x} - \rho_{yx}\omega_{2y} \qquad (3.89)$$

Seeking an asymptotic solution, we shall assume that as $\rho_{max} \to \infty$ the $z_x^1, \ldots, \omega_{2y}$ tend to finite nonzero limiting values. We set

$$\tan \gamma = (\tan \gamma)_0 + \mu m + O(\mu^2) \qquad (3.90)$$

---

* It may happen that this assertion contradicts (3.43), in accordance with which the vectors $\mathbf{j}$ and grad ω_2 are antiparallel in the optimal regime $\rho_{max} = \infty$. It is only an apparent contradiction, however, since (3.43) does not hold in the optimal range of the tensor. One can avoid differences in the expressions by introducing the vector $\mathbf{k} = (-\omega_{1y}, \omega_{1x})$, a vector of a fictitious current density in the conjugate-variables problem. In the optimal range $\rho_{max} = \infty$, the vectors $\mathbf{j}$ and $\mathbf{k}$ are parallel; for the scalar case this follows from (2.100) and (3.43), and for the tensor case from (3.101) and (3.119).

This relation is an expansion in powers of the parameter μ of the root of (3.86); the expansion is with respect to the parameter $\mu = \rho_{min}/\rho_{max}$, which occurs explicitly* in the coefficients of (3.86).

Therefore, for example,

$$(\tan \gamma)_0 \neq \tan \gamma_0, \qquad \text{but } (\tan \gamma_0) \to \tan \gamma_0 \text{ as } \mu \to 0$$

For $\cos 2\gamma$ and $\sin 2\gamma$, we obtain the expansions

$$\cos 2\gamma = (\cos 2\gamma)_0 - m(\sin 2\gamma)_0[1 + (\cos 2\gamma)_0]\mu + O(\mu^2) \quad (3.91)$$

$$\sin 2\gamma = (\sin 2\gamma)_0 + m(\cos 2\gamma)_0[1 + (\cos 2\gamma)_0]\mu + O(\mu^2) \quad (3.92)$$

It is necessary to retain the terms of order μ in these expressions because of the structure of (3.88) and (3.89). In this connection, we should point out that, in accordance with the remark made above, $(\cos 2\gamma)_0$ and $(\sin 2\gamma)_0$ are also expanded in series of the form

$$(\cos 2\gamma)_0 = \cos 2\gamma_0 + O(\mu), \qquad (\sin 2\gamma)_0 = \sin 2\gamma_0 + O(\mu) \quad (3.93)$$

To calculate the values of $(\tan \gamma)_0$ and m it is necessary to use the first of (3.87) for the points at which $(\mathbf{j}, \text{grad } \omega_2)_0 > 0$, and the second for the points at which $(\mathbf{j}, \text{grad } \omega_2)_0 < 0$. After some simple calculations, we obtain: for the region where $(\mathbf{j}, \text{grad } \omega_2)_0 < 0$,

$$(\tan \gamma)_0 = \frac{z_y^2}{z_x^2}, \qquad m = -\frac{(\text{grad } z^2)^2(\text{grad } \omega_2, \text{grad } z^2)}{(\omega_{2y}z_x^2 - \omega_{2x}z_y^2)(z_x^2)^2} \quad (3.94)$$

for the region where $(\mathbf{j}, \text{grad } \omega_2)_0 > 0$,

$$(\tan \gamma)_0 = \frac{\omega_{2y}}{\omega_{2x}}, \qquad m = \frac{(\text{grad } \omega_2)^2(\text{grad } \omega_2, \text{grad } z^2)}{(\omega_{2y}z_x^2 - \omega_{2x}z_y^2)(\omega_{2x})^2} \quad (3.95)$$

It now is easy to write down the required asymptotic equations. We consider the case $(\mathbf{j}, \text{grad } \omega_2)_0 < 0$. We substitute the expansions (3.91) and (3.92) into (3.88) without introducing the series (3.93) for $(\cos 2\gamma)_0$ and $(\sin 2\gamma)_0$ and the corresponding expansions for $z_x^1, \ldots, \omega_{2y}$. Taking into account (3.94), we find that the coefficients of ρ_{max} on the right-hand side

---

* See the record to the preceding footnote.

of (3.88) vanish identically. In the remaining terms, we can go to the limit $\mu = 0$; then $z_x^1, \ldots, \omega_{2y}$ take on the corresponding limiting values, which, as before, we denote by $z_x^1, \ldots, \omega_{2y}$. With regard to (3.89), the coefficients on the right-hand sides of $\rho_1 = \rho_{max}$ are not zero; dividing both sides of each of these equations by ρ_{max}, we pass to the limit $\mu = 0$. We then obtain the equations that describe the optimal process in the region $(\mathbf{j}, \text{grad } \omega_2) < 0$:

$$z_x^1 = \rho_{min} z_y^2 + \rho_{min} z_x^2 K, \qquad z_y^1 = -\rho_{min} z_x^2 + \rho_{min} z_y^2 K + c^{-1} VB \quad (3.96)$$

$$\left. \begin{array}{c} \rho_{min}\omega_{1x} = -z_x^2 \dfrac{\omega_{2y} z_x^2 - \omega_{2x} z_y^2}{(\text{grad } z^2)^2} \\[2mm] \rho_{min}\omega_{1y} = -z_y^2 \dfrac{\omega_{2y} z_x^2 - \omega_{2x} z_y^2}{(\text{grad } z^2)^2} \end{array} \right\} \quad (3.97)$$

For the region where $(\mathbf{j}, \text{grad } \omega_2)_0 > 0$, we obtain the equations

$$z_x^1 = -\frac{\omega_{2y} z_x^2 - \omega_{2x} z_y^2}{(\text{grad } \omega_2)^2}(\rho_{max}\omega_{2x} - 2\rho_{min}\omega_{2y}K + \rho_{min}\,\omega_{2x}K^2) \quad (3.98)$$

$$\left. \begin{array}{c} z_y^1 = -\dfrac{\omega_{2y} z_x^2 - \omega_{2x} z_y^2}{(\text{grad } \omega_2)^2}(\rho_{max}\omega_{2y} - 2\rho_{min}\omega_{2x}K + \rho_{min}\omega_{2y}K^2) + c^{-1}VB \\[2mm] \rho_{max}\omega_{1x} = -\omega_{2y} + K\omega_{2x}, \qquad \rho_{max}\omega_{1y} = \omega_{2x} + K\omega_{2y} \end{array} \right\} \quad (3.99)$$

Here, we have introduced the notation

$$K = \frac{(\text{grad } \omega_2, \text{grad } z^2)}{\omega_{2y} z_x^2 - \omega_{2x} z_y^2} \quad (3.100)$$

In particular, it follows from (3.98) and (3.99) that $(\mathbf{j}, \text{grad } \omega_2) \to 0$ as $\rho_{max} \to \infty$ [see the remark after (3.87)].

3.10. Asymptotic Solution in the Range $(\mathbf{j}, \text{grad } \omega_2) < 0$

We now consider (3.96) and (3.97) in more detail. This last pair of equations shows that

$$z^2 = h(\omega_1) \quad (3.101)$$

where $h(\omega_1)$ is an arbitrary function. Equations (3.97) are equivalent to the system

$$\omega_{2x} = \rho_{\min}(\omega_{1y} - \omega_{1x}K), \qquad \omega_{2y} = -\rho_{\min}(\omega_{1x} + \omega_{1y}K) \qquad (3.102)$$

The set of equations (3.96), (3.102), and (3.100) can be obtained in a simpler fashion. We assume from the very start that $\rho_1 = \rho_{\max} = \infty, \rho_2 = \rho_{\min}$. The basic equations of the problem can then be written in the form

$$0 = z_y^2 \cos \gamma - z_x^2 \sin \gamma$$

$$z_y^1 \cos \gamma - z_x^1 \sin \gamma = -\rho_{\min}(z_y^2 \sin \gamma + z_x^2 \cos \gamma) + c^{-1}VB \cos \gamma \qquad (3.103)$$

The first of these equations expresses the vanishing of the component of $\mathbf{j}$ along the $\boldsymbol{\alpha}(\rho_1 = \infty)$ axis, while the second is the differential Ohm's law in the direction of the $\boldsymbol{\beta}(\rho_2 = \rho_{\min})$ axis.

An equivalent way of expressing the system (3.103) is obtained by adding the equation

$$\tan \gamma = \frac{z_y^2}{z_x^2} \qquad (3.104)$$

to the system

$$z_x^1 = \rho_{\min}z_y^2 + \rho_{\min}z_x^2 K, \qquad z_y^1 = -\rho_{\min}z_x^2 + \rho_{\min}z_y^2 K + c^{-1}VB \qquad (3.105)$$

in which K is an as yet arbitrary function.

We shall treat this function formally as a control and pose the original optimization problem [see Section 3.5 for the system (3.105)]. The adjoint system has the form

$$\omega_{2x} = \rho_{\min}(\omega_{1y} - \omega_{1x}K), \qquad \omega_{2y} = -\rho_{\min}(\omega_{1x} + \omega_{1y}K) \qquad (3.106)$$

and the condition of stationarity with respect to the control K leads to the equation

$$z^2 = h(\omega_1) \qquad (3.107)$$

Equations (3.105), (3.106), and (3.107) coincide with (3.96), (3.102), and (3.100) which is what we wanted to prove.*

The system (3.96), (3.102), and (3.100) can be represented in a more convenient form. For this, we eliminate the function z^2 from (3.96) by means of (3.101), obtaining

$$z_x^1 = \rho_{min} h'(\omega_1)(\omega_{1y} + \omega_{1x}K)$$

$$z_y^1 = \rho_{min} h'(\omega_1)(\omega_{1y}K - \omega_{1x}) + c^{-1}VB \tag{3.108}$$

Elimination of z^1 from this system leads to the equation

$$h'(\omega_1)[\Delta\omega_1 + \omega_{1x}K_y - \omega_{1y}K_x] + h''(\omega_1)(grad\ \omega_1)^2 = (c\rho_{min})^{-1}VB_x(x) \tag{3.109}$$

On the other hand, the system (3.102) generates the equation

$$\Delta\omega_1 = \omega_{1x}K_y - \omega_{1y}K_x \tag{3.110}$$

This equation in conjunction with (3.109) leads to

$$2h'(\omega_1)\Delta\omega_1 + h''(\omega_1)(grad\ \omega_1)^2 = (c\rho_{min})^{-1}VB_x(x) \tag{3.111}$$

If the function $h(\omega_1)$ is known, then (3.110) and (3.111) together with the corresponding boundary conditions determine the functions ω_1 and K.

Let us consider the boundary conditions on the electrodes. The requirements

$$z^1 = z_{\pm}^1, \qquad \omega_2 = \omega_{2\pm} \qquad for\ |x| < \lambda, y = \pm\delta$$

lead to [see (3.102) and (3.108)]

$$h'(\omega_1)(\omega_{1y} + \omega_{1x}K) = 0, \qquad \omega_{1y} - \omega_{1x}K = 0$$

which hold along the electrodes. Rejecting the trivial possibility $h'(\omega_1) = 0$ (this would mean, in particular, that $j_y = 0$ on the electrodes) we conclude that

$$\omega_{1y} = 0, \qquad K = 0 \tag{3.112}$$

along the electrodes.

---

* Our derivation leads to (3.96), (3.102), and (3.100), and not to (3.98), (3.99), and (3.100), since it follows from these last three equations that

$$\omega_{2x} \to 0, \qquad \omega_{2y} \to 0, \qquad z_x^2 \to 0, \qquad z_y^2 \to 0$$

in the limit $\rho_{max} \to \infty$.

Thus, the lines of the current $\mathbf{j}$ (and with them the principal directions $\boldsymbol{\beta}$) must be normal to the electrodes. Symmetry considerations show that (3.112) also hold along the x axis.

The insulating walls of the channel are flow lines of the current $\mathbf{j}$; this circumstance is expressed by (2.101). The symmetry of the problem shows that the segment AA (see Fig. 3.3) is also a current line; if

$$\omega_1 = \omega_{1-} \qquad \text{along } B'C'C'B'$$
$$\omega_1 = \omega_{1+} \qquad \text{along } BCCB \tag{3.113}$$

then

$$\omega_1 = \tfrac{1}{2}(\omega_{1+} + \omega_{1-}) \qquad \text{along } AA \tag{3.114}$$

It remains to write down the connections between the values of the functions ω_1 and $h(\omega_1)$ and the parameters of the external circuit. These conditions have the form [see (2.101) and (3.16)]

$$z'_+ - z'_- = R(z^2_+ - z^2_-)$$
$$\omega_{2+} - \omega_{2-} + 1 = R(\omega_{1+} - \omega_{1-}) \tag{3.115}$$

To transform the last two equations, we use (3.102), (3.108), and (3.114). We consider the current line $L(\omega_1 = \text{const})$ joining the electrodes (Fig. 3.14), and we calculate the line integral*

$$\int_{L(\omega_1)} \omega_{2x}\, dx + \omega_{2y}\, dy$$

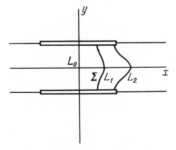

FIGURE 3.14

---

* It is assumed that there exist continuous contour lines of the function $\omega_1(x, y)$. In Section 3.14, the existence of such lines will be proved.

taken along this current line. Taking into account (3.102), we obtain

$$\int_{L(\omega_1)} \omega_{2x}\,dx + \omega_{2y}\,dy = \rho_{min}\int_{L(\omega_1)} \omega_{1y}\,dx - \omega_{1x}\,dy$$

Substitution of this result in the second of Eqs. (3.115) yields

$$\rho_{min}\int_{L(\omega_1)} \omega_{1y}\,dx - \omega_{1x}\,dy = R(\omega_{1+} - \omega_{1-}) - 1 \tag{3.116}$$

Similarly, using Eqs. (3.108), we obtain

$$\rho_{min}h'(\omega_1)\int_{L(\omega_1)} \omega_{1y}\,dx - \omega_{1x}\,dy = R[h(\omega_{1+}) - h(\omega_{1-})] - c^{-1}\int_{L(\omega_1)} VB\,dy$$

$$\tag{3.117}$$

A comparison of (3.116) and (3.117) leads to

$$h'(\omega_1) = \frac{1}{R(\omega_{1+} - \omega_{1-}) - 1}\left\{R[h(\omega_{1+}) - h(\omega_{1-})] - c^{-1}\int_{L(\omega_1)} VB\,dy\right\}$$

$$\tag{3.118}$$

$$\omega_{1-} \le \omega_1 \le \omega_{1+}$$

which determines the function $h(\omega_1)$ implicitly. If we add (3.116) taken for the parameter value* $\omega_1 = 1/2(\omega_{1+} + \omega_{1-})$ to the condition (3.118) and

---

* If the condition (3.118) is satisfied, it is sufficient to take (3.116) for only one value of the parameter ω_1 [e.g., $\omega_1 = (1/2)(\omega_{1+} + \omega_{1-})$] since its validity for other values of the parameter then follows automatically.

To prove this, we integrate both sides of (3.111) over the region Σ bounded by two current lines (see Fig. 3.14)

$$L_0(\omega_1 = (1/2)(\omega_{1+} + \omega_{1-})), \qquad L_1(\omega = \omega_1^0)$$

$$(1/2)(\omega_{1+} + \omega_{1-}) < \omega_1^0 \le \omega_{1+}$$

and two segments of the electrodes; using (3.118) and the condition (3.116) taken for the parameter value $\omega_1 = (1/2)(\omega_{1+} + \omega_{1-})$, we may represent the result in the form

$$\iint_\Sigma [h'(\omega_1^0) + h'(\omega_1)]\Delta\omega_1\,dx\,dy = 0$$

This equation is valid for any region Σ of the above type; it follows from this that the integrand of the last integral admits the representation

$$\frac{\partial}{\partial y}(u\omega_{1x}) - \frac{\partial}{\partial x}(u\omega_{1y})$$

and the function u vanishes on the electrodes. The required result then follows readily.

include the remaining boundary conditions, then we shall have a complete system of relations which, in conjunction with (3.111) and (3.110), suffice to determine the functions ω_1 and K.

The basic requirement $(\mathbf{j}, \operatorname{grad} \omega_2)_0 \leq 0$ tends to the inequality $(\mathbf{j}, \operatorname{grad} \omega_2) \leq 0$; in the limit from (3.101) and (3.102) we obtain the equivalent inequality

$$h'(\omega_1) \geq 0, \qquad \omega_1 \in [\omega_{1-}, \omega_{1+}] \tag{3.119}$$

It should be noted that the function ω_1 does not take values in G lying outside the interval $[\omega_{1-}, \omega_{1+}]$ under the imposed conditions.

Suppose that there are closed lines among the contour lines (Fig. 3.15). Then one and the same value $\omega = \omega_{1+}$ corresponds to the lines *abef* and *bedc*. For the line *abef*, (3.117) holds; using the single-valuedness of the potential $z^1(x, y)$ and arguing as in the derivation of (3.117), we obtain the equation

$$\rho_{\min} h'(\omega_{1+}) \oint_{bcdeb} \omega_{1y}\, dx - \omega_{1x}\, dy + \oint_{bcdeb} c^{-1} VB\, dy = 0$$

Similarly, the equation analogous to (3.116), namely

$$\rho_{\min} \oint_{bcdeb} \omega_{1y}\, dx - \omega_{1x}\, dy = \oint_{bcdeb} \omega_{2x}\, dx + \omega_{2y}\, dy$$

shows that the integral $\oint_{bcdeb} \omega_{1y}\, dx - \omega_{1x}\, dy$ can be nonzero only if $\omega_2(x, y)$ is a multiple-valued function. However, this is impossible since the integrals

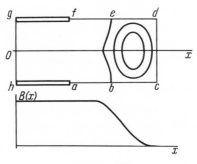

FIGURE 3.15

$\int_{abcf} d\omega_2$ and $\int_{abcdef} d\omega_2$ are both equal to the same quantity $\omega_{2+} - \omega_{2-}$. Therefore, the integral $\oint_{bcdeb} \omega_{1y} dx - \omega_{1x} dy$ is zero, and with it the integral $\oint_{bcdeb} VB \, dy$, which is contradictory under the assumptions made about the functions $V(y)$ and $B(x)$.

If we consider current lines situated entirely in the region of constancy of the function $B(x)$, then the proof needs to be somewhat modified. It then becomes simpler to introduce the new dependent variable f in accordance with (3.125) (see below); this variable satisfies (3.127), which satisfies Laplace's equation in the region $B(x) = $ const. It follows from this that the contour lines of the function ω_1 and, hence, of the function z^2 do not have closed loops within the region; i.e., lines join the electrodes, which is what we wanted to prove.

For other arrangements of closed contour lines, the proof is similar. Thus, all of the contour lines of ω_1 join the electrodes. We shall need this important conclusion later.

Summarizing what we have said, we arrive at the following problem.

Problem A. Determine a function $\omega_1(x, y)$ which satisfies (3.112), the boundary conditions (3.112) and (3.113), the condition

$$- \rho_{\min} \int_{-\delta}^{\delta} \omega_{1x}(0, y) \, dy = R(\omega_{1+} - \omega_{1-}) - 1 \qquad (3.120)$$

and the additional requirement (3.118), in conjunction with the inequality (3.119), then defines the function $h(\omega_1)$. The integral on the right-hand side of (3.118) is taken along the current line $L(\omega_1)$ corresponding to the parameter value $\omega_1 = $ const which serves as the argument of the function $h'(\omega_1)$ on the left-hand side of this equation. The current lines $\omega_1 = \omega_{1+}$ and $\omega_1 = \omega_{1-}$ are critical; along them, the function $h'(\omega_1)$ takes the value zero.

Once the function $\omega_1(x, y)$ has been found, (3.110) in conjunction with the Cauchy condition $K(x, \pm\delta)|_{|x|<\lambda} = 0$ [see (3.112)] determines the function K.

If the function $B(x)$ is equal to a constant B_0 for some range of values of the argument, then the problem may be simplified somewhat.

Consider current lines L lying entirely in the region in which $B(x) = B_0 = $ const. Equation (3.118) shows that the function $h'(\omega_1)$ takes one and the same constant value on these current lines,* and (3.111) reduces to

---

* Equation (3.118) applies to such current lines since these lines connect the electrodes.

Laplace's equation in the same region. However, (3.110) then admits the integral

$$K = F(\omega_1) \tag{3.121}$$

where $F(\omega_1)$ is an arbitrary function. Since the region $B(x) = B_0 = $ const encompasses the electrodes (Fig. 3.16), along which $K = 0$, we must set $F \equiv 0$ in (3.121), i.e., $K \equiv 0$.

This result shows that the optimal control is isotropic and equal to $\rho_{\min} = $ const in the region occupied by the current lines situated entirely in the region $B(x) = B_0 = $ const. Anisotropy of the control is characterized by nonzero values of the function K; the control is anisotropic where there are current lines which lie, at least partially (to arbitrarily small extent), within the region $B(x) \neq $ const. In Fig. 3.16, we have emphasized the current line separating the regions of isotropic and anisotropic control. The current lines $BCCB$ and $B'C'C'B'$ are critical.

The assumption of the existence of critical current lines is confirmed by an investigation of (3.118). The first factor on the right-hand side of this equation is always negative [to prove this, multiply both sides of (3.110) by ω_1 and integrate over the region $CCC'C'$ (Fig. 3.16); the integral on the right-hand side is zero by virtue of the boundary conditions (3.112) and the fact that $\omega_1 = \omega_{1\pm} = $ const along CC and $C'C$; equating the integral on the left-hand side to zero, we obtain [see (3.116)]

$$(\omega_{1+} - \omega_{1-})[1 - R(\omega_{1+} - \omega_{1-})] = \rho_{\min} \iint_G (\mathrm{grad}\,\omega_1)^2 \, dx \, dy$$

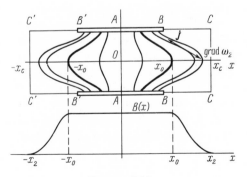

FIGURE 3.16

It follows that $\omega_{1+} - \omega_{1-} > 0$ and simultaneously that $1 - R(\omega_{1+} - \omega_{1-}) > 0$, which is what we wanted to prove.]*

The application of the previous argument to the region between the two lines $L_1(\omega_1 = \omega_1^0)$ and $L_2(\omega_1 = \omega_1^1)$ (see Fig. 3.14), and the use of the remark in the last footnote, together with the inequality $1 - R(\omega_{1+} - \omega_{1-}) > 0$, however, imply that one always has $\omega_1^1 > \omega_1^0$; i.e., the function ω_1 increases as one moves from one current line to the next from the left to the right.

The sign of the second factor on the right-hand side of (3.118) depends on the choice of the current line L. Suppose the function $B(x)$ is given by the graphs in Fig. 3.16, and $V = V(y)$. If the current line lies entirely in the region where $B(x) = B_0 = \text{const}$, then the expression in the braces in (3.118) is negative (it is clear that the term $R[h(\omega_{1+}) - h(\omega_{1-})]$ does not exceed $c^{-1}B_0 \int_{-\delta}^{\delta} V(y)\, dy$ if the graph of $B(x)$ is that of Fig. 3.16, since this term does not exceed the considered expression even when $B(x) = B_0$ everywhere in the channel). If the current line joining the electrodes in accordance with our assumption has an appreciable part lying in the region $B(x) = 0$, however, then the expression in braces in (3.118) is positive. This possibility must be rejected, since it contradicts the inequality (3.119). It follows that, within the adopted scheme, the current lines do not penetrate sufficiently far into the region of decrease of the field $B(x)$, and one can have the situation in which the entire region occupied by currents is bounded by the critical current lines $BCCB$ and $B'C'C'B'$ [along which $h'(\omega_1) = 0$] and by the electrodes. It is remarkable that the distribution of the resistivity outside this region has no effect since the condition $h'(\omega_1) = 0$ on the critical current line is sufficient to determine this line irrespective of what happens on its other side. For example, nothing stops us from assuming that the channel outside the region $CCC'C'$ is filled with a homogeneous isotropic insulator $\rho = \infty$, so that there are no currents outside $CCC'C'$ at all.

The coefficient of the principal part of (3.111) vanishes on the critical current lines. Sufficiently close to these lines, the control is always anisotropic since the critical lines necessarily go outside the region $B(x) = B_0 = \text{const}$. However, these small neighborhoods are not regions of rapid variation of ω_1. Along the critical lines, the derivatives ω_{1x} and ω_{1y} take finite values [this follows from (3.116) and the next-to-last footnote, or from (3.117) with allowance for the fact that on the critical lines the factor $H'(\omega_1)$ and the expression on the right-hand side vanish].

Of course, our arguments cannot be taken as proof of existence of a solution to Problem A. The proof will be given below; we first give a different, more convenient formulation of the problem.

---

* Another proof is based on the use of (3.111) in addition to (3.118) and (3.116).

Since $B(x)$ is an even function, the function $h(\omega_1)$ [see (3.118)] can be assumed to be an odd function of x; the first derivative $h'(\omega_1)$ is then even in x, and the second derivative $h''(\omega_1)$ is odd. At the same time, ω_1 is an odd function of x, and instead of the region $CCC'C'$ (see Fig. 3.3) we can consider its right-hand half $CBAABC$; the boundary AA of this new region can be assumed to be an insulator, along which [see (3.114)]

$$\omega_1 = 0 \tag{3.122}$$

We now integrate (3.118) with respect to ω_1 from zero to ω_{1+} to obtain [$L(\omega_1)$ denotes the current line parametrized by ω_1]

$$I = 2c^{-1} \int_0^{\omega_1} d\omega_1 \int_{L(\omega_1)} VB \, dy \tag{3.123}$$

$$h'(\omega_1) = \frac{1}{c(1 - 2R\omega_{1+})} \left[\int_{L(\omega_1)} VB \, dy - 2R \int_0^{\omega_{1+}} d\omega_1 \int_{L(\omega_1)} VB \, dy \right] \tag{3.124}$$

Next, we use Weierstrass's condition $h'(\omega_1) \geq 0$, and introduce the function $f(\omega_1)$ by the equation*

$$f'(\omega_1) = \sqrt{h'(\omega_1)} \tag{3.125}$$

We shall regard the function f as a new dependent variable; we may assume that (see Fig. 3.3)

$$f = 0 \qquad \text{along } AA$$
$$\tag{3.126}$$
$$f = f_+ \qquad \text{along } BCCB$$

Taking into account the equation

$$\Delta f = \sqrt{h'(\omega_1)} \, \Delta\omega_1 + \frac{h''(\omega_1)}{2\sqrt{h'(\omega_1)}} (\text{grad } \omega_1)^2$$

we then represent (3.111) in the form

$$\Delta f - \frac{1}{2c\rho_{\min} f'(\omega_1)} VB_x(x) = 0 \tag{3.127}$$

---

* As before, the prime denotes the derivative with respect to ω_1.

where $h'(\omega_1)$ must be eliminated by means of (3.125) and (3.124) [the connection between the constants ω_{1+} and f_+ is found from the correspondence between the boundary conditions $(3.113)_2$, (3.122), and (3.126) when (3.125) is integrated]. On the electrodes, the boundary condition for f has the form

$$f_y = 0 \tag{3.128}$$

and the conditions on the insulators are given by (3.126). The constant f_+ is determined from condition (3.120) which now takes the form

$$\rho_{min} \int_{L(\omega_{1+})} f_n(x, y) \, dt = f'(\omega_{1+})(1 - 2R\omega_{1+}) \tag{3.129}$$

and (3.124) and (3.125) and our remark made above about the connection between f_+ and ω_{1+}, must be taken into account.

On the critical line $h = h_+$, it follows from (3.125) that

$$f'(\omega_{1+}) = 0 \tag{3.130}$$

In investigating the posed problem, we may assume $R = 0$ without loss of generality; for, introducing the function

$$\tilde{B}(x) = B(x) - \frac{2R \displaystyle\int_0^{\omega_{1+}} d\omega_1 \int_{L(\omega_1)} VB \, dy}{\displaystyle\int_{L(\omega_1)} V \, dy} \tag{3.131}$$

and setting

$$f = \sqrt{1 - 2R\omega_{1+}}\, \tilde{f} \tag{3.132}$$

along with replacing $B(x)$ by $\tilde{B}(x)$ and assuming $R = 0$, we obtain a boundary-value problem for the determination of $\tilde{f}(x, y)$ which is identical with the problem (3.124)–(3.130). Of course, what we have said is true only under the condition that $\tilde{B}(x) \geq 0$ everywhere in the channel.

Under the assumptions made above about the functions $B(x)$ and $V(y)$ and the nature of the current lines L, the function $\tilde{B}(x)$ and $B(x)$ differ by a constant term (we denote it by $-b$). If we replace $B(x)$ by $B_0 = $ const in the calculation of this term in accordance with (3.131), but leave the current lines L unchanged, then the term under consideration is not increased. We obtain (Fig. 3.17)

$$\tilde{B}(x < x_2) \geq B_0(1 - 2R\omega_{1+}) > 0$$

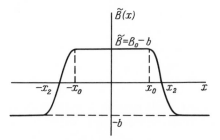

FIGURE 3.17

since the expression in the parentheses is positive and does not exceed unity as was shown earlier. On the other hand, for sufficiently large x, the function $\tilde{B}(x)$ preserves the constant negative value $-b$, so that, as a whole, $\tilde{B}(x)$ is given by a graph of the type shown in Fig. 3.17. The transition $B \to \tilde{B}$, $R \to 0$ leaves the lines of the current $\mathbf{j}$ invariant as well as the optimal value of the functional I. Let us prove the last assertion. Setting

$$B = \tilde{B} + b \tag{3.133}$$

and taking into account (3.131), we obtain (after a simple calculation)

$$b = \frac{2R \displaystyle\int_0^{\omega_{1+}} d\omega_1 \int_{L(\omega_1)} V\tilde{B}\, dy}{(1 - 2R\omega_{1+}) \displaystyle\int_{L(\omega_1)} V\, dy}$$

However, in accordance with (3.124), (3.125), (3.131), and (3.132), we have

$$d\omega_1 = \frac{df}{\sqrt{h'(\omega_1)}} = \frac{\sqrt{1 - 2R\omega_{1+}}\, df}{\left[\dfrac{1}{c}\displaystyle\int_{L(f)} V\tilde{B}\, dy\right]^{1/2}} = \frac{(1 - 2R\omega_{1+})d\tilde{f}}{\left[\dfrac{1}{c}\displaystyle\int_{L(\tilde{f})} V\tilde{B}\, dy\right]^{1/2}} \tag{3.134}$$

Integrating this equation, we obtain

$$\omega_{1+} = \frac{m}{1 + 2mR}, \qquad m = \int_0^{\tilde{f}_+} \frac{d\tilde{f}}{\sqrt{\dfrac{1}{c}\displaystyle\int_{L(\tilde{f})} V\tilde{B}\, dy}}$$

It follows that

$$1 - 2R\omega_{1+} = \frac{1}{1 + 2mR}$$

and [see (3.134) and (3.135)]

$$b = \left[2R \int_0^{\tilde{f}_+} \frac{d\tilde{f}}{\sqrt{\frac{1}{c} \int_{L(\tilde{f})} V\tilde{B}\,dy}} \int_{L(\tilde{f})} V\tilde{B}\,dy \right] \left[\left[\int_{L(\tilde{f})} V\,dy \right]^{-1} \right]$$

In what follows, we shall assume $V = \text{const}$ in order to simplify the calculations. We introduce dimensionless variables (we do not as yet fix the constant B^0)

$$x = \lambda\bar{x}, \qquad y = \lambda\bar{y}, \qquad \tilde{f} = \frac{\sqrt{VB^0\lambda}}{\rho_{\min}\sqrt{c}} \bar{\tilde{f}}, \qquad I = \frac{VB^0\lambda}{c\rho_{\min}} \bar{I}$$

$$B = B^0\bar{B}, \qquad \tilde{B} = B^0\bar{\tilde{B}}, \qquad b = B^0\bar{b}, \qquad R = \rho_{\min}\bar{R}$$

(3.135)

and let $\bar{I}(\bar{B}, \bar{R})$ be the optimal value of $\bar{I}$ corresponding to the field $\bar{B}$ and the load $\bar{R}$.

The expression for $\bar{b}$ becomes

$$\bar{b} = \bar{R}\frac{\lambda}{\delta} \int_0^{\tilde{f}_+} d\bar{\tilde{f}} \sqrt{\int_{L(\tilde{f})} \bar{\tilde{B}}\,dy} = \tfrac{1}{2}\bar{I}(\bar{\tilde{B}}, 0)\bar{R}\frac{\lambda}{\delta}$$

where we have taken (3.123)–(3.125) into account, along with the assumption $B = \tilde{B}$, $R = 0$.

On the other hand, (3.131), (3.133), and (3.123) imply

$$\bar{b} = \tfrac{1}{2}\bar{I}(\bar{B}, \bar{R})\bar{R}\frac{\lambda}{\delta}$$

Comparing this with the previous expression, we obtain

$$\bar{I}(\bar{\tilde{B}}, 0) = \bar{I}(\bar{B}, \bar{R})$$

which is what we wanted to prove.

For a given function $B(x)$ and fixed dimensions of the channel, the condition $\tilde{B}(x) \geq 0$ gives an upper bound on the values of R for which the transition $(B, R) \to (\tilde{B}, 0)$ has the above invariance property. Let us illustrate what we have said by the example of a channel which does not have horizontal sections of insulating walls (in Fig. 3.18 the channel is bounded on the left and right, respectively, by the insulators AA and CC, and at the top and bottom by the electrodes AC).

It is readily verified that the function

$$\omega_1(x) = \frac{1}{2(R + \rho_{\min}\delta/\lambda)} \frac{x}{\lambda} = \omega_{1+} \frac{x}{\lambda}$$

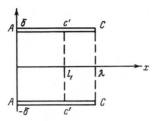

FIGURE 3.18

satisfies (3.111) for

$$h'(\omega_1) = \frac{2V\delta}{c} \frac{R + \rho_{min}\delta/\lambda}{\rho_{min}\delta/\lambda} \left[B(x) - 2R\frac{\omega_{1+}}{\lambda} \int_0^\lambda B(x)\, dx \right] \quad (3.136)$$

as well as the boundary conditions (3.112) and (3.120) for $\omega_1(0) = 0$, $\omega_{1+} = -\omega_{1-}$.

The single y component of the vector $\mathbf{j}$ is

$$j_y = \frac{\omega_{1+}}{\lambda} h'(\omega_1)$$

and the functional I is given by

$$I = 2 \int_0^\lambda j_y\, dx = 2 \int_0^{\omega_{1+}} h'(\omega_1)\, d\omega_1 = \frac{2V\delta}{c} \frac{\dfrac{1}{\lambda} \displaystyle\int_0^\lambda B(x)\, dx}{R + \rho_{min}\delta/\lambda} \quad (3.137)$$

If we now take the equation

$$1 - 2R\omega_{1+} = \frac{\rho_{min}\delta/\lambda}{R + \rho_{min}\delta/\lambda}$$

into account, we readily see that (3.137) is not changed by setting $R = 0$ in the denominator and replacing $B(x)$ by the function [see (3.131)]

$$\tilde{B}(x) = B(x) - 2R\frac{\omega_{1+}}{\lambda} \int_0^\lambda B(x)\, dx$$

Thus, we have invariance of the functional I under the substitution of $(B, R) \rightarrow (\tilde{B}, 0)$.

On the other hand, (3.137) is the optimal value of the functional I if $h'(\omega_1) \geq 0$ everywhere in the channel, i.e., if $j_y \geq 0$ or

$$B(x) - \frac{R}{R + \rho_{\min}\delta/\lambda} \frac{1}{\lambda} \int_0^\lambda B(x) \, dx \geq 0$$

We have pointed out above that the invariance $(B, R) \to (\tilde{B}, 0)$ holds only if this condition is satisfied.

For given (nonconstant) function $B(x)$ (of the type shown in Fig. 3.5) and for known λ, δ, and $\rho_{\min}$, the last inequality is an upper bound on the parameter R: $R \leq R_0$ where R_0 is determined by the condition

$$B(\lambda) - \frac{R_0}{R_0 + \rho_{\min}\delta/\lambda} \frac{1}{\lambda} \int_0^\lambda B(x) \, dx = 0$$

If $R > R_0$, then the value (3.137) of the functional is not optimal since the condition $h'(\omega_1) \geq 0$ is violated for ω_1 values sufficiently close to ω_{1+} (or for values of x sufficiently close to λ). If this condition is to be satisfied everywhere in the channel, we must obviously reduce λ; for every given $R_1 > R_0$ we can find a λ_1 such that a channel with $\lambda < \lambda_1$ will be optimal and the corresponding value of I will be given by (3.127); this value of λ_1 can be found from the condition

$$B(\lambda_1) - \frac{R_1}{R_1 + \rho_{\min}\delta/\lambda_1} \frac{1}{\lambda_1} \int_0^{\lambda_1} B(x) \, dx = 0$$

or, equivalently, from the condition $\tilde{B}(\lambda_1) = 0$.

This argument shows that a channel of given length x_c may be nonoptimal for the pair $B(x)$ and R if the parameter R is sufficiently large. Direct verification of optimality [i.e., of the inequality $h'(\omega_1) \geq 0$ for the pair $B(x)$ and R] is difficult, since it requires calculation of the right-hand side of (3.124) (this can be readily done only in the one-dimensional case considered above). Having the function $B(x)$, it is much easier to specify a constant value $b > 0$ [see (3.133)] such that the inequality $\tilde{B} = B - b \geq 0$ is violated nowhere in the channel and, using the invariance of the optimal value of the current under the transition $(B, R) \to (\tilde{B}, 0)$ discussed above, calculate this value for the pair $(\tilde{B}, 0)$. The corresponding resistance R is then found from the expression

$$R = \frac{2bV\delta}{cI(\tilde{B}, 0)}$$

which follows readily from (3.123), (3.131), and (3.133).

With the specification of each different b, we obtain (invariance) a sequence of optimal values of I corresponding to the values of R found from the last expression. If we choose an ascending sequence of constants b, then for some $b = b_0 = B(x_c)$ the difference $B - b$ becomes zero at the end of the channel $x = x_c$ [recall that the function $B(x)$ is given by a graph of the type shown in Fig. 3.16, and one can have the case when $x_2 > x_c$]; the corresponding value of R, equal to R_0, gives an upper bound on the set of values of R such that for the pair $(B(x), R)$, a channel of length x_c may be optimal, i.e., the condition $h'(\omega_1) \geq 0$ may be satisfied everywhere in the channel. For $R > R_0$, it is necessary to take a shorter channel and repeat the procedure we have just described, etc. An example of such a calculation will be given in Section 3.13.

Taking into account our remarks and changing the length of the channel when necessary, we shall next consider the problem (3.124)-(3.130) for $R = 0$, $B(x) \geq 0$. Equation (3.127) can then be written in the form (instead of $\tilde{f}$ and $\tilde{B}$ we write f and B)

$$M(f) \equiv -\Delta f + \frac{1}{2c\rho_{\min}} \frac{VB_x(y)}{\sqrt{\dfrac{1}{c} \displaystyle\int_{L(f)} VB \, dy}} = 0 \tag{3.138}$$

and the boundary condition (3.129) takes the form

$$\rho_{\min} \int_{L(f_+)} f_n(x, y) \, dt = \sqrt{\frac{1}{c} \int_{L(f_+)} VB \, dy} \tag{3.139}$$

In these equations, we have introduced the new parametrization $L(f)$ of the family of current lines.

Equation (3.138) can be integrated in the region $CBAABC$ (see Fig. 3.3) subject to the boundary conditions (3.126), (3.128), and (3.139) and the condition [see (3.130)]

$$\int_{L(f_+)} VB \, dy = 0 \tag{3.140}$$

which serves to determine the critical current line. We shall call the problem posed in this manner Problem B.

For $R = 0$, the functional is equal to [see (3.123)-(3.125)]

$$I = 2c^{-1/2} \int_0^{f_+} df \left(\int_{L(f)} VB \, dy \right)^{1/2} \tag{3.141}$$

3.11. Further Investigation of the Asymptotic Case: Existence of Solution

Before we consider the existence of a solution of Problem B for the rectangular region $KBAABK$ (Fig. 3.19), we note that the operator M defined by (3.138) is a potential operator on the set of twice continuously differentiable functions that satisfy the boundary conditions (3.126), (3.128), and (3.139). To see this, we consider the variational equation corresponding to (3.138). This equation is derived in Appendix A.5 [Eqs. (24), (26)]; it has the form

$$-\Delta\delta f_P - \frac{V(y_P)B_x(x_P)}{4c^2\rho_{\min}\left[c^{-1}\displaystyle\int_{L(f)} VB\,dy\right]^{3/2}}\int_{L(f)} V(y_M)B_x(x_M)\frac{\delta f_P - \delta f_M}{(f_n)_M}\,dt_M = 0$$

(3.142)

Here, δf_P denotes the variation δf at the point P; the integrations are performed along the current line $L(f)$ which passes through the point P; M denotes the point of integration, and $(f_n)_M$ the value of the derivative of f along the normal to the current line at the point M (see Fig. A.3 in Appendix A). The following variational condition corresponds to condition (3.139) [Eqs. (28) and (29) of Appendix A]:

$$\int_{L(f)} (\delta f_M)_n\,dt_M = 0$$

(3.143)

In Appendix A, it is shown that the integration in this equation can be performed along any current line of the family $L(f)$. The left-hand side of (3.142) defines a linear operator (which we denote by DM_f), which we shall consider on the class of functions that satisfy the condition (3.143)

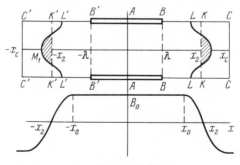

FIGURE 3.19

on *BCCB* and the boundary conditions

$$\left.\begin{array}{ll} \delta f = 0 & \text{along } AA \\[6pt] \delta f = \delta f_+ = \text{const} & \text{along } BCCB \end{array}\right\} \tag{3.144}$$

$$\delta f_y = 0 \qquad \text{along electrodes } AB \tag{3.145}$$

Note that the line *BCCB*, which bounds the region on the right (see Fig. 3.19), is not as yet assumed to be critical ($x_2 > x_c$); this assumption will not be made until the end of the present section.*

The conditions (3.144) and (3.145) are obtained by varying the conditions (3.126) and (3.128).

The operator DM_f defined in this manner is symmetric for every f, i.e., it satisfies the condition

$$\iint_G vDM_f u \, dx \, dy = \iint_G uDM_f v \, dx \, dy \tag{3.146}$$

To prove this, we calculate the left-hand side:

$$\iint_G vDM_f u \, dx \, dy$$

$$= -\iint_G v_P \Delta u_P \, dx_P \, dy_P$$

$$- \iint_G v_P \frac{B_x(x_P) u_P \, dx_P \, dy_P}{4c^2 \rho_{\min} \left[c^{-1} \displaystyle\int_{L(f)} VB \, dy \right]^{3/2}} \int_{L(f)} \frac{V(y_M) B_x(x_M)}{(f_n)_M} \, dt_M$$

$$+ \iint_G \frac{v_P V(y_P) B_x(x_P) \, dx_P \, dy_P}{4c^2 \rho_{\min} \left[c^{-1} \displaystyle\int_{L(f)} VB \, dy \right]^{3/2}} \int_{L(f)} \frac{V(y_M) B_x(x_M)}{(f_n)_M} u_M \, dt_M \tag{3.147}$$

The second term on the right-hand side of (3.147) is symmetric with respect to u and v; we transform the first and the third term, obtaining

$$\iint_G v_P \Delta u_P \, dx_P \, dy_P = \oint_\Gamma v_P \frac{\partial u_P}{\partial n} \, dt - \iint_G (\text{grad } v_P, \text{grad } u_P) \, dx_P \, dy_P$$

$$= -\iint_G (\text{grad } v_P, \text{grad } u_P) \, dx_P \, dy_P \tag{3.148}$$

---

* Figure 3.19 illustrates the case $x_2 < x_c$.

[the latter by virtue of the boundary conditions (3.143)–(3.145) satisfied by the functions u and v]. In the third term on the right-hand side of (3.147), we perform the outer integration in terms of the curvilinear coordinates f and q; the area element in these coordinates is $(dt = H_q \, dq)$

$$H_f H_q \, df \, dq = \frac{df \, dt}{|f_n|} = \frac{df \, dt}{f_n}$$

[the last equation holds because of the monotonic growth of f with respect to n (see above)]. We obtain

$$\iint_G v_P \frac{V(y_P) B_x(x_P) \, dx_P \, dy_P}{4c^2 \rho_{\min} \left[c^{-1} \int_{L(f)} VB \, dy \right]^{3/2}} \int_{L(f)} \frac{V(y_M) B_x(x_M)}{(f_n)_M} u_M \, dt_M$$

$$= \int_0^{f_+} \frac{df}{4c^2 \rho_{\min} \left[c^{-1} \int_{L(f)} VB \, dy \right]^{3/2}} \int_{L(f)} \frac{V(y_P) B_x(x_P) v_P}{(f_n)_P} \, dt_P$$

$$\times \int_{L(f)} \frac{V(y_M) B_x(x_M) u_M}{(f_n)_M} \, dt_M \tag{3.149}$$

Equations (3.148) and (3.149) show that (3.147) is symmetric with respect to u and v, and the symmetry of the operator DM_f is proved.

The condition (3.146) is necessary and sufficient for M to be a potential operator. The potential $\Phi(f)$ of this operator is determined by the expression

$$\Phi(f) - \Phi(f_0) = \int_0^1 dt \iint_G (f - f_0) M(f_0 + t(f - f_0)) \, dx \, dy \tag{3.150}$$

where $f_0(x, y)$ is any function in the region of definition $d(M)$ of M, i.e., it is a twice continuously differentiable function satisfying the boundary conditions (3.126), (3.128), and (3.139).

Note that the contour lines $L(f)$ of the functions f belonging to $d(M)$ cannot have closed branches in the part of the region G where $B_x(x) \neq 0$. This assertion follows from the very form of the operator M, which is not defined in the common center of all closed branches. We can therefore assume that functions in $d(M)$ do not have extrema in G; an extremum is attained only on the boundary of G.

The functionals $\Phi(f)$ and $I(f)$ [see (3.141)] have a common domain of definition*; we show that they take on stationary values for the same element f. The condition of stationarity of the functional $\Phi(f)$ on the set $d(M)$ is given by (3.138) (this may be verified directly). To obtain the condition of stationarity of $I(f)$, we represent this functional in an equivalent form, setting

$$f = f_+ u \tag{3.151}$$

where u is a twice continuously differentiable function satisfying the conditions

$$\left.\begin{array}{ll}
u = 0 & \text{along } AA \\[1em]
u = 1 & \text{along } BCCB \\[1em]
u_y = 0 & \text{on electrodes } AB \\[1em]
\rho_{\min} f_+ \displaystyle\int_{L(f_+)} u_n(x, y)\, dt = \sqrt{\dfrac{1}{c} \displaystyle\int_{L(f_+)} VB\, dy}
\end{array}\right\} \tag{3.152}$$

which are analogous to the conditions (3.126), (3.128), and (3.134).

We have

$$I(f_+ u) = 2c^{-1/2} f_+ \int_0^1 du \left(\int_{L(u)} VB\, dy \right)^{1/2}$$

The functional $I(f)$ will be completely defined if we know the connection between u and f_+. This connection is determined by the requirement of stationarity of the functional $\Phi(f)$. Making the substitution (3.151) in this functional, the condition of stationarity of $\Phi(f) = \Phi(f_+ u)$ with respect to f_+ becomes†

$$\frac{\partial \Phi(f_+ u)}{\partial f_+} = f_+ \iint_G (\nabla u)^2\, dx\, dy - \frac{1}{\rho_{\min}} \int_0^1 du \sqrt{\frac{1}{c} \int_{L(u)} VB\, dy} = 0$$

---

* As this domain, we can take the set $d(\Phi) = d(I)$ of square-integrable functions which are monotonic in the open region G, have square-integrable first derivatives, and satisfy the boundary conditions (3.126).

† See below; the remark following (3.156).

whence

$$f_+ = \frac{1}{\rho_{min}} \frac{\displaystyle\int_0^1 du \sqrt{\frac{1}{c}\int_{L(u)} VB\,dy}}{\displaystyle\iint_G (\nabla u)^2\,dx\,dy} \tag{3.153}$$

$$I(f_+ u) = \frac{2}{\rho_{min}} \frac{\left(\displaystyle\int_0^1 du \sqrt{\frac{1}{c}\int_{L(u)} VB\,dy}\right)^2}{\displaystyle\iint_G (\nabla u)^2\,dx\,dy} = I_1(u) \tag{3.154}$$

We shall now consider the functional $I_1(u)$ on functions $u(x, y)$ satisfying the first three of the conditions (3.152). We have

$$\frac{\rho_{min}}{2}\,\delta I_1(u) = \frac{2\displaystyle\int_0^1 du \sqrt{\frac{1}{c}\int_{L(u)} VB\,dy}}{\displaystyle\iint_G (\nabla u)^2\,dx\,dy} \left\{ \delta\int_0^1 du \sqrt{\frac{1}{c}\int_{L(u)} VB\,dy} \right.$$

$$\left. - \frac{\displaystyle\int_0^1 du \sqrt{\frac{1}{c}\int_{L(u)} VB\,dy}}{\displaystyle\iint_G (\nabla u)^2\,dx\,dy}\left[\oint_\Gamma \frac{\partial u}{\partial n}\,\delta u\,dt - \iint_G \Delta u\,\delta u\,dx\,dy\right]\right\}$$

$$\delta\int_0^1 du \sqrt{\frac{1}{c}\int_{L(u)} VB\,dy}$$

$$= \int_0^1 (d\,\delta u)\sqrt{\frac{1}{c}\int_{L(u)} VB\,dy} + \int_0^1 du\,\delta\sqrt{\frac{1}{c}\int_{L(u)} VB\,dy}$$

$$= \left[\sqrt{\frac{1}{c}\int_{L(u)} VB\,dy}\,\delta u\right]_{u=0}^{u=1}$$

$$+ \int_0^1\left[\delta\sqrt{\frac{1}{c}\int_{L(u)} VB\,dy} - \delta u\frac{d}{du}\sqrt{\frac{1}{c}\int_{L(u)} VB\,dy}\right]du$$

In accordance with (23) and (29) of the Appendix*

$$\delta \sqrt{\frac{1}{c} \int_{L(u)} VB\, dy} = \frac{1}{2\sqrt{c \int_{L(u)} VB\, dy}} \left[\delta u \int_{L(u)} \frac{V(y_M)B_x(x_M)}{(u_n)_M} dt_M \right.$$

$$\left. - \int_{L(u)} \frac{\delta u_M V(y_M)B_x(x_M)}{(u_n)_M} dt_M \right]$$

$$\frac{d}{du} \sqrt{\frac{1}{c} \int_{L(u)} VB\, dy} = \frac{1}{2\sqrt{c \int_{L(u)} VB\, dy}} \int_{L(u)} \frac{V(y_M)B_x(x_M)}{(u_n)_M} dt_M$$

Using these equations, we find

$$\delta \int_0^1 du \sqrt{\frac{1}{c} \int_{L(u)} VB\, dy} = \left[\sqrt{\frac{1}{c} \int VB\, dy} \, \delta u \right]_{u=0}^{u=1}$$

$$- \int_0^1 \frac{du}{2\sqrt{c \int_{L(u)} VB\, dy}}$$

$$\times \int_{L(u)} \frac{\delta u_M V(y_M)B_x(x_M)}{(u_n)_M} dt_M$$

$$\left. \frac{\rho_{\min}}{2} \delta I_1(u) = \frac{2\int_0^1 du \sqrt{\frac{1}{c}\int_{L(u)} VB\, dy}}{\iint_G (\nabla u)^2\, dx\, dy} \left\{ \left[\sqrt{\frac{1}{c}\int_{L(u)} VB\, dy}\, \delta u \right]_{u=0}^{u=1} \right. \right.$$

$$- \int_0^1 \frac{du}{2\sqrt{c\int_{L(u)} VB\, dy}} \int_{L(u)} \frac{\delta u_M V(y_M)B_x(x_M)}{(u_n)_M} dt_M$$

$$- \frac{\int_0^1 du \sqrt{\frac{1}{c}\int_{L(u)} VB\, dy}}{\iint_G (\nabla u)^2\, dx\, dy}$$

$$\left. \times \left[\oint_\Gamma \frac{\partial u}{\partial n} \delta u\, dt - \iint_G \Delta u\, \delta u\, dx\, dy \right] \right\} \qquad (3.155)$$

* For the notation, see the paragraph after (3.142).

Hence, taking into account the notation (3.153) and the boundary conditions $(3.152)_{1-3}$, we obtain (3.138) as a necessary condition of stationarity of $I_1(u)$. The third condition of (3.152) is a natural condition for the functional $I_1(u)$; the fourth condition of (3.152) is contained in (3.153), for it follows from (3.138) that the difference

$$\rho_{\min}f_+ \int_{L(u)} \frac{\partial u}{\partial n} dt - \sqrt{\frac{1}{c} \int_{L(u)} VB\, dy}$$

does not depend on u; (3.153) shows that the difference is zero—to see this, we recall that

$$\iint_G (\nabla u)^2 \, dx\, dy = \int_0^1 du \int_{L(u)} \frac{\partial u}{\partial n} dt$$

What we have said here corresponds to the specification of the domains of definition $d(\Phi) = d(I)$ of the functional Φ and I given earlier in the next-to-last footnote.

In this definition, it was not required that the functions belonging to $d(\Phi) = \delta(I)$ satisfy the natural boundary conditions (3.128) and (3.139).

Simultaneously, it is helpful to provide an expanded expression for the functional $\Phi(f)$ defined by (3.150). We set $\Phi(f) = \Phi_1(f) - \Phi_2(f)$; taking into account the condition (3.126) (we shall assume that the functions f and f_0 take on the same value f_+ on $BCCB$), along with condition (3.128), we find

$$\Phi_1(f) - \Phi_1(f_0) = -\int_0^1 dt \iint_G (f - f_0)\Delta(f_0 + t(f - f_0)) \, dx\, dy$$

$$= -\iint_G (f - f_0)\Delta f_0 \, dx\, dy$$

$$-\frac{1}{2} \iint_G (f - f_0)\Delta(f - f_0) \, dx\, dy$$

$$= -\oint_\Gamma (f - f_0)\frac{\partial f_0}{\partial n} dt - \frac{1}{2}\oint_\Gamma (f - f_0)\frac{\partial(f - f_0)}{\partial n} dt$$

$$+ \iint_G (\nabla(f - f_0), \nabla f_0) \, dx \, dy$$

$$+ \tfrac{1}{2} \iint_G (\nabla(f - f_0))^2 \, dx \, dy$$

$$= \tfrac{1}{2} \iint_G (\nabla f)^2 \, dx \, dy - \tfrac{1}{2} \iint_G (\nabla f_0)^2 \, dx \, dy$$

$$\rho_{\min} \sqrt{c} \left[\Phi_2(f) - \Phi_2(f_0) \right]$$

$$= - \int_0^1 dt \iint_G (f - f_0) \frac{V(y) B_x(x)}{2 \sqrt{\displaystyle \int_{L(f_0 + t(f - f_0))} VB \, dy}} \, dx \, dy$$

$$= - \int_0^1 dt \int_0^{f_+} \frac{d(f_0 + t(f - f_0))}{2 \sqrt{\displaystyle \int_{L(f_0 + t(f - f_0))} VB \, dy}}$$

$$\times \int_{L(f_0 + t(f - f_0))} \frac{(f - f_0)_M V(y_M) B_x(x_M)}{(f_0 + t(f - f_0))_{n_M}} \, dt_M$$

$$+ \int_0^1 dt \int_0^{f_+} \frac{d(f_0 + t(f - f_0))}{2 \sqrt{\displaystyle \int_{L(f_0 + t(f - f_0))} VB \, dy}}$$

$$\times \int_{L(f + t(f - f_0))} \frac{(f - f_0)_P V(y_M) B_x(x_M)}{(f_0 + t(f - f_0))_{n_M}} \, dt_M$$

$$- \int_0^1 dt \int_0^{f_+} (f - f_0)_P \frac{d(f_0 + t(f - f_0))}{2 \sqrt{\displaystyle \int_{L(f_0 + t(f - f_0))} VB \, dy}}$$

$$\times \int_{L(f_0 + t(f - f_0))} \frac{V(y_M) B_x(x_M)}{(f_0 + t(f - f_0)_{n_M}} \, dt_M$$

$$= \int_0^1 dt \int_0^{f_+} d(f_0 + t(f - f_0)) \frac{\partial}{\partial t} \sqrt{\int_{L(f_0 + t(f - f_0))}} VB \, dy$$

$$- \int_0^1 dt \int_0^{f_+} (f - f_0)_P \, d \sqrt{\int_{L(f_0 + t(f - f_0))}} VB \, dy$$

This transformation was already used in the derivation of (3.155). Note that in the last integral we can omit the symbol P appended to the function $f - f_0$. We then have

$$\int_0^1 dt \int_0^{f_+} d(f_0 + t(f - f_0)) \frac{\partial}{\partial t} \sqrt{\int_{L(f_0 + t(f - f_0))}} VB \, dy$$

$$= \int_0^1 dt \frac{\partial}{\partial t} \int_0^{f_+} d(f_0 + t(f - f_0)) \sqrt{\int_{L(f_0 + t(f - f_0))}} VB \, dy$$

$$- \int_0^1 dt \int_0^{f_+} \sqrt{\int_{L(f_0 + t(f - f_0))}} VB \, dy \, d(f - f_0)$$

$$= \int_0^{f_+} \sqrt{\int_{L(f)}} VB \, dy \, df - \int_0^{f_+} \sqrt{\int_{L(f_0)}} VB \, dy \, df_0$$

$$+ \int_0^1 dt \int_0^{f_+} (f - f_0) \, d \sqrt{\int_{L(f_0 + t(f - f_0))}} VB \, dy$$

Substituting this into the preceding equation and using the expression obtained earlier for $\Phi_1(f)$, we find

$$\Phi(f) = \tfrac{1}{2} \iint_G (\Delta f)^2 \, dx \, dy - \frac{1}{\rho_{\min}\sqrt{c}} \int_0^{f_+} \sqrt{\int_{L(f)}} VB \, dy \, df \quad (3.156)$$

The expression given earlier [prior to (3.153)] for $\partial \Phi / \partial f_+$ follows from (3.156). In (3.156) we make the substitution (3.151) eliminating f_+ from (3.156) by means of (3.153), we obtain

$$\Phi(f_+u) = \Phi_1(u) = -\frac{1}{4\rho_{\min}\sqrt{c}} I_1(u) \quad (3.157)$$

This equation shows that a minimum of $\Phi_1(u)$ corresponds to a maximum of the functional $I_1(u)$.

We now consider the existence of a solution to Problem B, retaining our previously made assumption that the broken line $BCCB$ (see Fig. 3.19) is not critical (i.e., the abscissa x_c of the vertical segment CC is less than x_2). The operator DM_f defined by the left-hand side of (3.142) is the derivative of the operator M_f. We have already shown that the operator DM_f is symmetric; let us now consider the question of its positivity. We have

$$\iint_G u\, DM_f u\, dx\, dy$$

$$= \iint_G (\text{grad } u)^2\, dx\, dy - \frac{1}{4c^2 \rho_{\min}} \int_0^{f_+} \frac{df}{\left[\dfrac{1}{c} \displaystyle\int_{L(f)} VB\, dy\right]^{3/2}}$$

$$\times \left\{ \int_{L(f)} \frac{V(y_M)B_x(x_M)}{|f_n|_M} u_M^2\, dt_M \int_{L(f)} \frac{V(y_M)B_x(x_M)}{|f_n|_M}\, dt_M \right.$$

$$\left. - \left[\int_{L(f)} \frac{V(y_M)B_x(x_M)}{|f_n|_M} u_M\, dt_M \right]^2 \right\} \tag{3.158}$$

The use of Schwarz's inequality, for the function $F(t)$ of constant sign, implies

$$\int Fu^2\, dt \times \int F\, dt - \left[\int Fu\, dt \right]^2 \geq 0$$

the expression in the braces in (3.158) is nonnegative; the right-hand side of (3.158) is the difference of two nonnegative numbers. Therefore, the operator DM_f is not in general positive; for fixed G it becomes positive if the functions $V(y)$ and $B(x)$ satisfy certain additional conditions. Suppose $V(y) > 0$; our earlier assumption that $x_2 > x_c$ (see Fig. 3.19) shows that Weierstrass's condition is satisfied. The sign of the expression (3.158) obviously depends on the nature of the decrease of the function $B(x)$ in the interval $(0, x_c)$. If $x_0 > x_c$, then the second term on the right-hand side in (3.158) is zero for all f, from which it follows that DM_f is positive. Moreover, on the set $d(DM_f)$ of functions satisfying the boundary conditions (3.143)-(3.145), this operator is also uniformly positive definite (for the proof of this assertion we follow the arguments in Ref. 110, pp. 340-342).

If $x_2 > x_c > x_0$, then to obtain conditions on $B(x)$ sufficient for the uniform positivity of the operator DM_f on the set $d(DM_f)$, we must construct an upper bound on the coefficient of $-(4c^2 \rho_{\min})^{-1}$ on the right-hand side of (3.158) which is uniform with respect to f. To obtain such a bound, we assume that

$$|f_n| \geq |f_n|_{\min} > 0 \tag{3.159}$$

Let $|B_x|_{max}$ be the maximum of the modulus of the derivative of the function $B(x)$ in the interval $[0, x_c]$, and let $V(y) = V = \text{const.}$ The coefficient of $-(4c^2\rho_{min})^{-1}$ in (3.158) is bounded above by

$$K_{max} \iint_G u^2 \, dx \, dy$$

where

$$K_{max} = \frac{2V^{1/2}|B_x|_{max}^2(x_c - \lambda + \delta)}{[2c^{-1}B(x_c)\delta]^{3/2}|f_n|_{min}} \tag{3.160}$$

Now let λ_1 be the first eigenvalue of the operator $-\Delta$ subject to the boundary conditions (3.143) along $BCCB$, conditions (3.114) along AA and $BCCB$, and conditions (3.145) along the electrodes (see Fig. 3.19). In accordance with the definition,

$$\iint_G (\text{grad } u)^2 \, dx \, dy \geq \lambda_1 \iint_G u^2 \, dx \, dy \tag{3.161}$$

Suppose

$$\frac{K_{max}}{4c^2\rho_{min}} < \lambda_1 \tag{3.162}$$

We then have

$$\iint_G (\text{grad } u)^2 \, dx \, dy - \frac{K_{max}}{4c^2\rho_{min}} \iint_G u^2 \, dx \, dy \geq \mu \iint_G u^2 \, dx \, dy$$

where

$$\mu = \lambda_1 - \frac{K_{max}}{4c^2\rho_{min}} > 0$$

Under the conditions imposed, the operator DM_f is positive definite, which is what we wanted to prove.

Now, using some well-known theorems (Ref. 109, pp. 307 and 308) we can assert that if the inequality (3.161) holds, Problem B is equivalent to the problem of minimizing the functional $\Phi(f)$ [see (3.156)] on $d(M)$ or the problem of maximizing the functional $I(f)$ [see (3.136)] on the same set.

In the formulation of this result, we can replace the set $d(M)$ by the set $d(\Phi) = d(I)$ if we can show that the functional $d(\Phi)$ is upper semicontinuous (Ref. 109, p. 308) in a Banach space $\bar{W}_2^1(G)$ with respect to which $d(M)$ and $D(\Phi) = d(I)$ are everywhere dense sets.

As the space $\bar{W}_2^1(G)$, we take the closure in the norm

$$\|f\|_{\bar{W}_2^1(G)}^2 = \|\nabla f\|_{L_2(G)}^2 + \|f\|_{L_2(G)}^2 + \|f\|_{L_2(\Gamma)}^2$$

of the set of sufficiently smooth functions that are monotonic in the open region G and satisfy the boundary conditions (3.126). We prove that if the function $B(x) > 0$ is continuously differentiable and does not increase (see Fig. 3.19), then the functional $\Phi(f)$ is continuous in the space $\bar{W}_2^1(G)$, and continuity holds irrespective of whether or not the inequality (3.162) is satisfied.

We have

$$\Phi(f_k) - \Phi(f) = \frac{1}{2} \iint_G (\nabla f_k)^2 \, dx \, dy - \frac{1}{2} \iint_G (\nabla f)^2 \, dx \, dy$$

$$+ \frac{1}{2\rho_{\min}\sqrt{c}} \int_0^1 dt \iint_G (f - f_k) \frac{V(y) B_x(x)}{\sqrt{\displaystyle\int_{L(f+t(f_k-f))} VB \, dy}} \, dx \, dy$$

$$\tag{3.163}$$

We estimate the first two terms on the right by means of the Cauchy–Schwarz inequality:

$$\iint_G [(\nabla f_k)^2 - (\nabla f)^2] \, dx \, dy$$

$$\leq \left\{ \iint_G [\nabla(f_k - f)]^2 \, dx \, dy \iint_G [\nabla(f_k + f)]^2 \, dx \, dy \right\}^{1/2}$$

whose right-hand side vanishes for $\|f_k - f\|_{\bar{W}_2^1(G)} \to 0$ since the norm $\|f_k + f\|_{\bar{W}_2^1(G)}$ is then bounded.

We now consider the inner integral in the last term on the right-hand side of (3.163). The function

$$\frac{V(y) B_x(x)}{\sqrt{\displaystyle\int_{L(f+t(f_k-f))} VB \, dy}}$$

belongs to $L_2(G)$ since the family $L(f + t(f_k - f))$ does not contain closed contour lines. Therefore, the last term on the right-hand side of (3.163) is a functional which is continuous in $L_2(G)$, and as a consequence of the preceding arguments it follows that $\Phi(f)$ is continuous in $\bar{W}_2^1(G)$.

Thus, if the inequality (3.163) is satisfied [or some other condition which guarantees uniform positivity of the operator DM_f with respect to $f \in d(M)$], Problem B is equivalent to the problem of maximizing the functional $I(f)$ on the set $d(I)$. We show that if the condition (3.162) is satisfied, the functional $\Phi(f)$ is essentially convex in the space $\bar{W}_2^1(G)$ for all f in $\bar{W}_2^1(G)$.

Consider the gradient M of the functional Φ. We have

$$(M, \varphi) = (\text{grad } \Phi(f), \varphi) = \frac{d}{dt} \Phi(f + t\varphi)\big|_{t=0}$$

$$= \iint_G (\nabla f, \nabla \varphi) \, dx \, dy - \frac{1}{\rho_{min}\sqrt{c}} \varphi \bigg|_{BCCB} \sqrt{\int_{L(f_+)} VB \, dy}$$

$$- \frac{1}{\rho_{min}\sqrt{c}} \int_0^{f_+} \frac{df}{2\sqrt{\int_{L(f)} VB \, dy}}$$

$$\times \int_{L(f)} \frac{V(y_M) B_x(x_M)}{(f_n)_M} \varphi(x_M, y_M) \, dt_M \qquad (3.164)$$

Arguments similar to those above show that the right-hand side of (3.164) is a functional over φ which is bounded in $\bar{W}_2^1(G)$ for all f in $\bar{W}_2^1(G)$. As we proved earlier, the derivative

$$\frac{d}{df} \text{grad } \Phi = DM_f$$

is a linear operator in $L_2(G)$.

By arguments similar to those given above with regard to this operator, we can show that the expression

$$(DM_f\varphi, \varphi_1) = \frac{d}{dt} (M(f + t\varphi), \varphi_1)\big|_{t=0}$$

is a bilinear functional over φ and φ_1 which is bounded in $\bar{W}_2^1(G)$ for any $f \in \bar{W}_2^1(G)$. If the inequality (3.162) is satisfied, then the derivative DM_f is uniformly positively bounded below; on the basis of a well-known theorem (Ref. 109, p. 310), we then conclude that the functional $\Phi(f)$ is increasing on $\bar{W}_2^1(G)$.

The proof follows from the following system of relations [see (3.143), (3.160)]:

$$\Phi(f) - \Phi(f_0) = \int_0^1 \int_0^1 t \, dt \, d\tau \int \int_G (f - f_0) DM_\xi (f - f_0) \, dx \, dy$$

$$\geq \frac{1}{2} \left[\|\nabla (f - f_0)\|_{L_2(G)}^2 - \frac{K_{max}}{4c^2 \rho_{min}} \|f - f_0\|_{L_2(G)}^2 + \alpha \|f - f_0\|_{L_2(\Gamma)}^2 \right]$$

$$(\alpha > 0, \ \xi = f_0 + t(f_1 - f_0))$$

From (3.161) and (3.162) it follows that $\Phi(f) \to \infty$ when $\|f\|_{\bar{W}_2^1(G)} \to \infty$. On the other hand, by virtue of the continuity of

$$\Phi(f) - \Phi(f_0) \leq \beta \|f - f_0\|_{\bar{W}_2^1(G)}^2, \qquad \beta > 0$$

it follows that the functional $\Phi(f)$ is bounded if the norm of the function f is bounded. Our results enable us to assert that under the conditions (3.160) and (3.162) the functional $\Phi(f)$ attains a lower bound in the space $\bar{W}_2^1(G)$ at the unique point f. Furthermore, any minimizing sequence converges weakly to f (Ref. 109, p. 312). The conditions (3.160) and (3.162) together with (3.159) reduce to the inequality (see Fig. 3.19)

$$\frac{(x_c - \lambda + \delta) V^{1/2} |B_x|_{max}^2}{2c^2 \rho_{min} [c^{-1} B(x_c) 2\delta]^{3/2} |f_n|_{min}} < \lambda_1 \tag{3.165}$$

For a fixed region $ACCA$, this inequality provides a lower bound for the minimum of the modulus of the gradient of f, thus defining a certain subset S_f of $d(M)$:

$$S_f : |f_n|_{min} > \frac{(x_c - \lambda + \delta) V^{1/2} |B_x|_{max}^2}{2c^2 \rho_{min} [c^{-1} B(x_c) 2\delta]^{3/2} \lambda_1} \tag{3.166}$$

The inequality (3.165) [or (3.166)] expresses a condition which is sufficient for the functional $I(f)$ to attain an upper bound on the corresponding uniqueness set S_f. In order to obtain more precise bounds on the uniqueness set, we consider the operators p and $Q(f)$ defined by the equations

$$P_u = -\Delta u$$

$$\left.\begin{array}{l} Q(f)u = \dfrac{V(y)B_x(x)}{4c^2\rho_{\min}\left[c^{-1}\displaystyle\int_{L(f)} VB\,dy\right]^{3/2}}\left[u\displaystyle\int_{L(f)}\dfrac{V(y_M)B_x(x_M)}{(f_n)_M}\,dt_M\right. \\[20pt] \left.-\displaystyle\int_{L(f)}\dfrac{V(y_M)B_x(x_M)}{(f_n)_M}u(x_M,y_M)\,dt_M\right] \end{array}\right\} \quad (3.167)$$

Obviously, $DM_ju = Pu - Q(f)u$. We shall consider the operator P on smooth functions satisfying the boundary conditions (3.143) along $BCCB$, (3.144) along AA and $BCCB$, and (3.145) along the electrodes (see Fig. 3.19), and we shall consider the operator $Q(f)$ on functions in $L_2(G)$ satisfying the conditions (3.144) along AA.

If $f_n > 0$ everywhere in G and the function $B(x)$ decreases monotonically and is positive in G, then both operators P and $Q(f)$ are positive definite and $d(P) \subset d(Q(f))$; in addition, every set of functions which is bounded in the energy norm of the operator P is compact in $L_2(G)$, and it therefore is bounded as well in the energy norm of the operator $Q(f)$. Therefore,[108] there exists an infinite set of eigenvalues

$$0 < \mu_1 \le \mu_2 \le \cdots \le \mu_n \le \cdots$$

of the equation

$$Pu - \mu Q(f)u = 0 \qquad (3.168)$$

The smallest eigenvalue μ_1 is determined by the equation

$$\mu_1 = \min_u \frac{(Pu, u)}{(Q(f)u, u)} \qquad (3.169)$$

The inequality (3.165) is now replaced by the condition

$$\mu_1 > 1 \qquad (3.170)$$

where

$$\mu_1 = 4c^2\rho_{\min}\left[\int\int_G (\nabla u)^2\,dx\,dy\right]\left\{\int_0^{f_+}\frac{df}{\left(c^{-1}\displaystyle\int_{L(f)} VB\,dy\right)^{3/2}}\right.$$

$$\times\left[\int_{L(f)}\frac{V(y_M)B_x(x_M)}{|f_n|_M}u_M^2\,dt_M\int_{L(f)}\frac{V(y_M)B_x(x_M)}{|f_n|_M}\,dt_M\right.$$

$$-\left[\int_{L(f)} \frac{V(y_M)B_x(x_M)}{|f_n|_M} u_M \, dt_M\right]^2\right]\right\}^{-1} \geq \mu_{1*}$$

$$= 4c^2\rho_{\min}\left[\iint_G (\nabla u)^2 \, dx \, dy\right]\left\{\int_0^{f_+} \frac{df}{\left(c^{-1}\int_{L(f)} VB \, dy\right)^{3/2}}\right.$$

$$\times \int_{L(f)} \frac{V(y_M)B_x(x_M)}{|f_n|_M} u_M^2 \, dt_M \int_{L(f)} \frac{V(y_M)B_x(x_M)}{|f_n|_M} \, dt_M\right\}^{-1}$$

$$= 4c^2\rho_{\min}\left[\iint_G (\nabla u)^2 \, dx \, dy\right]\left\{\iint_G \left(\int_{L(f)} \frac{V(y_M)B_x(x_M)}{|f_n|_M} \, dt_M\right)\right.$$

$$\times \frac{V(y_P)B_x(x_P)}{\left(c^{-1}\int_{L(f)} VB \, dy\right)^{3/2}} u_P^2 \, dx_P \, dy_P\right\}^{-1} \tag{3.171}$$

For the satisfaction of the inequality (3.170), it is sufficient that

$$\min_u \mu_{1*} > 1 \tag{3.172}$$

This is also a necessary condition since one can always restrict the set of admissible functions $u(x, y)$ to functions for which

$$\int_{L(f)} \frac{V(y_M)B_x(x_M)}{|f_n|_M} u_M \, dt_M = 0$$

for all $f \in (0, f_+)$. Using the assumptions made above about the nature of the functions $B(x)$ and $V(y)$, the inequality (3.172) yields a condition that determines the uniqueness set S_f. This condition has the form

$$\min_f \min_u \mu_{1*} \geq 4c^2\rho_{\min}\left(\max_f\left\{\frac{\int_{L(f)} \frac{V(y_M)|B_x(x_M)|}{|f_n|_M} \, dt_M}{\left(\frac{1}{c}\int_{L(f)} VB \, dy\right)^{3/2}}\right\}\right)^{-1}$$

$$\times \min_u \frac{\iint_G (\nabla u)^2 \, dx \, dy}{\iint_G V(y)|B_x(x)|u^2 \, dx \, dy} > 1$$

or

$$\max_f T(f) < \pi_1, \qquad T(f) = \frac{\displaystyle\int_{L(f)} \frac{V(y_M)|B_x(x_M)|}{|f_n|_M} dt_M}{\left(\dfrac{1}{c}\displaystyle\int_{L(f)} VB \, dy\right)^{3/2}} \qquad (3.173)$$

where π_1 is the smallest eigenvalue of the equation

$$\Delta u + \frac{\pi}{4c^2 \rho_{\min}} V(y)|B_x(x)|u = 0 \qquad (3.174)$$

subject to the boundary conditions (3.143)–(3.145).
   The functional T admits the estimates

$$\frac{\displaystyle\int_{L(f)} \frac{V(y_M)|B_x(x_M)|}{|f_n|_M} dt_M}{\left(\dfrac{1}{c} VB_{\max}2\delta\right)^{3/2}} \le T(f) \le \frac{\displaystyle\int_{L(f)} \frac{V(y_M)|B_x(x_M)|}{|f_n|_M} dt_M}{\left(\dfrac{1}{c} VB_{\min}2\delta\right)^{3/2}} \qquad (3.175)$$

whence

$$\frac{\displaystyle\int_{L(f)} \frac{V(y_M)|B_x(x_M)|}{|f_n|_M} dt_M}{\left(\dfrac{1}{c} VB_{\max}2\delta\right)^{3/2}} \le \frac{m(f)}{\left(\dfrac{1}{c} VB_{\max}2\delta\right)^{3/2}} \le \max_f T(f)$$

$$\le \frac{m(f)}{\left(\dfrac{1}{c} VB_{\min}2\delta\right)^{3/2}} \le \frac{|B_x|_{\max}2(x_c - \lambda + \delta)}{|f_n|_{\min}(c^{-1}B_{\min}2\delta)^{3/2} V^{1/2}}$$

where $m(f)$ denotes the maximum of the functional

$$\int_{L(f)} \frac{V(y_M)|B_x(x_M)|}{|f_n|_M} dt_M$$

The inequality (3.173) is satisfied if

$$\frac{|B_x|_{\max}2(x_c - \lambda + \delta)}{V^{1/2}|f_n|_{\min}(c^{-1}B_{\min}2\delta)^{3/2}} < \pi_1 \tag{3.176}$$

(3.173) is violated if the set of admissible functions includes a function f such that for any contour line $L(f)$ of this function one has

$$\int_{L(f)} \frac{V(y_M)|B_x(x_M)|}{|f_n|_M} dt_M > \pi_1\left(\frac{1}{c} VB_{\max}2\delta\right)^{3/2} \tag{3.177}$$

The inequality (3.176) is analogous to (3.165); it expresses a condition which is sufficient for the existence of a unique minimum of the functional $\Phi(f)$. We write this inequality in the following final form:

$$|f_n|_{\min} > \frac{|B_x|_{\max}2(x_c - \lambda + \delta)}{V^{1/2}(c^{-1}B_{\min}2\delta)^{3/2}\pi_1} \tag{3.178}$$

The inequality (3.177) expresses a restriction on the set $d(M)$ of functions f. The condition is sufficient for the derivative of the operator DM_f not to be positively uniformly bounded below with respect to f. For the contour line $L(f = \tau)$ the integration of the inequality (3.177) with respect to τ from 0 to f_+ yields

$$\iint_G V(y)|B_x(x)| \, dx \, dy > \pi_1 f_+\left(\frac{1}{c} VB_{\max}2\delta\right)^{3/2}$$

or, finally,

$$f_+ < \frac{\displaystyle\iint_G V(y)|B_x(x)| \, dx \, dy}{\pi_1(c^{-1}VB_{\max}2\delta)^{3/2}} \tag{3.179}$$

Suppose that the subset S_f of functions f in $d(M)$ is such that for given channel parameters $|B_x|_{\max}$ and $V = $ const the inequality (3.178) is satisfied; then, we have now shown that this subset contains a unique function $f^{(0)}$

which minimizes the functional $\Phi(f)$. For functions in the set S_f,

$$2x_c\delta = \int\int_G \frac{df\,dt}{|f_n|} \le \int\int_G \frac{df\,dt}{|f_n|_{\min}} \le \pi_1 \frac{f_+ V^{1/2}(c^{-1}B_{\min}2\delta)^{3/2}}{|B_x|_{\max}}$$

or

$$f_+ \ge \frac{2x_c\delta|B_x|_{\max}}{\pi_1(c^{-1}B_{\min}2\delta)^{3/2}V^{1/2}} \tag{3.180}$$

We fix the set S_f and increase the channel length to $2x_c$ (Fig. 3.19). The parameter τ_1 then decreases continuously,* and the right-hand sides of the inequalities (3.178) and (3.180) increase. From some x_c on, both (3.178) and (3.180) are violated for certain elements of S_f; at even larger x_c, the inequality (3.179) will be satisfied. Thus, to obtain conditions which guarantee the existence of a unique minimum of the functional $\Phi(f)$, we must modify the set S_f. We do this by increasing the lower limit of the values $|f_n|_{\min}$ (or f_+) of functions $f \in S_f$ to a value corresponding to the new (larger) value of x_c. The inequalities (3.178) and (3.180) show that the admissible values of $|f_n|_{\min}$ and f_+ are increased for the member functions of the new set S_f; it therefore is a part of the original set S_f. The minimal element $f^{(0)}$ varies continuously† in this process of restriction of the sets until one arrives at a set $S_{f(0)}$. At the corresponding value of x_c the first eigenvalue $\mu_1^{(0)}$ of (3.168) is attained; to achieve this, it is necessary to set $f = f^{(0)}$. At larger values of x_c there exists a new minimal element $f^{(1)}$; obviously, $S_{f^{(1)}} \subset S_{f^{(0)}}$, and, in addition, $f_+^{(1)} > f_+^{(0)}$. Hence, when the last footnote is taken into account, we may conclude that $I(f^{(1)}) \ge I(f^{(0)})$, so that the functional I does not decrease with increasing x_c.

If x_c is further increased, we may reach the first eigenvalue $\mu_1^{(1)}$ of (3.168), in which we must set $f = f^{(1)}$, etc.

The process of replacement of the minimal elements as the parameter x_c increases may be restricted by two circumstances. If the graph of the function $B(x)$ has the form shown in Fig. 3.19, then the Weierstrass condition requires that $x_c < x_2$. For $x_c = x_2$, we arrive at a singular case, which will be considered separately. If the limit as $x \to \infty$ of the decreasing function $B(x)$ tends asymptotically to zero or to some positive constant, then the rate of decrease of the parameter τ_1 may be insufficient,‡ and the inequalities

---

* This follows from the general theorems[86] on the continuous dependence of eigenvalues on the shape of the region.

† One can show, at the same time, that the maximum of the functional $I(f)$ does not decrease; $f_+^{(0)}$ does not decrease either since the maximum of $I(f)$ increases monotonically with $f_+^{(0)}$.

‡ This rate is determined by the rate of decrease of $B(x)$ as $x \to \infty$.

(3.178) and (3.180), corresponding to some minimal solution $f^{(i)}$, thus may not be violated. In such a case, the process of replacement of the minimal solutions does not go beyond step i. In particular, if $i = 0$, then the problem has a unique minimal solution for all $x_c > 0$.

3.12. Jacobi's Necessary Condition

Application of this condition (see Section 2.9) to the problem under consideration enables us to illustrate the process of replacement of minimal solutions described in the previous section. To this end, consider (3.103) which describe the distribution of the current and of the potential in the channel in the asymptotic case $\rho_{max} = \infty$ (see Section 3.8). Regarding the function $u = \tan \gamma$ as a control, we write (3.103) in the normal form (Section 2.1):

$$z_x^1 = \zeta^1$$

$$z_y^1 = u\zeta^1 - \rho_{min}(u^2 + 1)\zeta^2 + c^{-1}VB$$

$$z_x^2 = \zeta^2$$ (3.181)

$$z_y^2 = u\zeta^2$$

(The symbols ζ^1 and ζ^2 should not be confused with the same symbols used earlier for the Cartesian components of the vector $\mathbf{j}$; here, they have a different meaning.) In the usual manner, we can show that the increment δu of the control, compared with its optimal value, changes the functional $-I$ [see (3.12)] by the amount

$$\delta(-I) = -\iint_G \delta H \, dx \, dy$$ (3.182)

where

$$\delta H = -\frac{\delta u}{\zeta^2} \eta_1 [\rho_{min}(\zeta^2)^2 \delta u - \zeta^2 \delta\zeta^1 + \zeta^1 \delta\zeta^2] - \rho_{min}\eta_1(\delta u)^2 \delta\zeta^2$$ (3.183)

$$\eta_1 = \omega_{1x}, \qquad \zeta^2 = z_x^2 = h'(\omega_1)\omega_{1x}$$

The function ω_1 is a solution of the problem (3.111)-(3.114), and $\delta\zeta^1$ and $\delta\zeta^2$ denote the variations of ζ^1 and ζ^2 due to the variation δu of the control.*

Equations (3.182) and (3.183) are exact; unfortunately, in the general case, one cannot determine the sign of the increment $\delta(-I)$ since virtually nothing can be said about the variations $\delta\zeta^1$ and $\delta\zeta^2$ for any general variation δu. Information about these variations can be obtained by specializing δu in different ways, each of which leads to a different necessary condition for a minimum. Thus, assuming that δu is a strong variation concentrated within a narrow strip on the xy plane (local variation), we arrive at the inequality (3.119) as a necessary condition for a minimum† of $-I$ (Weierstrass's condition).

Another way of specializing δu is to assume that this variation is weak, i.e., small in absolute value. In the general case, δu is nonzero in all of the basic region. The variations $\delta\zeta^1$ and $\delta\zeta^2$ are quantities of the same order of magnitude as δu; they satisfy the equations [see (2.95)]

---

* Making use of (3.181) it is easy to prove that the optimal value of concentration m of material with $\rho_{\max} = \infty$ should be zero (see p. 122). Assuming the contrary and recalling the remark at the end of Section 3.4, we can write $\rho_2 = \rho_{\min}/(1 - m)$ instead of $\rho_{\min}$ in (3.103) and, consequently, in (3.181). If we now consider m as an additional control, then the expression (3.183) for δH will include the term

$$-\frac{\rho_{\min}\,\delta m}{(1 - m)^2}\,\eta_1\zeta^2 = -\frac{\rho_{\min}\,\delta m}{(1 - m)^2}\,h'(\omega_1)\omega_{1x}^2$$

which should be nonpositive; bearing in mind that $h' \geq 0$, we obtain $\delta m > 0$ and, consequently, $m = 0$.

† To see this, we note that under the assumption we have made, the integral (3.182) is taken only over the area of the strip, within which the variations $\delta\zeta^1$ and $\delta\zeta^2$ are related to δu by

$$\delta\zeta^1 = -\frac{\delta u}{[x_t + (u + \delta u)y_t]^2}\,y_t\{[x_t + (u + \delta u)y_t] + \rho_{\min}\zeta^2[y_t - 2(u + \delta u)x_t - u(u + \delta u)y_t]\}$$

$$\delta\zeta^2 = -\frac{\delta u}{x_t + (u + \delta u)y_t}\,y_t\zeta^2$$

which express the continuity of the variables z^1 and z^2 on the boundary of the strip.

These equations may now be used to eliminate the variations $\delta\zeta^1$ and $\delta\zeta^2$ from the expression (3.183) for δH; after some simple calculations we obtain

$$\delta H = -\frac{\rho_{\min}h'(\omega_1)\omega_{1x}^2(\delta u)^2}{[x_t + (u + \delta u)y_t]^2}$$

The requirement $\delta H \leq 0$ implies $h'(\omega_1) \geq 0$.

$$\delta z_x^1 = \delta \zeta^1$$

$$\delta z_y^1 = u\delta\zeta^1 - \rho_{min}(u^2 + 1)\delta\zeta^2 + (\zeta^1 - 2\rho_{min}u\zeta^2)\delta u$$

$$\delta z_x^2 = \delta\zeta^2$$ (3.184)

$$\delta z_y^2 = u\delta\zeta^2 + \zeta^2\delta u$$

which are obtained by varying (3.181).

On the other hand, the exact expression $\delta(-I)$ for the increment of the functional differs by a higher-order term from the second variation

$$\frac{1}{2}\delta^2(-I) = \int\int_G \frac{\eta_1}{\zeta^2}\,\delta u[\,\rho_{min}(\zeta^2)^2\delta u - \zeta^2\delta\zeta^1 + \zeta^1\delta\zeta^2]\,dx\,dy \quad (3.185)$$

of this functional. If the functional $-I$ is to attain a minimum, its second variation must be nonnegative; this is Jacobi's necessary condition.

We arrive at the auxiliary problem of minimizing the second variation (see Section 2.9), namely, to find a minimum of the functional (3.185) subject to the constraints (3.184) and the boundary conditions (see Fig. 3.19): on the electrodes

$$\delta z^1(x, \pm\delta) = \pm\delta z_+^1$$

on the insulators

$$\delta z^2|_{AA} = 0, \qquad \delta z^2|_{BCCB} = \delta z_+^2$$
                                                                        (3.186)
$$\delta z_+^1 = R\delta z_+^2$$

which are obtained by varying the boundary conditions (2.5) and with consideration of the symmetry of the problem with respect to the y axis.

The Euler equations of this problem are constructed in the usual manner by making use of Lagrange multipliers $\delta\xi_1, \delta\eta_1, \delta\xi_2, \delta\eta_2$, corresponding to (3.184). We obtain

$$\eta_1\delta u + \delta\xi_1 + u\delta\eta_1 = 0$$

$$-\frac{\eta_1}{\zeta^2}\zeta^1\delta u - \rho_{min}(u^2 + 1)\delta\eta_1 + \delta\xi_2 + u\delta\eta_2 = 0$$

$$2\rho_{min}\eta_1\zeta^2\delta u + \frac{\eta_1}{\zeta^2}(\zeta^1\delta\zeta^2 - \zeta^2\delta\zeta^1)$$ (3.187)

$$- (\zeta^1 - 2\rho_{min}u\zeta^2)\delta\eta_1 - \zeta^2\delta\eta_2 = 0$$

$$\delta\xi_i = -\delta\omega_{iy}, \qquad \delta\eta_i = \delta\omega_{ix} \qquad (i = 1, 2)$$

If we set

$$\delta z^2 = h_1(\omega_1) + h'(\omega_1)\delta\omega_1$$

where the function $h'(\omega_1)$ is defined by (3.118), and $h_1(\omega_1) = \delta h(\omega_1)$, then the variation of the function $h(\omega_1)$ is a small function of its argument [the derivative $h_1'(\omega_1)$ is equal to the variation of the right-hand side of (3.128)], and we can readily show that (3.184) and (3.187) reduce to

$$2(h_1' + h''\delta\omega_1)\Delta\omega_1 + 2h'\Delta\delta\omega_1 + 2h''(\nabla\omega_1, \nabla\delta\omega_1) + (h_1'' + h''\delta\omega_1)(\nabla\omega_1)^2 = 0$$

$$(3.188)$$

As one would expect, this equation is a variational equation based on (3.111); the corresponding boundary conditions yield similar variational equations. It is readily seen that the resulting boundary-value problem is equivalent to the boundary-value problem (3.143)-(3.145) for (3.142) which is a variational equation based on (3.133), i.e., a Jacobi equation.

We investigate the boundary-value problem (3.142)-(3.145) for the special case when the horizontal sections of the insulators BC in Fig. 3.19 are absent, and the region is the rectangle bounded by the electrodes AC and the insulators AA and CC (see Fig. 3.18). If $B(x) > 0$ for all $x \in [0, \lambda]$, and $V = \text{const}$, then the function

$$f_0(x) = \frac{1}{\rho_{\min}} \sqrt{\frac{V}{2c\delta}} \int_0^x \sqrt{B(x)}\, dx \qquad (3.189)$$

satisfies (3.133), along with the conditions (3.126), (3.128), and (3.134); at the same time, the constant f_+ has the value

$$f_+ = \frac{1}{\rho_{\min}} \sqrt{\frac{V}{2c\delta}} \int_0^\lambda \sqrt{B(x)}\, dx$$

and the lines $f = \text{const}$ are vertical straight lines joining the electrodes. Weierstrass's condition $[f_{0x}(x)]^2 \geq 0$ obviously is also satisfied.

The Jacobi equation (3.142) takes the form

$$\Delta\delta f_P + \frac{[B_x(x_P)]^2}{4B^2(x_P)}\left(\delta f_P - \frac{1}{2\delta}\int_{-\delta}^{\delta} \delta f_M\, dy_M\right) = 0 \qquad (3.190)$$

and the boundary condition (3.143) can be written as

$$\int_{-\delta}^{\delta} (\delta f)_x \, dy = 0 \tag{3.191}$$

The integrals in the last two equations are performed for fixed x, i.e., along the contour lines of f_0; in (3.191) any one of these contour lines may be used. We seek the solution δf_P in the form of the series

$$\delta f_P = \sum_{n=0}^{\infty} a_n(x_P) \cos \frac{n\pi}{\delta} y_P \tag{3.192}$$

this expression satisfies the conditions (3.145) on the electrodes. Substitution in (3.190) leads to the following equations for the coefficients (we omit the symbol P):

$$\frac{d^2 a_n}{dx^2} - \left(\frac{n\pi}{\delta}\right)^2 a_n + \frac{[B_x(x)]^2}{4B^2(x)} a_n = 0 \qquad (n = 1, 2 \ldots) \tag{3.193}$$

$$\frac{d^2 a_0}{dx^2} = 0 \tag{3.194}$$

From (3.144) and (3.191) we obtain the boundary conditions

$$a_n(0) = 0 \qquad (n = 0, 1, \ldots)$$

$$a_0(\lambda) = \delta f_+, \qquad a_n(\lambda) = 0 \qquad (n = 1, 2, \ldots) \tag{3.195}$$

$$a_0'(0) = 0$$

The imposed conditions are satisfied only by the trivial solution of (3.194) for which $\delta f_+ = 0$; with regard to (3.193), some of them may have nontrivial solutions in the general case. The existence of such solutions depends on the function $B(x)$ and the number n.

Each of the operators p_n and q defined by the equations

$$p_n \varphi = \left[-\frac{d^2}{dx^2} + \left(\frac{n\pi}{\delta}\right)^2 \right] \varphi \qquad (n = 1, 2, \ldots)$$

$$q\varphi = \frac{[B_x(x)]^2}{4B^2(x)} \varphi$$

is separately positive definite on functions φ satisfying the conditions (3.195). Suppose that the positive function $B(x)$ is such that every set of functions with bounded energy norm of the operator p_n is compact with respect to the energy norm of the operator q, i.e., with respect to the norm

$$\|\varphi\|_q = \int_0^\lambda \frac{[B_x(x)]^2}{4B^2(x)} \varphi^2 \, dx$$

Then, as is well known, the equation $p_n\varphi - \mu q\varphi = 0$ has an infinite set of eigenvalues

$$0 < \mu_1^{(n)} \le \mu_2^{(n)} \le \cdots$$

At the same time, one obviously has

$$\mu_i^{(k)} < \mu_i^{(k+1)}$$

so that $\mu_1^{(1)}$ is the smallest eigenvalue.

If $\mu_1^{(1)} > 1$ then the problem (3.193)-(3.195) has only the trivial solution for all n, and the corresponding value of the functional (3.185) is zero. For $\mu_1^{(1)} = 1$, the first of Eqs. (3.193) has a nontrivial solution satisfying (3.195); if $\mu_1^{(1)} < 1$, then one can find a function δf such that the corresponding value of the functional (3.185) is negative. Indeed, under the condition $\mu_1^{(1)} < 1$ there exists a nonzero function $a_{1*}(x)$ which satisfies (3.193) $(n = 1)$ in the interval $(0, l_1)$, $l_1 > \lambda$, and the conditions $a_{1*}(0) = a_{1*}(l_1) = 0$ at the ends of this interval.

We set

$$a_1(x) = \begin{cases} a_{1*}(x) & 0 \le x \le l_1 \\ 0 & l_1 \le x \le \lambda \end{cases}$$

The function

$$\delta f = a_1(x) \cos \frac{\pi y}{\delta} \tag{3.196}$$

satisfies (3.193) for $x \in (0, l_1)$, and $x \in (l_1, \lambda)$; the corresponding value of the functional (3.185) is zero. This last assertion follows from the equation

$$\tfrac{1}{2}\delta^2(-I) = -\oint_\Gamma \delta\omega_{1t} \, \delta z^1 \, dt - \oint_\Gamma \delta\omega_{2t} \, \delta z^2 \, dt$$

in which the integration is performed around the boundary of the rectangle $Ac'c'A$ (see Fig. 3.18); here it is necessary to take (3.107), (3.125), (3.118), (3.187), (3.144), (3.145), and (3.191) into account.

If $\mu_1^{(1)} < 1$, then the zero value of the functional (3.185) is not a minimum, since the function (3.196) has a discontinuous derivative with respect to x. For, if this function is minimal, then on the discontinuity line $x = l_1$ the Weierstrass-Erdmann conditions must be satisfied, which here have the form

$$[\delta\omega_{1t}]_-^+ = 0, \qquad [\delta\omega_{2t}]_-^+ = 0$$

$$\left\{\frac{\eta_1}{\zeta^2}\,\delta u[\,\rho_{\min}(\zeta^2)^2\delta u - \zeta^2\delta\zeta^1 + \zeta^1\delta\zeta^2] + \delta\omega_{1t}\delta z_n^1 + \delta\omega_{2t}\delta z_n^2\right\}_-^+ = 0$$

The expressions in the brackets to the right of the discontinuity line (for $x = l_1 + 0$) are zero; their values on the left can be calculated by means of the equations

$$\zeta^1 = 0, \qquad \zeta^2 = \frac{V}{c\rho_{\min}}\,B(x), \qquad \eta_1 = \frac{1}{2\rho_{\min}\delta}, \qquad u = 0$$

$$\delta u = -\frac{2\rho_{\min}\delta}{\sqrt{\dfrac{2V\delta}{c}B(x)}}\,a_{1*}(x)\frac{\pi}{\delta}\sin\frac{\pi y}{\delta}$$

$$\delta\omega_{2y} = -\frac{\rho_{\min}}{\sqrt{\dfrac{2V\delta}{c}B(x)}}\left[a_{1*x}(x) - \frac{B_x(x)}{2B(x)}a_{1*}(x)\right]\cos\frac{\pi y}{\delta}$$

$$\delta\omega_{1y} = -\frac{1}{\sqrt{\dfrac{2V\delta}{c}B(x)}}\,a_{1*}(x)\frac{\pi}{\delta}\sin\frac{\pi y}{\delta}$$

which readily follow from the foregoing.

The Weierstrass-Erdmann condition now leads to the requirement $B(l_1) \neq 0$

$$a_{1*x}(l_1) = 0$$

which cannot be satisfied unless the function $a_{1*}(x)$ vanishes identically on the interval $(0, l_1)$. Therefore, there exists a function δf which results in a negative value of the functional (3.185). This shows that if the second variation (3.185) is to be nonnegative, the first eigenvalue of the boundary-value problem (3.193)–(3.195) [or, which is the same thing, that of problem (3.142)–(3.145)] must be greater than or equal to unity.

Note that the corresponding value of δu is not local (in contrast to variation in a strip). In addition, this variation is essentially two dimensional since (3.193) have a nontrivial solution; (3.194) has only the trivial solution. In other words, the one-dimensional solution (3.189) satisfies Jacobi's condition with respect to one-dimensional variations but need not satisfy it with respect to two-dimensional variations.

For given λ and δ, the value of $\mu_1^{(1)}$ depends on $B(x)$. For applications, an interesting case arises when $B(x)$ is given by a graph of the type shown in Fig. 3.19, and $x_0 < \lambda < x_2$. At the same time, Weierstrass's condition $B(x) \geq 0$ is satisfied; with regard to Jacobi's condition, its satisfaction depends on l, λ, δ and the nature of the decrease of the function $B(x)$ for $x > x_0$. Thus, Jacobi's condition is satisfied if $x_0 = 0$ and $B(x)$ decreases linearly; but if $x_0 = 0$ and $B = B_0 e^{-\gamma x}$, $\gamma > 0$, then Jacobi's condition is violated for

$$\gamma > \gamma_0 = \frac{2\pi}{\lambda} \sqrt{1 + \frac{\lambda^2}{\delta^2}}$$

In the first of these cases, the rate of decrease of $B(x)$ for $x > x_0 = 0$ is bounded by Weierstrass's condition: $B \geq 0$ for $x \in [0, \lambda]$. In the second case, Weierstrass's condition is always satisfied ($e^{-\gamma x} > 0$ for all x), and the largest possible γ ($= \gamma_0$) is determined by Jacobi's condition. One needs to find optimal solutions under conditions when the function $B(x)$ decreases faster than is permitted by the Weierstrass and Jacobi conditions.

If the function $B(x)$ is nowhere increasing and changes sign from plus to minus at a point $x_2 \leq \lambda$, then the solution (3.189) is optimal only under the condition $x_2 = \lambda$. In other words, an insulating wall must be placed where the nonincreasing function $B(x)$ vanishes (this conclusion is also true for a channel of the type shown in Fig. 3.19 with horizontal sections BC of insulating walls).

If Weierstrass's condition is satisfied and the smallest ($n = 1$) eigenvalue of the boundary-value problem (3.193)–(3.195) is smaller than unity, Jacobi's condition is not satisfied. In the case $B(x) = B_0 e^{-\gamma x}$ described above, this is the case when $\gamma > \gamma_0$. For $\gamma < \gamma_0$, the Euler equation (3.138) has the unique solution (3.189) which satisfies the conditions (3.126), (3.128), and (3.129). For $\gamma = \gamma_0$, there is a bifurcation of this solution, and for $\gamma > \gamma_0$ there exist, in general, several solutions of (3.133). We therefore face the problem of continuing the solution (3.189), which is optimal for $\gamma < \gamma_0$, with respect to the parameter γ through the value $\gamma = \gamma_0$ in such a way that the continued solution remains optimal, i.e., such that the necessary conditions of Weierstrass and Jacobi are satisfied.

We shall seek small solutions of this problem, i.e., local continuations of the solution (3.189) with respect to the parameter γ, assuming that the function $B(x)$ is equal to $B_0 e^{-\gamma x}$, $x > 0$, and that the difference $(\gamma/\gamma_0) - 1$ is sufficiently small.

We set

$$f = f_0 + \varphi, \qquad \gamma = \gamma_0 + k \tag{3.197}$$

and consider the problem (3.138), (3.126), (3.128), and (3.139), assuming that $f_0 = f_0(x)$ is given by (3.189), $\varphi(x, y)$ is an unknown function, and k is a small (compared with γ_0) positive parameter.

In the new variables, the problem can be written as (see Fig. 3.18):

$$\Lambda(\varphi, k) \equiv L(f_0 + \varphi, \gamma_0 + k) = 0 \tag{3.198}$$

$$\left.\begin{array}{c} \varphi\big|_{AA} = 0, \qquad \varphi\big|_{CC} = \varphi_+, \qquad \varphi_y\big|_{AC} = 0 \\[2mm] \displaystyle\int_{-\delta}^{\delta} \varphi_x(0, y)\, dy = 0 \end{array}\right\} \tag{3.199}$$

The operator (Gâteaux derivative of the operator $-\Lambda$)

$$C = -\Lambda_\varphi(0, 0) = -L_\varphi(f_0, \gamma_0)$$

is a Fredholm operator. We write (3.198) in the form

$$C\varphi = R(\varphi, k), \qquad R(\varphi, k) = \Lambda(\varphi, k) - \Lambda_\varphi(0, 0)\varphi \tag{3.200}$$

The homogeneous boundary-value problem (3.199) for the equation

$$C\varphi \equiv -\Lambda_\varphi(0, 0)\varphi = 0 \tag{3.201}$$

has the nontrivial solution

$$z = a(x) \cos \frac{\pi y}{\delta} \tag{3.202}$$

where the function $a(x)$ satisfies (3.193), $n = 1$, and the corresponding conditions (3.195). We shall assume that the solution (3.202) is normalized:

$$\int_0^\lambda \int_{-\delta}^\delta z^2(x, y) \, dx \, dy = 1$$

For the existence of a solution to the inhomogeneous boundary-value problem

$$\Lambda_\varphi(0, 0)\varphi = g(x, y)$$

under the conditions (3.199), it is necessary and sufficient that

$$\int_0^\lambda \int_{-\delta}^\delta gz \, dx \, dy = 0$$

since the problem (3.201), (3.199) is identical with its adjoint problem.

We shall seek small solutions of (3.200) by the Lyapunov–Schmidt method (see Ref. 25). We represent (3.200) in the form

$$\tilde{C}\varphi = C\varphi + \xi z = F_{01}k + \sum_{i+j\geq 2} F_{ij}\varphi^i k^j + \xi z$$

$$\xi = \int_0^\lambda \int_{-\delta}^\delta \varphi z \, dx \, dy \tag{3.203}$$

where

$$F_{ij} = \frac{1}{i!j!} \frac{\partial^{i+j} R(0, 0)}{\partial \varphi^i \partial k^j}$$

is a power-law operator of order $i + j$, and ξ is a coefficient which is to be determined.[25] At the same time, we bear in mind that the function $z = a(x) \cos(\pi y/\delta)$ is a simple zero of the operator C [see (3.201)].

We shall seek a small solution $\varphi(x, y)$ of (3.203) in the form of the series

$$\varphi = \sum_{i=1}^\infty \varphi_{i0}\xi^i + \sum_{i=0}^\infty \xi^i \sum_{j=1}^\infty \varphi_{ij}k^j \tag{3.204}$$

which converges[25] for sufficiently small $|\xi|$ and k. Substitution of this series in (3.203) leads to the relation (bifurcation equation)

$$\sum_{i=2}^\infty L_{i0}\xi^i + \sum_{i=0}^\infty \xi^i \sum_{j=1}^\infty L_{0j}k^j = 0 \tag{3.205}$$

whose coefficients L_{ij} are determined by

$$L_{ij} = \int_0^\lambda \int_{-\delta}^\delta \varphi_{ij} z \, dx \, dy \qquad (3.206)$$

The functions Φ_{ij} are determined by the recursive system

$$\begin{aligned}
\tilde{C}\varphi_{01} &= F_{01} \\
\tilde{C}\varphi_{02} &= F_{02} + 2F_{11}\varphi_{01} + F_{20}\varphi_{01}^2 \\
\tilde{C}\varphi_{03} &= F_{03} + 3F_{12}\varphi_{01} + 3F_{21}\varphi_{01}^2 + F_{30}\varphi_{01}^3 + 2F_{11}\varphi_{02} + 2F_{20}\varphi_{01}\varphi_{02} \\
\tilde{C}\varphi_{10} &= z \\
\tilde{C}\varphi_{20} &= F_{20}\varphi_{10}^2 \\
\tilde{C}\varphi_{30} &= 2F_{20}\varphi_{10}\varphi_{20} + F_{30}\varphi_{10}^3 \\
\tilde{C}\varphi_{11} &= 2F_{11}\varphi_{10} + 2F_{20}\varphi_{10}\varphi_{01} \\
\tilde{C}\varphi_{21} &= 2F_{11}\varphi_{20} + 2F_{20}\varphi_{01}\varphi_{02} + 3F_{21}\varphi_{10}^2 + 3F_{30}\varphi_{01}\varphi_{10}^2
\end{aligned} \right\} \quad (3.207)$$

under boundary conditions of the type (3.199). Finding these functions and substituting them in (3.206), we arrive at equations that determine L_{ij}. We have

$$\begin{aligned}
L_{01} &= (F_{01}, z) \\
L_{02} &= (F_{02} + 2F_{11}(\Gamma F_{01}) + F_{20}(\Gamma F_{01})^2, z) \\
L_{03} &= (F_{03} + 3F_{12}(\Gamma F_{01}) + 3F_{21}(\Gamma F_{01})^2 + F_{30}(\Gamma F_{01})^3, z) \\
&\quad + ([2F_{11} + 2F_{20}(\Gamma F_{01})] \\
&\qquad \times \Gamma[F_{02} + 2F_{11}(\Gamma F_{01}) + F_{20}(\Gamma F_{01})^2], z) \\
L_{20} &= (F_{20}z^2, z) \\
L_{30} &= (2F_{20}z(\Gamma F_{20}z^2) + F_{30}z^3, z) \\
L_{11} &= (2F_{11}z + 2F_{20}z(\Gamma F_{01}), z)
\end{aligned} \right\} \quad (3.208)$$

Here we have introduced the notation

$$(u, z) = \int_0^\lambda \int_{-\delta}^\delta uz \, dx \, dy, \qquad \Gamma = C^{-1}$$

As is shown in Appendix A.6,

$$F_{01} = \sqrt{\frac{VB_0}{2c\delta}} \frac{e^{-\gamma_0 x/2}}{2\rho_{\min}} \left(1 - \frac{\gamma_0 x}{2}\right)$$

$$F_{02} = -\sqrt{\frac{VB_0}{2c\delta}} \frac{e^{-\gamma_0 x/2}}{8\rho_{\min}} x\left(1 + \frac{\gamma_0 x}{2}\right)$$

$$F_{11}\varphi k = \frac{\gamma_0 k}{2}\left[\varphi - \frac{1}{2\delta}\int_{-\delta}^{\delta} \varphi_M(x, y_M)\, dy_M\right]$$

$$F_{20}uv = \frac{\gamma_0^2 \rho_{\min}\sqrt{c}}{8\sqrt{2\,VB_0\delta}} e^{\gamma_0 x/2}\left\{\int_{-\delta}^{\delta} [(u - u_M)(v_M)_x + (v - v_M)(u_M)_x]\, dy_M\right.$$

$$+ \frac{\gamma_0}{4}\int_{-\delta}^{\delta} (u - u_M)(v - v_M)\, dy_M$$

$$\left. - \frac{3}{8}\frac{\gamma_0}{\delta}\int_{-\delta}^{\delta} (u - u_M)\, dy_M \int_{-\delta}^{\delta} (v - v_M)\, dy_M\right\}$$

$$\frac{4\,VB_0}{\gamma_0^2 \rho_{\min}^2 c} e^{-\gamma_0 x} F_{30}u^3 = \int_{-\delta}^{\delta} (u - u_M)[(u_M)_x]^2\, dy_M$$

$$- \frac{1}{2}\int_{-\delta}^{\delta} (u - u_M)^2 (u_M)_{xx}\, dy_M$$

$$+ \frac{\gamma_0}{4}\int_{-\delta}^{\delta} (u - u_M)^2 (u_M)_x\, dy_M$$

$$- \frac{3}{4}\frac{\gamma_0}{\delta}\int_{-\delta}^{\delta} (u - u_M)\, dy_M \int_{-\delta}^{\delta} (u - u_M)(u_M)_x\, dy_M$$

$$- \frac{3}{16}\frac{\gamma_0^2}{\delta}\int_{-\delta}^{\delta} (u - u_M)\, dy_M \int_{-\delta}^{\delta} (u - u_M)^2\, dy_M$$

$$+ \frac{5}{8}\frac{\gamma_0^2}{4\delta^2}\left[\int_{-\delta}^{\delta} (u - u_M)\, dy_M\right]^3$$

$$u = u(x, y), \qquad u_M = u(x, y_M)$$

$$\left. \right\} (3.209)$$

Hence, taking into account (3.208) and (3.202), we conclude that $L_{0i} = 0$, $i = 1, 2, \ldots$. In addition, we can verify that $L_{20} = 0$.

Therefore, the bifurcation equation (3.205) leads to the following characteristic equation[25]:

$$L_{30}\xi^2 + L_{11}k = 0$$

This equation has the two roots $\xi_{1,2} = \pm\sqrt{-(L_{11}/L_{30})k}$; both roots are real since we can show that

$$\xi_{1,2} = \pm\frac{2\lambda^2}{\pi}\left\{\frac{(1/2)\pi\sqrt{1+t}+2}{(1+t)[S(t)+T(t)]}\frac{VB_0k}{c\rho_{\min}^2}\right\}^{1/2}, \qquad t = \frac{\lambda^2}{\delta^2}$$

$$S(t) = \frac{1}{(2+t)(5+t)\pi^2}\left\{\left[\frac{35}{8}\pi^2(1+t) - 1 + t + \frac{21}{64}(1-t^2) + \frac{5}{4}\pi^2(5+t)^2\right.\right.$$

$$\left.\left. + 8\pi^2\frac{5+t}{1+t}(2+t)\right](e^{2\pi\sqrt{1+t}} - 1) - 72\pi^2(2+t)(e^{\pi\sqrt{1+t}} - 1)\right\}$$

$$T(t) = \frac{1+t}{16(1-t)}\left\{-\frac{e^{2\pi\sqrt{1+t}}-1}{2+t} + \left(\frac{1}{5+t} + \frac{1}{1-t}\right)\frac{2}{5+t}\right.$$

$$\times\left(5t+1-3\frac{e^{\pi\sqrt{1+t}} - \cos\pi\sqrt{1-3t}}{\sin\pi\sqrt{1-3t}}\sqrt{(1-3t)(1+t)}\right)$$

$$\times(e^{\pi\sqrt{1+t}}\cos\pi\sqrt{1-3t} - 1) + \left(3\sqrt{(1-3t)(1+t)}\right.$$

$$\left.\left. + (5t+1)\frac{e^{\pi\sqrt{1+t}} - \cos\pi\sqrt{1-3t}}{\sin\pi\sqrt{1-3t}}\right)e^{\pi\sqrt{1+t}}\sin\pi\sqrt{1-3t}\right\}$$

The graph of the function $S(t) + T(t)$ is shown in Fig. 3.20; clearly, $S(t) + T(t) > 0$ for all $t \geq 0$. The small solutions of (3.138) have

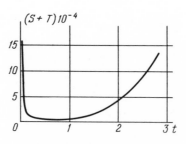

FIGURE 3.20

the form

$$
\begin{aligned}
f_{1,2} &= f_0 + \xi_{1,2}\sqrt{\frac{2}{\lambda\delta}}\,\sin\frac{\pi x}{\lambda}\cos\frac{\pi y}{\delta} \\
&= \frac{2}{\rho_{\min}}\sqrt{\frac{VB_0}{2c\delta}}\,\frac{1 - e^{-\gamma x/2}}{\gamma} \\
&\pm \frac{4\lambda}{\pi\rho_{\min}}\sqrt{\frac{VB_0}{2c\delta}}\left\{\frac{(1/2)\pi\sqrt{1+t}+2}{(1+t)[S(t)+T(t)]}\,k\lambda\right\}^{1/2}\sin\frac{\pi x}{\lambda}\cos\frac{\pi y}{\delta} \\
&= f_0 \pm \sqrt{k}\,\omega
\end{aligned}
\tag{3.210}
$$

Direct verification of fulfillment of Jacobi's condition is difficult; it is much simpler to calculate the increment of the total current due to the transition to the small solution (3.210).

The total current corresponding to the one-dimensional solution (3.189) is

$$
I = \frac{2VB_0}{c\rho_{\min}}\int_0^\lambda e^{-\gamma x}\,dx = \frac{2VB_0}{\gamma c\rho_{\min}}(1 - e^{-\gamma\lambda})
\tag{3.211}
$$

The component j_y of the vector $\mathbf{j}$ is given by ($V = \text{const}$)

$$
j_y = \sqrt{\frac{V}{c}}\int_{L(f)} B\,dy\,f_x
\tag{3.212}
$$

where the integration is carried out along the current line $L(f)$.

From Appendix A [see Eq. (29) and Section A.6] it is apparent that

$$
\sqrt{\frac{V}{c}}\int_{L(f)} B\,dy = \sqrt{\frac{V}{c}}\int_{L(f_0)} B\,dy + \frac{\sqrt{V}}{2\sqrt{c\int_{L(f_0)} B\,dy}}
$$

$$
\times\left\{\pm\rho_{\min}\sqrt{\frac{2ck\delta}{VB_0}}\int_{-\delta}^{\delta} B_x(x)\,e^{\gamma_0 x/2}(\omega - \omega_M)\,dy_M\right.
$$

$$+ \rho_{\min} \sqrt{\frac{2c\delta}{VB_0}} \, k \int_{-\delta}^{\delta} B_x(x) \, e^{\gamma_0 x/2} (\chi - \chi_M) \, dy_M$$

$$- \rho_{\min}^2 \frac{2c\delta}{VB_0} \, k \int_{-\delta}^{\delta} B_x(x) \, e^{\gamma_0 x} (\omega - \omega_M)(\omega_M)_x \, dy_M$$

$$+ \rho_{\min}^2 \frac{2c\delta}{VB_0} \frac{\gamma_0 k}{4} \int_{-\delta}^{\delta} B_x(x) \, e^{\gamma_0 x} (\omega - \omega_M)^2 \, dy_M$$

$$+ \tfrac{1}{2} \rho_{\min}^2 \frac{2c\delta}{VB_0} \, k \int_{-\delta}^{\delta} B_{xx}(x) \, e^{\gamma_0 x} (\omega - \omega_M)^2 \, dy_M \Bigg\}$$

$$- \frac{\sqrt{V}}{8\sqrt{c} \left[\displaystyle\int_{L(f_0)} B \, dy \right]^{3/2}} \rho_{\min}^2 \frac{2c\delta}{VB_0} \, k$$

$$\times \left[\int_{-\delta}^{\delta} B_x(x) \, e^{\gamma_0 x/2} (\omega - \omega_M) \, dy_M \right]^2 + o(k)$$

$$= \sqrt{\frac{V}{c} \int_{L(f_0)} B \, dy} \pm \sqrt{k} \, p + kq + o(k) \qquad (3.213)$$

Here, we have taken $B(x) = B_0 \, e^{-\gamma_0 x}$; in this equation we can obviously ignore the increment k of the exponent γ if we are interested in the increment of the total current due to the two-dimensional solution (3.210).

We write the increment δf in the form

$$\delta f = \pm\sqrt{k} \, \omega + k\chi, \qquad \chi = b \sin\frac{\pi x}{\lambda} \cos\frac{\pi y}{\delta} \qquad (3.214)$$

The first term on the right-hand side is determined by (3.210); the coefficient $b = 0(1)$ is unknown but must be retained for the necessary accuracy of the calculations; in what follows, we shall show that it does not need to be known. We can now write

$$f_x = (f_0)_x \pm \sqrt{k} \, \omega_x + k\chi_x \qquad (3.215)$$

The increment of the y component of the current density is equal to

$$
\left.
\begin{aligned}
\delta j_y &= \sqrt{\frac{V}{c}} \int_{L(f)} B\, dy\, f_x - \sqrt{\frac{V}{c}} \int_{L(f_0)} B\, dy\, (f_0)_x \\
&= \pm\sqrt{k}\left[\sqrt{\frac{V}{c}} \int_{L(f_0)} B\, dy\, \omega_x + p(f_{00})_x\right] \\
&\quad + k\left[\sqrt{\frac{V}{c}} \int_{L(f_0)} B\, dy\, \chi_x + p\omega_x + q(f_{00})_x\right] + o(k) \\
f_{00} &= \frac{2}{\rho_{\min}} \sqrt{\frac{VB_0}{2c\delta}}\, \frac{1 - e^{-\gamma_0 x/2}}{\gamma_0}
\end{aligned}
\right\} \tag{3.216}
$$

We have

$$
p = \frac{\sqrt{V}}{2\sqrt{c\int_{L(f_0)} B\, dy}}\, \rho_{\min} \sqrt{\frac{2c\delta}{VB_0}}\, (-\gamma_0) B(x)\, e^{\gamma_0 x/2} \int_{-\delta}^{\delta} (\omega - \omega_M)\, dy_M
$$

$$
= -\gamma_0 \rho_{\min} \omega \delta
$$

from which an expression for the coefficient of $\pm\sqrt{k}$ in (3.216) follows as

$$
\sqrt{\frac{2VB_0\delta}{c}}\, e^{-\gamma_0 x/2}\left(\omega_x - \frac{\gamma_0}{2}\omega\right) \tag{3.217}
$$

Integrating this expression (taken for $y = \delta$) with respect to x from 0 to λ, we obtain zero; the increment of the total current is therefore determined by the terms linear in k in (3.216).

In the expression for q [see (3.213)] we outline the term

$$
q_1 = \rho_{\min} \sqrt{\frac{2c\delta}{VB_0}}\, \frac{\sqrt{V}}{2\sqrt{c\int_{L(f_0)} B\, dy}} \int_{-\delta}^{\delta} B_x(x)\, e^{\gamma_0 x/2}(\chi - \chi_M)\, dy_M
$$

then compute $q_1(f_{00})_x$ and add it to $\sqrt{(V/c)}\int_{L(f_0)} B\, dy\, \chi_x$ [see (3.216)]; in the sum these terms yield an expression which differs from (3.217) by a constant factor, and the corresponding contribution to the total current therefore is zero. It remains to calculate the integral

$$
2k \int_0^{\lambda} (q - q_1)(f_{00})_x\, dx
$$

We have

$$q - q_1 = \frac{8\lambda^2}{\pi} \frac{(1/2)\pi\sqrt{1+t}+2}{\sqrt{1+t}\,[S(t)+T(t)]}$$

$$\times \sqrt{\frac{VB_0\delta}{2c}}\, e^{\gamma_0 x/2}\left(\frac{\gamma_0}{4} - \frac{\gamma_0}{4}\cos\frac{2\pi x}{\lambda} - \frac{\pi}{\lambda}\sin\frac{2\pi x}{\lambda}\right)$$

$$2k\int_0^\lambda (q - q_1)(f_{00})_x\, dx = 4\lambda^2 k\frac{VB_0}{\rho_{\min}c}\frac{(1/2)\pi\sqrt{1+t}+2}{S(t)+T(t)}$$

The increment of the current is equal to

$$\delta I = 4\lambda^2 k\frac{VB_0}{\rho_{\min}c}\frac{(1/2)\pi\sqrt{1+t}+2}{S(t)+T(t)}$$

The graph of the function $\delta I/Ik\lambda$ is shown in Fig. 3.21. It is apparent that the transition to the small solution is associated with an increase in the current taken from the electrodes (if $k\lambda = 2$, this increase is 5% for $t = 0.2$).

The increment of the current in the principal part (in the term of order k) does not depend on the choice of the small solution [i.e., on the choice of the sign in (3.210)]. A difference between these solutions can be manifested in the terms of higher order in k in the expression for the current density.

The physical cause of the increase in the current on the transition to the two-dimensional solution is obvious; for sufficiently rapid decrease of the external field $B(x)$ it is advantageous not to use the region of small $B(x)$ values with a view to reducing the effective resistance to currents generated in the central region of the channel, where $B(x)$ is large. The

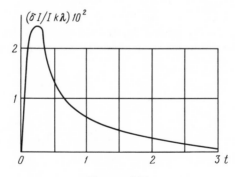

FIGURE 3.21

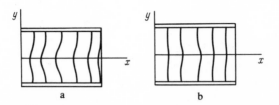

FIGURE 3.22

reduction of the resistance is achieved by the fact that current lines which come from the central region of the channel arrive on a larger part of the area of the electrodes. The solution (3.210) creates precisely such conditions of current flow; the corresponding lines of **j** are shown in Fig. 3.22a and b, the former corresponding to the upper sign in (3.210) and the latter to the lower sign.

To conclude this section, we consider the case when the function $B(x)$, which is positive for $0 < x < \lambda$, vanishes at the point $x = \lambda$. We set, for example, $B(x) = (1 - x/\lambda)^m$, $0 \le x \le \lambda$, $m \ge 0$; graphs of $B(x)$ for different values of the parameter are given in Fig. 3.23 a–c.

The solution (3.189) satisfies Weierstrass's condition; let us verify the fulfillment of Jacobi's condition. For $n = 1$, (3.193) has the form

$$\frac{d^2 a_1}{dx^2} + \left[\frac{m^2}{4(\lambda - x)^2} - \frac{\pi^2}{\delta^2} \right] a_1 = 0 \tag{3.218}$$

The general integral of this equation can be expressed in terms of the modified Bessel functions

$$a_1 = \sqrt{\lambda - x} \left[CI_{\sqrt{1-m^2}/2}\left(\frac{\pi}{\delta}(\lambda - x) \right) + DK_{\sqrt{1-m^2}/2}\left(\frac{\pi}{\delta}(\lambda - x) \right) \right] \tag{3.219}$$

We introduce the operator $d^n M_f$ defined by the left-hand side of (3.193) but taken with the minus sign; the operator DM_f on the functions (3.192)

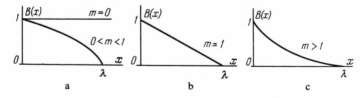

FIGURE 3.23

defined by the left-hand side of (3.142) is equal to

$$DM_f \delta f = -\frac{d^2 a_0}{dx^2} + \sum_{n=1}^{\infty} (d^n M_f a_n) \cos \frac{n\pi}{\delta} y$$

Positive definiteness of the operator DM_f on functions satisfying the boundary conditions (3.145), (3.191), and (3.195) will be established if the operators $-d^2/dx^2$ and $d^n M_f$ are positive definite on functions satisfying the corresponding boundary conditions (3.215). For the operator $-d^2/dx^2$ is obviously positive definite; with regard to the remaining operators, it suffices to establish positive definiteness for $d^1 M_f$.

It is natural to require that the increment of the y component of the current density [see (3.216)] be integrable over the interval $[0, \lambda]$.*

Based on (3.216) and (3.189) of this section and on the equations of the Appendix (see A.5 and A.6), which determine δj_y, we find that for $0 \le m < 1$ (see Fig. 3.23a) the solution (3.219) generates a singularity of order $(\lambda - x)^{-1-\sqrt{1-m^2}}$ at the point $x = \lambda$ in the expression for δj_y. This singularity is due to the second term in (3.219); it is nonintegrable, and the corresponding term must be rejected $(D = 0)$. The remaining term $\sqrt{\lambda - x} \, I_{\sqrt{1-m^2}/2}((\pi/2)(\lambda - x))$ does not vanish at $x = 0$ for any $0 \le m < 1$, and at $x = \lambda$ this term is an integrable singularity; thus, the operator $d^1 M_f$ is positive definite for the parameter values $0 \le m < 1$.

We now set $m = 1$ (see Fig. 3.23b) and consider the interval $[0, \lambda - \varepsilon)$, $\varepsilon > 0$. The function a_1 [see (3.219)] does not vanish at the ends of this interval if C and D are not equal to zero; for $\varepsilon = 0$ this function generates an integrable singularity at the end $x = \lambda$, so that in all cases $C = D = 0$ and $d^1 M_f$ again is positive definite.

We take $m > 1$ (see Fig. 3.23c) and consider the interval $[0, \lambda - \varepsilon)$, $\varepsilon > 0$. The boundary conditions (3.195) lead to the equation

$$v_m(\varepsilon) \equiv I_{i\sqrt{m^2-1}/2}\left(\frac{\pi\lambda}{\delta}\right) K_{i\sqrt{m^2-1}/2}\left(\frac{\pi\varepsilon}{\delta}\right) - I_{i\sqrt{m^2-1}/2}\left(\frac{\pi\varepsilon}{\delta}\right) K_{i\sqrt{m^2-1}/2}\left(\frac{\pi\lambda}{\delta}\right) = 0$$

For fixed $\varepsilon > 0$, there exists an infinite set of numbers $m > 1$ satisfying this equation†; let $m_0(\varepsilon)$ be the smallest of these numbers. We now establish the nature of the function $m_0(\varepsilon)$. We have

$$\frac{\partial v_m}{\partial \varepsilon}\bigg|_{m=m_0} \frac{d\varepsilon}{dm_0} + \frac{\partial v_m}{\partial m}\bigg|_{m=m_0} = 0 \tag{3.220}$$

---

* One can show that this requirement is equivalent to convergence of the integral (3.185), which determines the second variation of the functional I.

† A. Gray and G. B. Mathews, *A Treatise on Bessel Functions and Their Applications to Physics,* London 1922.

The following equation holds*:

$$\int_\lambda^\varepsilon v_m^2(x)\,\frac{dx}{x} = -\frac{2\varepsilon}{m}\left(v_m\frac{\partial^2 v_m}{\partial m\,\partial x} - \frac{\partial v_m}{\partial m}\frac{\partial v_m}{\partial x}\right)_{x=\varepsilon}$$

Setting $m = m_0(\varepsilon)$, we find

$$\int_\lambda^\varepsilon v_{m_0}^2(x)\,\frac{dx}{x} = \frac{2\varepsilon}{m_0}\frac{\partial v_m}{\partial m}\bigg|_{m=m_0(\varepsilon)}\frac{\partial v_m}{\partial \varepsilon}\bigg|_{m=m_0(\varepsilon)}$$

Together with (3.220), this results in

$$\frac{d\varepsilon}{dm_0} = -\frac{2\varepsilon}{m_0}\frac{\left(\dfrac{\partial v_m}{\partial m}\bigg|_{m=m_0}\right)^2}{\displaystyle\int_\lambda^\varepsilon v_{m_0}^2(x)\,\frac{dx}{x}}$$

The denominator of the right-hand side of the last equation is negative ($\varepsilon < \lambda$); thus, $m_0(\varepsilon)$ is an increasing function (Fig. 3.24). Suppose the parameters $m > 1$ and $\varepsilon < \lambda$ are chosen such that $m < m_0(\varepsilon)$. We fix m and let ε tend to zero [the channel length $2(\lambda - \varepsilon)$ then tends to 2λ]. If at some $\varepsilon = \varepsilon_0$ (Fig. 3.24) we have $m_0(\varepsilon_0) = m$, then for $\varepsilon < \varepsilon_0$ it follows that $m_0(\varepsilon) < m$; for $\varepsilon \le \varepsilon_0$ Jacobi's condition is violated, and we must then choose a solution of the original equation (3.138) which differs from (3.189), in the same way as in the preceding discussion.† If the equation $m_0(\varepsilon) = m$, for a given $m > 1$, is not satisfied as ε tends to zero, $\varepsilon = 0$ included, then one must investigate the behavior of the function (3.219) in the limit $x \to \lambda$ for $m > 1$. At $x = \lambda$, this function generates a nonintegrable singularity, and we must set $C = D = 0$, as in the case $m = 1$ considered above.

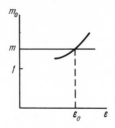

FIGURE 3.24

The choice between these two cases is obviously dictated by the behavior of the function $m_0(\varepsilon)$ in the limit $\varepsilon \to 0$: if $m_0(0) = 1$, then the first case holds; if $m_0(0) > 1$, then the second case is possible. Leaving out a detailed investigation of the behavior of the function $m_0(\varepsilon)$, we restrict ourselves to the remarks made above, which show that the operator DM_f remains positive definite even when the function $B(x)$ vanishes on the boundary of the channel and Jacobi's condition is satisfied simultaneously. Hence, by analogy with Section 3.11, we conclude that there exists a solution of the boundary-value problem B, although the inequalities of the type (3.162) [see also (3.160)] must be replaced by others.

3.13. Some Qualitative and Numerical Results

To illustrate the above theory, in this section we give the results of some calculations of the optimal and nonoptimal distributions of the current lines in the channel of an MHD generator of finite length $2x$ (see Fig. 3.19). In all cases it is assumed that the function $B(x) = B(-x) > 0$ is constant for $x \in [0, x_0]$, and for $x > x_0$ decreases monotonically to the value zero, which is taken at the point $x = x_2$, and for $x > x_2$, where x_2 may be greater than, equal to, or less than x_c it also decreases monotonically (Fig. 3.19 illustrates the case when $x_2 < x_c$).

On the basis of the results of Section 3.10 and using the dimensionless variables introduced there [see (3.135)],* we describe first the general form of the curve $\bar{I}(\bar{R})$, which expresses the optimal current taken from the electrodes as a function of the resistance $\bar{R}$ of the external load for given field $\bar{B}(\bar{x})$. We shall assume that Jacobi's condition is satisfied. We take a channel of definite length $2\bar{x}_{c_1}$, with $\bar{x}_{c_1} \leq \bar{x}_2$; the curve $\bar{I} = \bar{I}_1(\bar{R})$ begins at the point M_1 (Fig. 3.25). With increasing $\bar{R}$, the function $\bar{I} = \bar{I}_1(\bar{R})$ decreases monotonically; for relatively small $\bar{R}$, the value of $\bar{I}(\bar{R})$ can be calculated using the invariance of $\bar{I}$ under the replacement of the pair $(\bar{B}, \bar{R})$ by the pair $(\bar{B}, 0)$ (see Section 3.10). At the same time, as was pointed out in Section 3.10, it is convenient to specify, not $\bar{R}$, but an effective field $\bar{\bar{B}}(\bar{x})$, which differs from the given field $\bar{B}(\bar{x})$ by a constant: $\bar{B}(\bar{x}) = \bar{\bar{B}}(\bar{x}) + \bar{b}$; if $\bar{I}$ is the corresponding current, then the resistance $\bar{R}$ is found from (see Section 3.10)

$$\bar{R} = \frac{2\bar{b}\delta}{\bar{I}\lambda} \tag{3.221}$$

---

* We take the constant B^0 in (3.135) equal to the maximal value B^0 of the function $B(x)$ determined by Fig. 3.5.

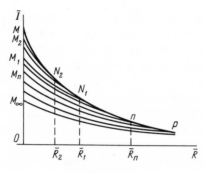

FIGURE 3.25

The critical value $\bar{R} = \bar{R}_1(\bar{b} = \bar{b}_1)$ after which the above invariance cannot be used corresponds to vanishing of the effective field $\tilde{\bar{B}}(\bar{x})$ at the point $\bar{x}_{c_1}$: $\tilde{\bar{B}}(\bar{x}_{c_1}) = 0$ (the point N_1 in Fig. 3.25). For $\bar{R} > \bar{R}_1(\bar{b} > \bar{b}_1)$, optimization requires a reduction in the channel length, i.e., the filling of definite sections adjoining the ends $|\bar{x}| = |\bar{x}_{c_1}|$ of the channel with insulating matter (which is equivalent to moving the ends toward the center of the channel). On each occasion, with the specification of $\bar{b} \geq \bar{b}_1$, we must choose the channel length $2\bar{x}_c$ (we shall call this the *critical length*) in such a manner that the equation $\tilde{\bar{B}}(\bar{x}_c) = \bar{B}(\bar{x}_c) - \bar{B} = 0$ holds for the given $\bar{b}$; the corresponding values of $\bar{R}$ are found in accordance with (3.221).* The points for which $\bar{R} > \bar{R}_1$ fill the part N_1P of the curve $\bar{I}_1(\bar{R})$ (Fig. 3.25) to the right of N_1.

We now repeat the foregoing discussion, increasing the original length of the channel to a value $2\bar{x}_{c_2} > 2x_{c_1}$, and leaving the original field $\bar{B}(\bar{x})$ as it was (as before, we assume $\bar{x}_{c_2} < \bar{x}_2$). The curve $\bar{I} = \bar{I}_2(\bar{R})$ begins at the point M_2 (Fig. 3.25); obviously, $\bar{I}_2(0) > \bar{I}_1(0)$. Furthermore, the smallest critical value $\bar{b} = \bar{b}_2$ is less than $\bar{b}_1$; the corresponding value of $\bar{I}_2$ is not less than $\bar{I}_1(\bar{R}_1)$; (3.221) shows that $\bar{R}_2 < \bar{R}_1$. The point N_2 on the curve $\bar{I}_2(\bar{R})$ corresponds to the vanishing of the effective field $\bar{B}(\bar{x})$ at $\bar{x} = \bar{x}_{c_2}$. For $\bar{b} > \bar{b}_2(\bar{R} > \bar{R}_2)$, the length of the optimal channel (the critical length) is reduced, and $\bar{I}_2(\bar{R})$ decreases monotonically; for $\bar{R} > \bar{R}_1$ we have the same conditions as in the preceding case, when the original length of the channel was $2\bar{x}_{c_1}$. Therefore, for $\bar{R} > \bar{R}_1$, the curves $\bar{I}_2(\bar{R})$ and $\bar{I}_1(\bar{R})$ coincide.

Thus, the common part N_1P of the curves $\bar{I}_1(\bar{R})$ and $\bar{I}_2(\bar{R})$ is universal in the sense that it belongs to all curves corresponding to channels of length greater than $2\bar{x}_{c_1}$ for one and the same original field $\bar{B}(\bar{x})$.

---

* Obviously, $\bar{x}_{c_1} > \bar{x}_c$ and $\bar{R} > \bar{R}_1$.

Continuing this argument, we find that the section $N_2 P$ of the curve $\bar{I}_2(\bar{R})$ is universal, belonging to all curves constructed for channels of length greater than $2\bar{x}_{c_2}$, etc. As a result, we obtain the limiting universal curve $MN_2 N_1 P$ (Fig. 3.25); the point M obviously corresponds to the current $\bar{I}$ taken from the electrodes for $\bar{R} = 0$ from a generator of length $2\bar{x}_2$. The coordinates $(\bar{I}_n, \bar{R}_n)$ of every point n of this curve are such that the current $\bar{I}_n$ is taken from a short-circuited channel of corresponding critical length $2\bar{x}_{c_n}$ for which

$$\bar{\bar{B}}(\bar{x}_{c_n}) = \bar{B}(\bar{x}_{c_n}) - \frac{\lambda}{2\delta} \bar{I}_n \bar{R}_n = 0$$

As the point n moves along the universal curve from the position M to infinity (the position P) the corresponding critical length $2\bar{x}_{c_n}$ decreases monotonically from $2\bar{x}_2$ to $2\bar{x}_0$. Considering a channel of length $2\bar{x}_{c_n}$, we can construct the curve $\bar{I}(\bar{R})$ for $\bar{R} < \bar{R}_n$. This part of the curve moves away from the point n to the left (Fig. 3.25) and ends at the point M_n corresponding to the short-circuited channel of length $2\bar{x}_{c_n}$. The points lying on $M_n n$ correspond to optimal values of the current taken from a channel of length $2\bar{x}_{c_n}$ for $\bar{R}$ values in the interval $[0, \bar{R}_n]$. Thus, the universal curve MnP generates a one-parameter family of curves of the type $M_n nP$, which consists of a section $M_n n$ below the universal curve, and a section nP of the universal curve. No two curves $M_n n$ of this family have common points: assuming otherwise, we should have channels of different length loaded with the same resistances and generating the same optimal current, which is impossible if the function $B(x)$ is given by a graph of the type shown in Fig. 3.19.*

As $\bar{R} \to \infty$ the universal curve tends asymptotically to the axis $\bar{R}$; if the point n goes to infinity ($\bar{R}_n \to \infty$), moving along the universal curve, then

$$\frac{\lambda}{2\delta} \bar{I}_n \bar{R}_n \to \bar{B}(x_0) = 1$$

and the curve $M_n n$ tends to the limiting position $M_\infty P$ corresponding to a channel of length $2\bar{x}_0$.

Having at our disposal the family of optimal curves constructed for given field $\bar{B}(\bar{x})$, we can find the optimal current taken from a channel of fixed original length $2\bar{x}_c < 2\bar{x}_2$ for different values of the load $\bar{R}$. Above all,

---

* Touching of the curves of the families can also be ruled out since it contradicts Jacobi's condition.

it is necessary to find the point of the universal curve corresponding to the case when the length $2\bar{x}_c$ is critical. This point lies on the intersection of the universal curve and the hyperbola $\bar{I}\bar{R} = 2(\delta/\lambda)\bar{B}(\bar{x}_c)$. Suppose n is such a point and $\bar{I}_n$ and $\bar{R}_n$ are the corresponding critical parameters (see Fig. 3.25). The critical current curve will be the curve corresponding to the family passing through the point n, i.e., the curve $M_n nP$; the channel length is $2\bar{x}_c$ for points of the branch $M_n n(\bar{R} < \bar{R}_n)$ and is less than $2\bar{x}_c$ for points of the branch nP; as $\bar{R} \to \infty$, the optimal length of the channel tends to $2\bar{x}_0$.* It is easy to see that there can exist only one point of intersection of the hyperbola $\bar{R} = 2(\delta/\lambda)\bar{B}(\bar{x}_c)$ and the universal curve. For sufficiently large $\bar{R}$, the hyperbola does not pass above the universal curve since in the limit $\bar{R} \to \infty$ the limiting relation

$$\bar{I}\bar{R} \to \frac{2\delta}{\lambda} \bar{B}(\bar{x}_0) \ge \frac{2\delta}{\lambda} \bar{B}(\bar{x}_c)$$

is satisfied for the points of this curve. For sufficiently small $\bar{R}$, the hyperbola lies below the universal curve. This proves the possible existence of at least one point of intersection; if $\bar{x}_0 < \bar{x}_2 < \bar{x}_c$, the point is unique since otherwise, proceeding along the section $M_1 1$ corresponding to the channel of length $2\bar{x}_c$ (Fig. 3.26, where ——— represents the universal curve, and ---- represents the hyperbola), and then along section 12 of the universal curve, we should arrive at the point 2, corresponding to a channel length $2\bar{x}_{c_2} < 2\bar{x}_{c_1}$, which contradicts the hypothesis.

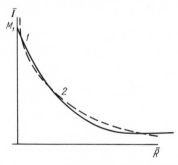

FIGURE 3.26

---

* It follows from these arguments that if the functions $\bar{B}_1(x)$ and $\bar{B}_2(\bar{x})$ both are of the type shown in Fig. 3.19 and differ only for $\bar{x} > \bar{x}_*$, then the corresponding universal curves and families will differ only for $\bar{R} < \bar{R}_*$, where the load $\bar{R}_*$ corresponds to the case when the length $2\bar{x}_*$ is critical.

The restriction $2\bar{x}_c < 2\bar{x}_2$ adopted in the foregoing arguments cannot be dropped without violating the conditions of optimality; therefore, a change of the parameter $\bar{x}_c$ from $\bar{x}_0$ to $\bar{x}_2$ corresponds to the curves of the family generated by the universal curve.

It is interesting to establish the shape of the universal curve and the curves of the family on the variation of the parameters $\bar{x}_0$ and $\bar{x}_2$ of the function $\bar{B}(\bar{x})$ (see Fig. 3.19). Suppose $\bar{x}_0$ is fixed and $\bar{x}_2 \to \bar{x}_0$, the interval $(\bar{x}_0, \bar{x}_2)$ of values of the parameter $\bar{x}_c$ contracts to $\bar{x}_0$. The universal curve together with the curves of the family move downward (Fig. 3.25), eventually coalescing with the limiting curve $M_\infty P$, which remains unchanged in this limit. If $\bar{x}_0 \geq 1$, then the equation of the limiting curve is obtained from (3.8), in which it is necessary to set $x_c = x_0$; expressed in the dimensionless variables (3.135), we have

$$\bar{I} = 2 \frac{\delta}{\lambda} \frac{\alpha^*}{1 + \bar{R}\alpha^*} \tag{3.222}$$

It is easy to show (see Appendix A.1) that as $\bar{x}_0$ increases from 1 to infinity, the parameter α^* increases monotonically from λ/δ to $\alpha^*_\infty = K(\kappa'_\infty)/K(\kappa_\infty)$, $\kappa^{-1}_\infty = \cosh(\pi\lambda/2\delta)$. In particular, for $\bar{x}_0 = \infty$ $[\bar{B}(x) = \text{const} = 1]$, the limiting curve (3.222) takes the highest possible position on the plane $(\bar{I}, \bar{R})$ which is determined by the equation

$$\bar{I} = 2 \frac{\delta}{\lambda} \frac{\alpha^*_\infty}{1 + \bar{R}\alpha^*_\infty} \tag{3.223}$$

while for $\bar{x}_0 = 1$ it takes the lowest possible position

$$\bar{I} = \frac{2}{1 + \bar{R}(\lambda/\delta)} \tag{3.224}$$

in the class of curves corresponding to functions $\bar{B}(\bar{x})$ with a region of constancy that encompasses the electrode region $|\bar{x}| < 1$. If $\bar{x}_0 < 1$, then the equation of the limiting curve $M_\infty P$ is obtained from (3.137) by replacing λ by x_0 and by assuming $B(x) = B_0 = \text{const.}^*$ In terms of the dimensionless variables,

$$\bar{I} = 2 \frac{\bar{x}_0}{1 + \bar{R}(\lambda/\delta)\bar{x}_0} \tag{3.225}$$

These arguments show that the universal curves $\bar{I}(\bar{R})$ together with the families they determine are situated in the part of the $(\bar{I}, \bar{R})$ plane

---

* See the arguments that follow (3.137).

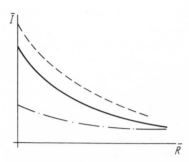

FIGURE 3.27

bounded by the upper limiting universal curve (3.223) and the lower limiting universal curve (3.224) or (3.225) (Fig. 3.27, where ---- represents the upper limiting universal curve, —— represents the universal curve, and —·— represents the lower limiting universal curve). The limiting universal curves correspond to functions $\bar{B}(\bar{x})$ that have no decreasing sections; therefore, the curves of the corresponding families coincide with the limiting curves themselves. This is confirmed by the fact that the hyperbola $\bar{I}\bar{R} = 2(\delta/\lambda)\bar{B}(\bar{x}_c)$, $\bar{x}_c < \bar{x}_2$, in the limiting cases $\bar{x}_2 = \bar{x}_0$ and $\bar{x}_0 = \infty$, has the equation $\bar{I}\bar{R} = 2(\delta/\lambda)$; this last equation determines a hyperbola that for any finite $\bar{R}$ lies above the upper limiting universal curve (3.223) and coincides with it only asymptotically in the limit $\bar{R} \to \infty$.

What we have presented here can be illustrated by a number of quantitative results for channels of different lengths and different loads. The function $\bar{B}(\bar{x})$ was given by

$$\bar{B}(\bar{x}) = \begin{cases} 1, & |\bar{x}| \le 1 \\ e^{-(|\bar{x}|-1)/\alpha}, & |\bar{x}| \ge 1 \end{cases} \tag{3.226}$$

The parameter α was taken equal to 1.5; the values of the parameters λ/δ and $2\bar{x}_c$ were varied. Numerical solutions of (3.138) were obtained under the boundary conditions (3.126), (3.128), and (3.139) with allowance for the invariance property described in Section 3.10; the functional I was calculated in accordance with (3.141). A BÉSM-4 computer was used for the calculations.

Figure 3.28a shows the optimal curves of the current $\bar{I}(\bar{R})$ corresponding to the parameter values $\lambda/\delta = 1$, $\bar{x}_c = 2.5$; the results of the calculations are given in Table 3.1. Curve P is a section of the universal curve of the family; $M_\infty P$ is the limiting curve of the family corresponding to $\bar{x}_c = 1$ or, equivalently, the lower limiting curve corresponding to the value $\alpha = \infty$; this curve is defined by (3.224).

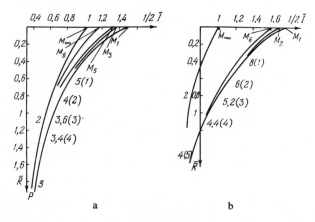

FIGURE 3.28

The critical values $2\bar{x}_c$ (the parameters of the family) are given next to the points of the universal curve from which the sections $M_1 1, \ldots, M_5 5$ of the curves of the family initiate.

In Fig. 3.28b and in Table 3.2 we give the analogous curves and calculated data for $\lambda/\delta = \frac{1}{2}$, $\bar{x}_c = 4$.

It is of interest to compare the constructed optimal curves with the curves $\bar{I}(\bar{R})$ corresponding to nonoptimal control regimes. In Fig. 3.29, we have plotted the dependences $\bar{I}(\bar{R})$ obtained from the numerical solution of the problem (3.15)–(3.16) for $\rho = \rho_{\min} = \text{const}$, $\alpha = 1.5$, $\lambda/\delta = 1$; the curves are parametrized in terms of the channel length $2\bar{x}_c$.* Each of the

Table 3.1

$\bar{x}_c$	1.2	1.2	1.2	1.2	1.2	1.2	1.2*	1.4*	1.5
$\bar{b}$	0	0.1	0.15	0.2	0.25	0.776	0.872*	0.766*	0
$\bar{R}$	0	0.093	0.149	0.210	0.281	2.958	5.910*	2.634*	0
$1/2\bar{I}$	1.189	1.070	1.010	0.951	0.891	0.262	0.148*	0.290*	1.353
$\bar{x}_c$	1.50	1.5	1.5	1.5*	1.7*	1.8	1.8	1.8	1.8
$\bar{b}$	0.2	0.4	0.617	0.715*	0.627*	0	0.2	0.35	0.485
$\bar{R}$	0.186	0.502	1.243	1.983*	1.282*	0	0.179	0.390	0.695
$1/2\bar{I}$	1.075	0.797	0.496	0.361*	0.489*	1.416	1.119	0.897	0.698
$\bar{x}_c$	1.8*	2.0	2.0	2.0*	2.50	2.5	2.5	2.5*	
$\bar{b}$	0.584*	0	0.25	0.512*	0	0.1	0.2	0.368	
$\bar{R}$	1.059*	0	0.237	0.777*	0	0.078	0.177	0.410	
$1/2\bar{I}$	0.552*	1.431	1.053	0.659*	1.490	1.286	1.132	0.902	

---

* For an appropriate choice of the parameters, these curves are described by (3.8).

Table 3.2

$\bar{x}_c$	2.0	2.0	2.0	2.0	2.0	2.0*	2.2
$\bar{b}$	0	0.1	0.2	0.3	0.4	0.512*	0
$\bar{R}$	0	0.136	0.310	0.539	0.857	1.417*	0
$1/2\bar{I}$	1.650	1.471	1.292	1.113	0.934	0.723*	1.699
$\bar{x}_c$	2.2	2.2	2.2	2.2	2.2*	2.6	2.6*
$\bar{b}$	0.1	0.2	0.3	0.4	0.449*	0	0.344*
$\bar{R}$	0.132	0.302	0.531	0.848	1.052*	0	0.647*
$1/2\bar{I}$	1.511	1.323	1.130	0.943	0.852*	1.749	1.063*
$\bar{x}_c$	3.0	3.0*	4.0	4.0	4.0*	3.0	
$\bar{b}$	0	0.263*	0	0.06	0.135*	0.13	
$\bar{R}$	0	0.429	0	0.072	0.180*	0.173	
$1/2\bar{I}$	1.776	1.225	1.785	1.660	1.492*	1.506	

constructed curves lies to the left of the corresponding curve of the family shown in Fig. 3.28a; in particular, the curve M_1P in Fig. 3.28a characterizes the gain in the current compared with the curve of Fig. 3.29a, which corresponds to $2\bar{x}_c = 5$; the gain is maximal (5%) for $\bar{R} = 0$, equal to 3.5% for $\bar{R} = 1$, and tends monotonically to zero as $\bar{R} \to \infty$.*

If we construct the envelope of the curves in Fig. 3.29 (the envelope is not shown in Fig. 3.29a), we obtain the dependence $\bar{I}(\bar{R})$ for a channel of original length $2\bar{x}_c = 5$ which has been partly optimized by choosing the optimal length $2\bar{x}_c$ for each value of the load $\bar{R}$. Comparing the envelope with the curve M_1P in Fig. 3.28a, we see the influence of the partial

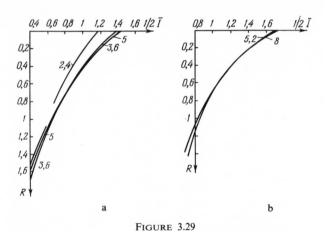

a   b

FIGURE 3.29

---

* The relative gain in the power is twice the relative gain in the current.

optimization: The relative increment of the current is reduced (in particular, for $\bar{R} = 1$ it is now 3.2%).

In Fig. 3.29b, we have plotted the nonoptimal dependences of $\bar{I}(\bar{R})$ for $\delta = \delta_{min} = $ const, $\alpha = 1.5$, $\lambda/\delta = 1/2$; the parameter of the curves is again the length $2\bar{x}_c$ of the channel. Comparison of the curves of Figs. 3.29b and 3.28b corresponding to $2\bar{x}_c = 8$ shows that the gain in the current is maximal (5.5%) for $\bar{R} = 0$, equal to 4.7% for $\bar{R} = 1$, and tends monotonically to zero as $\bar{R} \to \infty$; recalling the foregoing, we find that the gain increases as λ/δ decreases. With allowance for the partial optimization, the gain is 5.5% for $\bar{R} = 0$ and 1.8% for $\bar{R} = 1$.

Partial optimization can be implemented in practice by introducing into a channel filled with a uniformly conducting medium ($\delta = \delta_{min}$) a number of nonconducting baffles (Section 3.1) placed along the flow from the channel ends $x = \pm x_c$ to the positions $x = \pm x_{cn}$ (Fig. 3.30a, b) corresponding to the optimal length of the channel for each value of $\bar{R}$ (Fig. 3.28a, b). One could get even closer to the exact optimal solution by introducing additional horizontal baffles into the channel.

Let us consider the case when two horizontal baffles are placed along the x axis at the levels $\pm l$ (Fig. 3.30b) in a channel of optimal length $2\bar{x}_c = 5$ corresponding to a resistance $\bar{R} = 0.410$ (Table 3.1). Figure 3.31 illustrates the dependence $\bar{I}(\bar{l}) = \bar{I}(l/\lambda)$. It can be seen that a maximum of the current is attained for $\bar{l} = 0.34$; the relative increment of the current given by the exact optimal solution is reduced to 3% when the two baffles are introduced. An increase in the number of baffles improves the result even more.

In the foregoing discussions, we have assumed that Jacobi's condition is satisfied; if this condition is violated, then it cannot be restored by local variations of the resistivity (including the use of horizontal baffles). As an illustration, we can take the example in Section 3.12. If the channel is

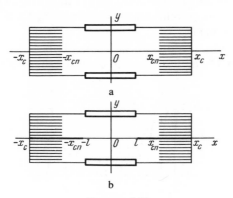

FIGURE 3.30

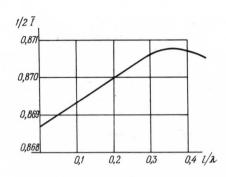

FIGURE 3.31

bounded by (short-circuited) horizontal electrodes and vertical insulators, then for $B(x) \geq 0$ the solution (3.189) satisfies Weierstrass's condition irrespective of the length of the channel.

At the same time, if $B = B_0\, e^{-\gamma x}$, then for

$$\gamma\lambda > 2\pi[1 + (\lambda/\delta)^2]^{1/2}$$

the solution (3.189) does not satisfy Jacobi's condition. This condition is satisfied by the solution (3.210), the transition to this solution being accompanied by an increase in the current taken from the electrodes. On the other hand, the introduction of one or several nonconducting baffles into such a channel, or a reduction of the scalar conductivity in some part of the channel, only reduces the current taken from the electrodes. The increase in the current associated with the solution (3.210) is due to the effect of the entire region of tensor resistivity. If this region is replaced by a sequence of baffles (whose shape is determined as in Fig. 3.32), then one can assert that the increase in the current is due to the combined action of the charges precipitated on the baffles.

To conclude this section, we give graphs that represent the current lines in the channel for the optimal control range (Fig. 3.32) and the

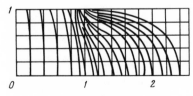

FIGURE 3.32

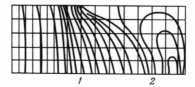

FIGURE 3.33

nonoptimal control range (Fig. 3.33)

$$\rho = \rho_{\min} = \text{const}$$

without baffles. In both cases, we have chosen a channel of length $2\bar{x}_c = 5$ with load resistance $\bar{R} = 0.237$ in the field (3.226) for $\alpha = 1.5$ (only a quarter of the channel is shown in either figure).

It is evident from the figures that the nonoptimal range exhibits well-developed current vortices, whereas there are no vortices at all in the optimal range.

3.14. On a Class of Shape Optimization Problems: An Application of Symmetrization

The results of the preceding sections open up the possibility of a new approach to various problems of optimization of the shape of bodies. Traditional examples of such problems are the isoperimetric problems of mathematical physics.[132]

Isoperimetric problems can be naturally generalized by introducing a differential constraint which differs from the Euler equation of an isoperimetric problem.

In this section, we study the problem of maximizing the current I taken from an MHD channel; in contrast to Section 3.2, we assume that the resistivity $\rho(x, y)$ is given and constant, and we take the shape of the lateral insulating walls CC and $C'C'$ of the channel as the control; we shall assume that the area of the channel is fixed:

$$\iint_G dx\, dy = S \tag{3.227}$$

Before we turn to the consideration of this problem, we return to the problem in Section 3.2 and represent it in a form more suited to the present purpose. It is easy to show that the boundary-value problem (3.15)–(3.16) is equivalent to the variational problem

$$\min_{z^2} \Phi = \min_{z^2} \left[\iint_g \rho(\text{grad } z^2)^2 \, dx \, dy + 2R(z_+^2)^2 - 2z_+^2 \varepsilon_1 \right.$$

$$\left. + \frac{2}{c} \iint_g z^2 VB'(x) \, dx \, dy \right] \qquad\qquad (3.228)$$

$$\varepsilon = \frac{1}{c} \int_{CC} VB \, dy$$

under the additional conditions $z^2(0, y) = 0$, $z^2|_{BCCB} = z_+^2$; the integration in (3.228) is extended to the region g, the right-hand half $CBAABC$ of the channel (see Fig. 3.19), and the integral in the expression for ε_1 is taken along the boundary CC in the direction from bottom to top. The condition $z_y^2(x, \pm\delta) = 0$ for $0 \le x \le \delta$ and the last condition in (3.16) are the natural conditions for the problem (3.228).

Introducing the dimensionless variable w by $z^2 = wz_+^2$, we obtain

$$\min_{w, z_+^2} \left\{ (z_+^2)^2 \left[\iint_g \rho(\text{grad } w)^2 dx \, dy + 2R \right] - 2z_+^2 \varepsilon_1 \right.$$

$$\left. + \frac{2}{c} z_+^2 \iint_g w VB'(x) \, dx \, dy \right\}$$

instead of the problem (3.228). Eliminating z_+^2 by means of the stationarity condition

$$z_+^2 = \frac{\varepsilon_1 - \dfrac{1}{c} \displaystyle\iint_g w VB'(x) \, dx \, dy}{\displaystyle\iint_g \rho(\text{grad } w)^2 \, dx \, dy + 2R} \qquad\qquad (3.229)$$

we arrive at the variational problem

$$\max_w \frac{\left[\varepsilon_1 - \dfrac{1}{c} \displaystyle\iint_g w VB'(x) \, dx \, dy \right]^2}{\displaystyle\iint_g \rho(\text{grad } w)^2 \, dx \, dy + 2R} \qquad\qquad (3.230)$$

under the additional conditions

$$w(0, y) = 0, \qquad w|_{BCCB} = 1 \tag{3.231}$$

The optimization problem posed in Section 3.2 can now be formulated as follows: to determine a function $\rho(x, y)$ under the restrictions (3.18) in accordance with the extremal requirement

$$\max_{\rho} z_+^2 = \max_{\rho} \frac{\varepsilon_1 - \dfrac{1}{c} \displaystyle\iint_g w VB'(x) \, dx \, dy}{\displaystyle\iint_g \rho (\operatorname{grad} w)^2 \, dx \, dy + 2R} \tag{3.232}$$

The functional z_+^2 depends on ρ not only explicitly but also implicitly through the function w, which satisfies the conditions (3.231) and the extremal requirement (3.230).

Let us consider the special case $B(x) = B_0 = $ const. In this case, the problems (3.230)–(3.231) are combined by the single extremal requirement

$$\max_{\rho,w} z_+^2 = \max_{\rho,w} \frac{\varepsilon_1}{\displaystyle\iint_g \rho (\operatorname{grad} w)^2 \, dx \, dy + 2R} \tag{3.233}$$

subject to the constraints (3.18) and (3.231). The problem of Section 3.2 for $B(x) = $ const is reduced to the simplest problem (3.233): the differential constraints (3.15) and the last condition (3.16) now are the Euler equations (with respect to the variable w) of the functional z_+^2, i.e., of the same functional that is maximized with respect to the variable ρ.

It is easy to see, in particular, that if the control ρ satisfies the restrictions (3.18), then a maximum of (3.233) is attained if $\rho = \rho_{\min}$ everywhere in the channel. In other words, for a homogeneous field B, the condition $\rho = \rho_{\min}$ is sufficient to maximize the total current. Simultaneously, our discussion also establishes the existence of an optimal control ρ in the case $B(x) = $ const.

The situation changes if $B(x) \neq $ const. In this case, (3.15) and the last condition in (3.16) are generated by the problem (3.230), i.e., the problem of maximizing a functional that differs from z_+^2 with respect to w. The maximum of the functional z_+^2 with respect to ρ is to be obtained subject to the constraints (3.15), (3.16), and (3.18). This is not a problem of the simplest kind.

We now return to the problem of the best shape of the lateral insulating walls of the channel for given (constant) resistivity function of the medium: $\rho(x, y) = \rho_{\min} = $ const. Obviously, this problem is equivalent to that of maximizing the functional (3.229) with respect to the shape of the wall CC for a function w given by the condition (3.230); it will be assumed that $\rho(x, y) = \rho_{\min} = $ const. If $B(x) = $ const, we arrive at the variational problem

$$\max_{CC,w} = \max_{CC,w} \frac{\varepsilon_1}{\rho_{\min} \displaystyle\iint_g (\operatorname{grad} w)^2 \, dx \, dy + 2R} \tag{3.234}$$

subject to the conditions (3.231). Note that the condition $j_n = 0$ expresses the insulating property of the unknown boundary $CC(C'C')$; this property can be ascribed to the material medium adjoining the channel along the boundaries CC and $C'C'$. A change in the shape of the channel can be interpreted as the creation of inclusions of insulating ($\rho = \infty$) material of the surrounding medium and the creation of inclusions of material with $\rho(x, y) = \rho_{\min}$ in the surrounding medium within the overall volume of the channel.

We shall investigate the influence of an insulating inclusion of the shape of an infinitesimally thin strip of short length a on the functional I. The expression for the increment ΔI produced by such an inclusion can be readily obtained by means of (2.102) and (2.103) in which u must be replaced by ρ and U by P. Passing to the limit as $P \to \infty$, integrating over the ellipse with semiaxes a and b (oriented, respectively, along the directions of $\boldsymbol{\alpha}$ and $\boldsymbol{\beta}$), and setting $b = 0$ in the final result, we obtain [see (2.100)]

$$\Delta I = \pi a^2 j_\beta \omega_{2\beta} = -\rho_{\min} \pi a^2 j_\beta k_\beta$$

$$(\mathbf{k} = (-\omega_{1y}, \omega_{2x}))$$

Since $\boldsymbol{\beta}$ is any direction at the given point, it is necessary that

$$\mathbf{j} \uparrow\uparrow \mathbf{k} \tag{3.235}$$

everywhere in the region in order to maximize I. The condition (3.235) is necessary for optimality. However, on the unknown boundaries CC and $C'C'$ we must have*

---

* These relations can be obtained by analogy with the last Weierstrass–Erdmann condition (2.42) under the assumption of sufficient smoothness of the boundaries CC and $C'C'$; it must be kept in mind that the function ω_1 has constant value [see (2.101)] along these boundaries.

$$j_t k_t = \gamma = \text{const} \qquad (3.236)$$

where **t** is the unit tangent vector to the unknown boundary, and γ is the constant Lagrangian multiplier corresponding to the constraint (3.227). We shall assume that the parameter S in (3.227) is chosen in such a way that the constant γ is zero.

The vector **j** is determined from the boundary-value problem (3.15)–(3.16), and the vector **k** by the boundary-value problem (2.100)–(2.101). These problems differ in the type of inhomogeneity: in the first case it occurs in the equations and in the second in the boundary condition. One can show that the vector **k** is a "fictitious" current density in the adjoint problem (2.100)–(2.101). If $B(x) = \text{const}$, then the condition (3.235) is satisfied [see (3.25)]. However, if $B(x) \neq \text{const}$, this condition cannot be satisfied almost everywhere in the channel for a continuously differentiable free boundary $CC(C'C')$. To see this, it is sufficient to compare Fig. 3.4, which shows the vector lines of $-\mathbf{k}$, with Fig. 3.34b–e, which represent the vector lines of **j** for the field $B(x)$ defined by the graph in Fig. 3.34a for different positions of the free boundaries CC and $C'C'$.

We see that as the length of the channel is increased, current vortices appear and develop, and these prevent the fulfillment of condition (3.235) within the region and of condition (3.236) on the free boundary. This is manifested especially clearly in the case $\gamma = 0$: the current density **j** vanishes only *at the critical points* (in Fig. 3.34c–e at the center of the vortex), and the condition $j_t k_t = 0$ *cannot be satisfied along the complete free boundary* $CC(C'C')$.

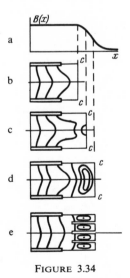

FIGURE 3.34

If the free boundary has sufficiently deep indentations (Fig. 3.34e), the large vortex is broken up into a number of smaller ones, the number of critical points is increased, but conditions (3.235) and (3.236) still cannot be satisfied.

These conditions can be satisfied by an appropriate modification of the formulation of the problem; namely, by introducing a tensor resistivity, as we did in Section 3.8. The description of the resistivity by means of a symmetric tensor may be assumed to be the limiting case of a distribution for which a large number of parallel, infinitesimally thin insulating baffles oriented in a suitable manner are introduced into a channel with $\rho(x, y) = \rho_{min} = $ const (obviously, these baffles define the directions of one of the principal axes of the tensor $\wp$). The baffles can be regarded as isolated branches of the boundary of the region.

The main advantage of the new formulation is that in the general case $B(x) \neq$ const, it essentially eliminates the differential constraints, and the problem again becomes an isoperimetric one. Indeed, in Section 3.11 we have shown that this formulation of the problem reduces it to the problem of maximizing the functional (3.154) subject to the constraints $(3.152)_1$ and $(3.152)_2$ or, equivalently, subject to the constraints (3.231); of course, the requirement (3.227) must be included here. This problem, when formulated for tensor function $\wp(x, y)$ in the general case $B(x) \neq$ const, has exactly the same structure as the analogous problem (3.234), (3.231), and that of problem (3.227) posed in the case $B(x) = $ const for the scalar function* $\rho(x, y) = \rho_{min} = $ const; for $B(x) = $ const, the problem (3.154) and (3.231) reduces to (3.234) and (3.231). We have seen that for $B(x) = $ const, condition (3.235) is satisfied for a scalar resistivity; for $B(x) \neq$ const, this condition can be satisfied only when a tensor resistivity is admitted. Essentially, the main reason for this is the circumstance that the introduction of a tensor resistivity transforms the optimization problem with differential constraints into an isoperimetric problem.

Returning to the problem (3.154) and (3.231), we note that, in accordance with the results of the preceding sections, attainment of a maximum

---

* In order to make the comparison correctly, it is necessary to set $R = 0$ in (3.234). What we have said also remains true for $R \neq 0$; in this case, following Section 3.10, $B(x)$ in (3.154) must be replaced by

$$B(x) - \frac{cIR}{2V\delta}$$

(we assume $V = $ const); the variational problem (3.234) can readily be reduced to an analogous form.

requires satisfaction of the condition

$$\int_{L(w)} VB \, dy \geq 0$$

almost everywhere in g. This condition is equivalent to condition (3.235); it appeared as a consequence of our consideration of local variations—strip-shaped inclusions—described in Section 3.12 [in the footnote between (3.183) and (3.184)]. Such variations are compatible with the constraint (3.227). In Section 3.12, we also considered weak global variations of the current line that satisfy (3.227); they led to Jacobi's necessary condition.

Variations of each of these types characterize definite properties of the optimal solution, but they are not well suited to prove its existence. The isoperimetric nature of the problem (3.154), (3.231), and (3.227) makes it possible to use an essentially different type of variation for this purpose: the geometrical operation known as Steiner symmetrization. A detailed description of this operation can be found in Pólya and Szegö's book[132]; here we give only the definition of the symmetrization and list its main properties.

Let B be some body in three-dimensional space and P be a plane, which will be referred to as the symmetrization plane. Symmetrization of B with respect to P consists of constructing another body B^* as follows:

(I) B^* is symmetric with respect to P.

(II) Any straight line perpendicular to P that intersects either B or B^* intersects the other of these bodies; the chords cut out on this line by the two bodies have the same length.

(III) The intersection of this straight line with B^* consists of only one segment which may degenerate to a point (the intersection with B may consist of several segments). This segment is divided in two by a plane P.

Consider a plane Q which is perpendicular to P and which intersects this one by a straight line l. The plane figures which appear as intersections of Q with the bodies B and B^* will be called, respectively, b and b^*. We shall say that the figure b (which may be either multiconnected or represented as a sum of several figures of different order of connection) is transformed to the figure b^* by symmetrization with regard to the straight line l. This line is called an axis of symmetrization.

The figure b^* is symmetric with regard to the axis l; any straight line perpendicular to l intersects the boundary of b^* in no more than two symmetrically placed points.

An example of symmetrization is given by Fig. 3.35: the figure B is obtained as a result of symmetrization of A with regard to the y axis, and the figure C appears as a result of symmetrization of B with regard to the x axis. The result of two successive symmetrizations depends on their order.

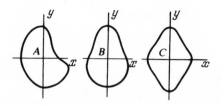

FIGURE 3.35

Symmetrization with regard to two perpendicular axes results in a simple connected figure bounded by symmetric curves monotonic with reference to those axes.

The importance of symmetrization for problems of mathematical physics is associated with the fact that this operation does not influence the values of a number of important functionals; the values of some others are changed in some fully determinate directions. A detailed description of these symmetrization properties can be found in Ref. 132; for our purposes, it suffices to mention only some of these properties:

1. Symmetrization of a body about any plane leaves its volume unchanged and does not increase the area of the surface. Symmetrization of a planar figure about any straight line lying in its plane leaves the area unchanged and does not increase the perimeter of the figure.
2. Symmetrization does not increase the Dirichlet integral of a positive function which vanishes* on the boundary. Let us clarify this result. Let D be a region in the (x, y) plane and $f(x, y)$ a function that is positive on D and vanishes on its boundary. We consider the points (x, y, z); (x, y) belongs to D and $0 \le z \le f(x, y)$.

The set of such points forms a certain body B; this body is bounded by the base D and the "upper surface" $z = f(x, y)$.

Under symmetrization of B with respect to the plane (y, z) the area of its surface is not increased; actually, this applies only to the area of the upper surface since the area of the base remains unchanged. Thus,

$$\iint_D [1 + f_x^2 + f_y^2]^{1/2} \, dx \, dy \ge \iint_{D^*} [1 + (f_x^*)^2 + (f_y^*)^2]^{1/2} \, dx \, dy.$$

Here, D^* is the symmetrized base, and $z = f^*(x, y)$ is the symmetrized upper surface. In the last inequality we replace f by εf, where ε is a small

---

* The result is not affected if the function is defined only to within a constant term.

positive constant, and expand both sides in a series in powers of ε. Remembering that D and D^* have the same area, and going to the limit $\varepsilon \to 0$, we obtain

$$\iint_D (f_x^2 + f_y^2)\, dx\, dy \geq \iint_{D^*} [(f_x^*)^2 + (f_y^*)^2]\, dx\, dy,$$

which is what we wanted to prove.

We now turn to problem (3.154), (3.231), and (3.227) and examine the effect of symmetrization on the functional in (3.154), namely,

$$\varphi = \frac{\left(\displaystyle\int_{-1}^{1} dw \sqrt{\frac{1}{c} \int_{L(w)} B(x)\, dy} \right)^2}{\displaystyle\iint_G (\nabla w)^2\, dx\, dy}$$

(we may assume that $V = \text{const}$ without loss of generality).

The function w is odd with respect to x; it is defined in all of the working region G of the channel $CCC'C'$ (see Fig. 3.16). As before, we assume that the contour lines of the function are continuous and connect the electrodes.

We take a strip of width 2δ and construct two identical nonoverlapping regions G and G_1 within the strips (see Fig. 3.36). We denote by E the region consisting of G, G_1, and the part D of the strip between them; the region E is bounded by the curves $B'C'C'B'$ and $B_1'C_1'C_1'B_1'$ and two segments of the straight lines $y = \pm\delta$. We shall assume that the curves $B'C'C'B'$, $BCCB$, $B_1C_1C_1B_1$, and $B_1'C_1'C_1'B_1'$ are variable curves situated outside the rectangles $B'BBB'$ and $B_1B_1'B_1'B_1$ but inside the strip. In particular, it is not necessary to require that these curves contain the horizontal sections $B'C'$, BC, B_1C_1, $B_1'C_1'$; the configurations shown in Fig. 3.36 using dashed curves are admissible. We shall assume that the area of each of the regions G and G_1 is given by (3.227).

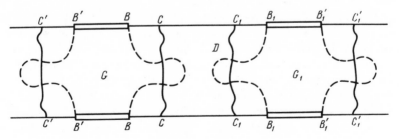

FIGURE 3.36

We define the function w in E as follows: $w = -1$ on $B'C'C'B'$ and $B_1'C_1'C_1'B_1'$, $w = 1$ on $BCCB$ and $B_1C_1C_1B_1$ and in the region D between them; i.e., at the points of the region E which lie neither in G nor in G_1. In the open region G, the function w is taken to be twice continuously differentiable with respect to the two variables x, y, and we assume that the derivatives w_x and w_y do not vanish simultaneously; the contour lines $w = \text{const}$ join the upper and lower electrodes $B'B$ to one another, and the parameter w on these lines increases monotonically from -1 on $B'C'C'B'$ to 1 on $BCCB$. In the region G_1, we use the even extension of w. As a whole, w is continuous in E, and the functional

$$\frac{\left(\int_{-1}^{1} dw\,\sqrt{\frac{1}{c}\int_{L(w)} B(x)\,dy}\right)^2}{\iint_G (\nabla w)^2\,dx\,dy} + \frac{\left(\int_{-1}^{1} dw\,\sqrt{\frac{1}{c}\int_{L(w)} B(x)\,dy}\right)^2}{\iint_{G_1} (\nabla w)^2\,dx\,dy}$$

$$= \frac{2\left(\int_{-1}^{1} dw\,\sqrt{\frac{1}{c}\int_{L(w)} B(x)\,dy}\right)^2}{\iint_G (\nabla w)^2\,dx\,dy} \tag{3.237}$$

is equal to $2\rho_{\min}I$. The set of functions w that are defined on E and have the above properties are denoted by $W(E)$.

We introduce rectangular coordinates x, y, z in such a way that E lies in the plane $z = -1$. We denote by B the body which is bounded below by E, above by the surface $z = w(x, y)$, and on the sides by the two planes $y = \pm\delta$. In accordance with what we have said above, the sections of the body B formed by the planes $z = \text{const}$ are simply connected planar regions.

We symmetrize the body B with respect to the plane yz; under this operation, B, E, G, B_1, D, w, I are transformed to B^*, E^*, G^*, G_1^*, D^*, w^*, I^*. The region E^* is obtained by symmetrization of E about the y axis; under this process, mes E = mes E^*.

The region D is the projection of the horizontal (lying in the plane $z = 1$) part of the surface of B onto the plane $z = -1$; under the symmetrization of B, this part of the surface is symmetrized about the axis parallel to the y axis lying in the plane $z = 1$, and its projection D is symmetrized about the axis parallel to the y axis lying in the plane $z = -1$. As a result, it can be shown that mes D = mes D^*; therefore, mes G = mes G^*, and mes G_1 = mes G_1^*; i.e., the symmetrization does not violate conditions (3.227). Under the symmetrization, the surface of the body B does not increase; only the parts of B that have the regions G and G_1 as horizontal projections can decrease. Making the transition from the area of the surface to the Dirichlet integral, as we did above, we conclude that the denominator of the right-hand side of (3.237) does not increase as a result of the symmetrization with respect to the yz plane.

The behavior of the numerator in (3.237) under a symmetrization of this type depends on the nature of the function $B(x)$. Consider Fig. 3.37, which shows a quarter of the channel; suppose that the current line $w = 2_0$, which has the shape *fabcde* before symmetrization, transforms into the line *fade* after the symmetrization. If the numerator of (3.237) is not to decrease as a result, it is sufficient that

$$\int_{ad} B(x)\, dy \geq \int_{abcd} B(x)\, dy$$

or

$$\int_{ad} B(x_3 - x_2 + x_1)\, dy \geq \int_{ab} B(x_3)\, dy + \int_{bc} B(x_2)\, dy + \int_{cd} B(x_1)\, dy$$

and for this, in turn, it is sufficient that for all $x_1 \leq x_2 \leq x_3$

$$B(x_3 - x_2 + x_1) \geq B(x_3) - B(x_2) + B(x_1)$$

Setting $x_2 = \lambda x_1 + (1 - \lambda)x_3, 0 \leq \lambda \leq 1$, in the last inequality, we obtain the equivalent condition

$$B[(1 - \lambda)x_1 + \lambda x_3] + B[\lambda x_1 + (1 - \lambda)x_3] \geq B(x_1) + B(x_3)$$

This condition shows that the function $B(x)$ is convex upward. Thus, if $B(x)$ is convex upward, i.e., represented by a graph of the type shown in Fig. 3.38, then $I^* \geq I$. In other words, symmetrization with respect to the yz plane does not reduce the current drawn from the electrodes.

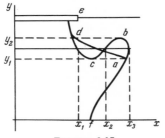

FIGURE 3.37

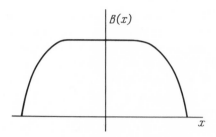

FIGURE 3.38

We now consider the influence of symmetrization with respect to the xz plane. As before, we show that the denominator in (3.237) does not increase (in the proof, we use the requirement that the curvilinear sections $C'C'$, CC, C_1C_1, $C_1'C_1'$ of the boundaries of G and G_1 lie outside the rectangles $BB'B'B$ and $B_1B_1'B_1'B_1$ in Fig. 3.36; we do so because, using this requirement, we can keep the electrodes at a distance 2δ from one another after symmetrization). The numerator, however, is not changed by symmetrization with respect to xz. To prove this, suppose that the sections of the current lines situated in the electrode region $|x| \leq \lambda$ are monotonic in the corresponding quadrants and that meanderings of the current lines can occur only for $|x| > \lambda$.

After symmetrization, the integral (Fig. 3.39)

$$\int_{fabc} B(x) \, dy = \int_f^a B(x) \, dy + \int_a^b B(x) \, dy + \int_b^c B(x) \, dy$$

is replaced by the equivalent integral:

$$\int_{fc} B(x) \, dy = \int_{fc} B(x) \, d(y_1 - y_2 + y_3)$$

$$= \int_b^c B(x) \, dy - \int_b^a B(x) \, dy + \int_f^a B(x) \, dy$$

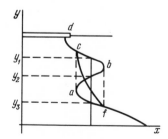

FIGURE 3.39

The integrals $\int B(x)\,dy$, which are taken along the current lines (or parts of them) situated in the electrode region $|x| < \lambda$, do not change as a result of the additional condition imposed upon the comparison curves.

As a result of symmetrization with respect to the two symmetry planes (yz and xz), the current lines in the channel are monotonic in the corresponding quadrants. The symmetrization preserves (3.227) and does not decrease the functional $I(I^* \geq I)$ if $B(x)$ is convex upward; under this condition, it is natural to seek the region which maximizes I in the class of regions that are symmetric with respect to the x and y axes and are bounded by monotonic curves. We see then that the application of symmetrization has made it possible to rule out regions with meandering boundaries as long as these meanderings lie outside the electrode region.

Functions w in $W(G)$ belong the class $C_2(G)$ and therefore form a set which is compact in $C_1(G)$. From any sequence $w_n \in W(G)$ one can thus select a subsequence that converges uniformly together with its first derivatives to a continuously differentiable function w; at the same time, by virtue of continuity, $I[w_n] \to I[w] = \max I$. The function w is monotonic in x for fixed y and, conversely, monotonic in y for fixed x since the functions w_n have this property.

Furthermore, as the limit of a sequence of elements $w_n \in C_2(G)$ the function w has generalized derivatives of second order belonging to $L_2(G)$. It therefore satisfies the Euler equation

$$f_+\Delta w = \frac{VB_x(x)}{2c\rho_{\min}\sqrt{\dfrac{1}{c}\displaystyle\int_{L(w)} VB\,dy}}$$

where the constant f_+ is determined by (3.153). If the derivative $B_x(x)$ satisfies a Lipschitz condition, then w has continuous derivatives of second order, i.e., it is a classical solution of the Euler equation. This result can be obtained by methods similar to those used in Chapter 4 of Ref. 91.

4

Relaxation of Optimization Problems with Equations Containing the Operator $\nabla \cdot \mathbb{D} \cdot \nabla$: An Application to the Problem of Elastic Torsion

In the preceding chapter, we investigated the problem of optimal distribution of the specific resistance of the working fluid in a channel of an MHD power generator. This problem represents a typical example of optimization problems which arise when a control function $\mathbb{D}$ (either scalar or tensor) enters the main part of the elliptic operator $\nabla \cdot \mathbb{D} \cdot \nabla$ of the second order. As was pointed out in the Introduction, the first step in such problems consists of building the G-closure of the initially given set U of controls $\mathbb{D}$; i.e., an extension of this set to some wider set GU including, in addition to U, all possible composites assembled of the initially given components— elements of the set U. The problem of obtaining an invariant description of this wider set is also essential for the theory of composite materials. For a given microstructure of a composite, its effective characteristics may be obtained by the procedure of *homogenization* (see Refs. 403, 405, 406, 421, 427, 441, 525, and 526) which required the solution of some fairly complicated auxiliary boundary-value problems. On the other hand, for many applications, particularly for optimal control and optimal design problems, a description of the *whole set* of materials which may be obtained from some initially given compounds with the aid of the process of mixing is of central interest.[476-479,499,514] Above that, for design purposes, it is necessary to indicate at least one (possibly simple) microstructure which would actually represent any admissible composite material.

Though, at first glance, these two problems seem to provide complexities, they do allow an explicit solution in a number of cases. Particularly, such a solution can be obtained for the elliptic operator $\nabla \cdot \mathbb{D} \cdot \nabla$ in two dimensions. In this chapter we shall give this solution for an arbitrary set U of tensors $\mathbb{D}$ belonging to the space L_∞, where U is assumed to be closed and bounded with respect to the L_∞ norm. At the end of the chapter these results will be applied to the problem of design of an elastic bar of extremal torsional rigidity.

4.1. Problem Statement

We consider a process governed by the equations

$$\nabla \cdot \mathbf{M} = f, \qquad \mathbf{M} = \mathbb{D} \cdot \mathbf{e}, \qquad \mathbf{e} = \nabla w \tag{4.1}$$

or, equivalently, by a single second-order equation

$$Aw \triangleq \nabla \cdot \mathbb{D} \cdot \nabla w = f \tag{4.2}$$

where w denotes the dependent variable, $\mathbb{D}$ a symmetric, positive definite 2×2 tensor depending upon the coordinates (x, y) and characterizing the "conductivity" of a material medium, $f \in L_2(S)$ the sources' density, and S a two-dimensional region. Across the lines of discontinuity of the tensor $\mathbb{D}$, we shall assume that the normal component of $\mathbf{M}$ and the tangential component of $\mathbf{e}$ are continuous with

$$[M_n]_-^+ = 0, \qquad [e_t]_-^+ = 0$$

Typical problems described by (4.1) are from electrostatics, heat theory, the theory of elasticity, etc. Particularly, the MHD problem of Chapter 3 can also be formulated in this way.

Equation (4.2) now is to be solved in $S + \partial S$ under some prescribed boundary conditions along ∂S, its solution w being determined as the element of $W_2^1(S)$ satisfying the boundary conditions and the integral identity

$$\iint_S \nabla \eta \cdot \mathbb{D} \cdot \nabla w \, dx \, dy = - \iint_S \eta f \, dx \, dy, \qquad \forall \eta \in \mathring{W}_2^1(S) \tag{4.3}$$

The admissible set $U \triangleq \{\mathbb{D}^k\}$ of tensors $\mathbb{D}^k$ will be supposed to possess the following properties:

(i) The U set is a bounded closed set of $L_\infty(S)$:

$$U: -\infty < (d_{ij})_{\min} \leq d_{ij}^{(k)} \leq (d_{ij})_{\max} < \infty \quad (i, j = 1, 2)$$

(ii) The eigenvalues $\Lambda_1^{(k)}, \lambda_2^{(k)}$ of any tensor $D^k \in U$ satisfy the inequalities

$$0 < \alpha \leq \lambda_1^{(k)} \leq \lambda_2^{(k)} \leq \beta < \infty$$

where α and β denote some fixed constants.

(iii) If $D \in U$, then one also has $\mathbb{B}^T \cdot \mathbb{D} \cdot \mathbb{B} \in U$, where $\mathbb{B}$ denotes the plane rotation tensor. The latter condition means that the U set restricts only the invariant characteristics of the admissible tensors, and not the orientation of their principal axes in a plane.

We shall be interested in the material characteristics of conductive mixtures (composites) which may be assembled from the elements belonging to the initially given set U. Divide the S region into the parts occupied by different materials of the U set; each subdivision of this kind may include a finite or an infinite number of parts. Consider the sequence $\{\mathbb{D}^k\}$ of such subdivisions; without loss of generality we may speak of media with microstructure, meaning that from some number k onward, in each ε vicinity of any point of the S region, zones occupied by various materials of the U set may be found.

The *effective* tensor $\mathbb{D}^0$ of conductivity now is defined as a limit for $k \to \infty$ in the sense of G-convergence of operators[448,481,525,526]:

$$A_k \triangleq \nabla \cdot \mathbb{D}^k \cdot \nabla \underset{G}{\rightrightarrows} A_0 \triangleq \nabla \cdot \mathbb{D}^0 \cdot \nabla$$

Here, $\mathbb{D}^k \in U$, and $\mathbb{D}^0$ denotes some symmetric, positive definite 2×2 tensor belonging to some extended set GU ($U \in GU$). This new set is known[448,481] as the *G-closure* of U.

According to the definition of G-convergence,[448,481,525] the operator A_0 is called a G-limit of the sequence $\{A_k\}$ of operators, if for any $f \in L_2(S)$ and $g \in \mathring{W}_2^1(S)$ there holds the limiting equality

$$\lim_{k \to \infty} (g, A_k^{-1} f) = (g, A_0^{-1} f)$$

In other words, all weak limits w^0 (and only those) of the sequences $\{w^k\}$ of solutions to (4.2) with $\mathbb{D} = \mathbb{D}^k$ satisfy the relationship

$$\nabla \cdot \mathbb{D}^0 \cdot \nabla w^0 = f \qquad (w^k \xrightarrow[W_2^1(S)]{} w^0)$$

a fact which holds for *any* right-hand side $f \in L_2(S)$.* For this reason, the element $\mathbb{D}^0$ should be considered dependent only upon the U set and upon

---

* Here and below, the symbol $\to$ denotes a weak convergence in the corresponding space.

the particular subdivision of the S region. The sequence $\{e^k\} = \{\nabla w^k\}$ which is weakly convergent in $L_2(S)$ to the element $e^0 = \nabla w^0$,

$$e^k \xrightarrow[L_2(S)]{} e^0$$

now is put into correspondence with the sequence $\{\mathbf{M}^k\} = \{\mathbb{D}^k \cdot \nabla w^k\}$ which obviously is also weakly convergent in $L_2(S)$. It converges to the limit $\mathbf{M}^0 = \mathbb{D}^0 \cdot \nabla w^0$:

$$\mathbf{M}^k \xrightarrow[L_2(S)]{} \mathbf{M}^0$$

The tensor $\mathbb{D}^0$ is seen to represent the tensorial "coefficient of proportionality" between the weak limits $\mathbf{M}^0$ and e^0.

For the prescribed subdivision of S, one may calculate the characteristics of the $\mathbb{D}^0$ tensor with the aid of the procedure of homogenization.[421] This technique introduces the "slow" (macroscopic) and "fast" (microscopic) independent variables x, y and $x/\varepsilon, y/\varepsilon$, respectively, and provides the asymptotic expansion of the solution into power series in ε. For microstructures possessing certain regularity properties, the equation for the main part of this expansion (independent of ε) is obtained. The corresponding equation in "slow" variables represents the G-limiting form of the initial equation, and the $\mathbb{D}^0$ tensor which enters this limiting equation, simply describes the effective characteristics of the composite itself. To determine $\mathbb{D}^0$ for a periodic microstructure, one has to solve the corresponding boundary-value problem.[421]

In what follows, we are going to solve a different problem. Our goal will be to describe the *whole* set GU of effective tensors $\mathbb{D}^0$ generated (in a sense of G-convergence) by the elements of the initial set U. The term "description" presupposes determination of the set of invariants of the tensors $\mathbb{D}^0 \in GU$. This set will also be denoted by GU and called the G-closure of U. At the same time, we shall attempt to indicate at least one (possibly simple) particular microstructure providing (in the sense of G-convergence) any prescribed point of the GU set. In other words, it is required to find the form of those microinclusions which approximate any prescribed tensor $\mathbb{D}^0 \in GU$.

The problem just formulated is essential not only for the theory of composite materials; its solution also enables us to give a correct formulation of a broad class of problems of optimal design of conductive media.[477,479,499,514] The reason is that for weakly continuous functionals of solutions to such problems, we are ready to guarantee the existence of optimal controls belonging to GU (note, referring to Chapter 3, that such controls may *not* exist within the initial U set, in general).

The elements of the U set (initially given materials) are characterized (a) by the eigenvalues of the tensors $\mathbb{D}^k$ and (b) by the orientation of their principal axes. For this reason, it is natural to begin with some simple cases, namely:

I. Find the G-closure of the U set including the isotropic tensors $\mathbb{D}^k = d^{(k)}\mathbb{E}$ where $\mathbb{E}$ is a unit 2×2 tensor, and $d^{(k)} \in [\alpha, \beta]$.

II. Find the G-closure of the U set including the tensors $\mathbb{B}_k^T \cdot \mathbb{D}_1 \cdot \mathbb{B}_k$, where $\mathbb{D}_1 = \mathbb{D}_1(x, y)$ is a fixed tensor satisfying conditions (i), (ii), and $\mathbb{B}_k = \mathbb{B}_k(x, y)$ is a rotation tensor.

Problem I has been examined in Refs. 499 and 514, where it was shown that the eigenvalues λ_1, λ_2 of $\mathbb{D}^0$ satisfy the inequalities

$$\alpha \le \frac{\alpha\beta}{\alpha + \beta - \lambda_2} \le \lambda_1 \le \lambda_2 \le \beta \tag{4.4}$$

In the (λ_1, λ_2) plane, the corresponding GU region is bounded by the curve

$$\lambda_1 = \frac{\alpha\beta}{\alpha + \beta - \lambda_2} \tag{4.5}$$

and a segment of the diagonal $\lambda_1 = \lambda_2$ (Fig. 4.1). The curve corresponds to a medium with a layered microstructure.[499,514] The other points of the GU region will be approximated below with the aid of the solution to Problem II, which is also given in what follows. Note that some estimates of the GU region for Problem II have been obtained earlier in Ref. 476.

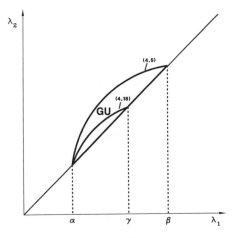

FIGURE 4.1

In an effort to give a complete description of the whole set GU, we shall proceed in the following way. First, we shall establish certain inequalities which determine some wider set Σ, containing GU in itself. Second, we shall try some specific microstructure and prove that it realizes any point of Σ; in this way, the sets Σ and GU will be shown to be the same.

The mentioned inequalities connect the tensors $\mathbb{D}^0$ and $\mathbb{D}^* = \lim_{\text{weak}} \mathbb{D}^k$ as well as $(\mathbb{D}^0)^{-1}$ and $(\mathbb{D}^{-1})^* = \lim_{\text{weak}} (\mathbb{D}^k)^{-1}$. These inequalities are

$$\lambda_i[\mathbb{D}^* - \mathbb{D}_0] \geq 0 \qquad (i = 1, 2) \qquad (4.6)$$

$$\lambda_i[(\mathbb{D}^{-1})^* - (\mathbb{D}^0)^{-1}] \geq 0 \qquad (i = 1, 2) \qquad (4.7)$$

Here, $\lambda_i[\,\cdot\,]$ denotes the ith eigenvalue of the tensor in the square brackets. The tensor $\mathbb{D}^*$ and $(\mathbb{D}^{-1})^*$ depend only upon the characteristics of the initial materials and their concentration rates, whereas the effective tensor $\mathbb{D}^0$ is also determined by the *form* of the inclusions; i.e., by the particular character of the microstructure of a composite.

In what follows, we shall also need the formula expressing the effective tensor $\mathbb{D}^0$ of a layered composite medium including only two materials with the conductivity tensors $\mathbb{D}_+$ and $\mathbb{D}_-$ [see (3.46)]:

$$\mathbb{D}^0 = \mathbb{D}^* - \frac{(\mathbb{D}^* - \mathbb{D}_-) \cdot \mathbf{nn} \cdot (\mathbb{D}_+ - \mathbb{D}^*)}{\mathbf{n} \cdot (\mathbb{D}_+ + \mathbb{D}_- - \mathbb{D}^*) \cdot \mathbf{n}} \qquad (4.8)$$

Here, $\mathbf{n}$ denotes a unit vector normal to the layers, and

$$\mathbb{D}^* = \lim_{\text{weak}} \mathbb{D}^k = m\mathbb{D}_+ + (1 - m)\mathbb{D}_- \qquad (4.9)$$

where m denotes the concentration of the $\mathbb{D}_+$ material. Note that (4.5) follows from (4.8) with $\mathbb{D}_+ = \alpha\mathbb{E}, \mathbb{D}_- = \beta\mathbb{E}$.

By analogy with Section 3.4 for the plane case we obtain the formula for the tensor $(\mathbb{D}^0)^{-1}$ of a layered medium

$$(\mathbb{D}^0)^{-1} = (\mathbb{D}^{-1})^* - \frac{[(\mathbb{D}^{-1})^* - \mathbb{D}_-^{-1}] \cdot \mathbf{tt} \cdot [\mathbb{D}_+^{-1} - (\mathbb{D}^{-1})^*]}{\mathbf{t} \cdot [\mathbb{D}_+^{-1} + \mathbb{D}_-^{-1} - (\mathbb{D}^{-1})^*] \cdot \mathbf{t}} \qquad (4.10)$$

Here, $\mathbf{t}$ denotes a unit vector parallel to the layers, and $(\mathbb{D}^{-1})^* = \lim_{\text{weak}} (\mathbb{D}^k)^{-1}$.

4.2. Weak Convergence in Energy and Estimates of the Set GU

In this section, we shall derive the inequalities (4.6) and (4.7). The procedure will be based on weak convergence in energy

$$N \triangleq \mathbf{M} \cdot \mathbf{e}$$

in some space $L_{p/2}(S)$, $p > 2$.

Consider the sequence $\{\mathbb{D}^i\}$ of "conductivities" ($\mathbb{D}^i \in U$) satisfying the conditions (i)–(iii) of the preceding section, and the corresponding sequences $\{\mathbf{e}^i = \nabla w^i\}$ and $\{\mathbf{M}^i\}$. The latter sequences may be assumed uniformly bounded in some $L_p(S)$, $p > 2$, and (presumably, some subsequences) weakly convergent in this space. We introduce the weak limits:

$$\mathbf{e}^i \xrightarrow[L_p(S)]{} \mathbf{e}^0, \qquad \mathbf{M}^i \xrightarrow[L_p(S)]{} \mathbf{M}^0 \tag{4.11}$$

Bearing in mind $(4.1)_2$, we may write the following expression for the energy

$$N^i \triangleq \mathbf{M}^i \cdot \mathbf{e}^i = \mathbf{e}^i \cdot \mathbb{D}^i \cdot \mathbf{e}^i \tag{4.12}$$

We shall prove that for $f \in L_2(S)$ [see $(4.1)_1$], one has the limiting relation

$$N^i \triangleq \mathbf{M}^i \cdot \mathbf{e}^i \xrightarrow[L_{p/2}(S)]{} \mathbf{M}^0 \cdot \mathbf{e}^0 \triangleq N^0 \tag{4.13}$$

According to the integral identity (4.3), we have

$$\iint_S \nabla w^i \cdot \mathbb{D}^i \cdot \nabla \eta^i \, dx\, dy = -\iint_S f \eta^i \, dx\, dy, \qquad \forall \eta^i \in \overset{\circ}{W}_2^1(S) \tag{4.14}$$

We shall take the test function η^i equal to $\xi(w^i - w^0)$, where $\xi(x, y)$ is some sufficiently smooth function, and w^i, w^0 denote the state variables corresponding to $\mathbf{e}^i$, $\mathbf{e}^0$, respectively ($\mathbf{e}^i = \nabla w^i$, $\mathbf{e}^0 = \nabla w^0$). Equation (4.14) is now reduced to

$$\iint_S \xi \nabla w^i \cdot \mathbb{D}^i \cdot (\nabla w^i - \nabla w^0) \, dx\, dy$$

$$= \iint_S (w^i - w^0) \nabla w^i \cdot \mathbb{D}^i \cdot \nabla \xi \, dx\, dy - \iint_S f \xi (w^i - w^0) \, dx\, dy \tag{4.15}$$

Recall that $f \in L_2(S)$. Then for any strictly internal domain S', $S' \in S$ all w^i belong to some bounded set of $W_p^1(S')$ for some $p > 2$.[483] The right-hand side of (4.15) obviously tends to zero, and, in view of arbitrariness of ξ, we obtain [see (4.1)$_2$]

$$\mathbf{M}^i \cdot (\mathbf{e}^i - \mathbf{e}^0) \xrightarrow[L_{p/2}(S')]{} 0$$

A more detailed analysis shows that for a smooth boundary S, one may substitute S' for S in this limiting equation, and, furthermore, one may use weaker restrictions for the density f of external sources. We hence arrive at the relationship (4.13):

$$N^i \triangleq \mathbf{M}^i \cdot \mathbf{e}^i \xrightarrow[L_{p/2}(S)]{} \mathbf{M}^0 \cdot \mathbf{e}^0 \triangleq N^0$$

The weak limits $\mathbf{M}^0$ and $\mathbf{e}^0$ are connected by some bounded linear operator L since the operation of transition to a weak limit is linear. The operator L can also be shown to be local; that is, $\mathbf{M}^0(x, y) = L\mathbf{e}^0(x, y)$ for any $(x, y) \in S$. The limiting function N^0 is positive as the weak limit of positive functions $N^i \triangleq \mathbf{M}^i \cdot \mathbf{e}^i = \mathbf{e}^i \cdot \mathbb{D}^i \cdot \mathbf{e}^i$ ($\mathbb{D}^i$ are positive definite symmetric tensors of rank two), and therefore L may be considered as the positive definite symmetric tensor $\mathbb{D}^0$ of rank two:

$$L \equiv \mathbb{D}^0, \qquad \mathbf{M}^0 = \mathbb{D}^0 \cdot \mathbf{e}^0$$

Weak convergence of the energy N allows one to obtain some estimates of $\mathbb{D}^0$. Define $\mathbb{D}^*$ as the weak limit of the sequence $\{\mathbb{D}^i\}$:

$$\mathbb{D}^i \xrightarrow[L_\infty(S)]{} \mathbb{D}^*$$

Note that the weak limit $\mathbb{D}^*$ and the effective tensor $\mathbb{D}^0$ of conductivity are essentially different tensors. In the following it will become apparent that the tensor $\mathbb{D}^*$ serves as a tool providing the required estimates of the tensor $\mathbb{D}^0$. The advantageous feature of $\mathbb{D}^*$ is that it can easily be calculated for the given compounds by a simple averaging over the element of microstructure. For example, two materials $\mathbb{D}_+$ and $\mathbb{D}_-$ represented with concentrations m and $1 - m$, respectively, produce $\mathbb{D}^* = m\mathbb{D}_+ + (1 - m)\mathbb{D}_-$.

The sequence $\{A^i\}$,

$$A^i \triangleq 2(\mathbf{M}^i \cdot \mathbf{e}^i - \mathbf{M}^i \cdot \mathbf{e}^0) = 2(\mathbf{e}^i - \mathbf{e}^0) \cdot \mathbb{D}^i \cdot \mathbf{e}^i$$

is weakly convergent to zero:

$$A^i \xrightarrow[L_{p/2}(S)]{} 0$$

Transforming A^i, we obtain

$$A^i \triangleq (\mathbf{e}^i - \mathbf{e}^0) \cdot \mathbb{D}^i \cdot (\mathbf{e}^i - \mathbf{e}^0) - \mathbf{e}^0 \cdot \mathbb{D}^i \cdot \mathbf{e}^0 + \mathbf{e}^i \cdot \mathbb{D}^i \cdot \mathbf{e}^i \xrightarrow[L_{p/2}(S)]{} 0 \quad (4.16)$$

The first term of the left-hand side is nonnegative; passing over to the weak limit, we get

$$\mathbf{e}^0 \cdot \mathbb{D}^0 \cdot \mathbf{e}^0 \le \mathbf{e}^0 \cdot \mathbb{D}^* \cdot \mathbf{e}^0$$

from which it follows that the eigenvalues of $\mathbb{D}^* - \mathbb{D}^0$ are nonnegative:

$$\lambda_i[\mathbb{D}^* - \mathbb{D}^0] \ge 0 \quad (i = 1, 2)$$

We have thus obtained inequality (4.6). To prove inequality (4.7), we begin with the limiting relationship

$$B^i \triangleq 2(\mathbf{M}^i \cdot \mathbf{e}^i - \mathbf{M}^0 \cdot \mathbf{e}^i) = (\mathbf{M}^i - \mathbf{M}^0) \cdot (\mathbb{D}^i)^{-1} \cdot (\mathbf{M}^i - \mathbf{M}^0)$$

$$- \mathbf{M}^0 \cdot (\mathbb{D}^i)^{-1} \cdot \mathbf{M}^0 + \mathbf{M}^i \cdot (\mathbb{D}^i)^{-1} \mathbf{M}^i \xrightarrow[L_{p/2}(S)]{} 0$$

In view of positive definiteness of $(\mathbb{D}^i)^{-1}$ and of the weak convergence of the expression

$$\mathbf{M}^i \cdot (\mathbb{D}^i)^{-1} \cdot \mathbf{M}^i = \mathbf{M}^i \cdot \mathbf{e}^i \xrightarrow[L_{p/2}(S)]{} \mathbf{M}^0 \cdot \mathbf{e}^0 = \mathbf{M}^0 \cdot (\mathbb{D}^0)^{-1} \cdot \mathbf{M}^0$$

we arrive at the inequality

$$\mathbf{M}^0 \cdot (\mathbb{D}^0)^{-1} \cdot \mathbf{M}^0 \le \mathbf{M}^0 \cdot (\mathbb{D}^{-1})^* \cdot \mathbf{M}^0$$

where, by the definition,

$$(\mathbb{D}^i)^{-1} \xrightarrow[L_\infty(S)]{} (\mathbb{D}^{-1})^*$$

The eigenvalues of the tensor $(\mathbb{D}^{-1})^* - (\mathbb{D}^0)^{-1}$ thus are nonnegative,

$$\lambda_i[(\mathbb{D}^{-1})^* - (\mathbb{D}^0)^{-1}] \ge 0 \quad (i = 1, 2)$$

which is the same as (4.7).

4.3. *G*-Closure of a Set *U* Consisting of Two Isotropic Plane Tensors of the Second Rank (Problem I)

The inequalities (4.6), (4.7) will be applied here for derivation of the inequalities (1.4) which characterize the GU set for this case.

The initial set U is represented by two isotropic tensors

$$\mathbb{D}_+ = \alpha(\mathbf{ii} + \mathbf{jj}), \qquad \mathbb{D}_- = \beta(\mathbf{ii} + \mathbf{jj}), \qquad 0 < \alpha \leq \beta < \infty$$

The tensor $\mathbb{D}^0$ is of the following general structure:

$$\mathbb{D}^0 = \lambda_1 \mathbf{aa} + \lambda_2 \mathbf{bb}, \qquad \lambda_1 \leq \lambda_2$$

where λ_1, λ_2 denote the eigenvalues and $\mathbf{a}, \mathbf{b}$ are a pair of orthogonal unit vectors.

Let m denote a concentration of the $\mathbb{D}_+$ material; (4.6), (4.7) show that

$$\lambda_2(\mathbb{D}^0) \leq \lambda_2(\mathbb{D}^*) \leq m\alpha + (1 - m)\beta$$

$$\lambda_2((\mathbb{D}^0)^{-1}) = \frac{1}{\lambda_1(\mathbb{D}^0)} \leq \lambda_2((\mathbb{D}^{-1})^*) \leq \frac{m}{\alpha} + \frac{1 - m}{\beta}$$

Combining these inequalities, we get

$$\alpha \leq \frac{\alpha\beta}{m\beta + (1 - m)\alpha} \leq \lambda_1(\mathbb{D}^0) \leq \lambda_2(\mathbb{D}^0) \leq m\alpha + (1 - m)\beta \leq \beta \quad (4.17)$$

which means in particular that $(\beta \geq \alpha)$

$$m \leq \frac{\beta - \lambda_2}{\beta - \alpha}$$

This estimate may be used to eliminate the parameter m from the second inequality of (4.17); we obtain

$$\alpha \leq \frac{\alpha\beta}{\alpha + \beta - \lambda_2} \leq \lambda_1 \leq \lambda_2 \leq \beta$$

which is the same as the inequalities (4.4). The equality sign [(4.5)] in the second inequality is achieved at the layered microstructure. This may readily be seen from (3.47) and (3.50). The principal directions **a** and **b** of the effective tensor $\mathbb{D}^0$ then coincide with the normal and tangent to the layers, respectively. Figure 4.1 represents a part of the plane (λ_1, λ_2) bounded by the segment of the curve (4.5) and that of a diagonal between the points (α, α) and (β, β). In the meantime this domain should be considered only as a Σ domain as introduced in Section 4.1, since we have not as yet indicated a microstructure which realizes *all* points of it. In particular, we still have given no indication of what microstructure actually realizes the points on the diagonal. The answer to that question will be postponed until the next section; now, we simply observe that if we were given a *continuous* set of materials whose conductivities belong to the interval $[\alpha, \beta]$ instead of a pair of materials with the conductivities α and β (Problem I of Section 4.1), then the problem of finding microstructures that imitate all points of the Σ domain would allow a simple solution. It would then be sufficient to assemble a layered composite of the components possessing the conductivities α and γ, where γ is any *intermediate* value of the conductivity $(\alpha < \gamma < \beta)$; this microstructure would then be set into correspondence with the curve (Fig. 4.1)

$$\lambda_1 = \frac{\alpha\gamma}{\alpha + \gamma - \lambda_2} \tag{4.18}$$

which sweeps over all of the Σ domain when the parameter γ assumes the values between α and β. In the absence of materials possessing the intermediate values of conductivity, the points of Σ can be imitated in a different way.

4.4. *G*-Closure of a Set U Consisting of Symmetric Plane Tensors of the Second Rank Possessing Different Orientations of the Principal Axes (Problem II)

The inequalities (4.6), (4.7) applied to Problem I provide[476] the characteristic of the corresponding G-closure in terms of inequalities (4.4) which restrict some domain at the (λ_1, λ_2) plane.

One of the boundaries (4.5) of this domain corresponds to a layered composite "of the first rank" assembled of the initially given isotropic compounds. To imitate the other points of GU, it has been suggested in Ref. 476 to introduce a layered composite "of the second rank" which is nothing but a sequence of layers made of one and the same composite of the first rank and different from one another only by orientation of the principal axes of the tensor $\mathbb{D}$. The calculation by (4.8) now shows that the $\mathbb{D}^0$ tensor for this case ($\mathbb{D}_- = \mathbb{B}^T \cdot \mathbb{D}_+ \cdot \mathbb{B}$) possesses the same second invariant as the common value of those for $\mathbb{D}_+$ and $\mathbb{D}_-$. It is remarkable that this statement is true irrespective of the angle of relative rotation of the principal axes of the tensors $\mathbb{D}_+$ and $\mathbb{D}_-$ (i.e., of the tensor $\mathbb{B}$) within the layers which constitute a composite. In other words, the eigenvalues λ of $\mathbb{D}^0$ depend on this rotation only when taken separately; however, this not the case for their product $\lambda_1 \lambda_2 = I_2(\mathbb{D}^0)$.

To prove this, we note that the following identity

$$\mathbb{D}^{-1} = \frac{1}{I_2(\mathbb{D})} \mathbb{O}^T \cdot \mathbb{D} \cdot \mathbb{O} \tag{4.19}$$

holds, for the 2×2 tensors $\mathbb{D}$, where $\mathbb{O} = \mathbf{ij} - \mathbf{ji}$ is the plane rotation tensor for an angle $\pi/2$; this tensor satisfies the obvious conditions

$$\mathbb{O} \cdot \mathbb{O} = \mathbb{O}^T \cdot \mathbb{O}^T = -\mathbb{E}, \qquad \mathbb{O} \cdot \mathbb{O}^T = \mathbb{O}^T \cdot \mathbb{O} = \mathbb{E}$$

Transforming (4.10) with the aid of (4.9) and (4.19), and bearing in mind that

$$I_2(\mathbb{D}_+) = I_2(\mathbb{D}_-), \qquad \mathbf{t} = \mathbb{O} \cdot \mathbf{n}$$

we arrive at the formula

$$\frac{1}{I_2(\mathbb{D}^0)} \mathbb{O}^T \cdot \mathbb{D}^0 \cdot \mathbb{O} = \frac{1}{I_2(\mathbb{D}_+)} \mathbb{O}^T \cdot \left[\mathbb{D}^* - \frac{(\mathbb{D}^* - \mathbb{D}_-) \cdot \mathbf{nn} \cdot (\mathbb{D}_+ - \mathbb{D}^*)}{\mathbf{n} \cdot (\mathbb{D}_+ + \mathbb{D}_- - \mathbb{D}^*) \cdot \mathbf{n}} \right] \cdot \mathbb{O}$$

Comparing this with (4.8), we obtain

$$I_2(\mathbb{D}^0) = \lambda_1 \lambda_2 = I_2(\mathbb{D}_+) = \text{const} \tag{4.20}$$

At the plane (λ_1, λ_2) of the eigenvalues of $\mathbb{D}^0$, layered composites are now seen to be corresponding to the points of a *curve* (4.20) (Fig. 4.2) rather than to some two-dimensional domain like that of (4.4) for problem I (Fig. 4.1).

This observation gives rise to the hypothesis that (4.20) is equally true for the general case when a composite possesses arbitrary microstructure but is assembled of compounds which differ from one another only by orientation of their principal axes of conductivity.

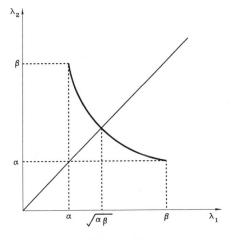

FIGURE 4.2

To prove this conjecture, consider the sequence $\{\mathbb{D}^k\}$ of 2×2 tensors of conductivity which differ only by a rotation of their axes; that is, $\mathbb{D}^k = \mathbb{B}_k^T \cdot \mathbb{D}_1 \cdot \mathbb{B}_k$ (for notation see Section 4.1).

We have

$$\mathbf{M}^k = \mathbb{D}^k \cdot \mathbf{e}^k = I_2(\mathbb{D}_1)\mathbb{O}^T \cdot (\mathbb{D}^k)^{-1} \cdot \mathbb{O} \cdot \mathbf{e}^k \tag{4.21}$$

which means that

$$\mathbb{O} \cdot \mathbf{e}^k = I_2^{-1}(\mathbb{D}_1)\mathbb{D}^k \cdot \mathbb{O} \cdot \mathbf{M}^k \tag{4.22}$$

Since $e^k = \nabla w^k$, one has the relationship

$$\nabla \cdot \mathbb{O} \cdot \mathbf{e}^k = 0 \tag{4.23}$$

On the other hand, with the partial solution $\mathbf{M}^x$ of the equation $\nabla \cdot \mathbf{M}^k = f$, we may represent the vector $\mathbf{M}^k$ in the following form:

$$\mathbf{M}^k = -\mathbf{i}\varphi_y^k + \mathbf{j}\varphi_x^k + \mathbf{M}^x$$

and, consequently,

$$\mathbb{O} \cdot \mathbf{M}^k = \nabla \varphi^k + \mathbb{O} \cdot \mathbf{M}^x$$

Making use of (4.22) and (4.23), we obtain

$$\nabla \cdot \mathbb{D}^k \cdot \mathbb{O} \cdot \mathbf{M}^k = \nabla \cdot \mathbb{D}^k \cdot (\nabla \varphi^k + \mathbb{O} \cdot \mathbf{M}^x) = 0 \tag{4.24}$$

The operator in (4.24) has the same G-closure as the operator appearing in the initial problem (4.2) (see Ref. 421). In fact, the explicit formulas for $\mathbb{D}^0$ (see Ref. 427) show that shifting of the inhomogeneous member f (which obviously depends upon the "slow" variables) into the main part of the operator (4.24) (the term $\mathbb{O} \cdot \mathbf{M}^x$) does not influence the final result. The reason is that for the solutions within the microstructure itself, both factors play the role of some normalizing uniform external fields.

This means that for the G-limit of (4.24), we should take the equation

$$\nabla \cdot \mathbb{D}^0 \cdot (\nabla \varphi^0 + \mathbb{O} \cdot \mathbf{M}^x) = 0$$

For the G-limiting relationship (4.22) we obtain

$$\mathbb{O} \cdot \mathbf{e}^0 = \frac{1}{I_2(\mathbb{D}_1)} \mathbb{D}^0 \cdot \mathbb{O} \cdot \mathbf{M}^0 \quad (\nabla \varphi^k + \mathbb{O} \cdot \mathbf{M}^x \xrightarrow[L_2(S)]{} \nabla \varphi^0 + \mathbb{O} \cdot \mathbf{M}^0) \quad (4.25)$$

On the other hand, we have

$$\mathbf{e}^0 = (\mathbb{D}^0)^{-1} \cdot \mathbf{M}^0$$

and

$$\mathbb{O} \cdot \mathbf{e}^0 = \mathbb{O} \cdot (\mathbb{D}^0)^{-1} \cdot \mathbf{M}^0 = \frac{1}{I_2(\mathbb{D}^0)} \mathbb{O} \cdot \mathbb{O}^T \cdot \mathbb{D}^0 \cdot \mathbb{O} \cdot \mathbf{M}^0$$

$$= \frac{1}{I_2(\mathbb{D}^0)} \mathbb{D}^0 \cdot \mathbb{O} \cdot \mathbf{M}^0$$

Comparing this equality with (4.25), we see that

$$I_2(\mathbb{D}^0) = \lambda_1 \lambda_2 = I_2(\mathbb{D}_1) = \alpha \beta$$

which is the required statement.*

We now have a material with some fixed eigenvalues α, β of the conductivity tensor at our disposal and we thus are free to intermingle its differently oriented pieces to obtain a set of composites whose eigenvalues λ_1, λ_2 fill in some segment of a hyperbola between the points (α, β) and

---

* This result has been obtained in Ref. 466 with the aid of a similar technique; see also Ref. 442. An alternate proof is given in Appendix A.7.

(β, α).* An example of such a medium is provided by a layered composite [see (4.8) and (4.9)] assembled of the compounds

$$\mathbb{D}_+ = \alpha \mathbf{nn} + \beta \mathbf{tt}, \qquad \mathbb{D}_- = \alpha \mathbf{tt} + \beta \mathbf{nn} \qquad (4.26)$$

In particular, this enables us to obtain an isotropic medium whose conductivity can only be equal to $\sqrt{\alpha\beta}$. It is also obvious that inner points of the hyperbolic segment correspond to a span $s(\mathbb{D}^0) \triangleq \lambda_2 - \lambda_1$ of the matrix $\mathbb{D}^0$ not exceeding $s(\mathbb{D}_1) \triangleq \beta - \alpha$ for the $\mathbb{D}_1$ matrix

$$s(\mathbb{D}_1) \geq s(\mathbb{D}^0) \qquad (4.27)$$

In other words, mixing decreases the degree of anisotropy of the material, and this property characterizes irreversibility of the mixing process. It should be noted in conclusion that for the coordinates $(\lambda_1^{-1}, \lambda_2)$ [or $(\lambda_1, \lambda_2^{-1})$] the hyperbolic segment is transformed into that of a straight line which passes through the origin.

Remark 4.1. The results of the present section will obviously remain the same if we supplement the U set by the materials whose tensors $\mathbb{D}^k$ have the same second invariant $I_2(\mathbb{D}_1)$ and $s(\mathbb{D}^k) \leq s(\mathbb{D}_1)$. All such materials may be obtained as mixtures of some differently oriented pieces of one and the same material $\mathbb{D}_1$.

The influence of additional materials may be essential if we somehow restrict the possibility of rotation of the principal axes of the compounds. For instance, if we are given two anisotropic materials with equal values of the second invariants and with the same (fixed) orientation of the corresponding principal axes, then the GU set is represented as a hyperbolic segment connecting the corresponding points of the (λ_1, λ_2) plane. In general, we cannot obtain an isotropic mixture in this way.

4.5. G-Closure of a Set U Consisting of Two Arbitrary Plane Symmetric Tensors of Rank Two

Suppose we possess two materials whose tensors of conductivity are given by

$$\mathbb{D}_+ = \alpha_+ \mathbf{a}_+ \mathbf{a}_+ + \beta_+ \mathbf{b}_+ \mathbf{b}_+, \qquad \alpha_+ \leq \beta_+ \qquad (4.28)$$

---

* This conclusion follows immediately from inequalities (4.6) and (4.7); physically it is quite obvious.

$$\mathbb{D}_- = \alpha_- \mathbf{a}_- \mathbf{a}_- + \beta_- \mathbf{b}_- \mathbf{b}_-, \qquad \alpha_- \leq \beta_- \tag{4.29}$$

Here $\alpha_+, \beta_+, \alpha_-, \beta_-$ are some fixed eigenvalues, and the orientation of the pairs $(\mathbf{a}_+, \mathbf{b}_+), (\mathbf{a}_-, \mathbf{b}_-)$ of eigenvectors is arbitrary. We also may always assume that $\alpha_+ \beta_+ \leq \alpha_- \beta_-$. To construct the G-closure, we first note that each of the two materials gives rise to a set of materials whose eigenvalues λ_1, λ_2 correspond to the hyperbolic segments

$$\lambda_1 \lambda_2 = \alpha_+ \beta_+, \qquad \alpha_+ \leq \lambda_1, \lambda_2 \leq \beta_+ \tag{4.30}$$

$$\lambda_1 \lambda_2 = \alpha_- \beta_-, \qquad \alpha_- \leq \lambda_1, \lambda_2 \leq \beta_- \tag{4.31}$$

The media (4.30) and (4.31) may actually be constructed from the layers of the initially given compounds (4.28), (4.29) (see Section 4.4).

We shall now prove that the hyperbolic segments (4.30) and (4.31) belong to the boundary of the set GU. Toward this purpose, establish the inequalities

$$\alpha_+ \beta_+ \leq \lambda_1 \lambda_2 \leq \alpha_- \beta_- \tag{4.32}$$

Assume first that

$$\alpha_+ \leq \alpha_-, \qquad \beta_+ \leq \beta_- \qquad \text{(Fig. 4.3)} \tag{4.33}$$

Next, consider some composite assembled of the materials (4.28) and (4.29) and possessing the eigenvalues λ_1, λ_2 of the tensor $\mathbb{D}^0$. If we replace

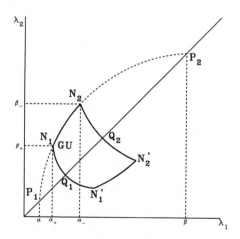

FIGURE 4.3

the material (4.28) with that of (4.29) in all of the inclusions, having preserved the orientation of the principal axes, that is, if we take α_- and β_- everywhere instead of α_+ and β_+, respectively, then according to (4.33), the eigenvalues of the $\mathbb{D}^0$ will obviously not decrease. This operation results in a material assembled of differently oriented pieces of the material (4.29), and the corresponding point in the (λ_1, λ_2) plane belongs to the hyperbolic segment $N_2 Q_2$ (Fig. 4.3). The right inequality (4.32) is thus proved, and the left one may be demonstrated in an analogous manner.

Consider now the case

$$\alpha_+ \leq \alpha_-, \qquad \beta_+ \geq \beta_- \qquad \text{(Fig. 4.4)} \tag{4.34}$$

[clearly, (4.33) and (4.34) cover all possible cases].

The set GU may only become larger if we extend the set U of initially given materials. In particular, this will be the case, provided that we replace the material (4.29) with the material $\hat{\mathbb{D}}_-$ characterized by eigenvalues $\hat{\alpha}_-, \hat{\beta}_-$ for which

$$\alpha_+ \leq \hat{\alpha}_-, \qquad \beta_+ \leq \hat{\beta}_-, \qquad \hat{\alpha}_- \hat{\beta}_- = \alpha_- \beta_-$$

The required material is always sure to exist, and the material (4.29) may be obtained as a mixture of the differently oriented samples of the new material (Problem II).

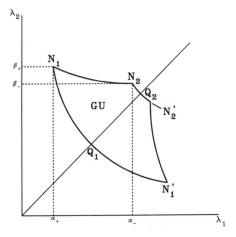

FIGURE 4.4

An alternate set of initial materials (call it $\hat{U}$) has thus expanded; on the other hand, the material $\hat{\mathbb{D}}_-$ satisfies (4.33) and consequently, the set $G\hat{U}$ (and GU even more so) satisfies (4.32).

The hyperbolic segments (4.30) and (4.31) constitute a part of the boundary of the set GU. To obtain the other parts, the media (4.28) and (4.29) should be combined within one composite. We shall see below that these parts are given by different analytical expressions for cases (4.33) and (4.34).

Consider the case (4.33) and recall that $\alpha_+ \leq \beta_+, \alpha_- \leq \beta_-$ and that, by assumption, $\lambda_1 \leq \lambda_2$. As for the case of two isotropic components (Section 4.3), the corresponding part of the boundary may be determined with the aid of (4.6), (4.7)

$$\lambda_2 \leq m\beta_+ + (1 - m)\beta_- = m(\beta_+ - \beta_-) + \beta_- \tag{4.35}$$

$$\lambda_1^{-1} \leq \frac{m}{\alpha_+} + \frac{1-m}{\alpha_-} = m\left(\frac{1}{\alpha_+} - \frac{1}{\alpha_-}\right) + \frac{1}{\alpha_-} \tag{4.36}$$

Bearing (4.33) in mind, we eliminate the parameter m from these inequalities; the result will be

$$\alpha_+ \leq \frac{\alpha_+\alpha_-(\beta_- - \beta_+)}{\alpha_-\beta_- - \alpha_+\beta_+ - (\alpha_- - \alpha_+)\lambda_2} \leq \lambda_1 \leq \lambda_2 \leq \beta_- \tag{4.37}$$

The latter inequalities generalize those of (4.17) and reduce to them if $\alpha_+ = \beta_+ = \alpha, \alpha_- = \beta_- = \beta$.

The relationship

$$\lambda_1 = \frac{\alpha_+\alpha_-(\beta_- - \beta_+)}{\alpha_-\beta_- - \alpha_+\beta_+ - (\alpha_- - \alpha_+)\lambda_2} \tag{4.38}$$

corresponds to a layered composite of the first rank (Section 4.4); the normal to the layers coincides with the principal axes $\mathbf{a}_+, \mathbf{a}_-$ of tensors $\mathbb{D}_+, \mathbb{D}_-$, associated with their *smallest* eigenvalues α_+, α_-.

The GU set is now formed (Fig. 4.3) as an intersection of the sets admitted by inequalities (4.32) and (4.37).

Consider now the case (4.34); for this one, elimination of the parameter m from (4.35), (4.36) is no longer possible and therefore the corresponding part of the boundary must be determined from additional considerations.

We will not give a detailed exposition here and only point out the final result: the component in question is given by the relationship[*]

$$\lambda_2 = \frac{\beta_+\beta_-(\alpha_- - \alpha_+)}{\alpha_-\beta_- - \alpha_+\beta_+ - (\beta_- - \beta_+)\lambda_1} \tag{4.39}$$

This equality is realized for a layered composite of the first rank; the normal to the layers coincides with the principal axes $\mathbf{b}_+, \mathbf{b}_-$ of tensors $\mathbb{D}_+, \mathbb{D}_-$ associated in this case with their *largest* eigenvalues β_+, β_-.

The GU set is now formed (Fig. 4.4) as an intersection of the sets admitted by (4.32) and by

$$\lambda_2 \le \frac{\beta_+\beta_-(\alpha_- - \alpha_+)}{\alpha_-\beta_- - \alpha_+\beta_+ - (\beta_- - \beta_+)\lambda_1} \tag{4.40}$$

Figures 4.3 and 4.4 represent GU sets for the cases (4.33) and (4.34), respectively; these sets are given by curvilinear quadrangles $N_1N_2Q_2Q_1$ symmetrically continued across the diagonal by the quadrangles $N_1'N_2'Q_2Q_1$. Note the difference in shape of the curves N_1N_2 and $N_1'N_2'$ corresponding to the cases (4.33) and (4.34): for the first of them the curve N_1N_2 is upward convex (Fig. 4.3); for the second, downward (Fig. 4.4). For the case (4.33), the segment N_1N_2 may be treated as a part of some longer curve $P_1N_1N_2P_2$ (Fig. 4.3) which is determined by (4.5) and connects the points P_1 and P_2 of a diagonal, these points corresponding to some isotropic initial compounds with some nonnegative conductivities α, β. The values of these parameters can be calculated comparing (4.5) and (4.38) and using (4.33). Having those isotropic media at our disposal, we are in a position to assemble a layered composite whose eigenvalues would correspond to any point of the segment N_1N_2.

The case (4.34) is different from the one just considered since there no longer exists a pair of isotropic materials which might give rise to an admissible set of composites.

The results of this section may be illustrated graphically in the plane generated either by the parameters $(\lambda_1, \lambda_2^{-1})$ or by $(\lambda_1^{-1}, \lambda_2)$ provided that we give up the earlier assumption $\lambda_1 \le \lambda_2$. Any anisotropic material whose $\mathbb{D}$ tensor possesses the eigenvalues d_1, d_2 is represented in each of those planes by two points passing to one another under the interchange of

---

[*] This result belongs to Dr. F. Murat (personal communication, 1985). The analysis of that case in Refs. 477, 479 contains an error. The author is indebted to Dr. Murat for that observation.

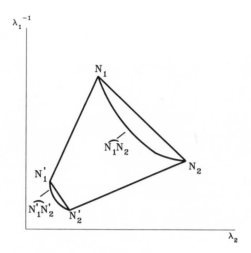

FIGURE 4.5

symbols d_1 and d_2: in coordinates $(\lambda_1, \lambda_2^{-1})$ these points will be (d_1, d_2^{-1}) and (d_2, d_1^{-1}), and in coordinates $(\lambda_1^{-1}, \lambda_2)$, the points (d_1^{-1}, d_2) and (d_2^{-1}, d_1). It is apparent that if the point (α, β^{-1}) belongs to the boundary of the GU set in the $(\lambda_1, \lambda_2^{-1})$ plane, then the same also holds for the point (β, α^{-1}).

In coordinates $(\lambda_1^{-1}, \lambda_2)$ the material $\mathbb{D}_1$ is represented by the points N_1 and N_1' [Fig. 4.5, case (4.33); Fig. 4.6, case (4.34)], the hyperbola (4.30) by a straight segment $N_1 N_1'$ passing through the origin; the material $\mathbb{D}_2$ is represented by the points N_2 and N_2' and the hyperbola (4.31) by a straight segment $N_2 N_2'$.

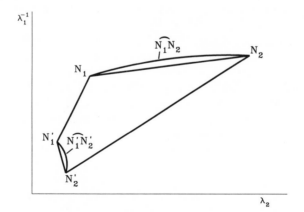

FIGURE 4.6

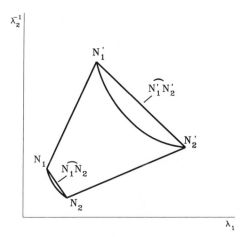

FIGURE 4.7

Connect now the points N_1 and N_2 as well as the points N_1' and N_2' by straight segments $N_1 N_2$ and $N_1' N_2'$. Equation (4.38) shows that the first of those segments represents layered first rank composite with normal to the layers oriented parallel to the principal axes $\mathbf{a}_+, \mathbf{a}_-$ of compounds associated with their smallest eigenvalues α_+, α_-. The eigenvalue λ_1 then corresponds to the principal direction of $\mathbb{D}_0$ normal to the layers. The segment $N_1' N_2'$ also represents a layered composite of the first rank [(4.39)], this time, however, the normal to the layers being oriented parallel to the principal axes $\mathbf{b}_+, \mathbf{b}_-$ of compounds associated with their largest eigenvalues

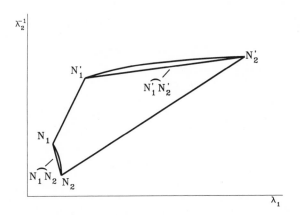

FIGURE 4.8

β_+, β_-. The normal then corresponds to the eigenvalue λ_2 of $\mathbb{D}_0$. The straight segment $N_1 N_2$ represents a part of the boundary of the GU set in case (4.33), not (4.34), and the segment $N_1' N_2'$ a part of the boundary in case (4.34), not (4.33). For each of these two cases, three sides of a quadrangle $N_1 N_2 N_2' N_1'$ represent a part of the GU boundary, and the fourth side $[N_1' N_2'$ in case (4.33), $N_1 N_2$ in case (4.34)] belongs to the inner part of GU. The reason is that the straight segments mentioned above are not transformed to one another as the coordinates λ_1 and λ_2 interchange.

In order to complement the quadrangle to the GU set, one should build the images of the straight segments $N_1 N_2$ and $N_1' N_2'$ resulting from that interchange operation. The segment $N_1 N_2$ would then be mapped to a curvilinear segment $\overset{\frown}{N_2' N_1'}$ (Figs. 4.5, 4.6), and the segment $N_1' N_2'$ to a curvilinear segment $\overset{\frown}{N_1 N_2}$. The set GU for both cases is represented as a sum of quadrangles $N_1 N_2 N_2' N_1'$ and $\overset{\frown}{N_1 N_2} \overset{\frown}{N_2' N_1'}$. This construction is also easily illustrated in coordinates $(\lambda_2^{-1}, \lambda_1)$ (Figs. 4.7, 4.8). In these coordinates, rectilinear and curvilinear segments exchange their places, and the GU set is built as a convex hull of an image of the quadrangle $\overset{\frown}{N_2 N_1} \overset{\frown}{N_1' N_2'}$.

4.6. The General Case: G-Closure of an Arbitrary Set of Symmetric Plane Tensors of Rank Two

The results obtained in the preceding section are easily generalized to the case of an arbitrary set U. First, we prove that the set GU may be determined as a sum of the GU sets borne by any pair of elements belonging to the initial set U. It suffices to consider the case when the set U consists of three materials represented by the points N_1, N_2, N_3 [Fig. 4.9 illustrates the case when for each of the three pairs $(N_1 N_2)$, (N_2, N_3), (N_3, N_1) inequalities (4.33) hold]. The aforementioned sum—the curvilinear polygon $N_1 N_2 N_3 Q_3 Q_2 Q_1$—obviously belongs to GU. We will show that there holds an inverse inclusion.

The polygon $N_1 N_2 N_3 Q_3 Q_2 Q_1$ is a part of a larger polygon $N_1 N_2 N_4 Q_3 Q_2 Q_1$ whose vertex N_4 is determined as a point of intersection of the (prolonged) segment $N_1 N_2$ and the ray $Q_3 N_3$. This new polygon may be treated as the set $G\bar{U}$ borne by two materials represented by the points N_1 and N_4. But it is apparent that $GU \subset G\bar{U}$ since the N_2 material can be obtained as a layered first rank composite assembled of N_1 and N_4 materials, and the material N_3 as a polycrystal of material N_4. This implies that a curvilinear open polygon $Q_3 Q_2 Q_1 N_1 N_2$ is a part of the boundary of $G\bar{U}$; at the same time it represents a part of the boundary of a sum of the GU sets borne by the pairs (N_1, N_2), (N_2, N_3), (N_3, N_1) of compounds.

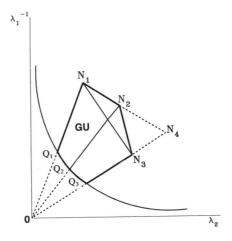

FIGURE 4.9

The same conclusion equally follows for an open curvilinear polygon $Q_1 Q_2 Q_3 N_3 N_2$ which renders the desired result.

The GU set corresponding to any initial set U of anisotropic materials is now constructed as the least set containing U and convex in coordinates $(\lambda_1, \lambda_2^{-1})$ and $(\lambda_1^{-1}, \lambda_2)$. All points of GU may be realized by some layered media which can be assembled only of those representatives of the U set which belong to the boundary of the GU set. Because the GU set includes media of arbitrary microstructure, it follows, in particular, that any such material is equivalent (in the sense of its effective properties) to some layered composite. The boundary points of GU are modeled by layered composites of the first rank whereas its inner points are represented by composites of the second rank.

The general G-closure procedure described here will be illustrated further if we represent the basic equations (4.1) in the form of (2.2) [see also (2.12)]. Namely, (4.1) are equivalent to the system $(w = z^1)$

$$\mathbf{M} = -\mathbf{i} z_y^2 + \mathbf{j} z_x^2 + \mathbf{M}^x$$

$$\mathbf{M} = \mathbb{D} \cdot (\mathbf{i} z_x^1 + \mathbf{j} z_y^1), \qquad \mathbb{D} = \alpha \mathbf{n} \mathbf{n} + \beta \mathbf{t} \mathbf{t}$$

where $\mathbf{M}^x$ denotes some partial solution of the equation

$$\nabla \cdot \mathbf{M} = f$$

With the coordinates $\mathbf{n} = \mathbf{i}y_t - \mathbf{j}x_t, \mathbf{t} = \mathbf{i}x_t + \mathbf{j}y_t$, the system takes on the form [of (2.120)]

$$z_n^1 = \frac{1}{\alpha}(-z_t^2 + M_n^x), \qquad z_n^2 = \beta z_t^1 - M_t^x \tag{4.41}$$

Assume that the set U consists of the two tensors (4.28) and (4.29) and let (4.33) hold. Then, according to Section 4.5, we should first construct a tensor corresponding to any point of the curve $N_1 N_2$ (Fig. 4.3); that is, corresponding to a layered microstructure with a normal to the layers parallel to the principal directions $\mathbf{a}_+, \mathbf{a}_-$ within each layer. This is achieved with the aid of averaging by Filippov's lemma (see Chapter 1 along with Refs. 43, 447, and 523) *under the additional condition that the orientation $\mathbf{n}$ and $\mathbf{t}$ of the principal axes should be preserved.* The result is

$$z_n^1 = \left(\frac{1}{\alpha}\right)^*(-z_t^2 + M_n^x), \qquad z_n^2 = \beta^* z_t^1 - M_t^x \tag{4.42}$$

$$\left(\frac{1}{\alpha}\right)^* = \frac{m}{\alpha_+} + \frac{1-m}{\alpha_-}, \qquad \beta^* = m\beta_+ + (1-m)\beta_- \tag{4.43}$$

The derivatives $z_t^1, z_t^2, z_n^1, z_n^2$ which appear in (4.42) represent the averaged values of the corresponding derivatives within the layers; the operation of averaging is performed over an elementary volume. It is essential that the averaged values of the *tangential* derivatives z_t^1, z_t^2 coincide with their (constant) values within the layers, and this is the reason for the success of Filippov's averaging procedure.

Equations (4.42) correspond to the tensor of conductivity

$$\mathbb{D} = \lambda_1 \mathbf{n}\mathbf{n} + \lambda_2 \mathbf{t}\mathbf{t}, \qquad \lambda_1 = 1/(1/\alpha)^*, \qquad \lambda_2 = \beta^*$$

According to the rule formulated earlier, the second step of the procedure requires formation of a layered microstructure from the components whose tensors of conductivity are given by

$$\mathbb{D}_+ = \lambda_1 \mathbf{n}\mathbf{n} + \lambda_2 \mathbf{t}\mathbf{t}, \qquad \mathbb{D}_- = \lambda_2 \mathbf{n}\mathbf{n} + \lambda_1 \mathbf{t}\mathbf{t} \tag{4.44}$$

where **n** and **t** denote the unit normal and unit tangent vectors to the new layers (of the second rank),* respectively.

The first of these materials corresponds to (4.42) and the second yields the equations

$$z_n^1 = \frac{1}{\beta^*}(-z_t^2 + M_n^\chi), \qquad z_n^2 = \frac{1}{(1/\alpha)^*}z_t^1 - M_t^\chi \qquad (4.45)$$

The second step of averaging results in a system

$$z_n^1 = \frac{1}{\lambda_{10}}(-z_t^2 + M_n^\chi), \qquad z_n^2 = \lambda_{20}z_t^1 - M_t^\chi \qquad (4.46)$$

where the eigenvalues λ_{10}, λ_{20} are given by

$$\lambda_{10}^{-1} = \mu\left(\frac{1}{\alpha}\right)^* + \frac{1-\mu}{\beta^*}, \qquad \lambda_{20} = \mu\beta^* + \frac{1-\mu}{(1/\alpha)^*} \qquad (4.47)$$

These formulas together with (4.43) describe the GU set. Each point of this set is determined by two parameters: m and μ. We see that the difference between the multidimensional and one-dimensional cases is that the former requires *two successive* Filippov procedures of averaging, connected with the introduction of layers of first and second rank.

No further comments are needed for the generalization of these results to the case of any initial set of materials.

4.7. Elastic Bar of Extremal Torsional Rigidity: On the Existence of Optimal Control

Suppose we have a prismatic bar of given simply connected cross section G bounded by a smooth curve Γ. The material of the bar is assumed to be linearly elastic and inhomogeneous; we shall assume that the elastic

---

* According to Section 4.5, we may take any directions of the principal axes of conductivity within the layers for the second step of the averaging procedure; these directions need not necessarily coincide with those of **n** and **t** specified by (4.44). Moreover, the microstructure of a composite may even be arbitrary in this case; it is only essential that it include pieces of the same material differing only in the orientation of their principal axes of conductivity or, more generally, pieces of materials with one and the same value of the second invariant $\lambda_1\lambda_2$ of the tensor $\mathbb{D}$. The set GU will remain the same under these conditions.

compliance* of the material $u(x, y)$ belongs to some subset of the set U of the space $L_\infty(G)$ of measurable positive functions u defined by the conditions

$$U: \quad 0 < u_{min} \leq \text{vrai max } u(x, y) \leq u_{max} < \infty \qquad (4.48)$$

Namely, adding the restriction (u_0 is fixed and α is an integer)

$$\iint_G u^\alpha \, dx \, dy = u_0^\alpha \text{ mes } G, \qquad \alpha \geq 1 \qquad (4.49)$$

to (4.48), we obtain a definition of the subset U_α. Equation (4.49) may be regarded as a formal estimate of the cost of the material used. (The symbols u_{min}, u_{max}, u_0 denote constants.) If the angle of twist per unit length of the bar (the degree of torsion) is taken equal to unity, the components of the displacement vector $\mathbf{u}$ will be determined by the equations (Fig. 4.10)

$$u_1 = -x_3 x_2, \qquad u_2 = x_3 x_1, \qquad u_3 = z^1(x_1, x_2) \qquad (4.50)$$

The function $z^1(x_1, x_2)$, the warping of the cross section of the rod, is to be determined from the equilibrium equation

$$\frac{\partial t_{13}}{\partial x_1} + \frac{\partial t_{23}}{\partial x_2} = 0 \qquad (4.51)$$

in which the stress components t_{13} and t_{23} are related to the corresponding strains

$$\varepsilon_{13} = \frac{1}{2}\left(\frac{\partial u_3}{\partial x_1} + \frac{\partial u_1}{\partial x_3}\right), \qquad \varepsilon_{23} = \frac{1}{2}\left(\frac{\partial u_3}{\partial x_2} + \frac{\partial u_2}{\partial x_3}\right) \qquad (4.52)$$

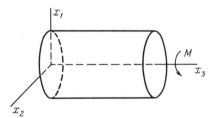

FIGURE 4.10

---

* The compliance is the reciprocal of the shear modulus: $u = \mu^{-1}$.

by the equations

$$\varepsilon_{13} = \tfrac{1}{2}ut_{13}, \qquad \varepsilon_{23} = \tfrac{1}{2}ut_{23} \tag{4.53}$$

which express Hooke's law.

The equation of equilibrium is satisfied by introducing the Prandtl function $z^2(x_1, x_2)$ by means of the equations

$$t_{13} = \frac{\partial z^2}{\partial x_2}, \qquad t_{23} = -\frac{\partial z^2}{\partial x_1} \tag{4.54}$$

Based on the relations (4.50), (4.52), and (4.53), we obtain a system of equations for the two functions z^1 and z^2 (for the independent variables x_1 and x_2, we adopt the notation x and y):

$$z_x^1 = uz_y^2 + y, \qquad z_y^1 = -uz_x^2 - x \tag{4.55}$$

If we set

$$\zeta^1 = -z_y^2 = -t_{13}, \qquad \zeta^2 = z_x^2 = -t_{23} \tag{4.56}$$

then the system (4.55) has the normal form

$$\begin{aligned}
z_x^1 &= -u\zeta^1 + y, & z_y^1 &= -u\zeta^2 - x \\
z_x^2 &= \zeta^2, & z_y^2 &= -\zeta^1
\end{aligned} \tag{4.57}$$

The boundary condition

$$z^2 = \text{const} \qquad \text{on } \Gamma \tag{4.58}$$

expresses the absence of external forces on the lateral surface of the rod. Without loss of generality, the constant in (4.58) can be assumed equal to zero.

The torsional rigidity, i.e., the torque per degree of torsion, is expressed by the integral

$$I = 2 \iint_G z^2 \, dx \, dy \tag{4.59}$$

The magnitude of this integral depends on the distribution of the compliance $u(x, y)$ over the rod section: $I = I[u]$. Our aim will be to find a distribution $u(x, y) \in U_\alpha$ such that the torque I of the rod is a maximum (Problem I) or a minimum (Problem II).

Before we turn to the proof of the existence of solutions of the posed optimization problems, we formulate the boundary-value problem (4.57)–(4.58) in the form of an integral identity. We define the function $z^2(x, y) \in \overset{\circ}{W}{}^1_2(G)$ by the identity

$$\iint_G u(\operatorname{grad} z^2, \operatorname{grad} \varphi)\, dx\, dy = 2 \iint_G \varphi\, dx\, dy \qquad (4.60)$$

which holds for all $\varphi \in \overset{\circ}{W}{}^1_2(G)$, where $\overset{\circ}{W}{}^1_2(G)$ is the space of functions that that vanish on Γ and which have square-integrable generalized derivatives of first order in S. For sufficiently smooth functions u, the identity (4.60) is equivalent to the boundary-value problem

$$\operatorname{div} u \operatorname{grad} z^2 = -2, \qquad z^2|_\Gamma = 0 \qquad (4.61)$$

which coincides with (4.57)–(4.58). In other words, we have to determine a generalized solution of this boundary-value problem; if the function $u \in U$ is given, then such a solution exists and is unique.[91]

We first consider minimizing I, assuming $u \in U$. Thus, let $v_n \in U$ be admissible controls, and $z^2_n \in \overset{\circ}{W}{}^1_2(G)$ be the corresponding solutions of the problem (4.60). The sequence v_n is bounded in $L_\infty(G)$; we may assume that it converges weakly to v in this space. The limit element $v \in U$ because the set U is weakly compact in itself.

Let z^2 denote the solution of the problem (4.60) corresponding to the control v. We then have

$$I[v] = 2 \iint_G z^2\, dz\, dy$$

$$= \iint_G v(\operatorname{grad} z^2)^2\, dx\, dy = \iint_G v_n(\operatorname{grad} z^2_n, \operatorname{grad} z^2)\, dx\, dy$$

$$= 2 \iint_G v_n(\operatorname{grad} z^2_n, \operatorname{grad} z^2)\, dx\, dy - \iint_G v(\operatorname{grad} z^2)^2\, dx\, dy$$

here, we have used the identity (4.60) for $\varphi = z^2$. Furthermore,

$$I[v_n] - I[v] = \iint_G v_n (\operatorname{grad} z_n^2)^2 \, dx \, dy - 2 \iint_G v_n (\operatorname{grad} z_n^2, \operatorname{grad} z^2) \, dx \, dy$$

$$+ \iint_G v (\operatorname{grad} z^2)^2 \, dx \, dy + \iint_G v_n (\operatorname{grad} z^2)^2 \, dx \, dy$$

$$- \iint_G v_n (\operatorname{grad} z^2)^2 \, dx \, dy = \iint_G v_n (\operatorname{grad}(z_n^2 - z^2))^2 \, dx \, dy$$

$$- \iint_G (v_n - v)(\operatorname{grad} z^2)^2 \, dx \, dy \qquad (4.62)$$

Hence, by virtue of weak convergence $v_n \to v$, and we conclude that

$$I[v] \le \lim_{n \to \infty} I[v_n] \qquad (4.63)$$

i.e., the functional $I[v]$ is weakly lower semicontinuous.

Suppose u_n is a minimizing sequence for $I[u]$:

$$\lim_{r \to \infty} I[u_n] = j$$

where $j = \inf_{v \in A} I[v]$. In accordance with the definition, for any $u \in A$ we have $I[u] \ge j$. On the other hand, $I[u] \le j$ by virtue of (4.63), so that

$$I[u] = j$$

i.e., the functional (4.59) attains a minimum. Setting $v_n = u_n$ and $v = u$ in (4.62) and using the inequality $0 < u_{\min} \le u$, yield

$$\lim_{n \to \infty} \iint_G (\operatorname{grad}(z_n^2 - z^2))^2 \, dx \, dy = 0$$

or, by virtue of the boundary condition,

$$\lim_{n \to \infty} \|z_n^2 - z^2\|_{\mathring{W}_2^1(G)} = 0$$

i.e., the approximate solutions z_n^2 of the optimization problem converge strongly to the minimizing function z.

Our proof remains in force if we restrict the class of admissible controls to the set U_1; for if the sequence v_n converges weakly to v in $L_\infty(G)$, then it converges weakly to the same element in $L_1(G)$ as well. The limit element $v \in U_1$: this follows from the definition of weak convergence in conjunction with the restriction (4.49). This proves the existence of a solution of the problem min for $\alpha = 1$.

Note that the proof does not follow for the case $\alpha > 1$ because the weak limit v [in the norm of $L_\alpha(G)$] of the sequence $v_n \in U_\alpha$ need not belong to U_α since the sphere (4.49) in the space $L_\alpha(G), \alpha > 1$, is not weakly compact in itself. The ball

$$\iint_G u^\alpha \, dx \, dy \leq u_0^\alpha \text{ mes } G, \qquad u_{\min} < u_0 < u_{\max} \tag{4.64}$$

does have this property* (Ref. 104), and the existence of a solution of Problem II for $\alpha > 1$ thus can be guaranteed under the weaker restriction (4.64), which now replaces the restriction (4.49).

The proof that a solution to Problem I exists encounters additional difficulties. Here, strictly speaking, we require only upper semicontinuity of $I[u]$, although it cannot be established on the basis of only the weak compactness of the set of controls. If we interchange the positions of v_n and v and also z_n^2 and z^2 in (4.62), we obtain

$$I[v] - I[v_n] = \iint_G v(\text{grad}\,(z_n^2 - z^2))^2 \, dx \, dy - \iint_G (v - v_n)(\text{grad}\, z_n^2)^2 \, dx \, dy$$

Upper semicontinuity would follow from this if the second integral on the right-hand side were to tend to zero as $n \to \infty$. But this does not follow from the weak convergence of v_n to v; therefore, to guarantee the existence of a solution, we should either characterize the sets of admissible controls in such a way that they are compact in the norms of the corresponding spaces, or resort to relaxation. We shall choose the latter approach.

---

* It is evidently more natural to replace equality by less than or equal to in the restrictions (4.49); however, we shall not do so since the modifications of the necessary conditions entailed by such a transition are obvious.

4.8. Necessary Conditions for Optimality

In accordance with the general theory, we introduce Lagrange multipliers ξ_1, η_1, ξ_2, η_2, κ, corresponding to (4.57) and (4.49), and also a multiplier Γ^* corresponding to the restriction (4.48) written in the equivalent form [see $(2.28)_*$]

$$(u_{max} - u)(u - u_{min}) - u_*^2 = 0 \tag{4.65}$$

Using these multipliers, we construct the function H, corresponding to Problem I:

$$H = -\xi_1 u \zeta^1 - \eta_1 u \zeta^2 + \xi_1 y - \eta_1 x + \xi_2 \zeta^2 - \eta_2 \zeta^1 + 2z^2$$

$$+ \kappa u^\alpha - \Gamma^*[(u_{max} - u)(u - u_{min}) - u_*^2] \tag{4.66}$$

and derive the necessary stationarity conditions

$$\xi_{1x} + \eta_{1y} = 0, \qquad \xi_{2x} + \eta_{2y} = -2 \tag{4.67}$$

$$\xi_1 u + \eta_2 = 0, \qquad \eta_1 u - \xi_2 = 0 \tag{4.68}$$

$$-\xi_1 \zeta^1 - \eta_1 \zeta^2 + \kappa \alpha u^{\alpha-1} + \Gamma^*(2u - u_{max} - u_{min}) = 0 \tag{4.69}$$

$$\Gamma^* u_* = 0$$

Equations (4.67) are satisfied by introducing functions ω_1 and ω_2 with

$$\begin{align}
\xi_1 &= -\omega_{1y}, & \xi_2 + x &= -\omega_{2y} \\
\eta_1 &= \omega_{1x}, & \eta_2 + y &= \omega_{2x}
\end{align} \tag{4.70}$$

Equations (4.68) are reduced to the form [see (4.55)]

$$\omega_{2x} = u\omega_{1y} + y, \qquad \omega_{2y} = -u\omega_{1x} - x$$

which results in the equation

$$\text{div } u \text{ grad } \omega_1 = -2 \tag{4.71}$$

Along the boundary, the following natural boundary condition is satisfied [see (2.40)]

$$\omega_{1y} y_t + \omega_{1x} x_t = 0$$

or

$$\omega_1|_\Gamma = \text{const} \tag{4.72}$$

The function ω_1 is determined to within a constant by (4.71); we can thus assume that the constant in the boundary condition (4.72) is zero.

Comparing the boundary-value problems (4.61), (4.71), and (4.72), and using uniqueness, we conclude that

$$z^2 \equiv \omega_1 \tag{4.73}$$

This result serves as the source of a number of distinctive features of these optimization problems.

We now turn to the formulation of Weierstrass's necessary condition. As before, we denote optimal quantities by lowercase letters, and the corresponding admissible quantities by uppercase letters. This yields the following expression for the increment of the functional $-I$:

$$\Delta(-I) = \iint_G E \, dx \, dy$$

where the Weierstrass function E is given by

$$E = H(\xi, \eta, \kappa, \zeta, u) - H(\xi, \eta, \kappa, Z, U)$$

$$= -\xi_1(u\zeta^1 - UZ^1) + \xi_2(\zeta^2 - Z^2) - \eta_1(u\zeta^2 - UZ^2)$$

$$- \eta_2(\zeta^1 - Z^1) + \kappa(u^\alpha - U^\alpha) \tag{4.74}$$

If the functional I is to attain a maximum, it is necessary and sufficient that $\Delta(-I)$ be made nonnegative for all admissible U, Z^1, and Z^2. It should be noted that variations in a small region (strip, disk) in accordance with (2.176) are now inadmissible since they violate restriction of the type (4.49).

We avoid this difficulty, by taking two regular* points P and Q of the measurable control $u(x, y)$ within the region and surrounding these points

---

* A point $(x, y) \in S$ is called *regular point* of the measurable function $u(x, y) \in U$ if for any neighborhood $O \in U$ of the point $u(x, y)$

$$\lim_{\text{mes } d \to 0} \frac{\text{mes}\,(u^{-1}(O) \cap d)}{\text{mes } d} = 1$$

Here, d is any set containing the point (x, y) and the symbol $u^{-1}(0)$ denotes the set of all points $(x, y) \in S$ for which $u(x, y) \in O$.

If a measurable function is defined in the domain S, then almost every point of S is a regular point of this function.[114]

with small regions d_p and d_Q, respectively. We then consider increments of u only within d_P and d_Q, and in such a way as to satisfy the equation

$$\iint_G U^\alpha \, dx \, dy - \iint_G u^\alpha \, dx \, dy = \iint_{d_p \cup d_Q} (U^\alpha - u^\alpha) \, dx \, dy = 0$$

This method of variation is obviously possible under the condition that the constant u_0 in (4.49) satisfy the strict inequality

$$u_{min} < u_0 < u_{max}$$

The inequality $\Delta(-I) \geq 0$ can now be written in the form

$$\iint_{d_p} E \, dx \, dy + \iint_{d_Q} E \, dx \, dy \geq 0$$

We divide both parts by mes $(d_p \cup d_Q)$ and contract d_P and d_Q, respectively, to the points P and Q, making each of these regions pass through a sequence of regularly contractible* set; we then obtain

$$\lambda E(P) + (1 - \lambda)E(Q) \geq 0, \qquad 0 \leq \lambda \leq 1$$

It follows that at least one of the numbers $E(P)$ and $E(Q)$ is nonnegative; however, since the points P and Q are equally arbitrary, both of these numbers must be nonnegative. Thus, if the functional I is to attain a maximum, we must have $E \geq 0$ at almost every point of the region.

The further operations in the derivation and transformation of the function E are carried out in accordance with the usual procedure.

We introduce variations in a strip; the variables Z^1 and Z^2 are eliminated by means of the conditions of continuity of the tangential derivatives

$$Z_t^1(x, y) = z_t^1(x, y), \qquad Z_t^2(x, y) = z_t^2(x, y)$$

which hold on the strip's boundary. It follows from these relations that

$$\zeta^1 - Z^1 = \frac{\Delta u}{U} x_t(\zeta^1 x_t + \zeta^2 y_t)$$

(4.75)

$$\zeta^2 - Z^2 = \frac{\Delta u}{U} y_t(\zeta^1 x_t + \zeta^2 y_t)$$

---

*See the footnote after (2.142).

here, $\Delta u = U - u$, and x_t and y_t are the direction cosines of the tangent to the strip boundary. Eliminating the variables Z^1 and Z^2 from the Weierstrass condition $E \geq 0$ and taking into account (4.57) and (4.70), we may write this condition in the form

$$\Delta u \left[(\text{grad } z^2)^2 - \frac{\Delta u}{U}(z_n^2)^2 - \kappa \sum_{i=1}^{\alpha} u^{\alpha-i}U^{i-1} \right] \geq 0 \qquad (4.76)$$

from which we obtain

$$u = u_{\max} \qquad \text{if } \kappa \sum_{i=1}^{\alpha} u^{\alpha-i}U^{i-1} - \frac{u_{\max}}{u_{\min}}(\text{grad } z^2)^2 \geq 0$$

$$\qquad\qquad\qquad\qquad\qquad\qquad\qquad\qquad\qquad\qquad\qquad (4.77)$$

$$u = u_{\min} \qquad \text{if } \kappa \sum_{i=1}^{\alpha} u^{\alpha-i}U^{i-1} - \frac{u_{\min}}{u_{\max}}(\text{grad } z^2)^2 \leq 0$$

for the limiting control regimes.

An intermediate control regime in this problem is impossible for such a regime is characterized by

$$u^{\alpha-1} = (\kappa\alpha)^{-1}(\text{grad } z^2)^2$$

If the parameter κ is eliminated from (4.76) by means of this equation, the condition (4.76) is clearly violated.

It is readily seen that the term in the Weierstrass inequality (4.76) depending on the inclination of the variation strip, is proportional to $(\Delta u)^2$. It follows that the necessary conditions for a weak maximum can be obtained by means of the Weierstrass function without considering those terms which depend on the inclination of the strip.* For Problem I, these conditions have the form

$$u = u_{\max} \qquad \text{if } \kappa\alpha u_{\max}^{\alpha-1} - (\text{grad } z^2)^2 \geq 0$$

$$\qquad\qquad\qquad\qquad\qquad\qquad\qquad\qquad\qquad (4.78)$$

$$u = u_{\min} \qquad \text{if } \kappa\alpha u_{\min}^{\alpha-1} - (\text{grad } z^2)^2 \leq 0$$

There is no smooth curve (switching line) dividing the zones occupied by materials with the limiting compliances $u = u_{\max}$ and $u = u_{\min}$. Assuming

---

* A similar situation also arises in the problem in Chapter 3.

the contrary, we arrive at a contradiction between (4.77) and the Weierstrass–Erdmann condition (2.62) (in what follows we assume $\alpha = 1$ to simplify the calculations)

$$-\kappa + (z_t^2)^2 + \frac{u_{\min}}{u_{\max}}(z_{n-}^2)^2 = 0 \qquad (4.79)$$

where $\mathbf{n}$ and $\mathbf{t}$ represent the normal and the tangent to the switching line. In (4.79) z_{n-}^2 denotes the limiting value of the normal derivative taken from the $u_{\min}$ side of the line, the value z_t^2 is continuous across it.

To demonstrate the contradiction, add (4.79) and the second inequality of (4.77); the result is

$$\left(\frac{u_{\min}}{u_{\max}} - 1\right)(z_t^2)^2 \geq 0 \qquad (4.80)$$

it is consistent only if $z_t^2 = 0$, which means that $z^2 = \text{const}$ along the switching line. The latter condition makes the problem overdetermined since the condition must be satisfied along the line together with (4.79). The same contradiction may be derived by analyzing (4.77) together with the continuity conditions

$$u_{\min}z_{n-}^2 = u_{\max}z_{n+}^2, \qquad z_{t-}^2 = z_{t+}^2 \qquad (4.81)$$

valid along the switching line.

Within the "forbidden" interval (Fig. 4.11)

$$\kappa \frac{u_{\min}}{u_{\max}} \leq (\nabla z^2)^2 \leq \kappa \frac{u_{\max}}{u_{\min}}$$

none of the stationary ranges is optimal. On the other hand, the value $(\nabla z^2)^2$ has a jump across the switching line. In general, this jump is less than the width of the "forbidden" interval. The use of the second inequality of (4.77) taken at the switching line, and of (4.81) yield

$$\kappa \leq \frac{u_{\min}}{u_{\max}}(\nabla z_-^2)^2 = \frac{u_{\max}}{u_{\min}}(\nabla z_+^2)^2 + (z_t^2)^2\left(\frac{u_{\min}}{u_{\max}} - \frac{u_{\max}}{u_{\min}}\right)$$

This inequality and (4.77) show that simultaneous validity of both is possible only provided that $z_t^2 = 0$.

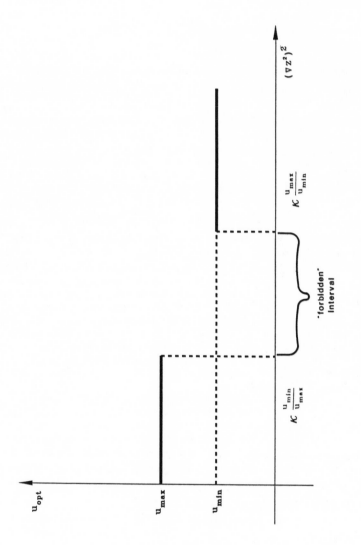

FIGURE 4.11

The minimum problem for the functional (4.59) was discussed in Ref. 202, where existence of the optimal solution was proven. The optimal compliance u now admits either the limiting values u_{min} and u_{max}, or any intermediate values. The latter domain connects those of the limiting values of compliance.

In fact, the Weierstrass condition corresponding to this problem yields [see (4.76) with $\alpha = 1$]:

$$\Delta u \left[(\text{grad } z^2)^2 - \frac{\Delta u}{U} (z_n^2)^2 - \kappa \right] \leq 0$$

which means that

$$
\left.
\begin{aligned}
u &= u_{max} \quad \text{if } \kappa - (\text{grad } z^2)^2 \leq 0 \\
u &= u_{min} \quad \text{if } \kappa - (\text{grad } z^2)^2 \geq 0 \\
u &= u_{int} \quad \text{if } \kappa - (\text{grad } z^2)^2 = 0
\end{aligned}
\right\}
\qquad (4.82)
$$

If, however, we have only two materials $u = u_{max}$ and $u = u_{min}$ at our disposal, then the intermediate control ceases to exist and only the first two possibilities in (4.82) remain admissible. For this case, one again has a contradiction since the u_{max} and u_{min} domains cannot be divided by some smooth switching line. To prove this, we observe that along such line the relationship

$$(\nabla z_+^2)^2 - (\nabla z_-^2)^2 = - (z_{n-}^2)^2 \left[1 - \left(\frac{u_{min}}{u_{max}} \right)^2 \right]$$

must hold, due to (4.81). This relationship violates (4.82) unless the switching line is orthogonal to the line $z^2 = \text{const}$.

To find a way out of this situation, one must observe that in the process of approximation to the optimal solution, the u_{max} and u_{min} domains are divided by a strongly oscillating line, so that adjacent regions of different compliance alternate more and more rapidly (see Section 3.3). In other words, in the optimal range, one now has regions where the layers of maximum and minimum compliance replace one another infinitely often, corresponding to the formation of an anisotropic composite.

4.9. The Relaxed Optimal Torsion Problem

The introduction of laminar composites as admissible materials allows a reformulation of the problem so as to eliminate contradictions which arise within the necessary conditions for optimality. For example, Problem I of Section 4.7 now requires the choice of tensor $\mathbb{D}^0$

$$\mathbb{D}^0 = \lambda_1 \mathbf{a}_1 \mathbf{a}_1 + \lambda_2 \mathbf{a}_2 \mathbf{a}_2$$

instead of scalar function u. The eigenvalues λ_1, λ_2 ($\lambda_1 \leq \lambda_2$) of $\mathbb{D}^0$ are connected by the inequality [see (4.4)]

$$u_{\min} \leq \frac{u_{\max} u_{\min}}{u_{\max} + u_{\min} - \lambda_2} \leq \lambda_1 \leq \lambda_2 \leq u_{\max} \tag{4.83}$$

This tensor should be chosen so as to maximize the functional (4.59) subject to the conditions [see (4.61)]

$$\nabla \cdot \mathbb{D} \cdot \nabla z^2 = -2, \qquad z^2|_\Gamma = 0$$

and [see (4.49)]

$$u_0 = \frac{1}{\text{mes } G} \iint_G \lambda_2 \, dx \, dy \tag{4.84}$$

the latter constraint expressing the prescribed amount of compounds.

We choose the elements of $\mathbb{D}^0$ as our controls: the angle between the x axis and the eigenvector $\mathbf{a}_1$ of $\mathbb{D}^0$ as well as the eigenvalues λ_1 and λ_2 of $\mathbb{D}^0$ subject to the conditions (4.83). The necessary conditions for optimality can be obtained by analysis of the increment of an augmented functional

$$\Delta I = \Delta \iint_G [2z^2 + \omega(\nabla \cdot \mathbb{D}^0 \cdot \nabla z^2 + 2) + \kappa(\lambda_2 - u_0)] \, dx \, dy$$

where ω and $\kappa = \text{const}$ denote the Lagrange multipliers. The problem is self-adjoint ($\omega = -z^2$), and the necessary condition for a maximum implies that

$$-\nabla z^2 \cdot \Delta \hat{\mathbb{D}} \cdot \nabla z^2 + \kappa \Delta \lambda_2 \leq 0 \tag{4.85}$$

where

$$\Delta \hat{\mathbb{D}} = \Delta \mathbb{D}^0 - \frac{\Delta \mathbb{D}^0 \cdot \mathbf{n} \mathbf{n} \cdot \Delta \mathbb{D}^0}{\mathbf{n} \cdot (\mathbb{D}^0 + \Delta \mathbb{D}^0) \cdot \mathbf{n}}$$

Retention of only the linear terms of the control variation provides the stationarity condition

$$-\nabla z^2 \cdot \delta \mathbb{D}^0 \cdot \nabla z^2 + \kappa \delta \lambda_2 \le 0 \qquad (4.86)$$

where the variation $\delta \mathbb{D}^0$ is to be calculated under the assumption that the second inequality of (4.83) is a strict equality:

$$\lambda_1 = \frac{u_{max} u_{min}}{u_{max} + u_{min} - \lambda_2}$$

$$(4.87)$$

$$\lambda_1 = [mu_{max}^{-1} + (1-m)u_{min}^{-1}]^{-1}, \qquad \lambda_2 = mu_{max} + (1-m)u_{min}$$

It should be apparent that this assumption does not influence the final result.

We obtain

$$\delta \mathbb{D}^0 = \left(\frac{d\lambda_1}{d\lambda_2}\mathbf{a}_1\mathbf{a}_1 + \mathbf{a}_2\mathbf{a}_2\right)\delta\lambda_2$$

$$+ \left[\lambda_1\left(\frac{d\mathbf{a}_1}{d\alpha}\mathbf{a}_1 + \mathbf{a}_1\frac{d\mathbf{a}_1}{d\alpha}\right) + \lambda_2\left(\frac{d\mathbf{a}_2}{d\alpha}\mathbf{a}_2 + \mathbf{a}_2\frac{d\mathbf{a}_2}{d\alpha}\right)\right]\delta\alpha \qquad (4.88)$$

$$= \left(\frac{\lambda_1^2}{u_{max}u_{min}}\mathbf{a}_1\mathbf{a}_1 + \mathbf{a}_2\mathbf{a}_2\right)\delta\lambda_2 + (\lambda_1 - \lambda_2)(\mathbf{a}_1\mathbf{a}_2 + \mathbf{a}_2\mathbf{a}_1)\delta\alpha$$

Aside from the indicated relationship between λ_1 and λ_2 we also made use of

$$\mathbf{a}_1 = \mathbf{i}\cos\alpha + \mathbf{j}_y\sin\alpha, \qquad \mathbf{a}_2 = -\mathbf{i}\sin\alpha + \mathbf{j}_y\cos\alpha$$

for the eigenvectors.

With the aid of (4.88) the following stationarity conditions are obtained from (4.86)

$$\delta\alpha(\lambda_1 - \lambda_2)(\mathbf{a}_1 \cdot \boldsymbol{\tau})(\mathbf{a}_2 \cdot \boldsymbol{\tau}) = 0 \qquad (4.89)$$

$$\delta\lambda_2\left[\frac{\lambda_1^2}{u_{max}u_{min}}(\mathbf{a}_1 \cdot \boldsymbol{\tau})^2 + (\mathbf{a}_2 \cdot \boldsymbol{\tau})^2 - \kappa\right] \ge 0 \qquad (4.90)$$

where $\boldsymbol{\tau} = \nabla z^2$.

From (4.89) we deduce that there arise two isotropic ranges of control

$$\lambda_1 = \lambda_2 = u_{\max} \qquad \text{if } \kappa - \frac{u_{\max}}{u_{\min}} (\mathbf{a}_1 \cdot \boldsymbol{\tau})^2 - (\mathbf{a}_2 \cdot \boldsymbol{\tau})^2 \geq 0$$

$$\lambda_1 = \lambda_2 = u_{\min} \qquad \text{if } \kappa - \frac{u_{\min}}{u_{\max}} (\mathbf{a}_1 \cdot \boldsymbol{\tau})^2 - (\mathbf{a}_2 \cdot \boldsymbol{\tau})^2 \leq 0$$

$$(4.91)$$

where $\mathbf{a}_1$ and $\mathbf{a}_2$ represent two arbitrary orthogonal unit vectors. As $u_{\min} < u_{\max}$, both inequalities are valid only if the vectors $\mathbf{a}_1$ and $\boldsymbol{\tau}$ are collinear. We thus arrive at the conditions

$$\lambda_1 = \lambda_2 = u_{\max} \qquad \text{if } \kappa - \frac{u_{\max}}{u_{\min}} |\boldsymbol{\tau}|^2 \geq 0 \qquad (4.92)$$

$$\lambda_1 = \lambda_2 = u_{\min} \qquad \text{if } \kappa - \frac{u_{\min}}{u_{\max}} |\boldsymbol{\tau}|^2 \leq 0 \qquad (4.93)$$

which coincide with the Weierstrass conditions (4.77) obtained in Section 4.8 for the nonrelaxed version of the same problem. At this time, however, the isotropic ranges (4.92) and (4.93) are divided, not by a switching line, but by an anisotropic zone which arises from the condition

$$(\mathbf{a}_1 \cdot \boldsymbol{\tau})(\mathbf{a}_2 \cdot \boldsymbol{\tau}) = 0 \qquad (4.94)$$

expressing the collinearity of $\boldsymbol{\tau}$ with one of the eigenvectors of $\mathbb{D}^0$.

Equations (4.90) and (4.94) show that within an anisotropic zone, two stationary ranges of control are possible:

$$\kappa = \frac{\lambda_1^2}{u_{\max} u_{\min}} |\boldsymbol{\tau}|^2 = \frac{1}{u_{\max} u_{\min}} |\mathbb{D}^0 \cdot \boldsymbol{\tau}|^2, \qquad (\mathbf{a}_2 \cdot \boldsymbol{\tau}) = 0 \qquad (4.95)$$

and

$$\kappa = |\boldsymbol{\tau}|^2, \qquad (\mathbf{a}_1 \cdot \boldsymbol{\tau}) = 0 \qquad (4.96)$$

To choose an optimal range between these two, we use the Weierstrass necessary condition (4.85). Let the strip of variation be distributed along $\boldsymbol{\tau}$ (this position of a strip can easily be shown to be the "most dangerous" from the viewpoint of the Weierstrass condition). The Weierstrass inequality (4.85) can then be rewritten in the form

$$|\tau|^2\left(\frac{(D^0_{nn} + \Delta D^0_{nn})(D^0_{tt} + \Delta D^0_{tt}) - (D^0_{nt} + \Delta D^0_{nt})^2}{D^0_{nn} + \Delta D^0_{nn}} - \lambda_i\right) - \kappa(\Lambda_2 - \lambda_2) \geq 0$$

(4.97)

[Note here that λ_i: $(\mathbf{a}_i \cdot \boldsymbol{\tau}) = |\boldsymbol{\tau}|$.] The left-hand side of the latter inequality should be nonpositive for any admissible tensor $\mathbb{D}^0 + \Delta\mathbb{D}^0$ whose eigenvalues Λ_1 and Λ_2 are connected by (4.83).

Having observed that

$$(D^0_{nn} + \Delta D^0_{nn})(D^0_{tt} + \Delta D^0_{tt}) - (D^0_{nt} + \Delta D^0_{nt})^2 = \Lambda_1\Lambda_2$$

we see that the left-hand side of (4.97) is a minimum if the admissible tensor $\mathbb{D}^0 + \Delta\mathbb{D}^0$ is such that $D^0_{nn} + \Delta D^0_{nn} = \Lambda_2$, since $\Lambda_1 \leq D^0_{nn} + \Delta D^0_{nn} \leq \Lambda_2$. The Weierstrass test would therefore be satisfied if

$$|\tau|^2(\Lambda_1 - \lambda_i) - \kappa(\Lambda_2 - \lambda_2) \geq 0$$

(4.98)

Now examine the ranges (4.95) and (4.96) separately. For the range (4.95), (4.98) takes on the form $(\lambda_i = \lambda_1)$

$$|\tau|^2(\Lambda_2 - \lambda_2)^2 \cdot \frac{\Lambda_1\lambda_1^2}{(u_{max}u_{min})^2} \geq 0$$

which shows that the range in question is optimal.

For the range (4.96), (4.98) reduces to $(\lambda_i = \lambda_2)$

$$(\Lambda_2 - \Lambda_1)|\tau|^2 \leq 0$$

and the Weierstrass test is violated because of (4.83).

Recalling the description of the layered microstructure which gave rise to the anisotropic material described by $\mathbb{D}^0$, we see that for the control range (4.95) the layers are arranged perpendicular to the vector $\boldsymbol{\tau}$. Equation (4.95) and the state equation now show that within the anisotropic zone, the problem reduces to the first-order partial differential equation

$$\frac{\partial}{\partial x}(\cos\alpha) + \frac{\partial}{\partial y}(\sin\alpha) = -\frac{2}{\sqrt{u_{max}u_{min}}}$$

For the $u = u_{min}$ and $u = u_{max}$ zones of isotropy, we obtain the Poisson equation whose right-hand sides are equal to $-2/u_{min}$ and $-2/u_{max}$, respectively.

In Ref. 202, where the problem of minimum torsional rigidity was considered, it turned out to be unnecessary to introduce anisotropic zones of compliance. The necessary conditions for optimality for this problem are given by (4.82); these conditions were derived assuming that intermediate compliance u_{int} was admissible. This assumption is, however, not necessary since the same problem of minimum rigidity can equally well be stated for only two admissible materials, possessing the compliance values u_{min} and u_{max}. This problem [initially ill-posed; see Section (4.8)] can be relaxed by the same procedure as before, and the necessary optimality condition (4.96) will now be related to the anisotropic zone which arises instead of the intermediate range of isotropic control.

For the minimum rigidity problem, the regularizing orientation of the layers is along the vector τ which thus is collinear with the eigenvector a_2; for this reason, only the largest eigenvalue λ_2 now is essential. The material behaves isotropically with compliance λ_2 which may take on any intermediate value between u_{min} and u_{max} due to the variable parameter m of the concentration.

For the maximum rigidity problem, it was the lowest eigenvalue λ_1 which turned out to be the only essential one.

4.10. Necessary Conditions for Optimality within the Layers of Laminated Composites

In the preceding section, it was shown that the Weierstrass necessary condition is satisfied for composites. More exactly, this condition is fulfilled when the strain is averaged over an elementary volume, and satisfies Hooke's law for an anisotropic composite body. The local stresses and strains, however, observed within the compounds which constitute a composite, may vary considerably from their averaged values since they obey Hooke's law for the compounds themselves. The question now arises whether the strains within the compounds would belong to the "forbidden" interval (see Section 4.8) for which none of the stationary ranges is optimal. The answer is negative. We shall illustrate this by means of the problem of maximum torsional rigidity.

As was noted in Section 4.8, the value of $(\nabla z^2)^2$ should not belong to the "forbidden" interval (Fig. 4.11)

$$\kappa \frac{u_{min}}{u_{max}} < (\nabla z^2)^2 < \kappa \frac{u_{max}}{u_{min}}$$

We now express the values ∇z_+^2 and ∇z_-^2 (corresponding to the zones $u = u_{max}$ and $u = u_{min}$, respectively) for compounds by making use of the averaged value ∇z^2 of Section 3.4 and recalling that $\nabla z^2 \cdot \mathbf{t} = 0$ ($\mathbf{t}$ denotes a unit vector tangent to the layers); we obtain

$$\nabla z_+^2 = \left[1 - \frac{(1-m)(u_{max} - u_{min})}{mu_{min} + (1-m)u_{max}} \right] \nabla z^2 = \frac{u_{min}}{\tilde{D}} \nabla z^2$$

$$\nabla z_-^2 = \left[1 + \frac{m(u_{max} - u_{min})}{mu_{min} + (1-m)u_{max}} \right] \nabla z^2 = \frac{u_{max}}{\tilde{D}} \nabla z^2$$

$$\tilde{D} = mu_{min} + (1-m)u_{max}$$

In accordance with the definition (4.87) of the eigenvalue λ_1, we may write

$$(\nabla z_+^2)^2 = \frac{\lambda_1^2}{u_{max}^2}(\nabla z^2)^2, \qquad (\nabla z_-^2)^2 = \frac{\lambda_1^2}{u_{min}^2}(\nabla z^2)^2 \qquad (4.99)$$

The stationarity condition for the anisotropic range (the regularized formulation) is given by (4.95)

$$(\nabla z^2)^2 = \kappa \frac{u_{max}u_{min}}{\lambda_1^2}$$

The use of this expression for the elimination of ∇z^2 from (4.99) yields

$$(\nabla z_+^2)^2 = \kappa \frac{u_{min}}{u_{max}}$$

$$(\nabla z_-^2)^2 = \kappa \frac{u_{max}}{u_{min}}$$

These formulas show that the stresses may take on constant absolute values within the compounds, and that these values are exactly the boundary values of the "forbidden" interval at each point where there arises an anisotropic range of control. For this range the value of an averaged stress $|\nabla z^2|$ varies only due to the variations of concentration.

The Weierstrass necessary test is thus shown to be satisfied both for the averaged strain and for the "microscopic" strain within different compounds.* At the same time, the obtained result provides a solution to the problem posed in Section 3.3 where mention was made of an infinitely often alternation of weak and strong materials. The switching line dividing those two now is seen to be a generalized curve: it is composed of an infinite number of switching lines dividing all of the layers of the microstructure.

As a corollary we note that the investigation of the optimality conditions for composites suggests two sequences tending to vanish: the dimensions of a "strip of variation" and the width of the layers which constitute the composite. The Weierstrass condition for the averaged strain is supplied by variation in a strip which tends to its center more slowly than the width of the layers tends to vanish. The conditions checked in this section correspond to the inverse situation, when the strip vanishes more rapidly than the width of layers.

4.11. Some Numerical Results

The regularized problem of torsion has been solved numerically (the calculations were performed by N. A. Lavrov). It has been observed that for both maximum and minimum rigidities the optimal control can be considered as substantially isotropic since only one eigenvalue (either λ_1 or λ_2) of the tensor $\mathbb{D}^0$ is really at work. The results are represented in Table 4.1 and Figs. 4.12–4.15.

Table 4.1. Numerical Results

Case	u_+	u_-	m_0	Figure	$100(I_{max} - I_0)/I_0$	$100(I_0 - I_{min})/I_0$
1	1.5	0.5	0.25	4.13a,b	43	16
2	1.5	0.5	0.50	4.14a,b	63	17
3	1.5	0.5	0.75		53	10
4	4.5	0.5	0.50		266	

---

* This conclusion is in accordance with the remark made in Section 3.3 about the use of limiting ranges of control "to their limit," i.e., until the Weierstrass necessary test is satisfied as an equality.

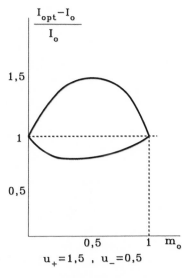

FIGURE 4.12

Figures 4.13 and 4.14 illustrate the optimal distributions of control within the square cross section of a bar for different values of the parameters $u_{\min}$, $u_{\max}$, and m_0. The latter determines the averaged concentration of a weak material. Figures 4.13a and 4.14a represent bars of maximum rigidity, and Figs. 4.13b and 4.14b those of minimum rigidity. The zones of strong isotropic material are indicated by a minus sign, and those of weak isotropic

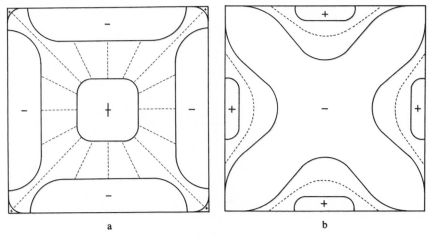

a   b

FIGURE 4.13

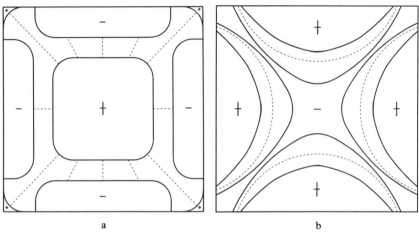

a   b

FIGURE 4.14

material by a plus sign. These two zones are divided by anisotropic layers of the isotropic compounds. For a strong bar, the zones of strong compound are distributed where the gradient of the Prandtl function z^2 is maximal, i.e., near the middle of each side of a square. At the corners and near the center of the cross section, it is economic to use the weak compound. For a weak bar, the compounds are in a sense inversely distributed.

Table 4.1 and Fig. 4.12 represent a relative increase (decrease) of rigidity of an optimal bar compared to that of a homogeneous bar of the same geometry made of the same materials taken in the same proportion. The

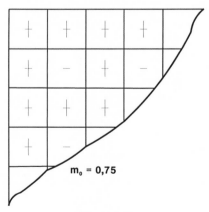

FIGURE 4.15

latter bar was modeled by a doubly periodic grating of nonhomogeneous cells as shown in Fig. 4.15.

The optimal bar of maximum rigidity was also compared with one whose square cross section has an inner part (a congruent square) made of weak isotropic material, with the rest of the area occupied by a strong isotropic compound. The relative loss in rigidity was 4% for $u_{max} = 1.5$, $u_{min} = 0.5$, and $m_0 = 0.5$.

5

Relaxation of Some Problems in the Optimal Design of Plates

In this chapter we shall consider some problems in the optimal design of thin elastic plates. In recent years, these problems have received attention in connection with the broad possibilities of optimization provided by the variation of both the thickness and the elastic moduli of a plate.

The behavior of elastic plates bent by a transverse load or subject to a plane state of stress is described by an equation including the fourth-order differential operator $\nabla\nabla \cdot \cdot \mathbb{D} \cdot \cdot \nabla\nabla$. This operator depends upon the tensor $\mathbb{D}$ of rank four, characterizing the rigidity of a plate. The tensor $\mathbb{D}$ depends upon both the thickness of a plate and its elastic moduli, and for this reason the corresponding optimization problems have many common features. Of these, the most important one is that both problems represent the same qualitative behavior as the less difficult second-order problems discussed before. Namely, optimal control for these problems may *not* exist if the admissible set U of controls is not G-closed. As before, in this connection, there arises the necessity of working out the procedure of G-closure of the initially given admissible sets U of controls.

Support for this approach to the problem may be found in Refs. 400, 402, 433–436, 452, 465, 472, 474, 475, and 489–492. Mainly, these papers are concerned with the numerical investigation of the problem of minimization of the elastic energy of a plate due to a suitable distribution of its thickness h. The authors observed divergence of numerical algorithms connected with frequent switches of a control function (the plate thickness) from its maximum to its minimum value, and vice versa; this sequence did not exhibit convergence to some definite limit. In Refs. 402, 433–436, 474, and 490–492, heuristic considerations concerning the optimal thickness distribution were given. According to Refs. 490 and 491 it should be characterized by some fine cellular structure. This viewpoint found confirmation in Ref. 474 where it was shown that the Legendre and Weierstrass conditions are violated in the corresponding problems if stationary distribution of the thickness is assumed smooth. Rather, the material of a plate should be concentrated in stiffeners which would decrease the value of the energy functional.

The solution of the general G-closure problem for plates has not yet been obtained; major difficulties arise here from the high dimensionality of the space of invariants for the tensor $\mathbb{D}$ of fourth rank (this dimensionality is five for the plane case). In addition, for the problem involving the thickness distribution, it is specified that all components of $\mathbb{D}$ be proportional to h^3, and this requirement presents additional difficulties for the G-closure problem. Finally, for that case there are well-founded doubts concerning the applicability of Kirchhoff's theory of plates; this theory requires additional motivation when the thickness changes rapidly and lines of discontinuity of the thickness follow one another too frequently* (see Ref. 536).

Nevertheless, the G-closure problem remains substantial for a vast class of problems of optimal control of the elastic moduli, and its solution represents a necessary element of the correct formulation of such problems. In this chapter, such a solution will be obtained for a number of important special cases; at the end of the chapter, we shall characterize some peculiarities of the problem of optimal thickness distribution.

The difficulties connected with the nonexistence of optimal controls may be avoided if we extend the set of controls to some smoother functional space than L_∞. In Ref. 135 it was shown, for instance, that the optimal control exists within some compact set of L_1. For various optimization problems in the theory of elasticity (Refs. 469 and 470), existence theorems were stated for controls belonging to W_2^1.

The proofs of these theorems are based upon the compactness of the admissible control sets; this property relieves one from the necessity of constructing the G-closure of these sets.

It should be mentioned, however, that the requirement that the control belong to smooth functional classes, often involves additional constraints of the admissible class of media. It is more natural to look for controls $\mathbb{D}$ within some bounded sets of L_∞. For each control belonging to these sets, there usually exists a well-defined solution to the corresponding boundary value problem of the theory of elasticity. An approach developed in Refs. 469 and 470 seems to be reasonable in problems involving thin-walled structures where the thickness h is taken as the control. On the other hand, if an optimal structure consisting of two different materials is required, then the admissible values of $\mathbb{D}$ obviously belong to L_∞ and do not form a compact set within this space.

---

* It should be mentioned, however, that for the axisymmetric (one-dimensional) plate problem, the G-closure was constructed for the case when the set U is given by ($U: 0 < h_{\min} \leq h \leq h_{\max} < \infty$) (see Refs. 436, 437, and 492).

5.1. Formulation of the Optimization Problem and of the Set of Controls

The distribution of momenta $\mathbb{M}$, the shear forces $\mathbf{Q}$, and displacement w of a plate S located by a normal force $q(x, y)$ are governed by the equilibrium equations (see Section 2.1)

$$\nabla \cdot \mathbb{M} = \mathbf{Q}, \qquad \nabla \cdot \mathbf{Q} = -q \tag{5.1}$$

and by Hooke's law (the symbol $\cdot\cdot$ denotes a double convolution)

$$\mathbb{M} = \mathbb{D} \cdot\cdot \, \mathbf{e} \tag{5.2}$$

which connects the tensors $\mathbb{M}$ of momenta and

$$\mathbf{e} = -\nabla\nabla w \tag{5.3}$$

of strain (see the Introduction). These equations are complemented by some definite boundary conditions along the contour ∂S of the plate.

The displacement $w \in W_2^2(S)$ satisfies the integral identity

$$\iint_S \nabla\nabla\eta \cdot\cdot \, \mathbb{D} \cdot\cdot \, \nabla\nabla w \, dx \, dy = \iint_S q\eta \, dx \, dy \tag{5.4}$$

where η denotes an arbitrary element of $\mathring{W}_2^2(S)$, and the components d_{ijkl} of the tensor $\mathbb{D}$ are assumed to belong to some bounded set of $L_\infty(S)$. The load q is taken to belong to $L_2(S)$.

For sufficiently smooth solutions w, (5.4) reduces to

$$L(D)w \triangleq \nabla\nabla \cdot\cdot \, \mathbb{D} \cdot\cdot \, \nabla\nabla w = q \tag{5.5}$$

The tensor $\mathbb{D}$ of fourth rank characterizes the elastic moduli of materials; for two independent variables, the tensor $\mathbb{D}$ is determined by six parameters. Since the tensor is self-adjoint,* it allows a dyadic expansion over symmetric eigentensors $\mathbb{a}_i$ of second rank (see Ref. 539); for two independent variables, we have three such eigentensors: $\mathbb{D} = d_i\mathbb{a}_i\mathbb{a}_i$ (hereafter unless stated otherwise, the numeration convention will be used from 1 to 3).

---

* Let $\mathbb{D}$ be a tensor of fourth rank. If $\mathbb{a} \cdot\cdot \, \mathbb{D} \cdot\cdot \, \mathbb{b} = \mathbb{b} \cdot\cdot \, \mathbb{D} \cdot\cdot \, \mathbb{a}$ for all tensors $\mathbb{a}, \mathbb{b}$ of second rank, then $\mathbb{D}$ is symmetric. If, moreover, $\mathbb{D}$ satisfies the condition $\mathbb{a} \cdot\cdot \, \mathbb{D} = 0$ for all antisymmetric tensors $\mathbb{a}$ of second rank, then $\mathbb{D}$ is self-adjoint. The tensor of elastic moduli possesses this property provided an elastic potential exists.

The parameters d_i—the eigenvalues of the self-adjoint tensor $\mathbb{D}$ of fourth rank—are always real; they are determined from the equations (no summation) $\mathbb{D} \cdot\cdot \, \mathfrak{a}_i = d_i \mathfrak{a}_i$ together with the eigentensors $\mathfrak{a}_i$.

The eigentensors $\mathfrak{a}_i$ and $\mathfrak{a}_\kappa$ corresponding to different eigenvalues d_i and d_κ are orthogonal: $\mathfrak{a}_i \cdot\cdot \, \mathfrak{a}_\kappa = 0$ $(d_i \neq d_\kappa)$; these tensors may also be normalized: $\mathfrak{a}_i \cdot\cdot \, \mathfrak{a}_i = 1$ (no summation). The latter equations also hold if $d_i = d_\kappa$; i.e., one always has

$$\mathfrak{a}_i \cdot\cdot \, \mathfrak{a}_\kappa = \delta_{i\kappa} \tag{5.6}$$

If $d_i = d_\kappa$, then the subspace $(\mathfrak{a}_i, \mathfrak{a}_\kappa)$ has no distinguished eigentensors.

The set of linearly independent tensors $\mathfrak{a}_i$ provides a basis in the space of symmetric tensors of second rank; in particular,

$$\mathfrak{e} = e_i \mathfrak{a}_i, \qquad e_i = \mathfrak{e} \cdot\cdot \, \mathfrak{a}_i \tag{5.7}$$

and the postulation of Hooke's law (5.2) allows the representation

$$\mathsf{M} = \mathbb{D} \cdot\cdot \, \mathfrak{e} = d_i e_i \mathfrak{a}_i \tag{5.8}$$

For the specific work of the strain, we have $\mathsf{M} \cdot\cdot \, \mathfrak{e} = d_i e_i^2$; the inequalities $d_i > 0$ express the positive definiteness of the specific work.

The triple of eigentensors $\mathfrak{a}_i$ is determined by nine parameters, of which only three are independent because of the six orthogonality conditions (5.6). Together with the three eigenvalues d_i, we obtain a totality of six independent parameters describing the tensor $\mathbb{D}$ for the two-dimensional case.

In particular, for an isotropic medium, we have

$$\mathsf{M} = 2(\lambda + \mu)e_1 \mathfrak{a}_1 + 2\mu(e_2 \mathfrak{a}_2 + e_3 \mathfrak{a}_3)$$

and, consequently,

$$\mathbb{D} = 2k\mathfrak{a}_1 \mathfrak{a}_1 + 2\mu(\mathfrak{a}_2 \mathfrak{a}_2 + \mathfrak{a}_3 \mathfrak{a}_3) \tag{5.9}$$

where $\lambda + \mu = k$, μ denote the dilatation modulus and the shear modulus, respectively, and where the eigentensors

$$\mathfrak{a}_1 = \frac{1}{\sqrt{2}}(\mathbf{ii} + \mathbf{jj}), \qquad \mathfrak{a}_2 = \frac{1}{\sqrt{2}}(\mathbf{ii} - \mathbf{jj}), \qquad \mathfrak{a}_3 = \frac{1}{\sqrt{2}}(\mathbf{ij} + \mathbf{ji}) \tag{5.10}$$

include a spherical tensor $\mathfrak{a}_1$ and two deviators $\mathfrak{a}_2, \mathfrak{a}_3$.

For a medium with cubic symmetry, we have

$$\mathbb{D} = k\mathfrak{a}_1 \mathfrak{a}_1 + \mu_2 \mathfrak{a}_2 \mathfrak{a}_2 + \mu_3 \mathfrak{a}_3 \mathfrak{a}_3 \tag{5.11}$$

where k may be considered as a (doubled) dilatation modulus, and μ_2, μ_3 as two (doubled) shear moduli.*

Isotropy of the tensor (5.9) can easily be demonstrated with the aid of the easily checked formulas $(\mathbf{i} \cdot \mathbf{i}^1 = \cos \alpha)$

$$\mathbb{a}_1^1 = \mathbb{a}_1, \qquad \mathbb{a}_2^1 = \mathbb{a}_2 \cos 2\alpha + \mathbb{a}_3 \sin 2\alpha, \qquad \mathbb{a}_3^1 = -\mathbb{a}_2 \sin 2\alpha + \mathbb{a}_3 \cos 2\alpha$$

where $\mathbb{a}_1^1, \mathbb{a}_2^1, \mathbb{a}_3^1$ are determined in the rotated basis $\mathbf{i}^1, \mathbf{j}^1$ by the same formulas as (5.10). We see that the deviators $\mathbb{a}_2, \mathbb{a}_3$ are transformed in quite the same way as if they were vectors rotated by the angle 2α around the fixed "direction" $\mathbb{a}_1$.

The tensor (5.9) thus is completely determined by only two parameters: k and μ.

The tensors (5.11) possess a lower rate of symmetry: they depend upon four parameters, namely, the dilatation modulus k, two shear moduli μ_2 and μ_3, and the angle φ which describes the orientation of the principal axes of the eigentensor $\mathbb{a}_2$ relative to fixed axes. We shall also make use of the following representation of the operator $\nabla\nabla$ in a basis $(\mathbb{a}_1, \mathbb{a}_2, \mathbb{a}_3)$:

$$\nabla\nabla = \frac{1}{\sqrt{2}}\left[\mathbb{a}_1\Delta + \mathbb{a}_2\left(\frac{\partial^2}{\partial x^2} - \frac{\partial^2}{\partial y^2}\right) + 2\mathbb{a}_3\frac{\partial^2}{\partial x\,\partial y}\right], \qquad \Delta = \frac{\partial^2}{\partial x^2} + \frac{\partial^2}{\partial y^2}$$

$$(5.12)$$

Note that the self-adjoint plane tensor $\mathbb{D}$ of fourth rank possesses five invariants, in general. These invariants include the three eigenvalues (d_1, d_2, d_3) and two angles between the principal axes of three eigentensors $\mathbb{a}_1, \mathbb{a}_2, \mathbb{a}_3$. The sixth parameter (noninvariant) which characterizes the tensor $\mathbb{D}$, is the angle between some principal axis and the x axis of the fixed coordinate system.

It is useful to introduce a representation of the tensor $\mathbb{D}$ in any orthogonal basis. If this basis is denoted by $\{\mathbb{a}_i\}$, we get $(i = 1, 2, 3)$

$$\mathbb{D} = d_{i\kappa}\mathbb{a}_i\mathbb{a}_\kappa, \qquad d_{i\kappa} = d_{\kappa i}$$

where the symmetry of the coefficients $d_{i\kappa}$ follows from the self-adjointness of $\mathbb{D}$.

We shall consider the following typical problem of optimal design: minimize the functional

$$I(w) = \iint_S \omega w \, dx \, dy \qquad (5.13)$$

---

* In what follows, we shall call k the dilatation modulus and μ the shear modulus.

where $\omega(x, y) \in L_2(S)$ is some weighting function. In addition to the basic equation (5.4) (with the corresponding boundary conditions), certain restrictions on the controls u may be prescribed [it is assumed that $\mathbb{D} = \mathbb{D}(u)$]. In particular, these restrictions may be of the form

$$\iint_S f(\mathbb{D}(u)) \, dx \, dy = \text{fixed} \tag{5.14}$$

where f denotes some given scalar function.

5.2. Weierstrass's Necessary Conditions for Optimality

Following the procedure of Chapter 2, we shall now derive necessary conditions for the optimization problem formulated above. For the time being, the restriction (5.14) will be neglected; it can, however, be easily taken into account.

We introduce the augmented functional corresponding to (5.5) and (5.13)

$$I = \iint_S [\omega w + \lambda(L(\mathbb{D})w - q)] \, dx \, dy \tag{5.15}$$

and assume that the tensor $\mathbb{D}$ depends upon some scalar control $u(x, y)$: $\mathbb{D} = \mathbb{D}(u)$.

The increment of the functional (5.15) caused by that of the control u is given by

$$\Delta I = \iint_S \{\Delta w[L(\mathbb{D})\lambda + \omega] + \varepsilon \cdot \cdot \Delta\mathbb{D} \cdot \cdot e + \varepsilon \cdot \cdot \Delta\mathbb{D} \cdot \cdot \Delta e\} \, dx \, dy \tag{5.16}$$

where the symbol $\Delta(\cdot)$ means the difference between the admissible and the optimal values of the argument, and λ represents the "adjoint displacement" satisfying

$$L(\mathbb{D})\lambda + \omega = 0$$

The latter equation differs from (5.5) only by $-\omega$ which now appears instead of q; the tensor $\varepsilon = -\nabla\nabla\lambda$ denotes the "adjoint strain."

Linear terms in the increment Δu of control are easily separated in the left-hand side of (5.16); they provide the stationarity condition

$$\Delta u \left(\varepsilon \cdot \cdot \frac{\partial \mathbb{D}}{\partial u} \cdot \cdot e \right) \geq 0 \qquad (5.17)$$

We now turn to the derivation of the necessary condition of Weierstrass; we observe that (5.16) contains an increment Δe together with that of $\Delta \mathbb{D}$, both increments generally having the same order of magnitude. To find out the connection between the two, we restrict ourselves to some special variation of control. This one will be performed within a narrow strip of variation S_δ of area δ; the strip will be allowed to tend to its center (x_0, y_0) with similarity being preserved. Within the strip, the strain is uniform up to terms of $0(\delta)$ (see footnote on p. 58); along the boundary of the strip one has the continuity conditions

$$[\mathbb{M}]_-^+ \cdot \cdot \mathbf{nn} = 0, \qquad [e]_-^+ \cdot \cdot (\mathbf{nt} + \mathbf{tn}) = 0, \qquad [e]_-^+ \cdot \cdot \mathbf{tt} = 0 \quad (5.18)$$

Here $\mathbf{n}$ and $\mathbf{t}$ denote the unit normal and the unit tangent to the boundary of the strip, respectively, $\mathbf{nn}, \mathbf{tt}, \mathbf{nt} + \mathbf{tn}$ being the corresponding dyadics. The symbol $(+)$ relates to the interior, and $(-)$ to the exterior of the strip S_δ.

Assuming that $\mathbb{M}_-, e_-$ coincide with the stationary values $\mathbb{M}(x_0, y_0)$ and $e(x_0, y_0)$, respectively (see Chapter 2), we may use (5.18) to calculate the value of Δe within the strip S_δ. Based on the last two equations of (5.18), we have

$$e_+ = e_- + \alpha \mathbf{nn}$$

where the parameter α may be evaluated with the aid of the first equation of (5.18)

$$(\mathbb{M}_+ - \mathbb{M}_-) \cdot \cdot \mathbf{nn} = [(\mathbb{D} + \Delta \mathbb{D}) \cdot \cdot e_+ - \mathbb{D} \cdot \cdot e_-] \cdot \cdot \mathbf{nn}$$

$$= [\Delta \mathbb{D} \cdot \cdot e + \alpha \mathbf{nn} \cdot \cdot (\mathbb{D} + \Delta \mathbb{D})] \cdot \cdot \mathbf{nn}$$

and

$$\alpha = - \frac{\mathbf{nn} \cdot \cdot \Delta \mathbb{D} \cdot \cdot e}{\mathbf{nn} \cdot \cdot (\mathbb{D} + \Delta \mathbb{D}) \cdot \cdot \mathbf{nn}}, \qquad \Delta e = e_+ - e_- = - \mathbf{nn} \frac{\mathbf{nn} \cdot \cdot \Delta \mathbb{D} \cdot \cdot e}{\mathbf{nn} \cdot \cdot (\mathbb{D} + \Delta \mathbb{D}) \cdot \cdot \mathbf{nn}}$$

The increment (5.16) now is written in the form

$$\Delta I = \iint_{S_\delta} \varepsilon \cdot \cdot \Delta \hat{\mathbb{D}} \cdot \cdot e \, dx \, dy \qquad (5.19)$$

where

$$\Delta \hat{\mathbb{D}} = \Delta \mathbb{D} - \frac{(\Delta \mathbb{D} \cdot \cdot \mathbf{nn}) \otimes (\Delta \mathbb{D} \cdot \cdot \mathbf{nn})}{\mathbf{nn} \cdot \cdot (\mathbb{D} + \Delta \mathbb{D}) \cdot \cdot \mathbf{nn}}$$

$$= \Delta \mathbb{D} - \frac{\Delta \mathbb{D} \cdot \cdot \mathbf{nnnn} \cdot \cdot \Delta \mathbb{D}}{\mathbf{nn} \cdot \cdot (\mathbb{D} + \Delta \mathbb{D}) \cdot \cdot \mathbf{nn}} \qquad (5.20)$$

and the symbol $\otimes$ denotes a dyadic product.

The Weierstrass necessary condition is given by

$$\varepsilon \cdot \cdot \Delta \hat{\mathbb{D}} \cdot \cdot \mathbf{e} \geq 0 \qquad (5.21)$$

The restriction (5.14) imposed upon the control may be taken into account with the aid of the (constant) Lagrange multiplier κ; the Weierstrass condition then is modified to

$$\varepsilon \cdot \cdot \Delta \hat{\mathbb{D}} \cdot \cdot \mathbf{e} + \kappa \Delta f \geq 0 \qquad (5.22)$$

where

$$\Delta f = f(\mathbb{D}(u + \Delta u)) - f(\mathbb{D}(u))$$

5.3. Contradictions within the Necessary Optimality Conditions

Consider the problem of optimal design of an elastic plate made of isotropic material whose dilatation modulus k is assumed fixed and with the shear modulus $\mu > 0$ treated as control. The set of admissible controls U will be given by

$$0 < \mu_- \leq \text{vraimax } \mu \leq \mu_+ < \infty, \qquad \mu \in L_\infty(S)$$

It was shown in Section 5.1 that the tensor $\mathbb{D}$ for an isotropic material is given by

$$\mathbb{D} = k\mathfrak{a}_1\mathfrak{a}_1 + \mu(\mathfrak{a}_2\mathfrak{a}_2 + \mathfrak{a}_3\mathfrak{a}_3)$$

where $\mathfrak{a}_1$ is a spherical eigentensor, and $\mathfrak{a}_2$, $\mathfrak{a}_3$ represent the deviator eigentensors

$$\mathfrak{a}_1 = \frac{1}{\sqrt{2}} (\mathbf{ii} + \mathbf{jj}), \qquad \mathfrak{a}_2 = \frac{1}{\sqrt{2}} (\mathbf{ii} - \mathbf{jj}), \qquad \mathfrak{a}_3 = \frac{1}{\sqrt{2}} (\mathbf{ij} + \mathbf{ji})$$

Consider the optimization problem of Section 5.1 without any restriction of the type (5.14) imposed upon the control. Following the procedure of Section 5.2 and assuming the normal **n** to the strip of variation is coincident with the unit vector **i**, the Weierstrass inequality (5.21) may be reduced to the form

$$\Delta\mu\left[\mathfrak{e}\cdot\cdot\left(\frac{k+\mu}{k+\mu+\Delta\mu}\,\mathfrak{a}_2\mathfrak{a}_2+\mathfrak{a}_3\mathfrak{a}_3\right)\cdot\cdot\mathfrak{e}\right]\geq 0 \qquad (5.23)$$

The condition **n** = **i** does not influence in the generality of the result, since all unit vectors are equivalent in the space of deviators $\mathfrak{a}_2\mathfrak{a}_2+\mathfrak{a}_3\mathfrak{a}_3$. Denoting the angles between the principal axes of $\mathfrak{a}_2$ and $\mathfrak{e}$, and between those of $\mathfrak{a}_2$ and $\mathfrak{e}$, by φ and ψ, respectively, the left-hand side of (5.23) may be written in the form

$$\Delta\mu|\text{dev }\mathfrak{e}||\text{dev }\mathfrak{e}|\left(\frac{k+\mu}{k+\mu+\Delta\mu}\cos 2\varphi\cos 2\psi+\sin 2\varphi\sin 2\psi\right)\geq 0$$

or

$$|\text{dev }\mathfrak{e}||\text{dev }\mathfrak{e}|\left[\Delta\mu\cos 2(\varphi-\psi)-\frac{(\Delta\mu)^2}{k+\mu+\Delta\mu}\cos 2\varphi\cos 2\psi\right]\geq 0 \qquad (5.24)$$

Here

$$|\text{dev }\mathfrak{e}|^2=\text{dev }\mathfrak{e}\cdot\cdot\text{dev }\mathfrak{e}=(\mathfrak{e}\cdot\cdot\mathfrak{a}_2)^2+(\mathfrak{e}\cdot\cdot\mathfrak{a}_3)^2$$

with an analogous result for $|\text{dev }\mathfrak{e}|$.

The stationarity condition for intermediate control shows that $2\varphi = 2\psi+\pi/2$. This condition does not, however, correspond to a minimum since (5.24) is then transformed to

$$-\frac{(\Delta\mu)^2}{k+\mu+\Delta\mu}\sin 4\varphi\tfrac{1}{2}|\text{dev }\mathfrak{e}||\text{dev }\mathfrak{e}|\geq 0$$

a condition which may be violated by suitable choice of φ.

We shall now show that the zones of the limiting ranges $\mu = \mu_+$ and $\mu = \mu_-$ of the control cannot generally be separated by a smooth switching line; this situation provides an analogue of what we had for the second-order problems of Chapters 3 and 4. It is apparent that the strongest Weierstrass condition is obtained when the strip of variation bisects the angle θ between the principal axes of e and ε, and $|\Delta\mu|$ is taken equal to $|\mu_+ - \mu_-|$. Based on (5.24), we then obtain

$$\mu = \mu_+ \quad \text{if } \tan^2 \theta \geq \frac{k + \mu_+}{k + \mu_-}$$

$$\mu = \mu_- \quad \text{if } \tan^2 \theta \leq \frac{k + \mu_-}{k + \mu_+} \tag{5.25}$$

These conditions show that there exists a "forbidden" interval of θ (see Section 4.8):

$$\frac{k + \mu_-}{k + \mu_+} \leq \tan^2 \theta \leq \frac{k + \mu_+}{k + \mu_-}$$

for which none of the stationarity ranges is optimal. The parameter θ has a jump across the switching line. This jump should exceed the width of the "forbidden" interval, a requirement which imposes a restriction upon the possible orientation of the normal to the switching line. This restriction overdetermines the problem. In fact, along the switching line we have [see (5.25)]

$$\left(\frac{\tan \theta_+}{\tan \theta_-}\right)^2 \geq \left(\frac{k + \mu_+}{k + \mu_-}\right)^2 = A^2 > 1$$

However, by virtue of (5.18) of continuity, we obtain

$$e_{3+} = e_{3-}, \quad (\mu_+ + k)e_{2+} = (\mu_- + k)e_{2-}(e_s = e \cdot \cdot \mathfrak{a}_s, \varepsilon_s = \varepsilon \cdot \cdot \mathfrak{a}_s)$$

and

$$\frac{\tan \theta_+}{\tan \theta_-} = \frac{e_{2+}\varepsilon_{2+} + e_{3+}\varepsilon_{3+}}{e_{2-}\varepsilon_{2-} + e_{3-}\varepsilon_{3-}} \cdot \frac{e_{2-}\varepsilon_{3-} - e_{3-}\varepsilon_{2-}}{e_{2+}\varepsilon_{3+} - e_{3+}\varepsilon_{2+}} = \frac{\tan \psi_{1-} \tan \psi_{2-} + A^2}{A(1 + \tan \psi_{1-} \tan \psi_{2-})} \tag{5.26}$$

Here $\psi_{1-} + \psi_{2-} = \theta_-$; $\psi_1(\psi_{2-})$ denotes an angle between the principal axis of $e_-(\varepsilon_-)$ and of the vector i. Inequality (5.26) generally is inconsistent with (5.25) (e.g., consider the case $\tan \psi_{1-} \tan \psi_{2-} > 1$).

5.4. The Effective Rigidity of a Laminated Composite

In this section, we shall derive the formulas which determine the effective tensor $\mathbb{D}^0$ of a laminated composite. The calculations will be based on the bending problem of a thin elastic plate.

Let the material of the plate be composed of infinitely often alternating layers of two different materials whose tensors of elastic constants (rigidities) are equal to $\mathbb{D}_+$ and $\mathbb{D}_-$, respectively. The concentration of the $\mathbb{D}_+$ material will be denoted by m.

We take a "physically small" area σ (Fig. 5.1) and define uniform strains e_+ and e_- as well as momenta M_+ and M_- within the layers σ_+ and σ_-:

$$M_+ = \mathbb{D}_+ \cdot\cdot \, e_+$$

$$M_- = \mathbb{D}_- \cdot\cdot \, e_-$$

(5.27)

The strain and moment averages (over σ) are given by

$$e^0 = me_+ + (1 - m)e_-$$

$$M^0 = mM_+ + (1 - m)M_-$$

(5.28)

Along the boundaries dividing the layers, one has the continuity conditions for the normal component of the moment and for the tangential component of the strain

$$[M \cdot\cdot \, nn]^+_- = 0, \qquad [e \cdot\cdot \, nt]^+_- = 0, \qquad [e \cdot\cdot \, tt]^+_- = 0 \qquad (5.29)$$

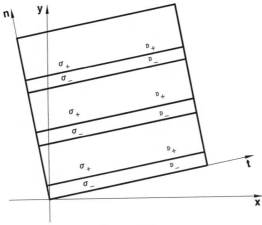

FIGURE 5.1

The last two equations together with the first equation of (5.28) show that the strain tensors within the layers differ only by their nn components

$$e_+ = e^0 + \alpha_1 nn$$

$$e_- = e^0 + \alpha_2 nn$$

To determine the unknown parameters α_1 and α_2, we use the first equation of (5.29), Eqs. (5.27), and the first equation of (5.28). We obtain

$$m\alpha_1 + (1 - m)\alpha_2 = 0$$

$$nn \cdot \cdot \mathbb{D}_+ \cdot \cdot (e^0 + \alpha_1 nn) - nn \cdot \cdot \mathbb{D}_- \cdot \cdot (e^0 + \alpha_2 nn) = 0$$

a linear system, whose solution is given by

$$\alpha_1 = \frac{1 - m}{\tilde{D}} nn \cdot \cdot (\mathbb{D}_- - \mathbb{D}_+) \cdot \cdot e^0$$

$$\alpha_2 = -\frac{m}{\tilde{D}} nn \cdot \cdot (\mathbb{D}_- - \mathbb{D}_+) \cdot \cdot e^0$$

where

$$\tilde{D} = nn \cdot \cdot [m\mathbb{D}_- + (1 - m)\mathbb{D}_+] \cdot \cdot nn \qquad (5.30)$$

The strains within the layers can now be written as

$$e_+ = \left[\mathbb{E} + \frac{1 - m}{\tilde{D}} nnnn \cdot \cdot (\mathbb{D}_- - \mathbb{D}_+) \right] \cdot \cdot e^0$$

$$\qquad (5.31)$$

$$e_- = \left[\mathbb{E} - \frac{m}{\tilde{D}} nnnn \cdot \cdot (\mathbb{D}_- - \mathbb{D}_+) \right] \cdot \cdot e^0$$

where $\mathbb{E}$ denotes a unit tensor of fourth rank.

The last equations permit the determination of an effective tensor $\mathbb{D}^0$ of the elastic moduli

$$\mathbb{M}^0 = \mathbb{D}^0 \cdot \cdot e^0$$

The use of the second equation of (5.28), Hooke's law (5.27), and (5.31), as well as the self-adjointness of $\mathbb{D}_+$ and $\mathbb{D}_-$ (see Section 5.1) yields

$$\mathbb{M}^0 = \left[m\mathbb{D}_+ + (1 - m)\mathbb{D}_- - \frac{m(1 - m)}{\tilde{D}} \mathbf{nn} \cdot \cdot (\mathbb{D}_- - \mathbb{D}_+) \right. $$
$$\left. \otimes (\mathbb{D}_- - \mathbb{D}_+) \cdot \cdot \mathbf{nn} \right] \cdot \cdot \mathbf{e}$$

that is,

$$\mathbb{D}^0 = m\mathbb{D}_+ + (1 - m)\mathbb{D}_- - \frac{m(1 - m)}{\tilde{D}} \mathbf{nn} \cdot \cdot (\mathbb{D}_- - \mathbb{D}_+) \otimes (\mathbb{D}_- - \mathbb{D}_+) \cdot \cdot \mathbf{nn}$$

This expression can also be represented in the form

$$\mathbb{D}^0 = \mathbb{D}^* - \frac{m(1 - m)}{\mathbf{nn} \cdot \cdot (\mathbb{D}_+ + \mathbb{D}_- - \mathbb{D}^*) \cdot \cdot \mathbf{nn}} \Delta\mathbb{D} \cdot \cdot \mathbf{nnnn} \cdot \cdot \Delta\mathbb{D} \qquad (5.32)$$

where

$$\mathbb{D}^* = m\mathbb{D}_+ + (1 - m)\mathbb{D}_-, \qquad \Delta\mathbb{D} = \mathbb{D}_+ - \mathbb{D}_-$$

In quite the same way, one may obtain the formula expressing the tensor $\mathbb{D}^0$ for a composite assembled of N different layers. Let the ith layer be characterized by a tensor $\mathbb{D}_i$ and a concentration m_i. The mentioned formula then may be written in the form

$$\mathbb{D}^0 = \sum_{i=1}^{N} m_i \mathbb{D}_i - \sum_{i=1}^{N} m_i \frac{\mathbb{D}_i \cdot \cdot \mathbf{nn}}{\bar{D}_i} \otimes \left[\mathbb{D}_i \cdot \cdot \mathbf{nn} - \frac{\sum_{\kappa=1}^{N} m_\kappa \frac{\mathbb{D}_\kappa \cdot \cdot \mathbf{nn}}{\bar{D}_\kappa}}{\sum_{\kappa=1}^{N} \frac{m_\kappa}{\bar{D}_\kappa}} \right]$$

where

$$\bar{D}_\kappa = \mathbf{nn} \cdot \cdot \mathbb{D}_\kappa \cdot \cdot \mathbf{nn}$$

Remark 5.1. For the torsional problem, the analogue of (5.32) is given by (3.46)

$$\mathbb{D}^0 = m\mathbb{D}_+ + (1 - m)\mathbb{D}_- - \frac{m(1 - m)(\mathbb{D}_+ - \mathbb{D}_-) \cdot \mathbf{n} \otimes (\mathbb{D}_+ - \mathbb{D}_-) \cdot \mathbf{n}}{\mathbf{n} \cdot [m\mathbb{D}_- + (1 - m)\mathbb{D}_+] \cdot \mathbf{n}}$$

5.5. The Problem of G-Closure: Estimation of the Set GU

The notion of G-closure of the initial set U of tensors of fourth rank is introduced in quite the same way as it was done in Section 4.1 (see also the Introduction) for tensors of second rank. As before, we shall first obtain the estimates of some sets Σ containing GU, these estimates being expressed in terms of weak limits of certain functions of the elements of the set U. The derivation will be based on weak convergence of the strain energy, and, an additional requirement for the fourth-order operator, also on weak convergence of the second strain invariant:

$$I_2(\mathbf{e}) \triangleq w_{xx}w_{yy} - w_{xy}^2 \longrightarrow w_{xx}^0 w_{yy}^0 - (w_{xy}^0)^2 \triangleq I_2(\mathbf{e}^0)$$

The set Σ will be described and it will be shown that its elements may be realized with the aid of laminated composites. In subsequent sections, G-closures will be built for a number of specific U sets. Namely, isotropic compounds with one and the same value of shear modulus μ and different values of the modulus of dilatation k will be considered. Also, the G-closure will be described for a set of materials possessing cubic symmetry and equal values of the dilatation modulus, the shear moduli of the compounds being assumed different.

5.5.1. Weak Convergence of the Strain Energy. Desiring to determine G-closure of the initial U set, we introduce the sequence $\{\mathbb{D}^i\}$ of controls [bounded in $L_\infty(S)$ space] and the corresponding sequences $\{\mathbf{e}^i\}$, $\{\mathbb{M}^i\}$ of deformations and moments, respectively. The latter may be treated as uniformly bounded in some $L_p(S)(p > 2)$ and (presumably, some subsequences) as weakly convergent in this space. Let $\mathbf{e}^0$ and $\mathbb{M}^0$ denote the weak limits

$$\mathbf{e}^i \xrightarrow[L_p(S)]{} \mathbf{e}^0, \qquad \mathbb{M}^i \xrightarrow[L_p(S)]{} \mathbb{M}^0 \tag{5.33}$$

The functions $\mathbb{M}^i$ and $\mathbf{e}^i$ are related by Hooke's law

$$\mathbb{M}^i = \mathbb{D}^i \cdot \cdot \, \mathbf{e}^i$$

with corresponding strain energy density

$$N^i \triangleq \mathbb{M}^i \cdot \cdot \, \mathbf{e}^i = \mathbf{e}^i \cdot \cdot \, \mathbb{D}^i \cdot \cdot \, \mathbf{e}^i$$

We now demonstrate that, for sufficiently smooth external loads, one has the limiting relation

$$N^i \triangleq \mathbb{M}^i \cdots e^i \xrightarrow[L_{p/2}(S)]{} \mathbb{M}^0 \cdots e^0 \triangleq N^0$$

Following the definition of the generalized solution (5.4) we have

$$\iint_S \nabla\nabla w^i \cdots \mathbb{D}^i \cdots \nabla\nabla \eta^i \, dx \, dy = \iint_S q\eta^i \, dx \, dy \quad \forall \eta^i \in \mathring{W}_2^2(S) \quad (5.34)$$

Assume that $\eta^i = \xi(w^i - w^0)$, where $\xi(x, y)$ is sufficiently smooth and w^i, w^0 denote deflections corresponding to the deformations e^i and e^0, respectively. We then obtain

$$\iint_S \xi\nabla\nabla w^i \cdots \mathbb{D}^i \cdots (\nabla\nabla w^i - \nabla\nabla w^0) \, dx \, dy$$

$$= -\iint_S (w^i - w^0)\nabla\nabla w^i \cdots \mathbb{D}^i \cdots \nabla\nabla \xi \, dx \, dy - 2\iint_S \nabla\nabla w^i \cdots \mathbb{D}^i \cdots \nabla\xi$$

$$\otimes \nabla(w^i - w^0) \, dx \, dy + \iint_S \xi q(w^i - w^0) \, dx \, dy \quad (5.35)$$

Recall that $q \in L_2(S)$. Then[483] for any strictly internal subregion S', $S' \in S$, all $w^i(\mathbb{D}^i)$ belong to some bounded set of $W_p^2(S')$ for some $p > 2$. The right-hand side of (5.35) now obviously tends to zero, and in view of the arbitrariness of ξ, one has

$$\mathbb{M}^i \cdots (e^i - e^0) \xrightarrow[L_{p/2}(S)]{} 0$$

or

$$N^i \triangleq \mathbb{M}^i \cdots e^i \xrightarrow[L_{p/2}(S)]{} \mathbb{M}^0 \cdots e^0 \triangleq N^0$$

The weak limits $\mathbb{M}^0$ and e^0 are obviously connected by some bounded linear operator L since the operation of transition to a weak limit is linear. It can also be shown that the operator L is local, i.e., $\mathbb{M}^0(x) = Le^0(x)$ for any $x \in S$. Now, because of the symmetry of the second-rank tensors $\mathbb{M}^0$ and e^0 the operator L turns out to be a self-adjoint (Section 5.1) tensor of fourth rank:

$$L \equiv \mathbb{D}^0, \qquad \mathbb{M}^0 = \mathbb{D}^0 \cdots e^0$$

Using weak convergence of the strain energy, we may now estimate the tensor $\mathbb{D}^0$. Define $\mathbb{D}^*$ as the weak limit of the sequence $\{\mathbb{D}^i\}$:

$$\mathbb{D}^i \xrightarrow[L_\infty(S)]{} \mathbb{D}^*$$

The sequence

$$A^i \triangleq 2(\mathbb{M}^i \cdot \cdot \, e^i - \mathbb{M}^i \cdot \cdot \, e^0) = 2(e^i - e^0) \cdot \cdot \, \mathbb{D}^i \cdot \cdot \, e^i$$

is weakly convergent to zero:

$$A^i \xrightarrow[L_{p/2}(S)]{} 0$$

Transforming A^i, we obtain

$$A^i \triangleq (e^i - e^0) \cdot \cdot \, \mathbb{D}^i \cdot \cdot \, (e^i - e^0) - e^0 \cdot \cdot \, \mathbb{D}^i \cdot \cdot \, e^0 + e^i \cdot \cdot \, \mathbb{D}^i \cdot \cdot \, e^i \xrightarrow[L_{p/2}(S)]{} 0 \tag{5.36}$$

The first term on the left-hand side is nonnegative since the same is true for the eigenvalues of $\mathbb{D}^i$; passing over to the weak limit in (5.36), we get

$$e^0 \cdot \cdot \, \mathbb{D}^0 \cdot \cdot \, e^0 \leq e^0 \cdot \cdot \, \mathbb{D}^* \cdot \cdot \, e^0$$

from which it follows that the eigenvalues of $\mathbb{D}^* - \mathbb{D}^0$ are also nonnegative:

$$\lambda_i[\mathbb{D}^* - \mathbb{D}^0] \geq 0 \qquad (i = 1, 2, 3) \tag{5.37}$$

In quite the same way, we may start with the limiting relationship

$$B^i \triangleq 2(\mathbb{M}^i \cdot \cdot \, e^i - \mathbb{M}^0 \cdot \cdot \, e^i)$$

$$= (\mathbb{M}^i - \mathbb{M}^0) \cdot \cdot \, (\mathbb{D}^i)^{-1} \cdot \cdot \, (\mathbb{M}^i - \mathbb{M}^0) - \mathbb{M}^0 \cdot \cdot \, (\mathbb{D}^i)^{-1} \cdot \cdot \, \mathbb{M}$$

$$+ \mathbb{M}^i \cdot \cdot \, (\mathbb{D}^i)^{-1} \cdot \cdot \, \mathbb{M}^i \xrightarrow[L_{p/2}(S)]{} 0$$

In view of positive definiteness of the tensor $(\mathbb{D}^i)^{-1}$ and of the weak convergence of the expression

$$\mathbb{M}^i \cdot \cdot \, (\mathbb{D}^i)^{-1} \cdot \cdot \, \mathbb{M}^i = \mathbb{M}^i \cdot \cdot \, e^i \xrightarrow[L_{p/2}(S)]{} \mathbb{M}^0 \cdot \cdot \, e^0 = \mathbb{M}^0 \cdot \cdot \, (\mathbb{D}^0)^{-1} \cdot \cdot \, \mathbb{M}^0$$

we arrive at the inequality

$$\mathbb{M}^0 \cdot \cdot (\mathbb{D}^0)^{-1} \cdot \cdot \mathbb{M}^0 \le \mathbb{M}^0 \cdot \cdot (\mathbb{D}^{-1})^* \cdot \cdot \mathbb{M}^0$$

where, by definition,

$$(\mathbb{D}^i)^{-1} \xrightarrow[L_\infty(S)]{} (\mathbb{D}^{-1})^*$$

The eigenvalues of the tensor $(\mathbb{D}^{-1})^* - (\mathbb{D}^0)^{-1}$ are obviously nonnegative, since

$$\lambda_i[(\mathbb{D}^{-1})^* - (\mathbb{D}^0)^{-1}] \ge 0 \qquad (i = 1, 2, 3) \tag{5.38}$$

5.5.2. Weak Convergence of the Second Strain Invariant. Here we shall prove an important feature of the tensor $e^i = -\nabla\nabla w^i$: its second invariant $I_2(e^i) = w^i_{xx} w^i_{yy} - (w^i_{xy})^2$ converges weakly in $L_{p/2}(S)(p > 2)$ to the second invariant $I_2(e^0) = w^0_{xx} w^0_{yy} - (w^0_{xy})^2$ of the tensor $e^0 = -\nabla\nabla w^0$. Observe that $2I_2(e^i) = e^i \cdot \cdot \mathbb{D}_{I_2} \cdot \cdot e^i$, where $\mathbb{D}_{I_2}$ denotes a self-adjoint tensor of the fourth rank which is represented by

$$\mathbb{D}_{I_2} = \mathfrak{a}_1 \mathfrak{a}_1 - \mathfrak{a}_2 \mathfrak{a}_2 - \mathfrak{a}_3 \mathfrak{a}_3$$

with basis

$$\mathfrak{a}_1 = \frac{1}{\sqrt{2}}(\mathbf{ii} + \mathbf{jj}), \qquad \mathfrak{a}_2 = \frac{1}{\sqrt{2}}(\mathbf{ii} - \mathbf{jj}), \qquad \mathfrak{a}_3 = \frac{1}{\sqrt{2}}(\mathbf{ij} + \mathbf{ji})$$

It will be shown that

$$e^i \cdot \cdot \mathbb{D}_{I_2} \cdot \cdot e^i \xrightarrow[L_{p/2}(S)]{} e^0 \cdot \cdot \mathbb{D}_{I_2} \cdot \cdot e^0 \tag{5.39}$$

Consider the sequence

$$e^i \cdot \cdot \mathbb{D}_{I_2} \cdot \cdot e^i - e^0 \cdot \cdot \mathbb{D}_{I_2} \cdot \cdot e^0 \tag{5.40}$$

and take some sufficiently smooth finite function $\xi = \xi(x, y)$ to construct the expression

$$A = \iint_S \xi e \cdot \cdot \mathbb{D}_{I_2} \cdot \cdot e \, dx \, dy = 2 \iint_S \xi[w_{xx}w_{yy} - (w_{xy})^2] \, dx \, dy$$

$$= \iint_S \xi \left[\frac{\partial}{\partial x}(w_x w_{yy} - w_y w_{xy}) - \frac{\partial}{\partial y}(w_x w_{xy} - w_y w_{xx}) \right] dx \, dy$$

Double integration by parts provides the equivalent form

$$A = \iint_S [(\xi_{xy}w_x - \xi_{xx}w_y)w_y + (\xi_{xy}w_y - \xi_{yy}w_x)w_x]\, dx\, dy \qquad (5.41)$$

Multiplying (5.40) by ξ and transforming the result by double integration, we get

$$\iint_S \xi(e^i \cdot\cdot \mathbb{D}_{I_2} \cdot\cdot e^i - e^0 \cdot\cdot \mathbb{D}_{I_2} \cdot\cdot e^0)\, dx\, dy$$

$$= \iint_S \{2\xi_{xy}(w_x^i w_y^i - w_x^0 w_y^0)$$

$$- \xi_{xx}[(w_y^i)^2 - (w_y^0)^2] - \xi_{yy}[(w_x^i)^2 - (w_x^0)^2]\}\, dx\, dy \qquad (5.42)$$

The right-hand side of the last equation tends to zero [since $w \in W_p^2(S)$] which proves the limit (5.39) because ξ is arbitrary.

Remark 5.2. In view of the well-known statical-geometrical analogy, one may establish the weak convergence of the second invariant of the stress tensor for the corresponding problem of extension of a plate.

Remark 5.3. The weak convergence of the bilinear forms

$$\mathbb{M} \cdot\cdot e = M_{nn}e_{nn} + M_{tt}e_{tt} + 2M_{nt}e_{nt}, \qquad I_2(e) = e_{nn}e_{tt} - e_{nt}^2$$

to the weak limits $\mathbb{M}^0 \cdot\cdot e^0$ and $I_2(e^0)$, respectively, holds because these forms satisfy the following conditions: (1) they are invariant; (2) across any boundary line dividing media with different rigidities, the jump of each form depends linearly upon the jumps of the components of the tensors $\mathbb{M}$ and e. The latter feature follows immediately from the continuity conditions

$$[M_{nn}]_-^+ = [e_{nt}]_-^+ = [e_{tt}]_-^+ = 0$$

which hold across the boundary line with normal $\mathbf{n}$ and tangent $\mathbf{t}$.

In view of linearity of the operation of passing to the weak limit (i.e., averaging over the element of microstructure) and due to the linearity of the jumps $[\cdot]_-^+$ of the forms $\mathbb{M} \cdot \cdot \mathbb{e}$ and $I_2(\mathbb{e})$ relative to the jumps $[M_{tt}]_-^+, [M_{nt}]_-^+, [e_{nn}]_-^+$, the sequences of those forms turn out to be weakly convergent.

The same argument holds for the second-order problem of Chapter 4 where the jump of the energy

$$[\mathbb{M} \cdot \mathbb{e}]_-^+ = M_n[e_n]_-^+ + [M_t]_-^+ e_t$$

depends linearly upon the jumps of $\mathbb{M}$ and $\mathbb{e}$ since the components M_n and e_t of these vectors are continuous.

5.5.3. Estimates of the Effective Tensors $\mathbb{D}^0$ in Terms of the Weak Limit Tensors.
The preceding sections provide the background for some important inequalities derived below.

We have

$$2(\mathbb{e}^i \cdot \cdot \mathbb{D}^i \cdot \cdot \mathbb{e}^i - \mathbb{e}^i \cdot \cdot \mathbb{D}^i \cdot \cdot \mathbb{e}^0) \xrightarrow[L_{p/2}(S)]{} 0 \tag{5.43}$$

$$2(\mathbb{e}^i \cdot \cdot \mathbb{D}_{I_2} \cdot \cdot \mathbb{e}^i - \mathbb{e}^i \cdot \cdot \mathbb{D}_{I_2} \cdot \cdot \mathbb{e}^0) \xrightarrow[L_{p/2}(S)]{} 0 \tag{5.44}$$

Multiplying (5.44) by some fixed constant α and subtracting the result from (5.43), we obtain

$$2[\mathbb{e}^i \cdot \cdot (\mathbb{D}^i - \alpha \mathbb{D}_{I_2}) \cdot \cdot \mathbb{e}^i - \mathbb{e}^i \cdot \cdot (\mathbb{D}^i - \alpha \mathbb{D}_{I_2}) \cdot \cdot \mathbb{e}^0] \xrightarrow[L_{p/2}(S)]{} 0$$

Denote $\tilde{\mathbb{D}}^i = \mathbb{D}^i - \alpha \mathbb{D}_{I_2}$ and transform the last relationship in the following way:

$$(\mathbb{e}^i - \mathbb{e}^0) \cdot \cdot \tilde{\mathbb{D}}^i \cdot \cdot (\mathbb{e}^i - \mathbb{e}^0) - \mathbb{e}^0 \cdot \cdot \tilde{\mathbb{D}}^i \cdot \cdot \mathbb{e}^0 + \mathbb{e}^i \cdot \cdot \tilde{\mathbb{D}}^i \cdot \cdot \mathbb{e}^i \xrightarrow[L_{p/2}(S)]{} 0$$

If the parameter α is chosen so as to leave the eigenvalues of $\tilde{\mathbb{D}}^i$ nonnegative for any number i:

$$\lambda_j[\tilde{\mathbb{D}}^i] \ge 0, \qquad j = 1, 2, 3 \tag{5.45}$$

then

$$\mathbb{e}^0 \cdot \cdot \tilde{\mathbb{D}}^0 \cdot \cdot \mathbb{e}^0 \le \mathbb{e}^0 \cdot \cdot \tilde{\mathbb{D}}^* \cdot \cdot \mathbb{e}^0$$

where

$$\tilde{\mathbb{D}}^0 = \mathbb{D}^0 - \alpha\mathbb{D}_{I_2}, \qquad \tilde{\mathbb{D}}^* = \lim_{\text{weak}} (\mathbb{D}^i - \alpha\mathbb{D}_{I_2})$$

The latter inequality shows that the eigenvalues of $\tilde{\mathbb{D}}^* - \tilde{\mathbb{D}}^0$ are nonnegative

$$\lambda_j[\tilde{\mathbb{D}}^* - \tilde{\mathbb{D}}^0] = \lambda_j[\mathbb{D}^* - \mathbb{D}^0] \geq 0 \qquad (j = 1, 2, 3) \qquad (5.46)$$

By analogy we arrive at the following estimate:

$$\lambda_j[(\tilde{\mathbb{D}}^{-1})^* - (\tilde{\mathbb{D}}^0)^{-1}] \geq 0 \qquad (j = 1, 2, 3) \qquad (5.47)$$

where

$$(\tilde{\mathbb{D}}^{-1})^* = \lim_{\text{weak}} (\tilde{\mathbb{D}}^i)^{-1} = \lim_{\text{weak}} (\mathbb{D}^i - \alpha\mathbb{D}_{I_2})^{-1}$$

The inequality (5.46) does not depend upon the parameter α and coincides with (5.37). As to (5.47), it does depend upon α, and it produces stronger estimates than (5.38) (the results are the same for $\alpha = 0$). The parameter α should be so chosen as to provide the strongest possible estimates; the requirement (5.45) should, however, be kept in mind.

The restrictions imposed on the elements of the U set form the basis for a relationship between the weak limits $\tilde{\mathbb{D}}^*$ and $(\tilde{\mathbb{D}}^{-1})^*$. The technique of derivation of such a relationship will be illustrated below by examples; the final estimates characterize the class of composite media which can be assembled from the given materials (compounds). For a number of cases, these estimates will be shown to be exact.

Remark 5.4. For problems of the second order connected with the operator $\nabla \cdot \mathbb{D} \cdot \nabla$, we used only the statement of Section 5.5.1 concerning the convergence of the strain energy. The corresponding estimates (see Section 4.2) can be obtained from (5.45)–(5.47) by setting $\alpha = 0$.

The set of tensors $\mathbb{D}^0$ satisfying (5.45)–(5.47) will be denoted by Σ. Clearly, $GU \subset \Sigma$; i.e., any corresponding pair of weak limits $\mathbb{M}^0, e^0$ are connected by the relationship (Hooke's law) $\mathbb{M}^0 = \mathbb{D}^0 \cdot \cdot e^0$ where $\mathbb{D}^0 \in \Sigma$. To demonstrate existence, it suffices to show that for any point of Σ, we are able to build a corresponding composite using only the initial set U of compounds. This is the same as establishing coincidence of the sets Σ and GU.

5.5.4. The Case with Partly Known $\mathbb{D}^0$. As stated in Section 5.1, the self-adjoint tensor $\mathbb{D}$ of fourth rank possesses five invariants. The corresponding region GU in the invariant space is therefore five-dimensional, its boundary being a four-dimensional manifold. Investigation of the G-closure problem would be substantially simplified if the dimensionality of the invariant space could be decreased in some way. In this section, we shall describe an important class of such cases.

Suppose that the elements of the admissible set U of materials are characterized by the tensors

$$\mathbb{D}^i = d_1 \mathfrak{a}_1 \mathfrak{a}_1 + d_2^i \mathfrak{a}_2^i \mathfrak{a}_2^i + d_3^i \mathfrak{a}_3^i \mathfrak{a}_3^i$$

having the same fixed eigenvalue d_1 and corresponding eigentensor $\mathfrak{a}_1$. Then the same will hold for the effective tensor; i.e., $\mathbb{D}^0$ will allow the following representation:

$$\mathbb{D}^0 = d_1 \mathfrak{a}_1 \mathfrak{a}_1 + d_2^0 \mathfrak{a}_2^0 \mathfrak{a}_2^0 + d_3^0 \mathfrak{a}_3^0 \mathfrak{a}_3^0 \qquad (5.48)$$

Consider the weakly convergent sequence of deformations $\{e^i\}$

$$e^i \to e^0$$

and the corresponding sequence of moments

$$\mathbb{M}^i = \mathbb{D}^i \cdot\cdot\, e^i \to \mathbb{M}^0 = \mathbb{D}^0 \cdot\cdot\, e^0 \qquad (5.49)$$

where

$$\mathbb{M}^i = d_1 e_1^i \mathfrak{a}_1 + d_2^i e_2^i \mathfrak{a}_2^i + d_3^i e_3^i \mathfrak{a}_3^i, \qquad e_1^i = e^i \cdot\cdot\, \mathfrak{a}_1, \qquad e_j^i = e^i \cdot\cdot\, \mathfrak{a}_j^i$$

$$(j = 2, 3) \quad (5.50)$$

Contracting the last tensor equation with the constant tensor $\mathfrak{a}_1$, we obtain

$$\mathbb{M}^i \cdot\cdot\, \mathfrak{a}_1 = d_1 e_1^i \to d_1 e_1^0 = \mathbb{M}^0 \cdot\cdot\, \mathfrak{a}_1$$

where

$$e_1^0 = e^0 \cdot\cdot\, \mathfrak{a}_1$$

We now expand the tensor $\mathbb{D}^0$ over the basis $\mathfrak{a}_1, \mathfrak{a}_2, \mathfrak{a}_3$, where $\mathfrak{a}_2, \mathfrak{a}_3$ are orthogonal to $\mathfrak{a}_1$ ($\mathfrak{a}_1 \cdot\cdot\, \mathfrak{a}_2 = \mathfrak{a}_1 \cdot\cdot\, \mathfrak{a}_3 = 0$), and hence arrive at the relationship

$$\mathbb{M}^0 = d_{ij}^0 \mathfrak{a}_i \mathfrak{a}_j \cdot\cdot\, e^0 \qquad (d_{ij}^0 = d_{ji}^0, i, j = 1, 2, 3)$$

Contracting this with $\mathfrak{a}_1$ and using (5.48) and (5.50) we get

$$d_1 e_1^0 = d_{1j}^0 e_j^0$$

which shows (in view of the arbitrariness of e^0) that

$$d_{11}^0 = d_1, \qquad d_{12}^0 = d_{13}^0 = 0$$

thus proving (5.48).

In particular, if the set U

$$\{U : \mathbb{D}^i = k\mathfrak{a}_1\mathfrak{a}_1 + \mu^i(\mathfrak{a}_2\mathfrak{a}_2 + \mathfrak{a}_3\mathfrak{a}_3)\}$$

$$\mathfrak{a}_1 = \frac{1}{\sqrt{2}}\,(\mathbf{ii} + \mathbf{jj}), \qquad \mathfrak{a}_2 = \frac{1}{\sqrt{2}}\,(\mathbf{ii} - \mathbf{jj}), \qquad \mathfrak{a}_3 = \frac{1}{\sqrt{2}}\,(\mathbf{ij} + \mathbf{ji}) \qquad (5.51)$$

of isotropic materials is characterized by the same dilatation modulus k and by different values μ^i of the shear modulus μ, then the effective tensor $\mathbb{D}^0$ can be represented as

$$\mathbb{D}^0 = k\mathfrak{a}_1\mathfrak{a}_1 + \mu_2^0\mathfrak{a}_2^0\mathfrak{a}_2^0 + \mu_3^0\mathfrak{a}_3^0\mathfrak{a}_3^0 \qquad (5.52)$$

where $\mathfrak{a}_2^0, \mathfrak{a}_3^0$ denote certain deviators; the latter tensor corresponds to some medium with cubic symmetry.

If the set U includes the media (5.11) with cubic symmetry and possessing the same value k of the dilatation modulus, then $\mathbb{D}^0$ allows the following general representation:

$$\mathbb{D}^0 = k\mathfrak{a}_1\mathfrak{a}_1 + \mu_2^0\mathfrak{a}_2^0\mathfrak{a}_2^0 + \mu_3^0\mathfrak{a}_3^0\mathfrak{a}_3^0$$

We see that the effective medium is a medium with cubic symmetry also. Now, if the elements of the initial class of isotropic media differ only by the values k of their dilatation moduli, then the effective tensor $\mathbb{D}^0$ allows the representation

$$\mathbb{D}^0 = k^0\mathfrak{a}_1\mathfrak{a}_1 + \mu(\mathfrak{a}_2\mathfrak{a}_2 + \mathfrak{a}_3\mathfrak{a}_3) \qquad (5.53)$$

which is typical for an isotropic medium. Here we have the case when the fixed eigenvalues of the initial compounds are double eigenvalues, a fact which does not influence the main limiting result.

In what follows, we give a number of applications of these results to some optimization problems involving bending of thin plates.

5.6. G-Closure of a Set U of Isotropic Materials with Equal Shear Moduli and Different Dilatation Moduli

Consider a plate made of two isotropic materials each of which has the same fixed value of the shear modulus μ and some different (but fixed) values k_+ and k_- of the dilatation moduli. The tensors of the elastic moduli of the compounds are given by

$$\mathbb{D}_+ = k_+ \mathbb{a}_1 \mathbb{a}_1 + \mu(\mathbb{a}_2 \mathbb{a}_2 + \mathbb{a}_3 \mathbb{a}_3)$$

$$\mathbb{D}_- = k_- \mathbb{a}_1 \mathbb{a}_1 + \mu(\mathbb{a}_2 \mathbb{a}_2 + \mathbb{a}_3 \mathbb{a}_3) \tag{5.54}$$

where $\mathbb{a}_1$ denotes a spherical tensor, and $\mathbb{a}_2, \mathbb{a}_3$ denote the deviators. This set of tensors constitutes the set U.

The set GU of $\mathbb{D}^0$ tensors (see Section 5.5.4)

$$GU = \{\mathbb{D}^0 : \mathbb{D}^0 = k^0 \mathbb{a}_1 \mathbb{a}_1 + \mu(\mathbb{a}_2 \mathbb{a}_2 + \mathbb{a}_3 \mathbb{a}_3), \qquad k_- \le k^0 \le k_+\}$$

represents a class of isotropic media with some intermediate value of the dilatation modulus k^0.

The results of Section 5.5 provide some estimates of k^0 expressed through the value k^* of the weak limit of the sequence $\{k^i\}$, i.e., through the value of k averaged over the element of the microstructure. To obtain these estimates, we make use of (5.46) and (5.47). First, calculate the tensors $\tilde{\mathbb{D}}^0, \tilde{\mathbb{D}}^*, (\tilde{\mathbb{D}}^{-1})^*$. We have

$$\tilde{\mathbb{D}}^0 = (k^0 - \alpha)\mathbb{a}_1 \mathbb{a}_1 + (\mu + \alpha)(\mathbb{a}_2 \mathbb{a}_2 + \mathbb{a}_3 \mathbb{a}_3)$$

$$\tilde{\mathbb{D}}^* = (k^* - \alpha)\mathbb{a}_1 \mathbb{a}_1 + (\mu + \alpha)(\mathbb{a}_2 \mathbb{a}_2 + \mathbb{a}_3 \mathbb{a}_3)$$

$$(\tilde{\mathbb{D}}^{-1})^* = \left(\frac{1}{k - \alpha}\right)^* \mathbb{a}_1 \mathbb{a}_1 + \frac{1}{\mu + \alpha}(\mathbb{a}_2 \mathbb{a}_2 + \mathbb{a}_3 \mathbb{a}_3)$$

where

$$\left(\frac{1}{k - \alpha}\right)^* = \lim_{\text{weak}} \left(\frac{1}{k^i - \alpha}\right)$$

The tensors $\tilde{\mathbb{D}}^* - \tilde{\mathbb{D}}^0$ and $(\tilde{\mathbb{D}}^{-1})^* - (\tilde{\mathbb{D}}^0)^{-1}$ are given by

$$\tilde{\mathbb{D}}^* - \tilde{\mathbb{D}}^0 = (k^* - k^0)\mathbb{a}_1 \mathbb{a}_1$$

$$(\tilde{\mathbb{D}}^{-1})^* - (\tilde{\mathbb{D}}^0)^{-1} = \left[\left(\frac{1}{k - \alpha}\right)^* - \frac{1}{k^0 - \alpha}\right]\mathbb{a}_1 \mathbb{a}_1$$

and the estimates (5.46) and (5.47) take on the form

$$k^* \geq k^0, \qquad \left(\frac{1}{k-\alpha}\right)^* \geq \frac{1}{k^0-\alpha}$$

or

$$\left[\left(\frac{1}{k-\alpha}\right)^*\right]^{-1} \leq k^0 - \alpha \leq k^* - \alpha \tag{5.55}$$

According to (5.45), the parameter α should belong to the interval

$$-\mu \leq \alpha \leq k_- \tag{5.56}$$

It remains to establish the relationship between the weak limits k^* and $[1/(k-\alpha)]^*$. We observe that the terms k^i of the weakly convergent sequence $\{k^i\}$ take on either the value k_+ or k_-; it follows that

$$(k_+ - k^i)(k^i - k_-) = [(k_+ - \alpha) - (k^i - \alpha)][(k^i - \alpha) - (k_- - \alpha)] = 0$$

Dividing this equation by $k^i - \alpha$ and passing over to the weak limit, we obtain

$$(k^* - \alpha) - (k_+ + k_- - 2\alpha) + (k_+ - \alpha)(k_- - \alpha)[1/(k-\alpha)]^* = 0$$

With this relationship we may eliminate the value $[(1/(k-\alpha))^*]^{-1}$ from (5.55); we then arrive at the following estimate:

$$\frac{(k_+ - \alpha)(k_- - \alpha)}{k_+ + k_- - 2\alpha - (k^* - \alpha)} + \alpha \leq k^0 \leq k^* \tag{5.57}$$

For each α satisfying (5.56), (5.57) outline a certain region on the k^0, k^* plane. Considering the α family of such regions, one can easily verify that the strongest estimate is provided by $\alpha = -\mu$, and the corresponding region is determined as the intersection of all those for which $\alpha \in [-\mu, k_-]$. For $\alpha = -\mu$, we obtain

$$\frac{(k_+ + \mu)(k_- + \mu)}{k_+ + k_- + \mu - k^*} - \mu \leq k^0 \leq k^* \tag{5.58}$$

The set Σ of admissible points (k^0, k^*) is shown in Fig. 5.2, where the boundary curve A corresponds to the first equation of (5.58) and the curve B to the second equation of (5.58). The A curve tends to that of B as the parameter μ increases to infinity.

It is easy to show that the A curve corresponds to the laminated medium assembled of the initially given materials. Substituting the tensors (5.54) into (5.32) and denoting the weak limit of k by $k^* = mk_+ + (1 - m)k_-$, we obtain

$$\mathbb{D}^0 = \left[\frac{(k_+ + \mu)(k_- + \mu)}{k_+ + k_- + \mu - k^*} - \mu \right] \mathfrak{a}_1 \mathfrak{a}_1 + \mu(\mathfrak{a}_2 \mathfrak{a}_2 + \mathfrak{a}_3 \mathfrak{a}_3) \qquad (5.59)$$

Note that the orientation of the layers is of no importance here since the term containing $\mathbf{n}$ in (5.32)

$$\frac{m(1 - m)}{\mathbf{nn} \cdot \cdot [m\mathbb{D}_- + (1 - m)\mathbb{D}_+] \cdot \cdot \mathbf{nn}} \Delta\mathbb{D} \cdot \cdot \mathbf{nnnn} \cdot \cdot \Delta\mathbb{D}$$

$$= \frac{m(1 - m)(k_+ + k_-)^2 (\mathfrak{a}_1 \cdot \cdot \mathbf{nn})^2 \mathfrak{a}_1 \mathfrak{a}_1}{[mk_- + (1 - m)k_+](\mathfrak{a}_1 \cdot \cdot \mathbf{nn})^2 + \mu[(\mathfrak{a}_2 \cdot \cdot \mathbf{nn})^2 + (\mathfrak{a}_3 \cdot \cdot \mathbf{nn})^2]}$$

$$= \frac{m(1 - m)(k_+ + k_-)^2}{mk_- + (1 - m)k_+ + \mu} \mathfrak{a}_1 \mathfrak{a}_1$$

is seen to be independent of this vector. This is obviously due to the exceptional nature of the spherical tensor $\mathfrak{a}_1$.

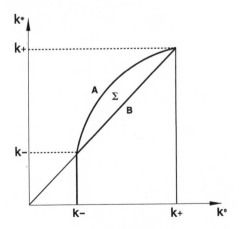

FIGURE 5.2

In close analogy with Section 4.4, the last observation suggests that any composite assembled of the components (5.54) corresponds to the points of a curve A regardless of the form of the inclusions which constitute the microstructure. In other words, (5.46) and (5.47) now provide rough estimates since the GU set is actually represented by a curve A, whereas the mentioned inequalities determine some wider domain.

To prove this hypothesis, we shall use the identity

$$\nabla\nabla \cdot\cdot \mathbb{D}_{I_2} \cdot\cdot \nabla\nabla w = 0 \tag{5.60}$$

The multiplication of (5.60) by the fixed value μ of the shear modulus and its combination with (5.5), along with (5.54) and (5.12) yield

$$\nabla\nabla \cdot\cdot (k + \mu)\mathfrak{a}_1\mathfrak{a}_1 \cdot\cdot \nabla\nabla w \equiv \tfrac{1}{2}\Delta(k + \mu)\Delta w = q$$

or, equivalently,

$$\Delta M = 2q$$
$$M = (k + \mu)\Delta w \tag{5.61}$$

Consider now the weakly [in $L_\infty(S)$] convergent sequence $\{\mathbb{D}^s\}$ of the elements of U [see (5.54)] and the corresponding sequence of deflections $\{w^s\}$ and deformations $\{e^s = -\nabla\nabla w^s\}$. The latter one is bounded in $L_2(S)$ and may be supposed weakly convergent in $L_2(S)$ to some tensor $e^0 = -\nabla\nabla w^0$:

$$-\nabla\nabla w^s = e^s \xrightarrow[L_2(S)]{} -\nabla\nabla w^0 = e^0$$

In particular,

$$-e^s \cdot\cdot \mathfrak{a}_1 = \frac{1}{\sqrt{2}}\Delta w^s \xrightarrow[L_2(S)]{} -e^0 \cdot\cdot \mathfrak{a}_1 = \frac{1}{\sqrt{2}}\Delta w^0 \tag{5.62}$$

Here, w^0 denotes a weakly limiting deflection which satisfies the equation

$$\nabla\nabla \cdot\cdot \mathbb{D}^0 \cdot\cdot \nabla\nabla w^0 = q$$

or, in view of (5.12) and (5.53),

$$\tfrac{1}{2}\Delta(k^0 + \mu)\Delta w^0 = q \tag{5.63}$$

and

$$\Delta M^0 = 2q$$

$$M^0 = (k^0 + \mu)\Delta w^0 \tag{5.64}$$

A comparison of $(5.61)_2$ and $(5.64)_2$ and the use of (5.4) imply

$$M^s \xrightarrow{L_2(S)} M^0$$

At the same time, the members of the sequence $\{M^s\}$ represent the generalized solutions of Poisson's equation (5.61). This sequence is therefore uniformly bounded in $W_2^1(S)$ and consequently *strongly* convergent in $L_2(S)$ to M^0

$$M^s \xrightarrow{L_2(S)} M^0 \qquad \text{(strongly)} \tag{5.65}$$

On the other hand, the weak limit equation (5.62) together with $(5.61)_2$ and $(5.64)_2$ show that

$$\Delta w^s = \frac{M^s}{k^s + \mu} \xrightarrow{L_2(S)} \Delta w^0 = \frac{M^0}{k^0 + \mu} \tag{5.66}$$

The use of the strong $L_2(S)$ convergence of $\{M^s\}$ to M^0, and passage to the weak limit in (5.66) result in

$$\frac{1}{k^0 + \mu} = \lim_{\text{weak}} \frac{1}{k^s + \mu} \tag{5.67}$$

It now remains to connect the weak limit

$$\lim_{\text{weak}} \frac{1}{k^s + \mu}$$

with the value

$$k^* = \lim_{\text{weak}} k^s$$

of the averaged dilatation modulus. But this has already been done before; making use of this previous result, we may reduce (5.67) to the form

$$k^0 = \frac{(k_- + \mu)(k_+ + \mu)}{k_+ + k_- + \mu - k^*} - \mu \tag{5.68}$$

It is now clear that the effective dilatation modulus k^0 is determined only by the elastic constants of the initial compounds and by their concentration rates; it does not depend on the microstructure of the composite, i.e., on the form of the inclusions which constitute the mixture. The physical reason is that the dilatation modulus connects the first invariants of moment and deformation tensors which only enter the basic equations (5.61) and their G-limits (5.64). This specific feature allows us to establish the *strong* $L_2(S)$-convergence of the sequence $\{M^s\}$ [(5.65)]. The presence of the tensor $\mathfrak{a}_1$ (which is invariant under rotation) within the basis $(\mathfrak{a}_1, \mathfrak{a}_2, \mathfrak{a}_3)$ is germane to problems of G-closure of the tensors of fourth rank and provides no analogy for the corresponding second-rank problems whose basic space includes only vectors.

5.7. *G*-Closure of a Set of Materials with Cubic Symmetry and Equal Dilatation Moduli

Consider now a set $U(\mu_2, \mu_3, \varphi)$ of materials of the type (5.11) possessing cubic symmetry and suppose that all of them have equal and fixed dilatation modulus k; all other parameters, i.e., the shear moduli μ_2 and μ_3 ($\mu_2 \le \mu_3$) and the angle φ between the principal axes of elasticity and some fixed set of axes, may vary and belong to some bounded sets of $L_\infty(S)$. For this set of materials, we now formulate the problem for the determination of G-closure of $U(\mu_2, \mu_3, \varphi)$.

As was shown in Section 5.5.4, all possible composites belonging to $GU = GU(\mu_{20}, \mu_{30}, \varphi_0)$ preserve their initial structure (5.11) with exactly the same dilatation modulus k as that of the compounds. The angle φ_0 is determined by the orientation of the principal axes of a composite and is clearly arbitrary. It remains to determine the set of possible values μ_{20}, μ_{30}; this set will be denoted by $GU(\mu_{20}, \mu_{30})$. As in the preceding section, it will be convenient to begin the analysis with the transformation of the basic equation (5.5). Multiplying (5.60) by $(-k)$ and combining the result with (5.5) we get

$$\nabla\nabla \cdot \cdot \, \tilde{\mathbb{D}} \cdot \cdot \, \nabla\nabla w = q \tag{5.69}$$

where the tensor

$$\tilde{\mathbb{D}} \triangleq \mathbb{D} - k\mathbb{D}_{I_2} = (\mu_2 + k)\mathfrak{a}_2\mathfrak{a}_2 + (\mu_3 + k)\mathfrak{a}_3\mathfrak{a}_3 \triangleq \tilde{\mu}_2\mathfrak{a}_2\mathfrak{a}_2 + \tilde{\mu}_3\mathfrak{a}_3\mathfrak{a}_3$$

$$\tag{5.70}$$

is acting in a subspace of deviators of a linear space of symmetric tensors of second rank. The basis $\mathfrak{a}_2, \mathfrak{a}_3$ of this subspace is quite similar to the vector basis in a plane (see Section 5.1) which makes it reasonable to use the results obtained in Section 4.5 for the problem of G-closure of a set of tensors of fourth rank relative to the operator (5.69) as well.

We shall have need of the estimates (5.46) and (5.47) for the eigenvalues $\lambda_i(\tilde{\mathbb{D}}^0)$ of the G-limiting tensor $\tilde{\mathbb{D}}^0$ (see Section 5.5.4)

$$\tilde{\mathbb{D}}^0 = \mathbb{D}^0 - k\mathbb{D}_{I_2} = (\mu_{20} + k)\mathfrak{a}_2^0\mathfrak{a}_2^0 + (\mu_{30} + k)\mathfrak{a}_3^0\mathfrak{a}_3^0 \qquad (5.71)$$

in terms of the weak limit tensors

$$\tilde{\mathbb{D}}^* = \lim_{\text{weak}} \tilde{\mathbb{D}}^s \quad \text{and} \quad (\tilde{\mathbb{D}}^{-1})^* = \lim_{\text{weak}} (\tilde{\mathbb{D}}^s)^{-1}$$

The estimates are given by the following formulas:

$$\lambda_i(\tilde{\mathbb{D}}^* - \tilde{\mathbb{D}}^0) \geq 0, \qquad \lambda_i((\tilde{\mathbb{D}}^{-1})^* - (\tilde{\mathbb{D}}^0)^{-1}) \geq 0 \qquad (i = 1, 2)$$

These inequalities show in particular that

$$\lambda_{\max}(\tilde{\mathbb{D}}^0) \triangleq \mu_{30} \leq \lambda_{\max}(\tilde{\mathbb{D}}^*) \leq (\tilde{\mu}_3)^*$$

$$\lambda_{\max}((\tilde{\mathbb{D}}^0)^{-1}) \triangleq \frac{1}{\tilde{\mu}_{20}} \leq \lambda_{\max}((\mathbb{D}^{-1*})) \leq \left(\frac{1}{\tilde{\mu}_2}\right)^*$$

or that

$$\left[\left(\frac{1}{\tilde{\mu}_2}\right)^*\right]^{-1} \leq \tilde{\mu}_{20} \leq \tilde{\mu}_{30} \leq (\tilde{\mu}_3)^* \qquad (5.72)$$

where $\tilde{\mu}_{i0} \equiv \mu_{i0} + k, i = 2, 3$.

Consider first the problem of G-closure of the set U of materials whose eigenvalues are fixed and equal to k, μ_2, μ_3, respectively, with $\mu_2 \leq \mu_3$. We shall actually be dealing with one and the same anisotropic material (5.11); the composite is assumed to be assembled of a number of differently oriented pieces thereof.

It will now be proved that the eigenvalues k, μ_{20}, μ_{30} of the composite satisfy the following conditions:

$$\mu_2 \leq \mu_{20} \leq \mu_{30} \leq \mu_3$$

$$(\mu_{20} + k)(\mu_{30} + k) = (\mu_2 + k)(\mu_3 + k) \tag{5.73}$$

The first inequality of (5.73) follows immediately from (5.68) and (5.72). To prove the last equation of (5.73), we represent an admissible tensor $(\tilde{D}^s)^{-1}$

$$(\tilde{D}^s)^{-1} = \frac{1}{\mu_2 + k}\, \mathfrak{a}_2^s \mathfrak{a}_2^s + \frac{1}{\mu_3 + k}\, \mathfrak{a}_3^s \mathfrak{a}_3^s$$

in the form

$$(\tilde{D}^s)^{-1} = \frac{1}{I_2(\tilde{D}^s)}\, \mathbb{O}^T \cdot \cdot \mathbb{D}^s \cdot \cdot \mathbb{O} \tag{5.74}$$

where the second invariant is given by

$$I_2(\tilde{D}^s) = (\mu_2 + k)(\mu_3 + k) = \tilde{\mu}_2 \tilde{\mu}_3 = \text{const}\,(s) = I_2(\tilde{D}) \tag{5.75}$$

and where the tensor

$$\mathbb{O} = \mathfrak{a}_2^s \mathfrak{a}_3^s - \mathfrak{a}_3^s \mathfrak{a}_2^s = \mathfrak{a}_2 \mathfrak{a}_3 - \mathfrak{a}_3 \mathfrak{a}_2 = \text{const}\,(s) \tag{5.76}$$

is independent of s.

The operator $\mathbb{O}$ acts on the space of symmetric tensors of second rank; it performs projection of any such tensor onto a subspace of deviators and then turns the axes of the projection by an angle $\pi/4$ (see Section 5.1). It is easily seen that $\mathbb{O} \cdot \cdot \mathbb{O}^T = \mathbb{E}, \mathbb{O} \cdot \cdot \mathbb{O} = -\mathbb{E}$, where $\mathbb{E} = \mathfrak{a}_2 \mathfrak{a}_2 + \mathfrak{a}_3 \mathfrak{a}_3$ represents a unit tensor of a deviator subspace.

Consider Hooke's law corresponding to (5.69)

$$\tilde{M}^s = -\tilde{D}^s \cdot \cdot \nabla\nabla w^s = -\tilde{D}^s \cdot \cdot \text{dev}\,\nabla\nabla w^s$$

The identity (5.74) together with (5.75) allow the following reformulation:

$$-\mathbb{O} \cdot \cdot \text{dev}\,\nabla\nabla w^s = \frac{1}{I_2(\tilde{D})}\, \tilde{D}^s \cdot \cdot \mathbb{O} \cdot \cdot \tilde{M}^s \tag{5.77}$$

From (5.12) and (5.76) we deduce that

$$\nabla\nabla \cdot\cdot \mathbb{O} \cdot\cdot \text{dev } \nabla\nabla w^s = 0 \qquad (5.78)$$

and also that

$$\mathbb{O} \cdot\cdot \tilde{\mathsf{M}}^s = \nabla\nabla \psi^s + \mathbb{O} \cdot\cdot \mathsf{M}^x \qquad (5.79)$$

where M^x denotes a partial solution of $\nabla\nabla \cdot\cdot \tilde{\mathsf{M}} = q$ which is determined by the right-hand side q and may be considered as independent of s.

Equation (5.78) can be derived from the relationships

$$\nabla\nabla \cdot\cdot (\tilde{\mathsf{M}}^s - \mathsf{M}^x) = 0, \qquad I_1(\tilde{\mathsf{M}}^s - \mathsf{M}^x) = \sqrt{2}(\tilde{\mathsf{M}}^s - \mathsf{M}^x) \cdot\cdot \mathfrak{a}_1 = 0$$

The application of the operator $\nabla\nabla \cdot\cdot$ to (5.77) subject to (5.78) and (5.79) yields

$$\nabla\nabla \cdot\cdot \tilde{\mathbb{D}}^s \cdot\cdot (\nabla\nabla \psi^s + \mathbb{O} \cdot\cdot \mathsf{M}^x) = 0 \qquad (5.80)$$

As in Section 4.4, we see then that the G-closure of the operators (5.80) and (5.69) coincides[427] because the shifting of the nonhomogeneous term $\mathbb{O} \cdot\cdot \mathsf{M}^x$ into the main part of the operator does not influence the final result. The G-limiting form of (5.80) is

$$\nabla\nabla \cdot\cdot \tilde{\mathbb{D}}^0 \cdot\cdot \mathbb{O} \cdot\cdot \tilde{\mathsf{M}}^0 = 0$$

and (5.77) becomes

$$-\mathbb{O} \cdot\cdot \text{dev } \nabla\nabla w^0 = \frac{1}{I_2(\tilde{\mathbb{D}})} \tilde{\mathbb{D}}^0 \cdot\cdot \mathbb{O} \cdot\cdot \tilde{\mathsf{M}}^0 \qquad (5.81)$$

On the other hand, we have

$$-\mathbb{O} \cdot\cdot \text{dev } \nabla\nabla w^0 = \mathbb{O} \cdot\cdot (\tilde{\mathbb{D}}^0)^{-1} \cdot\cdot \tilde{\mathsf{M}}^0 = \frac{1}{I_2(\tilde{\mathbb{D}}^0)} \tilde{\mathbb{D}}^0 \cdot\cdot \mathbb{O} \cdot\cdot \tilde{\mathsf{M}}^0$$

and a comparison with (5.81) shows that

$$I_2(\tilde{\mathbb{D}}^0) = \tilde{\mu}_{20}\tilde{\mu}_{30} = I_2(\tilde{\mathbb{D}}) = \tilde{\mu}_2\tilde{\mu}_3$$

which is the desired result.

All points of the set $GU(\tilde{\mu}_{20}, \tilde{\mu}_{30})$ may be represented by a laminated composite which may easily be deduced from (5.32).

Until now the results obtained in this section correspond to those given in Section 4.4 for the second-rank tensors $\mathbb{D}$ associated with the operator $\nabla \cdot \mathbb{D} \cdot \nabla$.

We now turn to the problem of G-closure for a set U including two materials whose $\mathbb{D}(\tilde{\mathbb{D}})$ tensors are given by the relationships

$$\mathbb{D}_+^s = k\mathbb{a}_1\mathbb{a}_1 + \mu_{2+}\mathbb{a}_{2+}^s\mathbb{a}_{2+}^s + \mu_{3+}\mathbb{a}_{3+}^s\mathbb{a}_{3+}^s$$

$$\tilde{\mathbb{D}}_+^s = (\mu_{2+} + k)\mathbb{a}_{2+}^s\mathbb{a}_{2+}^s + (\mu_{3+} + k)\mathbb{a}_{3+}^s\mathbb{a}_{3+}^s \triangleq \tilde{\mu}_{2+}\mathbb{a}_{2+}^s\mathbb{a}_{2+}^s + \tilde{\mu}_{3+}\mathbb{a}_{3+}^s\mathbb{a}_{3+}^s$$

$$\tag{5.82}$$

$$\mathbb{D}_-^s = k\mathbb{a}_1\mathbb{a}_1 + \mu_{2-}\mathbb{a}_{2-}^s\mathbb{a}_{2-}^s + \mu_{3-}\mathbb{a}_{3-}^s\mathbb{a}_{3-}^s$$

$$\tilde{\mathbb{D}}_-^s = (\mu_{2-} + k)\mathbb{a}_{2-}^s\mathbb{a}_{2-}^s + (\mu_{3-} + k)\mathbb{a}_{3-}^s\mathbb{a}_{3-}^s \triangleq \tilde{\mu}_{2-}\mathbb{a}_{2-}^s\mathbb{a}_{2-}^s + \tilde{\mu}_{3-}\mathbb{a}_{3-}^s\mathbb{a}_{3-}^s$$

$$\tag{5.83}$$

We may always assume that $\tilde{\mu}_{2+} \leq \tilde{\mu}_{3+}, \tilde{\mu}_{2-} \leq \tilde{\mu}_{3-}$ and that $\tilde{\mu}_{2+}\tilde{\mu}_{3+} \leq \tilde{\mu}_{2-}\tilde{\mu}_{3-}$. The notation $\mathbb{a}_{2+}^s, \ldots, \mathbb{a}_{3-}^s$ reflects the assumption that the orientation of the principal axes of eigentensors $\mathbb{a}_{2+}, \ldots, \mathbb{a}_{3-}$ of both materials may be different within different pieces of those materials which constitute the assembled composite.

This problem is quite similar to that discussed in Section 4.5. In a similar fashion, we must distinguish between two cases:

$$\tilde{\mu}_{2+} \leq \tilde{\mu}_{2-}, \qquad \tilde{\mu}_{3+} \leq \tilde{\mu}_{3-} \tag{5.84}$$

and

$$\tilde{\mu}_{2+} \leq \tilde{\mu}_{2-}, \qquad \tilde{\mu}_{3+} \geq \tilde{\mu}_{3-} \tag{5.85}$$

the last two inequalities being analogues of (4.33) and (4.34), respectively. The GU set is now built as the intersection of the set

$$\tilde{\mu}_{2+}\tilde{\mu}_{3+} \leq \tilde{\mu}_{20}\tilde{\mu}_{30} \leq \tilde{\mu}_{2-}\tilde{\mu}_{3-}$$

bounded in the $(\tilde{\mu}_{20}, \tilde{\mu}_{30})$ plane by two hyperbolas, and the sets

$$\tilde{\mu}_{2+} \leq \frac{\tilde{\mu}_{2+}\tilde{\mu}_{2-}(\tilde{\mu}_{3-} - \tilde{\mu}_{3+})}{\tilde{\mu}_{2-}\tilde{\mu}_{3-} - \tilde{\mu}_{2+}\tilde{\mu}_{3+} - (\tilde{\mu}_{2-} - \tilde{\mu}_{2+})\tilde{\mu}_{30}} \leq \tilde{\mu}_{20} \leq \tilde{\mu}_{30} \leq \tilde{\mu}_{3-} \tag{5.86}$$

if (5.84) holds, and

$$\tilde{\mu}_{2+} \leq \tilde{\mu}_{20} \leq \tilde{\mu}_{30} \leq \frac{\tilde{\mu}_{3+}\tilde{\mu}_{3-}(\tilde{\mu}_{2-} - \tilde{\mu}_{2+})}{\tilde{\mu}_{2-}\tilde{\mu}_{3-} - \tilde{\mu}_{2+}\tilde{\mu}_{3+} - (\tilde{\mu}_{3-} - \tilde{\mu}_{3+})\tilde{\mu}_{20}} \leq \tilde{\mu}_{3-} \quad (5.87)$$

if case (5.85) holds.

The last two inequalities are analogous to (4.37) and (4.40), respectively. The equality sign in the first inequality (5.86) corresponds to a laminated composite assembled of materials (5.82) and (5.83) oriented within the layers in such a way that the main axes of tensors $\mathfrak{a}_{2+}^s$, $\mathfrak{a}_{2-}^s$ would coincide with a normal and tangent to the layers, respectively. The equality sign in the next to the last inequality (5.87) corresponds to a laminated composite with the normal and tangent to the layers oriented along the principal axes of tensors $\mathfrak{a}_{3+}^s$, $\mathfrak{a}_{3-}^s$ of compounds. All of the results obtained in Sections 4.5 and 4.6 can also be extended to the present case; in particular, it is apparent that any composite belonging to the GU set formed by two initially given materials (5.82), (5.83) is equivalent to some laminated composite of the first or the second rank.

5.8. Elimination of Contradictions Arising within Necessary Conditions for Optimality

We shall illustrate the procedure by an example concerning the optimal distribution of the shear modulus of an isotropic elastic plate.

The corresponding problem (see Sections 5.2 and 5.3) is overdetermined and consequently ill-posed. Its regularization is also achieved by the introduction of laminated composites. The equivalent tensor $\mathbb{D}^0$ can be calculated with the aid of the general formula (5.32). We have

$$\mathbb{D}_+ = k\mathfrak{a}_1\mathfrak{a}_1 + \mu_+(\mathfrak{a}_2\mathfrak{a}_2 + \mathfrak{a}_3\mathfrak{a}_3)$$

$$\mathbb{D}_- = k\mathfrak{a}_1\mathfrak{a}_1 + \mu_-(\mathfrak{a}_2\mathfrak{a}_2 + \mathfrak{a}_3\mathfrak{a}_3)$$

where the eigentensors $\mathfrak{a}_i$ are determined by

$$\mathfrak{a}_1 = \frac{1}{\sqrt{2}}(\mathbf{ii} + \mathbf{jj}), \qquad \mathfrak{a}_2 = \frac{1}{\sqrt{2}}(\mathbf{ii} - \mathbf{jj}), \qquad \mathfrak{a}_3 = \frac{1}{\sqrt{2}}(\mathbf{ij} + \mathbf{ji})$$

Assuming $\mathbf{nn} = \mathbf{ii}$ (without loss of generality), we obtain

$$\mathfrak{a}_1 \cdot\cdot\, \mathbf{nn} = \frac{1}{\sqrt{2}}, \qquad \mathfrak{a}_2 \cdot\cdot\, \mathbf{nn} = \frac{1}{\sqrt{2}}, \qquad \mathfrak{a}_3 \cdot\cdot\, \mathbf{nn} = 0$$

The expression (5.32) for $\mathbb{D}^0$ now reduces to

$$\mathbb{D}^0 = k\mathfrak{a}_1\mathfrak{a}_1 + [m\mu_+ + (1-m)\mu_-]\mathfrak{a}_3\mathfrak{a}_3$$

$$+ \left[\frac{(\mu_+ + k)(\mu_- + k)}{k + m\mu_- + (1-m)\mu_+} - k\right]\mathfrak{a}_2\mathfrak{a}_2 \qquad (5.88)$$

This tensor characterizes a medium possessing cubic symmetry. The eigenvalues μ_2 and μ_3 of $\mathbb{D}^0$ corresponding to its "deviator" eigentensors $\mathfrak{a}_2$ and $\mathfrak{a}_3$ are obviously related by

$$\mu_2 + k = \frac{(\mu_+ + k)(\mu_- + k)}{(\mu_+ + k) + (\mu_- + k) - (\mu_3 + k)}; \qquad \mu_- \le \mu_3 \le \mu_+ \quad (5.89)$$

Figure 5.3 represents a family of curves (5.89) corresponding to different values of the dilatation modulus k. The curves 1, 2, 3 correspond to the increasing values of k: $k_1 < k_2 < k_3$; when $k \to \infty$, the curves tend to a segment of the diagonal $\mu_2 = \mu_3$.

As for the corresponding second-order problem (Chapter 4), we now consider this problem in its relaxed form. As controls, we choose the eigenvalue μ_3 of $\mathbb{D}^0$ and the angle α between the principal axis of $\mathfrak{a}_2$ and the x axis.

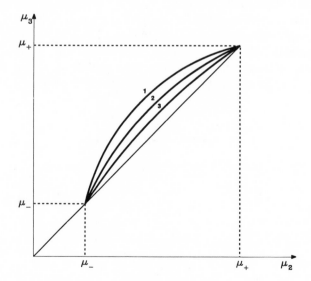

FIGURE 5.3

The stationarity conditions have the form

$$\Delta\mu_3\left(\mathbf{\iota} \cdot\cdot \frac{\partial \mathbb{D}^0}{\partial \mu_3} \cdot\cdot \, \mathbf{e}\right) \geq 0$$

$$\mathbf{\iota} \cdot\cdot \frac{\partial \mathbb{D}^0}{\partial \alpha} \cdot\cdot \, \mathbf{e}^0 = 0$$

(5.90)

The derivatives $\partial \mathbb{D}^0/\partial \mu_3$, and $\partial \mathbb{D}^0/\partial \alpha$ are calculated with the aid of (5.89) and the relationships

$$\mathbf{i} = \mathbf{i}_x \cos \alpha + \mathbf{i}_y \sin \alpha, \qquad \mathbf{j} = -\mathbf{i}_x \sin \alpha + \mathbf{i}_y \cos \alpha$$

where $\mathbf{i}_x, \mathbf{i}_y$ represent the unit vectors of the fixed axis system x, y.
The stationarity conditions then reduce to

$$\Delta\mu_3\left[\frac{(\mu_2 + k)^2}{(\mu_+ + k)(\mu_- + k)}\, e_2\varepsilon_2 + e_3\varepsilon_3\right] \geq 0$$

(5.91)

$$(\mu_2 - \mu_3)(e_2\varepsilon_3 + e_3\varepsilon_2) = 0$$

where

$$e_2 = \mathbf{e} \cdot\cdot \, \mathbf{\alpha}_2 = |\text{dev } \mathbf{e}| \cos 2\varphi, \qquad e_3 = \mathbf{e} \cdot\cdot \, \mathbf{\alpha}_3 = |\text{dev } \mathbf{e}| \sin 2\varphi$$

$$\varepsilon_2 = \mathbf{\iota} \cdot\cdot \, \mathbf{\alpha}_2 = |\text{dev } \mathbf{\iota}| \cos 2\psi, \qquad \varepsilon_3 = \mathbf{\iota} \cdot\cdot \, \mathbf{\alpha}_3 = |\text{dev } \mathbf{\iota}| \sin 2\psi$$

and φ, ψ denote the angles between the unit vector $\mathbf{i}$ and the principal axes of $\mathbf{e}$ and $\mathbf{\iota}$, respectively.
The formula (5.91) can be rewritten in the form

$$\Delta\mu_3\left[\frac{(\mu_2 + k)^2}{(\mu_+ + k)(\mu_- + k)}\cos 2\varphi \cos 2\psi + \sin 2\varphi \sin 2\psi\right] \geq 0 \quad (5.92)$$

$$(\mu_2 - \mu_3)\sin 2(\varphi + \psi) = 0 \qquad (5.93)$$

Optimal solutions may include both isotropic and anisotropic ranges of control; for the latter case, (5.93) shows that the principal axis $\mathbf{i}$ of $\mathbf{\alpha}_2$ bisects the angle θ between the principal axes of $\mathbf{e}$ and $\mathbf{\iota}$. The isotropic ranges are possible if [see (5.25)]

$$\mu_2 = \mu_3 = \mu_+ \qquad \text{if } \tan^2 \theta \geq \frac{k + \mu_+}{k + \mu_-} \qquad (5.94)$$

$$\mu_2 = \mu_3 = \mu_- \qquad \text{if } \tan^2 \theta \le \frac{k + \mu_-}{k + \mu_+} \tag{5.95}$$

From (5.93) it follows that $\varphi = -\psi$ (or $\varphi = \pi/2 - \psi$) for the anisotropic zone, and that, according to (5.92),

$$\frac{(k + \mu_2)^2}{(k + \mu_+)(k + \mu_-)} = \tan^2 \theta \tag{5.96}$$

It is clear now that the optimal solution includes the isotropic ranges (5.94) and (5.95) which are also characteristic for the nonregularized problem. At the same time, there now arises an anisotropic range (5.96) within the "forbidden" interval (see Section 5.3).

The Weierstrass test is applied in a manner similar to that of Section 4.9; it may easily be shown that it is satisfied.

5.9. Problems of Optimal Thickness Design for Plates

As mentioned earlier in this chapter, the optimal thickness problem has much in common with that of the optimal distribution of the elastic within a plate. There are, however, some particular features which make it well worthwhile to discuss this problem in more detail. We recall the complications which arise in the minimization problem for the volume

$$I = \iint_S h \, dx \, dy \tag{5.97}$$

of a Kirchhoff plate undergoing free vibrations with some prescribed frequency ω.

Such vibrations are described by the well-known equation[281]

$$\nabla^2 D \nabla^2 w + (1 - \nu) \left(2 \frac{\partial^2 D}{\partial x \, \partial y} \frac{\partial^2 w}{\partial x \, \partial y} - \frac{\partial^2 D}{\partial x^2} \frac{\partial^2 w}{\partial y^2} - \frac{\partial^2 D}{\partial y^2} \frac{\partial^2 w}{\partial x^2} \right) - h \omega^2 w = 0 \tag{5.98}$$

where w denotes the displacement, $D = D(h) = Eh^3/12\rho(1 - \nu^2)$ the rigidity of a plate, E and ν Young's modulus and the Poisson's ratio of the material, respectively, and ρ the density.

For definiteness, we assume the boundary conditions

$$w|_\Gamma = w_n|_\Gamma = 0 \tag{5.99}$$

corresponding to a fixed boundary.

Following the general procedure of Chapter 2, the exact expression for the increment ΔI of the functional (5.97) provided by the increment of thickness h has the form

$$- \Delta I = \iint_S \left\{ \Delta h(-1 - c^2\omega^2 w^2 - c^2\omega^2 w\Delta w) + c^2[D(H) - D(h)] \right.$$

$$\cdot \left[\left(\frac{\partial^2 w}{\partial x^2}\right)^2 + \left(\frac{\partial^2 w}{\partial y^2}\right)^2 + 2(1 - \nu)\left(\frac{\partial^2 w}{\partial x\, \partial y}\right)^2 + 2\nu\frac{\partial^2 w}{\partial x^2}\frac{\partial^2 w}{\partial y^2} \right.$$

$$\left. \left. - \frac{1}{D(h)}\left(M_x \frac{\partial^2 \Delta w}{\partial x^2} + M_y \frac{\partial^2 \Delta w}{\partial y^2} + 2M_{xy}(1 - \nu)\frac{\partial^2 \Delta w}{\partial x\, \partial y}\right) \right] \right\} dx\, dy$$

$$(5.100)$$

Here, M_x, M_y, M_{xy} denote the optimal values of the bending and torsional moments, respectively, w denotes the optimal displacement (its increment being Δw), and c is a constant.

It is seen from (5.100) that one has to know Δw only for these points of S where $\Delta h \neq 0$. To obtain an effective condition for a minimum, it is necessary to evaluate the increments Δw, $\partial^2 \Delta w/\partial x^2$, $\partial^2 \Delta w/\partial y^2$, $\partial^2 \Delta w/\partial x\, \partial y$ in terms of Δh. We shall arrive at the corresponding formulas by performing a local variation within a narrow strip with length $a\varepsilon$, and width $a\varepsilon^2$, and with $\varepsilon \to 0$.

Let $\mathbf{n}, \mathbf{t}$ be the unit vectors normal and tangential to the strip, respectively; we then have the relationships

$$\Delta w = 0, \qquad \frac{\partial^2 \Delta w}{\partial t^2} = 0, \qquad \frac{\partial^2 \Delta w}{\partial n\, \partial t} = 0, \qquad \Delta M_n = 0 \qquad (5.101)$$

valid up to terms $0(\varepsilon)$. They follow from the continuity of the displacement, the slope of normal to the strip and the normal component of momentum across the boundary of the strip. The limiting values of these quantities from without the strip are assumed to be coincident with their optimal values [see footnote following (2.118) on p. 58].

With the aid of Hooke's law (see the Introduction), the last equation of (5.101) may be represented in the form

$$\frac{\partial^2 \Delta w}{\partial n^2} = \frac{H^3 - h^3}{H^3}\frac{M_n}{D(h)}$$

This reduces the integrand in (5.100) to

$$\Delta h(-1 - c^2\omega^2 w^2) + \frac{c^2 E}{12\rho(1 - \nu^2)}(H^3 - h^3)\left[\Pi - \left(\frac{\partial^2 w}{\partial n^2} + \nu\frac{\partial^2 w}{\partial t^2}\right)\frac{H^3 - h^3}{H^3}\right]$$

$$(5.102)$$

where

$$\Pi = \left(\frac{\partial^2 w}{\partial x^2}\right)^2 + \left(\frac{\partial^2 w}{\partial y^2}\right)^2 + 2\nu\frac{\partial^2 w}{\partial x^2}\frac{\partial^2 w}{\partial y^2} + 2(1 - \nu)\left(\frac{\partial^2 w}{\partial x\,\partial y}\right)^2$$

is an (invariant) expression for the specific potential energy.

If we set the coefficient of $\Delta h = H - h$ in (5.102) equal to zero, we obtain the necessary condition

$$\frac{h^2 E c^2}{4\rho(1 - \nu^2)}\Pi = 1 + c^2\omega^2 w^2 \tag{5.103}$$

for stationarity of the intermediate control h. This control is, however, not optimal in the absence of pointwise restrictions on the admissible values of h. To show this, we consider the nonlinear (over Δh) part of (5.102):

$$\frac{1}{H^3}\frac{c^2 E}{12\rho(1 - \nu^2)}\left[(3h + \Delta h)(\Delta h)^2 H^3\Pi - \frac{M_n^2}{D^2(h)}(H^3 - h^3)^2\right] \tag{5.104}$$

The noninvariant term

$$-\frac{c^2 E}{12 H^3\rho(1 - \nu^2)}\frac{M_n^2}{D^2(h)}(H^3 - h^3)^2$$

depends on the orientation of the strip, and this term is a maximum for its *critical* position (see Section 2.3). Provided that the principal values M_1, M_2 of the momentum tensors have opposite signs, the critical position corresponds to the vanishing component M_n of this tensor, and the nonlinear part (5.104) of (5.102) becomes nonnegative implying nonoptimality of the stationary intermediate control. If the principal values of the tensor $\mathbb{M}$ are of the same sign, then the choice of that M_n whose principal value is the smaller of the two in absolute value, reduces (5.104) to the form

$$\frac{c^2 E(\Delta h)^2}{12 H^3\rho(1 - \nu^2)}\left[f_1(h, \Delta h)\left(\left(\frac{\partial^2 w}{\partial n^2}\right)^2 + 2\nu\frac{\partial^2 w}{\partial t^2}\frac{\partial^2 w}{\partial n^2}\right) + f_2(h, \Delta h)\left(\frac{\partial^2 w}{\partial t^2}\right)^2\right]$$

$$(5.105)$$

Here,

$$f_1(h, \Delta h) = -h^2[6h^2 + 8h\Delta h + 3(\Delta h)^2]$$

$$f_2(h, \Delta h) = (1 - \nu^2)[(\Delta h)^4 + 6h(\Delta h)^3] + 3(4 - 5\nu^2)h^2(\Delta h)^2$$

$$+ 2(5 - 9\nu^2)h^3\Delta h + 3(1 - 3\nu^2)h^4$$

The expression (5.105) can be made nonpositive for sufficiently large Δh. Smooth profiles satisfying (5.103) will thus be nonoptimal since they will obviously be improved if we distribute the material within some set of stiffeners rather than within some smooth profile of a plate. Note that our conclusion is valid regardless of the pointwise restriction upon h if the eigenvalues of the tensor M are of opposite signs; for the case of the same signs, it is required that the ratio $h_{max}/h_{min}(h_{min} \leq h \leq h_{max})$ be large enough. The mentioned set of stiffeners would obviously be dense within some two-dimensional part of a plate; we thus arrive at the situation where the necessity of relaxation becomes obvious. For a one-dimensional distribution of the thickness, e.g., for an axisymmetric plate, the relaxation means introduction of a laminated composite assembled of stiffeners possessing maximum and minimum thickness and distributed with some density to be determined. This model has actually been worked out in Refs. 436, 437, and 492. For the two-dimensional case, the problem remains open so long as we are still not in a position to describe the corresponding G-closure of the set U ($U: h_{min} \leq h \leq h_{max}$) of admissible values of the rigidity $D(h)$. In any case, the assumptions of continuous plate design provide final results whose validity is questionable.

Note that for the so-called sandwich plates whose rigidity D is proportional to h, not h^3, this difficulty disappears, and the necessary condition of the type (5.103) is also a sufficient condition for optimality.

6

Optimal Control of Systems Described by Equations of Hyperbolic Type

Hitherto, the general method of Chapter 2 has been illustrated by examples of elliptic optimization problems. The method also works effectively in hyperbolic optimization problems; the investigation of problems of this class presents a number of particular features which will be considered in the present chapter.

6.1. Quasi-linear First-Order Equation

We consider the optimization problem (2.88)-(2.90) of Example 6 in Section 2.2 and calculate the increment of the functional (2.90) due to a change of the control by the amount Δu compared with its optimal value u. We assume that the variation Δu is nonzero in the strip $abcd$ of the (t, x) plane of width h and length l (Fig. 6.1). We assume that the function $z(t, x)$ is continuous; it is easy to show that in the limit $h \to 0$ the increment can

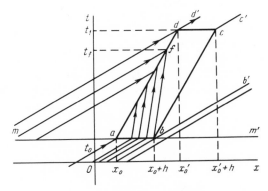

FIGURE 6.1

305

be written in the form

$$\Delta I = - \iint\limits_{abcd} \xi[(z_x + \Delta z_x)\delta f_1 + \delta f_0] \, dt \, dx + o(h) \tag{6.1}$$

Here, $\xi(t, x)$ is the function determined by problem (2.91)–(2.92), and the symbol δ has the meaning

$$\delta f_{0,1} = f_{0,1}(t, x, z, u + \Delta u) - f_{0,1}(t, x, z, u) \tag{6.2}$$

We transform the function

$$- \xi[(z_x + \Delta z_x)\delta f_1 + \delta f_0]$$

in the integrand of (6.1). To this end, we use the requirement of continuity of the tangential derivative of the function z on the boundary ad (or bc; in the limit $h \to 0$ this amounts to the same thing) of the variation strip. We shall ignore the difference between the values of z at the corresponding points of the boundaries ad and bc since allowance for this difference in the limit $h \to 0$ entails terms of higher order in h.

The condition of continuity of the tangential derivatives*

$$[f_0(t, x, z, u) + f_1(t, x, z, u)z_x]t_\tau + z_x x_\tau$$

$$= [f_0(t, x, z, u + \Delta u) + f_1(t, x, z, u + \Delta u)(z_x + \Delta z_x)]t_\tau + (z_x + \Delta z_x)x_\tau \tag{6.3}$$

makes it possible to eliminate Δz_x from the integrand of (6.1). As a result, this expression takes on the form

$$- \xi \frac{f_1(t, x, z, u)t_\tau + x_\tau}{f_1(t, x, z, u + \Delta u)t_\tau + x_\tau} (\delta f_0 + z_x \delta f_1) \tag{6.4}$$

By the usual argument, establish that if the functional I is to be minimized, (6.4) must be nonnegative.

Let us consider the individual factors in (6.4). The function $\xi(t, x)$ is determined by problem (2.91)–(2.92); solving this problem involves constructing the characteristic system of ordinary differential equations. For our purposes, it is sufficient to note that along the characteristic $x = x(t)$ we have

$$\frac{d\xi}{dt} = \xi(f_{1x} - f_{1z}z_x - f_{0z})_{x=x(t)} \tag{6.5}$$

---

* (t_τ, x_τ) denote the direction cosines of the direction τ.

This relation shows that the sign of ξ along the characteristic coincides with the sign of the initial value $\xi(T, x) = -\rho(x)$ [see (2.92)]. The integral surface $\xi = \xi(t, x)$ of (2.91) is divided into strips in which $\xi(t, x)$ is positive or negative depending on whether the value of $-\rho(x)$ is positive or negative along the corresponding segment of the straight line $t = T$.

We now consider the sign of the ratio

$$\frac{f_1(t, x, z, u)t_\tau + x_\tau}{f_1(t, x, z, u + \Delta u)t_\tau + x_\tau} \tag{6.6}$$

This ratio is equal to unity for $\Delta u = 0$; for $\Delta u \neq 0$, there exist directions (t_τ, x_τ) to which there correspond negative values of the ratio (6.6). Such directions must be ruled out, since variation in strips with forbidden directions immediately takes us out of the adopted class of comparison functions—continuous functions with piecewise continuous derivatives—and entails the appearance of discontinuities of the first kind in the solution $z(t, x)$ of the basic equation (2.88). It is important to note that these discontinuities are not due to the quasi-linearity of the equation but solely due to the orientation of the variation strip; therefore, what we have said also remains true in the case of a linear equation.

If discontinuous functions z are allowed for comparison, then the arguments leading to (6.4) lose their validity; (6.1) is already invalid for the quasi-linear case. Thus, the device used to obtain optimality conditions hitherto has made essential use of the assumption that the ratio (6.6) is positive.

We now consider the question of the occurrence of discontinuities in the solution. The situation here is very similar to the one that arises in the case of strong discontinuities in the solutions of quasi-linear equations. It is known, for example, that the solution of the quasi-linear equation

$$z_t + f(z)z_x = 0 \tag{6.7}$$

generated by smooth initial data can be discontinuous along a certain line on the (t, x) plane; if the slope of this line is U, then the condition on the discontinuity is given by the inequalities

$$f(z_-) \geq U \geq f(z_+) \tag{6.8}$$

where z_- and z_+ are the values of the solution to the left and right of the discontinuity, respectively. These inequalities show that the ratio (6.6), in which dx/dt is identified with U, is negative. The only difference is that in the quasi-linear case the discontinuity is formed as a result of the z-dependence of the phase velocity, whereas in the problem with control, fulfillment of the inequalities (6.8) is guaranteed by an appropriate external influence. These inequalities generate a nonuniqueness of the solution on the line where they are satisfied.* The characteristics which approach this line from different sides carry different Cauchy data onto the line, and a discontinuity is formed in the solution. The characteristics do not "collide" if (6.6) is positive; the characteristics on different sides of the discontinuity line of the control then "continue" each other, and the condition (6.3) guarantees a continuous extension of the solution through this line—one of the boundaries of the variation strip.

Note that if (6.6) is negative, the characteristics "collide" only on one boundary of the variaton strip (ad in Fig. 6.1); on the other boundary of the strip (bc in Fig. 6.1) the characteristics "diverge" from the boundary in opposite directions. A discontinuity in the solution arises only along ad.

Thus, if we admit only continuous comparison functions $z(t, x)$, then we must carry out the variation in such a way as not to violate the positivity condition of (6.6). In other words, the variation strip can be situated only within a certain angle, whose size depends on the admissible variation Δu of the control and, of course, on the form of the function f_1. It may happen, in particular, that the only possible variation is one in which the strip is perpendicular to the t axis ($t_\tau = 0$).† Under these conditions, we find that minimization of the functional I requires that the expression $-\xi(\delta f_0 + z_x \delta f_1)$ be nonnegative for all admissible Δu. This last result is Weierstrass's necessary condition for the optimization problem (2.88)–(2.90) in the class of continuous functions $z(t, x)$.

The possible occurrence of forbidden directions of the variation strip distinguishes hyperbolic optimization problems from the previously considered problems of elliptic type.

Let us consider a simple example which illustrates the new possibilities of optimization that arise when one relinquishes the continuity requirement of the function $z(t, x)$.

Suppose we are given the equation

$$z_t + uz_x = 0 \tag{6.9}$$

---

* In this case, on the boundary ad of the variation strip (see below).

† It is in fact the way used for the variation in Refs. 75, 149, although in the problems considered in these investigations this method is not the only one possible. The general case of variation is considered in Refs. 68, 96, 300, 301.

in which $u = u_0 = \text{const}$ for $-\infty < x < \infty$, $0 \le t < t_0$. The Cauchy data

$$z(0, x) = x \tag{6.10}$$

generate the solution

$$z = x - u_0 t \tag{6.11}$$

which holds for $-\infty < x < \infty$, $0 \le t < t_0$. At the time $t = t_0$, the characteristic velocity u changes abruptly from u_0 to the value u_1 on the interval $s(x_0 \le x \le x_0 + h)$; for values of x outside this interval, the velocity u remains equal to u_0.

The interval s, on which the value $u = u_1$ is maintained, moves as a whole with velocity U in the direction of increasing x; the values of u_0, U, and u_1 are related by

$$u_0 > U > u_1$$

(6.6) then is negative.

After the time $t_1 - t_0$ has elapsed, the perturbation u_1 is removed as abruptly as it occurred; at the time t_1 the interval s occupies the position $(x_0', x_0' + h)$, $x_0' = x_0 + U(t_1 - t_0)$, and the entire perturbed region occupies the parallelogram $abcd$ on the (t, x) plane (see Fig. 6.1). Subsequently, the characteristic velocity remains constant and equal to u_0 for any x.

Beginning at the time t_0, the solution is obtained from the Cauchy problem with the initial data [see (6.11)]

$$z(t_0, x) = x - u_0 t_0 \tag{6.12}$$

now, however, $u = u_1$ in $abcd$, while $u = u_0$ in the remaining part of the half-plane $t > t_0$. In the region to the left of the broken line $madd'$ (see Fig. 6.1), and also within the angle $b'bm'$, the solution is given, as before, by (6.11), i.e., it remains unperturbed. The perturbation of the solution is concentrated in the region $abcd$, and also in the half-strips $d'dcc'$ and $c'cbb'$. On the interval ab, the solution is continuous, and the Cauchy data (6.12) generate the solution

$$z = x - u_1(t - t_0) - u_0 t_0 \tag{6.13}$$

in the triangle abf; on the line af, this solution takes the value

$$z\big|_{af}^+ = U(t - t_0) + x_0 - u_1(t - t_0) - u_0 t_0$$

On the other hand, the solution (6.11) along af is equal to

$$z\,|_{af}^{-} = U(t - t_0) + x_0 - u_0 t$$

Along af the function z has a discontinuity of magnitude

$$\Delta z = z\,|_{af}^{+} - z\,|_{af}^{-} = (u_0 - u_1)(t - t_0)$$

This magnitude increases linearly with increasing t. Thus, the discontinuity "accumulates" on the left-hand boundary of s; at t_0 it is equal to zero, but by t_f it reaches the value $(u_0 - u_1)(t_f - t_0)$.

Perturbations created by the initial state do not penetrate into the region to the right of the broken line $b'bfdd'$; this may be called a "shadow zone." From the mathematical point of view, one can arbitrarily specify the values of the function $z(t, x)$ along bc; this determines the solution in the whole of the shadow zone. To specify the function z along bc, it is necessary to invoke additional physical or other arguments.

We assume that $z\,|_{bc} = 0$ (from which it follows that $z = 0$ in the shadow zone), and we estimate the increment Δz of the solution in different parts of the half-plane $t > 0$ due to the perturbation of the characteristic velocity.

We have:

- $\Delta z = 0$ in the strip $0 \le t \le t_0$, and also to the left of the broken line $madd'$ and within the angle $b'bm'$
- $\Delta z = (u_0 - u_1)(t - t_0)$ in the triangle abf
- $\Delta z = -z = -x + u_0 t$ inside the strip $d'dfbb'$

Inside the triangle abf,

$$t - t_0 \le t_f - t_0 \le Mh, \qquad M > 0$$

and therefore

$$|\Delta z| \le M|\Delta u|h, \qquad (t, x) \in \Delta_{abf}$$

A similar inequality holds for the discontinuity of the solution along af.

The increment of z in the half-strip $d'dfbb'$ cannot be estimated in this way; we see that it does not depend on h at all.

In the limit $h \to 0$, the parallelogram $abcd$ degenerates into the interval ad; the region bounded by this interval and the rays of characteristics passing through its ends is the shadow zone in which the main perturbation of the solution is concentrated.

The shadow zone opens up new possibilities for optimization. For example, suppose we need to minimize the functional

$$I = \int_{x_1}^{x_2} (z(T, x))^2 \, dx$$

subject to the constraints (6.9) and (6.10) by choice of the control $u(t, x)$, $0 < u_{min} \le u(t, x) \le U_{max}$, and discontinuous solutions are admitted. Clearly, if T is sufficiently large, then by an appropriate choice of the length of the strip within which $u = u_{min}$ (and without which $u = u_{max}$) and which forms an angle $\tan^{-1} U$, $u_{max} > U > u_{min'}$ with the t axis one can make the function $z(T, x)$ vanish on the interval (x_1, x_2), so that I attains an absolute minimum. There exists a nondenumerable set of controls having this property. Nothing is changed, for example, if we change the width of the strip; we can also take the function u inside the strip equal to, not u_{min}, but some other value $V < u_{max}$, it being only necessary that $V \le U < u_{max}$.

It is interesting to note that the concept of a variation of the control Δu that is discontinuous in the (t, x) plane is not stable against small perturbations of the control.* If we go over from the discontinuous control $u(x - Ut)$ to a continuous control that differs little from it (the dashed curve in Fig. 6.2), then the boundaries ad and bc of the parallelogram $abcd$ are replaced by narrow layers of width ε (Fig. 6.3). Within these layers, the slope of the characteristics varies continuously between the limits u_1 and u_0. Under these conditions, continuous initial data generate a continuous solution everywhere. In the limit $\varepsilon \to 0$, the family of characteristics tends to the limiting picture shown in Fig. 6.3. The limit solution is discontinuous on the characteristic dd', but along the pencil of characteristics penetrating into the shadow region there now propagates a constant† perturbation

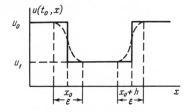

FIGURE 6.2

---

* This was pointed out to me by Dr. K. G. Guderley.

† This conclusion is valid under the assumption that the initial function $z(x, t_0)$ is continuous at the point $x = x_0 + h$.

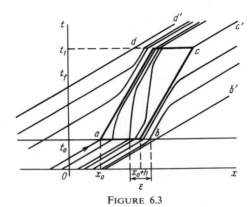

FIGURE 6.3

produced by the initial state $z(x_0 + h, t_0)$ at the point b. Instead of the shadow zone, we now obtain a region occupied by this constant perturbation, and all of our conclusions drawn earlier concerning the increment of the solution in this region must be modified appropriately.

In this case too the limit solution is one of the possible solutions in the shadow zone; on the other hand, there are problems in which one must assume from the very start that the variations are discontinuous, and cannot be regarded as limits of continuous variations, at the very least, not at either end of the variation strip.* A simple illustration thereof is a column of automobiles which has been stopped by a red traffic light whereas the automobiles that passed the crossing before the red signal continue their motion with the same velocity; between the two groups of automobiles, a shadow zone is formed.

Continuous variations of the phase velocity correspond to a different physical situation, and therefore the two types of variation must be regarded as essentially different.

In the case of (2.93), the situation is similar; the condition of occurrence of a discontinuity has the form

$$\frac{x_\tau^2 - u t_\tau^2}{x_\tau^2 - U t_\tau^2} < 0 \tag{6.14}$$

in this case, however, there is a complication due to the fact that (2.93) describes perturbations propagating in both directions along the x axis. We shall not consider this question here.

---

* Thus, in these problems a unique choice of solution is made in the shadow zone.

The case when the number of independent variables exceeds two entails only formal modifications in the arguments. Let us now consider the previous problem for the equation

$$z_t = f_0(t, x, z, u) + \sum_{i=1}^{m} f_i(t, x, z, u) z_{x_i}$$

with the boundary condition

$$z(0, x_1, \ldots, x_m) = z_0(x_1, \ldots, x_m)$$

The functional to be minimized has the form

$$I = \int_{-\infty}^{\infty} \rho(x_1, \ldots, x_m) z(T, x_1, \ldots, x_m) \, dx_1, \ldots, dx_m$$

Equation (2.91) has the analogue

$$\xi_t - \sum_{i=1}^{m} (\xi f_i)_{x_i} = -\xi \left(f_{0_z} + \sum_{i=1}^{m} f_{i_z} z_{x_i} \right)$$

the boundary condition (2.92) remains in force.

The control $u(t, x_1, \ldots, x_m)$ is varied within a thin (of width $h \to 0$) layer of the space $(t, x_1, \ldots, x_m)$ bounded by two parallel m-dimensional hyperplanes.

Introducing m independent parameters $q_1, \ldots, q_m$ as coordinates on these hyperplanes, we denote the corresponding direction cosines by $t_{q_i}, x_{1q_i}, \ldots, x_{mq_i}$. The analogue of (6.4) now is

$$-\xi \frac{d}{D} \left[\delta f_0 + \sum_{i=1}^{m} z_{x_i} \delta f_i \right]$$

where δf_0 and δf_i have their previous meaning [see (6.2)], and d is the characteristic determinant

$$d = \begin{vmatrix} 1 & -f_1(t, x, z, u) & -f_2(t, x, z, u) & \cdots & -f_m(t, x, z, u) \\ t_{q_1} & x_{1q_1} & x_{2q_1} & \cdots & x_{mq_1} \\ \cdots & \cdots & \cdots & \cdots & \cdots \\ t_{q_m} & x_{1q_m} & x_{2q_m} & \cdots & x_{mq_m} \end{vmatrix}$$

whereas D is a determinant which differs from d only in that the functions $f_i(t, x, z, u)$ are replaced by the functions $f_i(t, x, z, u + \Delta u)$.

Exactly as before, one can show that the ratio d/D must be positive if the optimal solution z is taken to belong to the class of continuous functions.

6.2. Example: A New Derivation of Bellman's Equation

Our results can be applied to the derivation of the partial differential equation which is satisfied by the optimal function (under certain conditions) in problems of minimization subject to constraints specified by a system of ordinary differential equations.

We consider the following problem: given a system of ordinary differential equations and initial conditions

$$\frac{dx^i}{dt} = f^i(t, x, u), \qquad x^i(t_0) = x_0^i \qquad (i = 1, \ldots, n) \tag{6.15}$$

where $u(t) = (u_1(t), \ldots, u_p(t))$ is a vector control function; a system of constraints expressed by the finite equalities*

$$R^j(t, x, u) = 0 \qquad (j = 1, \ldots, r \le p) \tag{6.16}$$

and assuming that the matrix $\|\partial R^j/\partial u^j\|$ has maximal rank, it is required to determine a control $u(t)$ that minimizes the functional $I = S_T(x(T), T)$, which is a function of the terminal point $(x(T), T)$ of the phase trajectory in the (x, t) space

$$S_T(x(T), T) = F(x^1(T), \ldots, x^n(T), T) \tag{6.17}$$

It is assumed that the function F is continuously differentiable with respect to all of its arguments.

It is well known that this problem can be solved by means of Bellman's differential equation[144] (the usual summation convention is adopted)

$$\frac{\partial S_T}{\partial t} = \max_{u \in A(t,x)} \left[-\frac{\partial S_T}{\partial x^i} f^i(t, x, u) \right] \tag{6.18}$$

---

* The case of constraints specified by finite inequalities can be reduced to the case considered here by the introduction of "auxiliary controls."

subject to the condition that the function S_T is differentiable. In (6.18), $A(t, x)$ is the set of admissible controls, i.e., controls satisfying (6.16).

The solution of (6.18) subject to the boundary condition (6.17) yields the function $S_T(x, t)$; its substitution into the expression

$$u = \gamma\left(x, t, \frac{\partial S_T}{\partial x^i}\right) \tag{6.19}$$

obtained from the condition[144]

$$\frac{\partial}{\partial u}\left[\frac{\partial S_T}{\partial x^i} f^i(t, x, u) + \Gamma_j R^j(t, x, u)\right] = 0 \tag{6.20}$$

leads to the relation

$$u = \varphi(x, t) \tag{6.21}$$

which synthesizes the optimal control.

The solution of the system of equations

$$\frac{dx^i}{dt} = f^i(t, x, \varphi(x, t)) \tag{6.22}$$

minimizes the functional (6.17) for arbitrary initial data $x^i(t_0) = x_0^i$.

The argument which leads to Bellman's equation (6.18) is well known. Essentially, this equation is nothing more than a differential formulation of the optimality principle.

The solution of (6.18) subject to given boundary conditions yields the synthesizing function (6.21). The substitution of this function, instead of u, into the right-hand side of (6.18) results in

$$\frac{\partial S_T}{\partial t} = -\frac{\partial S_T}{\partial x^i} f^i(t, x, \varphi(x, t)) \tag{6.23}$$

We now consider the problem posed at the very start and take some admissible control $U(x, t)$. Note that the function U is taken to depend on the two variables x and t, which are regarded as independent.

Substituting $U(x, t)$ instead of u in the system (6.15), we arrive at the differential equations of the admissible trajectories; for given initial state (x_0, t_0), the admissible trajectory is uniquely determined.

We consider the functional $S_T(x_0, t_0)$ on an admissible trajectory; since its value is determined solely by the final (at time $t = T$) position of the phase point, this value does not depend on the point of the admissible trajectory which we take as the initial point; this circumstance is expressed by the equation

$$\frac{dS_T(x_0, t_0)}{dt_0} = 0 \tag{6.24}$$

which holds along an admissible trajectory.

Differentiating in (6.24) and taking into account (6.15), we obtain

$$\frac{\partial S_T(x_0, t_0)}{\partial t_0} = -\frac{\partial S_T(x_0, t_0)}{\partial x_0^i} f^i(t_0, x_0, U(x_0, t_0))$$

or, dropping the subscripts 0,

$$\frac{\partial S_T(x, t)}{\partial t} = -\frac{\partial S_T(x, t)}{\partial x^i} f^i(t, x, U(x, t)) \tag{6.25}$$

We now note that as admissible trajectory we can choose any trajectory of the system (6.15) [in which u is replaced by $U(x, t)$], so that in accordance with (6.24) $S_T(x_0, t_0)$ is conserved along every trajectory of the system (6.15). But this means[86] that $S_T(x, t)$ as a function of two variables satisfies (6.25), regarded as a first-order partial differential equation.

It must be emphasized once more that the coordinates of the initial point of the admissible trajectory are the argument of the function $S_T(x, t)$, while the function S_T itself is determined as some combination of the coordinates of the terminal point of the trajectory. For the chosen control, the relative levels of $S_T(x, t)$ are preserved as one moves along the trajectory, so that, in particular, the minimum of $S_T(x(T), T)$ coincides with the minimum of $S_T(x_0, t_0)$. However, the values of $x(T)$ are unknown, whereas x_0 and t_0 are known. Therefore, instead of minimizing $S_T(x(T), T)$ we may minimize $S_T(x_0, t_0)$ at the given point (x_0, t_0) of the boundary of the admissible region of the (x, t) plane.

We arrive at the following optimization problem with (6.26) regarded as a first-order partial differential equation: to determine the control function $U(x, t)$ of two independent variables subject to the restrictions (6.16) in such a way that the solution $S_T(x, t)$, which satisfies the condition (6.17) at $t = T$, takes on the smallest possible value at the point (x_0, t_0). The control is sought in the class of piecewise continuous functions, and the solution $S_T(x, t)$ is assumed to belong to the class of continuous functions of two independent variables.

The solutions of this problem can be obtained directly in accordance with the method described above. The necessary condition for a minimum has the form

$$-\xi\frac{d}{D}\frac{\partial S_T}{\partial x^i}\,\delta f^i(t, x, u) \geq 0 \tag{6.26}$$

where d and D are the characteristic determinants (see the end of the previous section); the symbol δf^i has the same meaning as in (6.2), and $\xi(t, x)$ satisfies the equation (for clarity, the summation sign has been included here)

$$\xi_t + \sum_{i=0}^{n} (\xi f^i)_{x^i} = 0 \tag{6.27}$$

and the boundary condition

$$\xi(x, t_0) = \delta(x - x_0) \qquad (\delta \text{ is the delta function}) \tag{6.28}$$

Conditions (6.27)–(6.28) essentially distinguish only one characteristic, that which begins at the point (x_0, t_0); along this characteristic, $\xi > 0$. Therefore, the requirement $\xi(d/D)(\partial S_T/\partial x^i) \times \delta f^i(t, x, u) \geq 0$ is actually imposed and satisfied only along this characteristic, along which it is equivalent, to the inequality

$$-\frac{\partial S_T}{\partial x^i}\,\delta f^i(t, x, u) \geq 0 \tag{6.29}$$

i.e., the usual condition for a maximum. In accordance with the preceding section, the ratio d/D is assumed to be positive.

Under consideration of (6.29), (6.25) can be rewritten in the form

$$\frac{\partial S_T}{\partial t} = -\max_{U \in A(t, x)} \left[\frac{\partial S_T}{\partial x^i} f^i(t, x, U(x, t))\right]$$

which holds along the optimal trajectory. The well-known standard argument[144] then leads to Bellman's equation.

6.3. An Optimization Problem for the Plastic Torsion of a Rod

We consider an inhomogeneous rod in a state of plastic torsion. In the plane of the cross section G of the rod, we introduce Cartesian coordinates x_1 and x_2; the x_3 axis is directed along the rod generator (see Fig. 4.10). The components $t_{13}(x_1, x_2)$ and $t_{23}(x_1, x_2)$ of stress arising in the rod, which is twisted around the x_3 axis by couples with moment M, satisfy the yield condition

$$t_{13}^2 + t_{23}^2 = k^2 \tag{6.30}$$

Here, $k = k(x_1, x_2)$ is the yield point; for an inhomogeneous rod, this yield point may be a function of the points in the cross section.

In addition to condition (6.30), the equilibrium equation [see (4.51)]

$$\frac{\partial t_{13}}{\partial x_1} + \frac{\partial t_{23}}{\partial x_2} = 0 \tag{6.31}$$

must be satisfied, and the boundary condition

$$t_{23}\, dx_1 - t_{13}\, dx_2 = 0 \qquad \text{along } \Gamma \tag{6.32}$$

which expresses the absence of a load on the lateral surface of the rod.

In what follows, we shall write x, y instead of x_1, x_2. We satisfy (6.31) by means of a stress function $z(x, y)$, introducing it in accordance with the equations [see (4.54)]

$$t_{13} = z_y, \qquad t_{23} = -z_x$$

Equation (6.30) takes the form

$$z_x^2 + z_y^2 = k^2 \tag{6.33}$$

and the condition (6.32) is equivalent to

$$z = 0 \qquad \text{along } \Gamma \tag{6.34}$$

Bearing in mind the formulation of the optimization problem, we shall regard the yield point $k(x, y)$ as a control. We assume that $k(x, y)$ belongs to the set of measurable functions bounded by the inequalities

$$0 < k_{\min} \le \text{vrai max } k(x, y) \le k_{\max} < \infty \tag{6.35}$$

where k_{min} and k_{max} are constants. In addition, we shall assume that the average value of the yield point does not exceed some constant:

$$\frac{1}{\text{mes } G} \iint_G k(x, y) \, dx \, dy \le k_{av}, \qquad k_{min} < k_{av} < k_{max} \qquad (6.36)$$

this last restriction expresses an estimate of the available control resources [see the comment after (4.49)].

We assume that the function $z(x, y)$ is continuous everywhere except at lines on which characteristics of (6.33) carrying different initial data collide. The same lines may be discontinuity lines of the normal derivatives z_n even when the function z itself and the control k are continuous; the normal derivatives z_n on different sides of the discontinuity line differ only in sign.

One has the following optimization problem: to determine the control $k(x, y)$, subject to the indicated restrictions, in such a way as to maximize the functional

$$I = \iint_G z(x, y) \, dx \, dy \qquad (6.37)$$

which differs from the limiting moment M by a positive factor.

This is in the form in which the problem of optimal plastic torsion was posed in Ref. 303.*

We derive the necessary conditions of optimality. In accordance with the general scheme of Chapter 2, we represent (6.33) in the normal form:

$$z_x = \zeta, \qquad z_y = \sqrt{k^2 - \zeta^2} \qquad (6.38)$$

with corresponding H function:

$$H = \xi\zeta + \eta\sqrt{k^2 - \zeta^2} + z - \frac{\gamma}{\text{mes } G} k$$

Here, ξ, η, and γ are Lagrange multipliers. The stationarity conditions

$$\xi_x + \eta_y = -1, \qquad \xi - \eta\frac{\zeta}{\sqrt{k^2 - \zeta^2}} = 0$$

---

* The present exposition stems largely from this reference.

are equivalent to the relations

$$\xi = \omega z_x, \qquad \eta = \omega z_y, \qquad \text{div } \omega \text{ grad } z = -1 \qquad (6.39)$$

If we introduce a variation strip with normal **n** and tangent **t**, Weierstrass's condition can be written in the form

$$\Delta H = H(z, Z, K) - H(z, \zeta, k) = \omega z_n \Delta z_n - \frac{\gamma}{\text{mes } G} \Delta k$$

$$= \omega z_n (|\sqrt{K^2 - z_t^2}| \text{ sign } Z_n - |\sqrt{k^2 - z_t^2}| \text{ sign } z_n) - \frac{\gamma}{\text{mes } G} \Delta k \leq 0$$

Simultaneously, the continuity of z_t on the discontinuity line of the yield point (the boundary of the strip) has been used.

We require that $\Delta H = 0$ for $K = k$; this condition will be satisfied if sign $Z_n =$ sign z_n and we then have

$$\Delta H = \omega z_n (|\sqrt{K^2 - z_t^2}| - |\sqrt{k^2 - z_t^2}|) \text{ sign } z_n - \frac{\gamma}{\text{mes } G} \Delta k \leq 0 \quad (6.40)$$

In order to satisfy the last inequality for all **n** and **t**, we must have

$$\left(\omega |\nabla z| - \frac{\gamma}{\text{mes } G} \right) \Delta k \leq 0 \qquad (6.41)$$

The optimal function k can assume limiting values or intermediate values in accordance with the conditions

$$\left. \begin{array}{ll} k = k_{\min}, & \text{if } \omega |\nabla z| - \dfrac{\gamma}{\text{mes } G} \leq 0 \\[3mm] k_{\min} < k < k_{\max}, & \text{if } \omega |\nabla z| - \dfrac{\gamma}{\text{mes } G} = 0 \\[3mm] k = k_{\max}, & \text{if } \omega |\nabla z| - \dfrac{\gamma}{\text{mes } G} \geq 0 \end{array} \right\} \qquad (6.42)$$

Conversely, it is easy to see that inequality (6.41) implies inequality (6.40) for any orientation (**n**, **t**) of the strip: indeed, (6.41) is inequality (6.40) derived for the critical position of the strip when the vector grad z is directed along the normal **n**.

Conditions (6.42) can be transformed to a different form. Toward this end, we introduce orthogonal curvilinear coordinates $q_1 = s$, $q_2 = z$; the lines $z = $ const are contour lines of the stress function, and the lines $s = $ const $(0 \le s \le s_0)$ are their orthogonal trajectories. In these coordinates, (6.39) takes the form

$$\text{div}\,(\omega\nabla z) = \frac{1}{H_z H_s}\frac{\partial}{\partial z}\left(\omega\frac{H_s}{H_z}\right) = -1 \qquad (6.43)$$

where $H_z = |\mathbf{r}_z|$, $H_s = |\mathbf{r}_s|$ are Lamé coefficients. Note that $H_z = |\mathbf{r}_z| = |\text{grad}\,z|^{-1} = 1/k$.

Within the region we assume that there is a line Σ on which z is continuous but on which the normal derivatives z_n are discontinuous; let $z = z_0(s)$ be the equation of this line. We shall assume that the parameter z varies from 0 on Γ to $z_0(s)$ on Σ (Fig. 6.4) for fixed s and that one can reach any point (z, s) of G by moving along a line $s = $ const (i.e., along a vector line of grad z) which leaves the contour Γ and terminates on the discontinuity line Σ. Along Σ the natural boundary condition $\xi y_t - \eta x_t = \omega z_n = 0$ is satisfied; it thus follows that

$$\omega = 0 \qquad \text{along } \Sigma$$

Using the assumptions we have made, we integrate (6.43) with respect to z, for $s = $ const, from $z_0(s)$ to z; we obtain

$$\omega = -\frac{H_z}{H_s}\int_{z_0(s)}^{z} J(s,\rho)\,d\rho, \qquad J(s,z) = H_s H_z > 0$$

whence

$$\omega|\nabla z| = \omega H_z^{-1} = -\frac{1}{H_s}\int_{z_0(s)}^{z} J(s,\rho)\,d\rho = -\frac{1}{k(s,z)J(s,z)}\int_{z_0(s)}^{z} J(s,\rho)\,d\rho$$

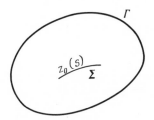

FIGURE 6.4

We use the function

$$C(s, z) = \int_z^{z_0(s)} J(s, \rho)\, d\rho - \frac{\gamma}{\text{mes } G} k(s, z) J(s, z)$$

$$= \int_z^{z_0(s)} \frac{H_s(s, \rho)}{k(s, \rho)}\, d\rho - \frac{\gamma}{\text{mes } G} H_s(s, z) \tag{6.44}$$

to write conditions (6.42) in the form

$$\left.\begin{array}{ll} k(s, r) = k_{\min}, & \text{if } C(s, z) \leq 0 \\[2mm] k_{\min} < k(s, r) < k_{\max}, & \text{if } C(s, z) = 0 \\[2mm] k(s, r) = k_{\max}, & \text{if } C(s, z) \geq 0 \end{array}\right\} \tag{6.45}$$

If the condition $C(s, z) = 0$ is satisfied at all points of some region, then

$$\frac{\partial C}{\partial z} = -J(s, z) - \frac{\gamma}{\text{mes } G} \frac{\partial H_s}{\partial z} = 0 \tag{6.46}$$

holds in that region. This last condition has a simple geometrical meaning. Namely, at the points at which it is satisfied the contour lines $z = \text{const}$ have constant curvature κ_z, i.e., are segments of circles of constant radius. To prove this, we first show that

$$\frac{\partial H_s}{\partial z} = -\kappa_z J$$

We then introduce unit vectors $\mathbf{i}_s$ and $\mathbf{i}_z$ along the axes q_1 and q_2:

$$\mathbf{i}_s = \frac{\mathbf{r}_s}{|\mathbf{r}_s|} = \frac{\mathbf{r}_s}{H_s}, \qquad \mathbf{i}_z = \frac{\nabla z}{|\nabla z|} = \frac{\mathbf{r}_z}{H_z} = k\mathbf{r}_z$$

With the assumption that they form a right-handed system, we can write

$$\frac{\partial H_s}{\partial z} = \frac{\partial}{\partial z} (\mathbf{i}_3, [\mathbf{i}_z, H_s \mathbf{i}_s]) = \left(\mathbf{i}_3, \frac{\partial}{\partial z}[\mathbf{i}_z, \mathbf{r}_s]\right)$$

$$= \left(\mathbf{i}_3, \left[\frac{\partial \mathbf{r}_z}{\partial z}, \mathbf{r}_s\right]\right) + \left(\mathbf{i}_3, \left[\mathbf{i}_z, \frac{\partial \mathbf{i}_z}{\partial z}\right]\right) = \left(\mathbf{i}_3, \left[\mathbf{i}_z, \frac{\partial \mathbf{r}_s}{\partial z}\right]\right)$$

$$= \left(\mathbf{i}_3, \left[\mathbf{i}_z, \frac{\partial \mathbf{r}_z}{\partial s}\right]\right) = \left(\mathbf{i}_3, \left[\mathbf{i}_z, \frac{\partial}{\partial s}\left(\frac{\mathbf{i}_z}{k}\right)\right]\right) = \frac{1}{k}\left(\mathbf{i}_3, \left[\mathbf{i}_z, \frac{\partial \mathbf{i}_z}{\partial s}\right]\right)$$

Here we have taken into account that the derivative of the unit vector $\mathbf{i}_z$ with respect to z is a vector oriented along $\mathbf{i}_s$, i.e., along $\mathbf{r}_s$. As a consequence of the Frenet formula

$$\frac{\partial \mathbf{i}_z}{\partial s} = -\kappa_z H_s \mathbf{i}_s$$

it follows that

$$\frac{\partial H_s}{\partial z} = -\kappa_z \frac{H_s}{k} = -\kappa_z J$$

which is what we wanted to prove.

By eliminating $\partial H_s / \partial z$ from (6.46), this equation may be written in the form

$$\frac{\partial C}{\partial z} = J\left(\frac{\gamma}{\operatorname{mes} G} \kappa_z - 1\right) = 0 \tag{6.47}$$

from which it follows that

$$\kappa_z = \frac{1}{\rho} \tag{6.48}$$

where the radius $\rho = \gamma / \operatorname{mes} G$. Obviously, C increases as a function of z if $\kappa_z > \rho^{-1}$, and decreases otherwise.

As before, we show that

$$\frac{\partial H_z}{\partial s} = \frac{\partial}{\partial s}\left(\frac{1}{k}\right) = -\kappa_s J$$

where κ_s is the curvature of the line $s = \text{const}$, i.e., the vector line of grad z. It follows that at points of the regions where $k = \text{const} = k_{\min}$ (or $k_{\max}$) the lines $s = \text{const}$ are straight.

We now consider the region corresponding to the intermediate control range, where $C(s, z) = 0$. In this region the lines $z = \text{const}$ are segments of circles of constant radius ρ. Suppose that, corresponding to different values of z, the centers lie on the curve

$$\mathbf{R} = \mathbf{R}(z) = (x_0(z), y_0(z)) \tag{6.49}$$

We have (Fig. 6.5)

$$\mathbf{R} = \mathbf{r} + \rho \mathbf{i}_z$$

where $\mathbf{r}$ is the radius vector of the point (s, z). Differentiating with respect to z yields

$$\mathbf{R}_z = \mathbf{r}_z + \rho \frac{\partial \mathbf{i}_z}{\partial z} = k^{-1} \mathbf{i}_z + \rho \frac{\partial \mathbf{i}_z}{\partial z}$$

and

$$(\text{grad } z, \mathbf{R}_z) = k(\mathbf{i}_z, \mathbf{R}_z) = 1$$

since the vector $\partial \mathbf{i}_z / \partial z$ is directed along $\mathbf{i}_s$.

We now rewrite our result in the form

$$k|\mathbf{R}_z| \cos \alpha = 1 \qquad (6.50)$$

where α denotes the angle between $\mathbf{i}_z$ and the tangent vector to the curve $\mathbf{R}(z)$ (Fig. 6.5). However, in view of the preceding expressions, we also have

$$\mathbf{R}_s = \mathbf{r}_s + \rho \frac{\partial \mathbf{i}_z}{\partial s} = \mathbf{i}_s H_s - \rho \kappa_z H_s \mathbf{i}_s = 0$$

so that the vector $\mathbf{R}$ does not depend on s; we mention in passing that this is obvious on the basis of its geometrical meaning. It follows that $|\mathbf{R}_z|$ is also independent of s. Equation (6.50) now shows that

$$k(s, z) \cos \alpha(s, z) = k(s', z) \cos \alpha(s', z) \qquad (6.51)$$

Our results enable us to draw a number of conclusions about the mutual disposition of the regions with different values of the optimal control.

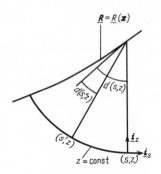

FIGURE 6.5

Suppose we trace the behavior along the line $s = s_0 = $ const, which begins at the point $z_0(s_0)$ of the line Σ. At this point $C \le 0$, so that $k = k_{\min}$ at the point itself and in a certain neighborhood of it in the direction of decrease of z [it is assumed that z increases from zero on Γ to some positive $z_0(s)$ on Σ]. In the region $k = k_{\min}$ the lines $s = $ const are straight. The function $C(s, z)$ can increase as z decreases; we assume that for $z = z_2$ this function vanishes. If the function $C(s, z)$ keeps its zero value as z decreases further, then an intermediate control regime is realized in the corresponding region. It is easy to see that the intermediate control region cannot come into contact with the boundary Γ of the basic region if this boundary is not a circle.

A further decrease of z for $s = s_0$ may lead to the appearance of a region with $k = k_{\max}$. This will be the case if the boundary Γ of G consists of a finite number of rectilinear sections. In accordance with (6.47), the function C increases in the immediate proximity of the boundary; that is at points of the boundary where $\partial C / \partial z < 0$, as the boundary is approached from within the region along the line $s = s_0 = $ const. There thus exists a point (s_0, z_1) on the line $s = s_0$ such that $C(s_0, z) > 0$ for $z < z_1$ and $C(s_0, z) \le 0$ for $z \ge z_1$. For $z < z_1$, we have $k = k_{\max}$.

A possible variant of the mutual disposition of the regions with different optimal control ranges is shown in Fig. 6.6.

The use of (6.44), (6.49), and (6.52) and of the given curve $\mathbf{R}(z)$ [see (6.50)] or one of the curves $z_2(s)$ and $z_1(s)$ bounding the region of intermediate control values, suffice to construct the system of lines $s = $ const and $z = $ const in all of G. We perform this construction for a square region, assuming that the ratio $k_{\max}/k_{\min}$ and the value of $\gamma/$mes G are known (the size of the square is not assumed to be fixed).

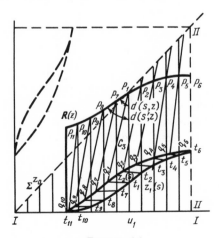

FIGURE 6.6

We construct a system of straight lines parallel to the edges of the square (obviously, for a homogeneous rod of square section, this will be a system of lines $s = $ const and $z = $ const). The discontinuity line Σ coincides with the diagonal of the square. We take an arbitrary point p_1 on Σ and assume that the curve $\mathbf{R}(z)$, oriented as shown in Fig. 6.6, passes through this point. The lines Σ and $z_2(s)$ bound the region where $k = k_{min}$; in this region, the lines $s = $ const are straight.

By definition of the curve $z_2(s)$,

$$\frac{1}{k_{min}} \int_{z_2(s)}^{z_0(s)} H_s(s, \rho)\, d\rho - \frac{\gamma}{\text{mes } G} H_s(s, z_2(s)) = 0 \qquad (6.52)$$

The integration here is performed along the straight line $s = s' = $ const; along this straight line, $d\rho = k_{min}|d\mathbf{r}|$. The parameter $H_s(s, z)$ is the absolute value of the derivative of the arc length of the line $z = $ const with respect to s. The introduction of the average value $\bar{H}_s(s')$ of this parameter over the segment $(z_2(s'), z_0(s'))$ of the straight line $s = s' = $ const results in the following form of (6.52):

$$\bar{H}_s(s')p(s') = \frac{\gamma}{\text{mes } G} H_s(s', z_2(s'))$$

Here $p(s')$ denotes the length of the segment $(z_2(s'), z_0(s'))$ of the straight line $s = s' = $ const which lies in the region $k = k_{min}$; this length has the value

$$p(s') = \frac{H_s(s', z_2(s'))}{\bar{H}_s(s')} \frac{\gamma}{\text{mes } G}$$

If we had a family of straight lines $s = $ const belonging to the region $k = k_{min}$, then we could approximately calculate the value of $\mu_s(s, z)$ and plot the point q_1 at the distance $\rho = \gamma/\text{mes } G$ from p_1. Having drawn the circular arc of radius $\gamma/\text{mes } G$ and of angular distance $\alpha(s', z) - \alpha(s, z)$ (see Fig. 6.6) with the center at p_1, we would build the point t_1 on the line $z_1(s)$ which separates zone of intermediate control from that of control $k = k_{max}$. The angle $\alpha(s, z)$ would then be determined from (6.51) which now accepts the form (Fig. 6.6)

$$k_{max} \cos \alpha(s, z) = k_{min} \cos \alpha(s', z)$$

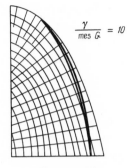

$$\frac{\gamma}{\text{mes } G} = 10$$

FIGURE 6.7

The line $z_1(s)$ separates the region of intermediate values of k from that where $k = k_{\max}$. Assume that the latter zone is adjacent to the domain's boundary. Because the latter is rectilinear the lines $s = $ const are straight lines perpendicular to the boundary. Consider the line $s = $ const (namely, $s = s_1$, the segment $u_1 t_1$) which comes to the point t_1. The vector grad z at the point t_1 is continuous; it is directed along this line and thus orthogonal to the arc $q_1 t_1$ at t_1. But the radius $p_1 t_1$ is also orthogonal to the same arc at the point t_1; consequently, the point p_1 lies on the prolongation of the segment $u_1 t_1$. This circumstance enables us to reconstruct the segment $q_1 p_1$ by a given segment $u_1 t_1$; the latter can easily be built if we know the point p_1 and the value of the parameter $\gamma/\text{mes } G$. In the region $k = k_{\min}$ the complete family of lines $s = $ const can be similarly constructed. The line $z_1(s)$ is obtained from the line $\mathbf{R}(z)$ by displacing it downward through a distance $\gamma/\text{mes } G$ (see Fig. 6.6), and the line $z_2(s)$ is recovered from $z_1(s)$ by an obvious geometrical construction.

The shape of the curve $\mathbf{R}(z)$ in Fig. 6.6 guarantees that the lines $z_1(s)$ and $z_2(s)$ join at the points t_{11} and t_6. It can be assumed that the edge of

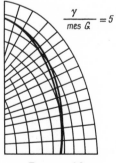

$$\frac{\gamma}{\text{mes } G} = 5$$

FIGURE 6.8

the square I-I passes through the point t_{11}; indeed, if it passed below this point, then it would be necessary to introduce an additional line separating regions with different values of k (k_{min} and k_{max}); along this line, we should either have a discontinuity of the function z (which is impossible), or (6.52) should not be satisfied.

Similarly, the straight line II-II can be taken as the second edge of the square.

Our solution describes the optimal distribution of the yield point for a quarter of the square. The solution for the complete square is constructed by symmetry. The constant ρ is chosen in such a way as to satisfy condition (3.38).

One can show (see Ref. 303) that the form of the curve $\mathbf{R}(z)$ chosen to construct the solution in Fig. 6.6 is not a good choice. If this curve lies entirely on one side of Σ, we would arrive at a contradiction with (6.52).

We emphasize that the size of the basic region in this construction was not known in advance. It was determined from the known parameters k_{max}/k_{min} and $\gamma/\text{mes } G$. We can assert that the contour of the region was chosen in accordance with the solution obtained. For regions bounded by curvilinear contours, such an approach presents great difficulties. The only (and trivial) exception is a circular region.

Even for an ellipse it is simpler to obtain the solution numerically by direct application of the method of successive approximation. In this way, solutions were obtained for the values $\gamma/\text{mes } G = 10$ (Fig. 6.7) and for $\gamma/\text{mes } G = 5$ (Fig. 6.8). As can be seen from these figures, the region occupied by material with the largest value of the yield point increases with decreasing parameter $\gamma/\text{mes } G$.

6.4. Hyperbolic Optimization Problems for Regions Whose Boundaries Contain Segments of Characteristics

We consider functions $z^1(x, y)$ and $z^2(x, y)$ satisfying the system of equations

$$z_x^1 = X(z^1, z^2, u), \qquad z_x^2 = \zeta^2$$

$$z_y^1 = \zeta^1, \qquad\qquad z_y^2 = Y(z^1, z^2, u) \tag{6.53}$$

with control $u(x, y)$. The characteristics of this system have projections onto the plane (x, y) which are straight lines parallel to the coordinate axes.

In the more general system

$$z_x^1 + \alpha(z^1, z^2)z_y^1 = f(z^1, z^2, u)$$

$$z_x^2 + \beta(z^1, z^2)z_y^2 = g(z^1, z^2, u)$$

(6.54)

the projections of the characteristic are lines with slopes α and β. If these parameters are constant, then (6.54) can be readily reduced to the form (6.53) by changing the independent variables.

In a number of problems, one needs to determine a solution of the system (6.53) in a region whose boundary contains sections parallel to the coordinate axes, i.e., characteristic sections. An example of such a region is shown in Fig. 6.9. It is well known that it is not always possible to specify arbitrary initial values of the corresponding functions on the characteristic sections; to guarantee the existence of a solution, the given initial values must satisfy compatibility relations that arise directly from the equations.

Thus, the initial values must satisfy the condition

$$\frac{dz^1}{dx} = X(z^1, z^2, u)$$

(6.55)

on the horizontal sections $y = \text{const}$ of the boundary Γ of G (Fig. 6.9) and on the vertical sections ($x = \text{const}$) the condition

$$\frac{dz^2}{dy} = Y(z^1, z^2, u)$$

(6.56)

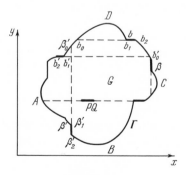

FIGURE 6.9

must be satisfied. These relations are ordinary differential equations which relate the initial values of the dependent variables on the characteristic sections of the boundary. In the investigation of optimization problems, the relations (6.54) and (6.55) must be taken into account in the derivation of necessary conditions, as prescribed by the general scheme of Chapter 2.

Another important property of characteristics is that discontinuities of the solutions of hyperbolic equations or of their derivatives frequently propagate along the characteristics. Equations (6.54) have this property if they are weakly nonlinear, i.e., if $\partial \alpha / \partial z^2 = \partial \beta / \partial z^1 = 0$ (Ref. 143, p. 440). The system (6.53) can be taken as an example of a weakly nonlinear system.

In the following we shall illustrate the influence of these features of hyperbolic equations on the formulation and solution of the corresponding optimization problems.

The example which follows is due to Jackson.[270] We shall follow his paper, modifying the arguments somewhat.

Suppose we want to find a continuous solution of the system (6.53) in the region G (see Fig. 6.9) if the values of z^1 are known at all points of the arc DAB of the boundary Γ with the exception of the horizontal (characteristic) sections, which are open (on the left) intervals of this arc; we shall denote the set of points of Γ at which the values of z^1 are given by Γ_{z^1}. Similarly, we shall assume that the values of z^2 are known on the part Γ_{z^2} of the arc ABC that does not contain vertical (characteristic) sections (which are open intervals at the bottom). These conditions and (6.53) are sufficient to determine z^1 and z^2 in the closed region $G \cup \Gamma$; the optimization problem consists of the determination of a control $u(x, y)$ which minimizes the functional

$$I = \oint_{\Gamma} (lz^1 + mz^2) \, dt$$

Here, t is the arc length of the boundary Γ, and l and m are given functions of t.

As before, we begin the derivation of the necessary conditions for a maximum by introducing the Lagrange multipliers $\xi_1, \ldots, \eta_2$ and the function H:

$$H = \xi_1 X + \eta_1 \zeta^1 + \xi_2 \zeta^2 + \eta_2 Y \tag{6.57}$$

To determine the multipliers, we have the system of equations

$$\xi_{1x} + \eta_{1y} = -\xi_1 X_{z^1} - \eta_2 Y_{z^1}, \qquad \eta_1 = 0$$

$$\xi_{2x} + \eta_{2y} = -\xi_1 X_{z^2} - \eta_2 Y_{z^2}, \qquad \xi_2 = 0 \tag{6.58}$$

and the boundary conditions

$$\xi_1 y_t = -l \qquad \text{at points of multiplicity } \Gamma/\Gamma_{z^1} \qquad (6.59)_1$$

$$\eta_2 x_t = m \qquad \text{at points of multiplicity } \Gamma/\Gamma_{z^2} \qquad (6.59)_2$$

It is, however, easy to see that the condition $(6.59)_1$ cannot be satisfied on the horizontal sections Γ/Γ_{z^1} of the arc, where $y_t = 0$, unless l vanishes identically on these sections. Similarly, if $m \neq 0$ on the vertical sections Γ/Γ_{z^2} of the arc, then $(6.59)_2$ cannot be satisfied on these sections.

It is easy to explain this difficulty. The point is that the derivation of necessary conditions for optimality must take into account (6.55) and (6.56) on the characteristic sections of the boundary. If the boundary is nonconvex and the characteristic sections of the boundary can be continued into the region (as, e.g., in Fig. 6.9), then (6.55) and (6.56) must also be taken into account on the extensions of the characteristic sections.*

For the region shown in Fig. 6.9, the constraint (6.55) must be taken into account on the horizontal sections of type b (i.e., $b_0 b_1 b_2$, etc.) and type b' (i.e., $b'_0 b'_1 b'_2$, etc.), and the constraint (6.56) on the vertical sections of type β (i.e., $\beta_0 \beta_1 \beta_2$, etc.) and type β' (i.e., $\beta'_0 \beta'_1 \beta'_2$, etc.). It is assumed that no two sections b and b' have the same ordinate, and that no two sections β and β' have the same abscissa.

The construction of all possible sections of both types, results in a partitioning of G into a number of partial regions G_i with boundaries Γ_i. In each of them, we determine the Lagrange multipliers $\xi_1, \ldots, \eta_2$, and the multipliers θ^b, $\theta^{b'}$, θ^β, $\theta^{\beta'}$ on the sections b, b', β, β', respectively. With these multipliers we then construct the functional

$$\Pi = I + \sum_i \iint_{G_i} [\xi_1(z_x^1 - X) + \eta_1(z_y^1 - \zeta^1) + \xi_2(z_x^2 - \zeta^2) + \eta_2(z_y^2 - Y)] \, dx \, dy$$

$$+ \sum \int_{b_0}^{b_2} \theta^b(z_x^1 - X) \, dx + \sum \int_{b'_2}^{b'_0} \theta^{b'}(z_x^1 - X) \, dx + \sum \int_{\beta_0}^{\beta_2} \theta^\beta(z_y^2 - Y) \, dy$$

$$+ \sum \int_{\beta'_2}^{\beta'_0} \theta^{\beta'}(z_y^2 - Y) \, dy \qquad (6.60)$$

---

* In what follows, we shall show that it suffices to take these relations into account only on some of the extensions of the characteristic sections of the boundary.

and calculate its first variation. When (6.58) are taken into consideration, the part of the variation generated by the integrals over G_i is given by

$$\sum_i \oint_{\Gamma_i} \xi_1 y_t \delta z^1 \, dt + \sum_i \oint_{\Gamma_i} (-\eta_2 x_t) \delta z^2 \, dt - \sum_i \iint_{G_i} \xi_1 \frac{\partial X}{\partial u} \delta u \, dx \, dy$$

$$- \sum_i \iint_{G_i} \eta_2 \frac{\partial Y}{\partial u} \delta u \, dx \, dy$$

or, equivalently,

$$\oint_{\Gamma/\Gamma_{z^1}} \xi_1 y_1 \delta z^1 \, dt + \sum_\beta \int_{\beta_0}^{\beta_1} (\xi_1^+ - \xi_1^-) \delta z^1 \, dy + \sum_{\beta'} \int_{\beta_1'}^{\beta_0'} (\xi_1^+ - \xi_1^-) \delta z^1 \, dx$$

$$- \oint_{\Gamma/\Gamma_{z^2}} \eta_2 x_t \delta z^2 \, dt - \sum_b \int_{b_0}^{b_1} (\eta_2^+ - \eta_2^-) \delta z^2 \, dy - \sum_{b'} \int_{b_1'}^{b_0'} (\eta_2^+ - \eta_2^-) \delta z^2 \, dy$$

$$- \sum_i \iint_{G_i} \xi_1 \frac{\partial X}{\partial u} \delta u \, dx \, dy - \sum_i \iint_{G_i} \eta_2 \frac{\partial Y}{\partial u} \delta u \, dx \, dy \qquad (6.61)$$

In the integrals over the sections b and b', we denote the limiting value of η_2 from the partial region below the section by η_2^+ and the limiting value of η_2 from the partial region above the section by η_2^-.

In the integrals along the sections β and β' we similarly denote the limiting value of ξ_1 from the partial region to the left of the section by ξ_1^+ and the limiting value of ξ_1 from the partial region to the right of the section by ξ_1^-.

The transformation of the variations of the last four integrals in (6.60) leads to the terms

$$\sum [\theta^b \delta z^1]_{b_0}^{b_2} - \sum \int_{b_0}^{b_2} \left[\left(\theta_x^b + \theta^b \frac{\partial X}{\partial z^1} \right) \delta z^1 + \theta^b \frac{\partial X}{\partial z^2} \delta z^2 + \theta^b \frac{\partial X}{\partial u} \delta u \right] dx$$

$$+ \sum [\theta^{b'} \delta z^1]_{b_2}^{b_0'} - \sum \int_{b_2'}^{b_0'} \left[\left(\theta_x^{b'} + \theta^{b'} \frac{\partial X}{\partial z^1} \right) \delta z^1 + \theta^{b'} \frac{\partial X}{\partial z^2} \delta z^2 + \theta^{b'} \frac{\partial X}{\partial u} \delta u \right] dx$$

$$+ \sum [\theta^\beta \delta z^2]_{\beta_0}^{\beta_2} - \sum \int_{\beta_0}^{\beta_2} \left[\left(\theta_y^\beta + \theta^\beta \frac{\partial Y}{\partial z^2} \right) \delta z^2 + \theta^\beta \frac{\partial Y}{\partial z^1} \delta z^1 + \theta^\beta \frac{\partial Y}{\partial u} \delta u \right] dy$$

$$+ \sum [\theta^{\beta'} \delta z^2]_{\beta_2}^{\beta_0'} - \sum \int_{\beta_2'}^{\beta_0'} \left[\left(\theta_y^{\beta'} + \theta^{\beta'} \frac{\partial Y}{\partial z^2} \right) \delta z^2 + \theta^{\beta'} \frac{\partial Y}{\partial z^1} \delta z^1 + \theta^{\beta'} \frac{\partial Y}{\partial u} \delta u \right] dy$$

$$(6.62)$$

We denote the part of the boundary Γ that does not contain horizontal or vertical sections by $\tilde{\Gamma}$. Varying the functional I and taking into account (6.61), we obtain the following natural boundary conditions along $\tilde{\Gamma}$:

$$\xi_1 y_t + l = 0 \qquad \text{on } \tilde{\Gamma}/\Gamma_{z^1}$$

$$\eta_2 x_t - m = 0 \qquad \text{on } \tilde{\Gamma}/\Gamma_{z^2}$$

(6.63)

The functions θ^b satisfy the equations

$$\theta_x^b + \theta^b \frac{\partial X}{\partial z^1} = \begin{cases} l(x) & \text{on } b_1 b_2 \\ 0 & \text{on } b_0 b_1 \end{cases}$$

(6.64)

and the initial conditions

$$\theta^b(b_2) = 0$$

(6.65)

The functions θ^β are determined from the equations

$$\theta_y^\beta + \theta^\beta \frac{\partial Y}{\partial z^2} = \begin{cases} m(y) & \text{on } \beta_1 \beta_2 \\ 0 & \text{on } \beta_0 \beta_1 \end{cases}$$

(6.66)

and the initial conditions

$$\theta^\beta(\beta_2) = 0$$

(6.67)

The multipliers $\theta^{b'}$ and $\theta^{\beta'}$ are determined by the equations and the conditions

$$\theta_x^{b'} + \theta^{b'} \frac{\partial X}{\partial z^1} = \begin{cases} l(x) & \text{on } b_1' b_2' \\ 0 & \text{on } b_0' b_1' \end{cases}$$

(6.68)

$$\theta^{b'}(b_0') = 0$$

(6.69)

$$\theta_y^{\beta'} + \theta^{\beta'} \frac{\partial Y}{\partial z^2} = \begin{cases} m(y) & \text{on } \beta_1' \beta_2' \\ 0 & \text{on } \beta_0' \beta_1' \end{cases}$$

(6.70)

$$\theta^{\beta'}(\beta_0') = 0$$

(6.71)

Equations (6.68)–(6.71) show that

$$\theta^{b'} = 0 \qquad \text{on } b_0' b_1'$$

$$\theta^{\beta'} = 0 \qquad \text{on } \beta_0' \beta_1'$$

(6.72)

i.e., the multipliers $\theta^{b'}$ and $\theta^{\beta'}$ vanish identically on the extensions of the characteristic sections of the boundary into the region. From (6.64)–(6.67), it is evident, however, that the multipliers θ^b and θ^β do not vanish on the extensions of the corresponding characteristic sections.

In the remaining terms of (6.60) [see (4.61) and (4.62)] we now set the coefficients of δz^1 equal to zero to obtain

$$- \theta^\beta \frac{\partial Y}{\partial z^1} + \xi_1^+ - \xi_1^- = 0 \qquad \text{along } \beta_0\beta_1 \qquad (6.73)$$

$$\theta^\beta \frac{\partial Y}{\partial z^1} - l(y) - \xi_1(y) = 0 \qquad \text{along } \beta_1\beta_2 \text{ on } BCD \qquad (6.74)$$

$$- \theta^{\beta'} \frac{\partial Y}{\partial z^1} + \xi_1^+ - \xi_1^- = 0 \qquad \text{along } \beta_1'\beta_0' \qquad (6.75)$$

$$\theta^{\beta'} \frac{\partial Y}{\partial z^1} - l(y) - \xi_1(y) = 0 \qquad \text{along } \beta_2'\beta_1' \text{ on } BAD \qquad (6.76)$$

In similar fashion, the coefficients of δz^2 yield the conditions

$$- \theta^b \frac{\partial X}{\partial z^2} - \eta_2^+ + \eta_2^- = 0 \qquad \text{along } b_0b_1 \qquad (6.77)$$

$$\theta^b \frac{\partial X}{\partial z^2} - m(x) + \eta_2(x) = 0 \qquad \text{along } b_1b_2 \text{ on } ADC \qquad (6.78)$$

$$- \theta^{b'} \frac{\partial X}{\partial z^2} - \eta_2^+ + \eta_2^- = 0 \qquad \text{along } b_1'b_0' \qquad (6.79)$$

$$\theta^{b'} \frac{\partial X}{\partial z^2} - m(x) + \eta_2(x) = 0 \qquad \text{along } b_2'b_1' \text{ on } ADC \qquad (6.80)$$

Equations (6.75) and (6.79) in conjunction with (6.72) show that the multipliers ξ_1 and η_2 are continuous on the characteristics $\beta_1'\beta_0'$ and $b_1'b_0'$, respectively; it follows from (6.64)–(6.67), (6.73), and (6.77) that these same multipliers generally differ on the characteristics b_0b_1 and $\beta_0\beta_1$. It now becomes clear that one can ignore (6.54) and (6.55) on the extensions $b_1'b_0'$ and $\beta_1'\beta_0'$ of the characteristic sections $b_2'b_1'$ and $\beta_2'\beta_1'$, altogether; indeed, one need not bother to introduce these extensions at all. Instead, one should extend the corresponding integrals in (6.60) only to the sections $b_2'b_1'$ and

$\beta_2'\beta_1'$ of the boundary. This follows from (6.72), and these, in turn, follow from the fact that the variations δz^1 and δz^2 at the ends b_0' and β_0' of $b_1'b_0'$ and $\beta_1'\beta_0'$ are free (the point b_0' belongs to the arc BCD and β_0' to the arc ADC).

Thus, one needs only extend those characteristic sections of the boundary which, when extended, intersect the boundary at points where the corresponding functions are specified (e.g., at the points b_0, where the values of z^1 are known, or at the points β_0, where the values of z^2 are given).

These results yield the following expression for the first variation of the functional I:

$$\delta I = -\iint_G \left(\xi_1 \frac{\partial X}{\partial u} + \eta_2 \frac{\partial Y}{\partial u} \right) \delta u \, dx \, dy - \sum \int_{b_0}^{b_2} \theta^b \frac{\partial X}{\partial u} \delta u \, dx$$

$$- \sum \int_{b_2'}^{b_1'} \theta^{b'} \frac{\partial X}{\partial u} \delta u \, dx - \sum \int_{\beta_0}^{\beta_2} \theta^\beta \frac{\partial Y}{\partial u} \delta u \, dy - \sum \int_{\beta_2'}^{\beta_1'} \theta^{\beta'} \frac{\partial Y}{\partial u} \delta u \, dy$$

$$(6.81)$$

The double integral in this formula expresses the variation of the functional for the case when the boundary of the region does not have characteristic sections. In the general case, the formula for δI contains terms due to small (weak) variations δu of the control concentrated on characteristic sections of the boundary or on extensions of some of them into the region (sections of the type PQ in Fig. 6.9). These types of control variations are essentially one-dimensional; two-dimensional variations occur in the first integral term of (6.81).

Any variation δu whose sign coincides with the sign of $\theta^\beta \partial X/\partial u$ on the sections $b_0 b_2$, with the sign of $\theta^{\beta'} \partial X/\partial u$ on $b_2' b_1'$, with the sign of $\theta^\beta \partial Y/\partial u$ on $\beta_0 \beta_2$, with the sign of $\theta^\beta \partial Y/\partial u$ on $\beta_2'\beta_1'$, and with the sign of $\xi_1 \partial X/\partial u + \eta_2 \partial Y/\partial u$ at almost all remaining points of the region, decreases the functional I. This property of (6.81) can be used as the basis for constructing a minimizing sequence of controls. The functions u obtained in this manner are in general different at points of discontinuity of the multipliers ξ_1 and η_2; i.e., in particular, on the extensions $b_0 b_1$ and $\beta_0 \beta_1$ of the characteristic sections of the boundary. In addition, discontinuities of u can arise, as usual, as a result of additional restrictions on the controls.

In the absence of such restrictions, (6.81) leads to the following stationarity conditions:

$$\theta^b \frac{\partial X}{\partial u} = 0 \qquad \text{along } b_0 b_2 \qquad (6.82)$$

$$\theta^{b'} \frac{\partial X}{\partial u} = 0 \qquad \text{along } b'_2 b'_1 \qquad\qquad (6.83)$$

$$\theta^{\beta} \frac{\partial Y}{\partial u} = 0 \qquad \text{along } \beta_0 \beta_2 \qquad\qquad (6.84)$$

$$\theta^{\beta'} \frac{\partial Y}{\partial u} = 0 \qquad \text{along } \beta'_2 \beta'_1 \qquad\qquad (6.85)$$

Generally, solution of these equations will be discontinuous at points of discontinuity of the multipliers ξ_1 and η_2.

Equations (6.58), (6.64), (6.66), (6.68), (6.70), (6.82)–(6.85) form the system of Euler equations, and (6.63), (6.65), (6.67), (6.69), (6.71) and (6.73)–(6.80) are the natural boundary conditions of the posed optimization problem.

Although the preceding discussion applied to weak variations of the controls, it shows that sections of the type PQ (see Fig. 6.9) play a particular role in the case of strong variation as well. In order to obtain the Weierstrass (Legendre) necessary conditions for this case, we must consider different types of strong (weak) variations of the control. First, we introduce a two-dimensional strong (weak) variation $\Delta u(x, y)$ concentrated in a small neighborhood of an interior point M_1 which does not belong to any of the sections $b_0 b_1$ or $\beta_0 \beta_1$ (see Fig. 6.9). The shape of the small region of variation in the neighborhood of the point M_1 here has no significance. In addition, we introduce variations $\Delta u(x)$ concentrated in a small neighborhood $(x - h, x + h)$ of a point M_2 belonging to the sections $b_0 b_2$ or $b'_2 b'_1$, and variations $\Delta u(y)$ concentrated in a small neighborhood of a point M_3 belonging to the sections $\beta_0 \beta_2$ or $\beta'_2 \beta'_1$. In the usual way,[270] we can show that the principal part of the increment ΔI of the functional due to weak variations of these types is given by (6.81), in which one need only to replace the weak variations δu by corresponding variations Δu. In addition to (6.57), we now introduce the functions

$$h_b = \theta^b X, \qquad h_{b'} = \theta^{b'} X, \qquad h_\beta = \theta^\beta Y, \qquad h_{\beta'} = \theta^{\beta'} Y$$

Weierstrass's necessary conditions then take the form

$$H(u) = \max_U H(U)$$

$$h_b = \max_U h_b(U), \qquad h_{b'} = \max_U h_{b'}(U)$$

$$h_\beta = \max_U h_\beta(U), \qquad h_{\beta'} = \max_U h_{\beta'}(U)$$

These relations are the analogue of Pontryagin's maximum principle for the problem we have considered.

6.5. The Optimal Shape of a Contour Surrounded by a Supersonic Gas Flow

Optimal control methods have found wide application in the theory of optimal aerodynamic shapes, which plays an important part in modern applied aerodynamics. Numerous investigations have been made into the variational problems of gas dynamics*; in this field, many important results have been achieved and specific methods have been developed for the investigation of extremal problems. Here, we cannot attempt to set forth even the basic facts in any detail. As a whole, these already constitute a fairly well-developed theory; the interested reader is referred to the literature cited in the footnote.

We shall consider only one direction in the investigation of variational problems in gas dynamics, namely, that for which the pioneering investigations were made by Guderley and Armitage[53] and Sirazetdinov.[148] We shall treat optimization problems for functionals of a fairly general form subject to constraints imposed in the form of the differential equations of gas dynamics. The role of controls is played by functions which determine the shape of the contour of the body or nozzle around which the flow takes place; somewhat arbitrary isoperimetric or other types of restrictions may be imposed on these functions.

In Refs. 53 and 148, it was proposed that the constraints be taken into account by suitably introduced Lagrange multipliers. Based on the usual procedures of the calculus of variations, the Euler equations and natural boundary conditions may be obtained; here, it is important to note that the Lagrange multipliers are discontinuous in a number of cases of practical interest. This last circumstance in connection with optimization problems of gas dynamics was discovered by Kraiko[79]; in a way, the circumstance is analogous to similar phenomena described in the previous section.

In two original papers, Refs. 53 and 148, as well as in subsequent publications by other authors,† the method was developed only as far as the derivation of necessary conditions of stationarity and study of the consequences of these conditions. In this fashion, many important results

---

* See Refs. 13–15, 79–81, and 166–170, along with two monographs, Refs. 160 and 171, which contain extensive bibliographies.
† For example, see Ref. 160.

were obtained concerning the optimal boundaries for various flow conditions; some of these will be given below. Some arguments which establish the nature of the extremum for a number of problems have been given by Shmyglevskii,[166,171] Fanselau,[240] Guderley *et al.*,[450] Fedorov,[443] and Lurie and Fedorov.[480]

In the development of investigations into optimal aerodynamic shapes, Nikol'skii's investigation[117] played an important part. His basic idea was to reduce the variational problem in the region of supersonic flow to a problem on the control contour—the boundary of this region formed by the characteristics of the equations of gas dynamics. In other words, his idea was to reduce the number of independent variables by one; for planar and axisymmetric flows, Nikol'skii's method thus reduces the optimization problems to one involving ordinary differential equations. Guderley and Hantsch[265] obtained the first exact solution of a variational problem of gas dynamics using a control contour. Extensive investigations in this direction, leading to deep results, have been made by Shmyglevskii.[166–171]

Unfortunately, the control contour method applies only in those cases where the size of the body is restricted and there are no irreversible processes (shock waves) in the flow. If the isoperimetric conditions have a more general nature (e.g., the area of the surface of the body or its volume may be specified), and if there are strong discontinuities, then it is necessary to use the procedures described in Refs. 53 and 148.

In the present section, we follow Ref. 53 to derive necessary conditions for stationarity in the general case, and we consider simplifications which make it possible to apply Nikol'skii's method in special problems.

We consider a steady-state vortex-free isentropic flow of a perfect gas. We shall assume that the flow is axisymmetric; then the velocity components $u(x, y)$, $v(x, y)$, the pressure $p(x, y)$, and the density $\rho(x, y)$ satisfy the following equations:

$$\frac{u^2 + v^2}{2} + \frac{\gamma}{\gamma - 1} \frac{p}{\rho} = \text{const} \qquad \text{(Bernoulli's integral)} \qquad (6.86)$$

$$(\rho u y)_x + (\rho v y)_y = 0 \qquad \text{(Continuity)} \qquad (6.87)$$

$$u_y - v_x = 0 \qquad \text{(Zero vorticity)} \qquad (6.88)$$

$$\frac{p}{\rho^\gamma} = \text{const} \qquad \text{(Constant entropy)} \qquad (6.89)$$

Here, γ is the adiabatic exponent; the flow is assumed to be supersonic ($|\mathbf{v}| > a$).

The velocity of sound a is determined by

$$a^2 = \frac{dp}{d\rho} = \frac{\gamma p}{\rho} \tag{6.90}$$

We assume that the gas moves through a nozzle of circular section with generator $y = f(x)$ (Fig. 6.10, where AC is the nozzle's contour, AB is a characteristic of the first family, and BC is a characteristic of the second family). The thrust of the nozzle is determined by

$$T = \int_{x_A}^{x_C} (p - p_0) ff' \, dx \tag{6.91}$$

where p_0 is the external pressure. On the nozzle contour, the normal component of the velocity is zero:

$$f'(x)u(x, f(x)) - v(x, f(x)) = 0, \qquad x_A \leq x \leq x_c \tag{6.92}$$

Some geometrical characteristics of the nozzle are determined by technical requirements. We introduce a restriction in the form

$$s = \int_{x_A}^{x_C} F(f, f', x) \, dx \tag{6.93}$$

where s is a given constant. The case $F = f(1 + f'^2)^{1/2}$ corresponds to the specification of the surface area of the nozzle.

We now formulate the following optimization problem: To determine a function $f(x)$ satisfying restriction (6.93) and the inequalities

$$d \leq f'(x) \leq b \tag{6.94}$$

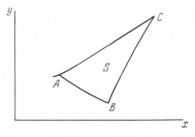

FIGURE 6.10

in such a way that the functional T takes on its largest possible value subject to the constraints (6.86)–(6.89) and (6.92). Inequalities (6.94) express additional requirements on the shape of the channel; it can be assumed that these requirements rule out the formation of shock waves and thus justify the use of (6.86)–(6.89) to describe the motion of the gas.

Before we turn to the derivation of necessary conditions, we reduce the equations of the problem to standard form. We shall assume that the functions u and v are continuous in an open region filled with streamlines. We set $z^1 = u$ and $z^2 = v$; (6.87) and (6.88) then are equivalent to the system

$$z_x^1 = \zeta^1, \qquad z_y^1 = \zeta^2$$

$$z_x^2 = \zeta^2, \qquad z_y^2 = \zeta^3 \tag{6.95}$$

$$\frac{\partial}{\partial x}(y\rho z^1) + \frac{\partial}{\partial y}(y\rho z^2) = 0$$

It is convenient to keep the last equation in the form (6.87).

The boundary condition (6.92) may be written in the system form

$$\frac{df}{dx} = w \tag{6.96}$$

$$wz^1 - z^2 = 0 \qquad \text{on } AC \tag{6.97}$$

the inequalities (6.94) may be replaced by the equivalent equation

$$(b - w)(w - d) - w_*^2 = 0$$

with additional control w_*.

Next, we introduce Lagrange multipliers $\xi_1(x, y), \ldots, \eta_2(x, y), \omega(x, y), \mu(x), \nu(x), c_1, c_2(x)$, along with the functional

$$\Pi = \int_{x_A}^{x_C} \{(p - p_0)fw + \mu(x)(f' - w) + c_1 F(f, w, x) + c_2(x)(wz^1 - z^2)$$

$$+ \nu(x)[(b - w)(w - d) - w_*^2]\} \, dx$$

$$+ \iint_S \left\{ \xi_1(z_x^1 - \zeta^1) + \eta_1(z_y^1 - \zeta^2) + \xi_2(z_x^2 - \zeta^2) + \eta_2(z_y^2 - \zeta^3) \right.$$

$$\left. + \omega \left[\frac{\partial}{\partial x}(y\rho z^1) + \frac{\partial}{\partial y}(y\rho z^2) \right] \right\} dx \, dy \qquad (6.98)$$

Here, the double integral is taken over the triangular region S (Fig. 6.10) bounded by the section AC of the boundary and the characteristics AB and CB of the different families passing through the points A and C. The region is the domain of dependence of AC: the distribution of the pressure along AC and, therefore, that of the thrust depend only on the hydrodynamic quantities within S.

On the characteristic sections of the boundary of S, the following conditions are satisfied:

$$y_t - x_t \tan(\alpha + \theta) = 0 \qquad \text{on } BC \qquad (6.99)$$

$$y_t + x_t \tan(\alpha - \theta) = 0 \qquad \text{on } AC \qquad (6.100)$$

In these conditions,

$$\theta = \arctan \frac{z^2}{z^1}, \qquad \alpha = \arcsin \frac{a}{[(z^1)^2 + (z^2)^2]^{1/2}} \qquad (6.101)$$

The characteristic AB and the values of the velocity components z^1 and z^2 along it are assumed known.* In the derivation of necessary conditions, it must be remembered that the pressure p and the density ρ are related to z^1 and z^2 by (6.86) and (6.89). From these equations, variations δp and $\delta \rho$ follow as

$$\delta p = -\rho(z^1 \delta z^1 + z^2 \delta z^2)$$

$$\delta \rho = a^{-2} \delta p \qquad (6.102)$$

The variation of the functional (6.98) must be equal to zero.

The expression for $\delta \Pi$ is made up of terms due to the variations δz^1, δz^2, $\delta \zeta^1$, $\delta \zeta^2$, $\delta \zeta^3$, as well as the variations δf, δw, and δx_C.

---

* More precisely, we assume that the characteristic of the second family emanating from A is known; the position of point B on this characteristic is not fixed.

The variations δz^1 and δz^2 generate the terms

$$\int_{x_A}^{x_C} \left[-\rho(z^1\delta z^1 + z^2\delta z^2)fw + c_2(w\delta z^1 - \delta z^2) \right] dx$$

$$+ \oint (\xi_1 y_t - \eta_1 x_t)\delta z^1\, dt + \oint (\xi_2 y_t - \eta_2 x_t)\, \delta z^2\, dt$$

$$+ \oint \omega y[\delta(\rho z^1)y_t - \delta(\rho z^2)x_t]\, dt$$

$$- \iint_S [\xi_1\delta\zeta^1 + \eta_1\delta\zeta^2 + \xi_2\delta\zeta^2 + \eta_2\delta\zeta^3 + \omega_x y\delta(\rho z^1) + \omega_y y\delta(\rho z^2)]\, dx\, dy$$

$$- \iint_S [(\xi_{1x} + \eta_{1y})\delta z^1 + (\xi_{2x} + \eta_{2y})\delta z^2]\, dx\, dy$$

When (6.97) and (6.102) are taken into account, this expression may be transformed to

$$\int_{x_A}^{x_C} \left\{ \left[-\rho z^1 fw + c_2 w - \xi_1 w + \eta_1 - \omega\frac{f\rho z^1 z^2}{a^2} - \omega\rho fw\left(1 - \frac{(z^1)^2}{a^2}\right) \right] \delta z^1 \right.$$

$$\left. + \left[-\rho z^2 fw - c_2 - \xi_2 w + \eta_2 + \omega\frac{f\rho z^1 z^2}{a^2} w + \omega\rho f\left(1 - \frac{(z^2)^2}{a^2}\right) \right] \delta z^2 \right\} dx$$

$$+ \left[\int_A^B + \int_B^C \right] \left\{ \xi_1\, dy - \eta_1\, dx + \omega y\rho\left[\left(1 - \frac{(z^1)^2}{a^2}\right) dy + \frac{z^1 z^2}{a^2}\, dx \right] \right\} \delta z^1$$

$$+ \left[\int_A^B + \int_B^C \right] \left\{ \xi_2\, dy - \eta_2\, dx + \omega y\rho\left[-\frac{z^1 z^2}{a^2}\, dy - \left(1 - \frac{(z^2)^2}{a^2}\right) dx \right] \right\} \delta z^2$$

$$- \iint_S \left\{ \xi_1\delta\zeta^1 + \eta_1\delta\zeta^2 + \xi_2\delta\zeta^2 + \eta_2\delta\zeta^3 \right.$$

$$+ y\rho\left[\omega_x\left(1 - \frac{(z^1)^2}{a^2}\right) - \omega_y\frac{z^1 z^2}{a^2} \right]\delta z^1$$

$$\left. + y\rho\left[\omega_y\left(1 - \frac{(z^2)^2}{a^2}\right) - \omega_x\frac{z^1 z^2}{a^2} \right]\delta z^2 \right\} dx\, dy$$

$$- \iint_S [(\xi_{1x} + \eta_{1y})\delta z^1 + (\xi_{2x} + \eta_{2y})\delta z^2]\, dx\, dy$$

The integral along AB is zero in this expression since the functions z^1 and z^2 are fixed on this section of the boundary*. In the double integral,

---

* This is certainly true if $\delta f'(x_A + 0) = 0$. For the general case, see Ref. 160 (p. 178).

the vanishing of the coefficients of $\delta\zeta^1$, $\delta\zeta^2$, $\delta\zeta^3$ yields the Euler equations

$$\xi_1 = 0, \qquad \xi_2 + \eta_1 = 0, \qquad \eta_2 = 0 \tag{6.103}$$

Based on these relations, the coefficients of δz^1 and δz^2 may be made to vanish also; we obtain

$$\eta_{1y} = -y\rho\left[\omega_x\left(1 - \frac{(z^1)^2}{a^2}\right) - \omega_y\frac{z^1z^2}{a^2}\right]$$

$$\tag{6.104}$$

$$\eta_{1x} = y\rho\left[\omega_y\left(1 - \frac{(z^2)^2}{a^2}\right) - \omega_x\frac{z^1z^2}{a^2}\right]$$

Subject to (6.97) and (6.103), the conditions on the boundary AC have the form

$$-\rho z^2 f + c_2\frac{z^2}{z^1} + \eta_1 - \omega\rho f\frac{z^2}{z^1} = 0$$

$$-\rho\frac{(z^2)^2}{z^1}f - c_2 + \eta_1\frac{z^2}{z^1} + \omega\rho f = 0$$

Multiplying the first of these equations by z^1/z^2 and adding it to the second, we arrive at the equation

$$\eta_1 = \rho f z^2 \qquad \text{on } AC \tag{6.105}$$

The elimination of η_1 from the first equation yields

$$\omega = \frac{c_2}{\rho f} \qquad \text{on } AC \tag{6.106}$$

Equations (6.105) and (6.106) are part of the system of natural boundary conditions on the boundary of the nozzle. The remaining conditions on AC will be obtained below.

The conditions along the characteristic BC are derived next. Subject to the constraint (6.99), the vanishing of the coefficient of δz^1 results in the condition

$$\eta_1 - \omega y\rho\left[\frac{z^1z^2}{a^2} + \left(1 - \frac{(z^1)^2}{a^2}\right)\tan(\alpha + \theta)\right] = 0$$

or, taking (6.101) into account,

$$\eta_1 + \omega y\rho\cot\alpha = 0 \qquad \text{on } BC \tag{6.107}$$

---

* This is certainly true if $\delta f'(x_A + 0) = 0$. For the general case, see Ref. 160 (p. 178).

The vanishing of the coefficient of δz^2 yields the same relation.

In Ref. 53 it is shown that (6.104) for a supersonic flow form a hyperbolic system; its characteristics coincide with the characteristics of the basic system (6.86)–(6.89). On the characteristics of the first family

$$\frac{dy}{dx} = \tan(\alpha + \theta)$$

the kinematic consistency condition has the form

$$d\eta_1 - y\rho \cot \alpha \, d\omega = 0$$

on the characteristics of the second family

$$\frac{dy}{dx} = \tan(\alpha - \theta)$$

the consistency condition is

$$d\eta_1 + y\rho \cot \alpha \, d\omega = 0$$

We now consider the conditions generated by the variations δf, δw, δw_*. Only the first integral in (6.98) is to be varied. It must be borne in mind that the variables p, z^1, z^2 depend on f when taken on the boundary AC:

$$p\big|_{AC} = p(z^1, z^2)\big|_{AC} = p[z^1(x, f(x)), z^2(x, f(x))]$$

Bearing this in mind, the corresponding variation is given by [see (6.102)]:

$$\int_{x_A}^{x_C} \left[-fw(\rho z^1 z^1_y + \rho z^2 z^2_y)\delta f + (p - p_0)w\delta f \right.$$

$$+ (p - p_0)f\delta w - \frac{d\mu}{dx}\delta f - \mu\delta w + c_1 F_f \delta f + c_1 F_w \delta w + c_2 z^1 \delta w$$

$$\left. + c_2(wz^1_y - z^2_y)\delta f - \nu(2w - d - b)\delta w - 2\nu w_* \delta w_* \right] dx$$

$$+ \mu\delta f \big|_{x_A}^{x_C} + [c_1 F + (p - p_0)fw]_{x=x_C}\delta x_C \qquad (6.108)$$

The vanishing of the coefficients of δf, δw, δw_* in the integrand implies

$$\left.\begin{array}{l}\dfrac{d\mu}{dx} + fw(\rho z^1 z_y^1 - \rho z^2 z_y^2) - (p - p_0)w - c_1 F_f - c_2(wz_y^1 - z_y^2) = 0 \\[12pt] (p - p_0)f - \mu + c_1 F_w + c_2 z^1 = \nu(2w - d - b), \qquad 2\nu w_* = 0\end{array}\right\} \quad (6.109)$$

Based on (6.95) and (6.97), we can readily show that

$$\rho z^1 z_y^1 + \rho z^2 z_y^2 = \rho z^1 \frac{dz^2}{dx}$$

$$c_2(wz_y^1 - z_y^2) = \frac{c_2}{\rho f}(wz_y^1 - z_y^2)\rho f$$

$$= \frac{c_2}{\rho f}[(wz_y^1 - z_y^2)\rho f + (\rho f)_y(wz^1 - z^2)]$$

$$= \frac{c_2}{\rho f}[w(\rho f z^1)_y + (\rho f z^1)_x] = \frac{c_2}{\rho f}\frac{d}{dx}(\rho f z^1)$$

on AC. Here, we have introduced the following notation for the total derivative with respect to x along AC:

$$\frac{d}{dx} = \frac{\partial}{\partial x} + w\frac{\partial}{\partial y}$$

Solving the second equation of (6.109) for μ and substituting the results in the first equation, we obtain

$$\rho f z^1 w \frac{dz^2}{dx} + f\frac{dp}{dx} - c_1 F_f + c_1 \frac{dF_w}{dx} - \frac{d}{dx}[\nu(2w - d - b)] + \rho f z^1 \frac{d}{dx}\left(\frac{c_2}{\rho f}\right) = 0$$

or, using (6.86),

$$-\rho f z^1 \frac{dz^1}{dx} - c_1 F_f + c_1 \frac{dF_w}{dx} - \frac{d}{dx}[\nu(2w - d - b)] + \rho f z^1 \frac{d}{dx}\left(\frac{c_2}{\rho f}\right) = 0$$

An integration then yields

$$\frac{c_2}{\rho f} = z^1 - \int_x^{x_C} \frac{1}{\rho f z^1}\left\{c_1 F_f - c_1 \frac{dF_w}{dx} + \frac{d}{dx}[\nu(2w - d - b)]\right\}dx + c_3 \quad (6.110)$$

We now use the remaining natural conditions which follow from (6.108) to determine the constant c_3 and the parameter x_C. The position of point A is fixed, and we therefore have $\delta f|_A = 0$. At the unknown point C, the use of (6.109) yields the conditions

$$\mu = (p - p_0)f + c_1 F_w + c_2 z^1 - \nu(2w - d - b) = 0 \qquad (6.111)$$

$$(p - p_0)fw + c_1 F = 0 \qquad (6.112)$$

Taking (6.110) into account, (6.111) may be written in the form

$$(p - p_0)f + c_1 F_w + \rho f z^1(z^1 + c_3) = \nu(2w - d - b) \qquad \text{at the point } C$$

$$(6.113)$$

It is worth noting that the use of (6.110) reduces (6.106) to the form

$$\omega = z^1 - \int_x^{x_C} \frac{1}{\rho f z^1} \left\{ c_1 F_f - c_1 \frac{dF_w}{dx} + \frac{d}{dx}[\nu(2w - d - b)] \right\} dx + c_3 \qquad \text{on } AC$$

$$(6.114)$$

The parameters c_1 and c_3 can be eliminated by means of (6.112) and (6.113).

As a result, we have obtained a complete system of necessary conditions for stationarity. It consists of (6.104), the conditions (6.105) and (6.114) on the nozzle contour AC, the condition (6.107) on the closing characteristic BC, and the conditions (6.112) and (6.113) at the point C.

We now take an arbitrary contour AC which satisfies the restrictions to verify the necessary conditions. Knowing the characteristic AB of the first family and the flow parameters thereon, the flow in the region S may be determined by the method of characteristics.[87]

If $\nu \equiv 0$ along AC (intermediate extremum), then the data obtained enable us to calculate the values of η_1 and ω on AC in accordance with (6.105) and (6.114). With these initial data, and the method of characteristics, the values of η_1 and ω inside the region, and, therefore, along BC, may be calculated. The optimality criterion for the chosen contour will be the identical vanishing of the expression $\eta_1 + \omega y \rho \cot \alpha$ along BC, as dictated by condition (6.107). In Ref. 53, the corresponding process of successive approximation is described.

This process becomes unnecessary if the restriction (6.93) is satisfied by a rectilinear contour AC with slope d or b. In this case, $w_* = 0$, $\nu \neq 0$ (boundary extremum). The function $\nu(x)$ can always be chosen in such a way that (6.107) holds along BC. This choice must, however, be made subject to the additional requirement that T be a maximum with respect to w. We shall not dwell on the derivation of the corresponding condition for $v(x)$. We merely note that in the general case the optimal contour is made up of boundary arcs and of intermediate extrema.

As a rule, the Lagrange multipliers η_1 and ω are determined numerically. Under certain conditions, however, an analytical solution for the multipliers may be obtained. Borisov and Shipilin[15] found a particular solution of (6.104). It has the form

$$\eta_1 = \rho y z^2, \qquad \omega = z^1 + c_3 \qquad (6.115)$$

It is remarkable that, in the absence of restrictions on the shape of the contour AC (i.e., for $c_1 = v = 0$), (6.115) also satisfies the boundary conditions (6.105) and (6.114). Moreover, (6.115) may be transformed into the corresponding expressions (6.105) and (6.114) for any function $y = f(x)$, whatever. In other words, the solution (6.115) of the system (6.104) which satisfies the boundary conditions

$$\eta_1 = \rho f_1 z^2, \qquad \omega = z^1 + c_3$$

on the curve $y = f_1(x)$ satisfies the conditions

$$\eta_1 = \rho f_2 z^2, \qquad \omega = z^1 + c_3$$

on any other curve $y = f_2(x)$.

Boundary conditions having this property are said to be matched (or mutually consistent),[45,94] and the set of matched boundary conditions forms a field with respect to the given system of equations. Thus, the functions (6.115) define a field for the system (6.104).

The system of necessary conditions has now been reduced to (6.107) on the closing characteristic BC and the conditions (6.112) and (6.113) at the point C. Equations (6.115) may now be used to replace (6.107) and the consistency condition

$$d\eta_1 = y\rho \cot \alpha \, d\omega = 0 \qquad \text{on } BC$$

by the equivalent equations

$$y\rho(z^2)^2 \tan \alpha = \text{const}, \qquad z^1 + z^2 \tan \alpha = -c_3 \qquad \text{on } BC$$

These equations were obtained by Shmyglevskii[166] in an investigation of the problem with fixed x_C and $f(x_C)$, when the conditions (6.112) and (6.113) are not imposed. Shmyglevskii used the control contour method; we see that the success of this method is due to the fact that under the given conditions, (6.115) determine a field for (6.104). The existence of a field makes it possible to eliminate the multipliers η_1 and ω from the expression for the first variation of the functional T; subsequently, this variation is represented as a sum of integrals along the generator AC of the nozzle and along the unknown characteristic BC, in complete agreement with the basic idea of the control contour method.

The functions (6.115) form a field whenever the integral on the right-hand side of (6.114) vanishes identically. For $\nu = 0$, this condition reduces to the identical vanishing of the differential Euler expression

$$\frac{dF_w}{dx} - F_f$$

for the function F. The well-known general form of this function is given by

$$F = \frac{dG(x, f)}{dx} \tag{6.116}$$

and the corresponding restriction has the form

$$s = G(x_C, f(x_C)) - G(x_A, f(x_A)) \tag{6.117}$$

i.e., for given x_A and $f(x_A)$ the constraint is imposed at the end point of the nozzle generator.

For a restriction of this type, the control contour method applies; obviously, nothing is changed if there are several such restrictions. Shmyglevskii imposed two restrictions in his paper: in one of them $G = x$, in the other $G = F$.

Borisov[14] constructed a field of Lagrange multipliers for a spatial problem without rotational symmetry.

As indicated above, in the general case, the optimal contour contains sections of interior and boundary extrema and therefore, in general, corners. Guderley and Armitage[160] constructed Weierstrass–Erdmann necessary conditions for these points. At these corners, there is flow around an angle greater than π (Prandtl–Meyer flow[87]); we shall not consider the cases when the angle is less than π and a compressible flow arises.

If the contour around which flow takes place has corners, then discontinuity lines of the Lagrange multipliers occur in the flow, and the gas-dynamic parameters (velocity, pressure, density) remain continuous on these lines. This important fact was established by Kraiko.[79] Indeed, the values of the multipliers on the characteristic de (Fig. 6.11) are determined by the conditions on dc and cf; in the general case, the values at the point d do not satisfy the natural boundary conditions on ad. It is therefore necessary to assume the existence of a line of discontinuity of the multipliers, emanating from d. It is readily seen that the Weierstrass–Erdmann conditions must be satisfied on such a line, and that they are expressed in the form of a system of linear homogeneous equations in the discontinuities of the multipliers because of the assumed continuity of the basic gas-dynamic parameters. It is a feature of hyperbolic problems that the determinant of this system can vanish nontrivially; it occurs on characteristics common to the original system and the system for the multipliers, characteristics which also include streamlines. In the elliptic case, such a situation is impossible; in the example of the problem in Chapter 3 we have seen that the discontinuity of the multipliers is due to discontinuity of the control; if the control is continuous, so are the multipliers.*

By virtue of the linearity and homogeneity of the equations for the discontinuities of the multipliers, the latter are zero or nonzero on all of the characteristics simultaneously.

In Fig. 6.11, the discontinuity line of the multipliers is the characteristic de.

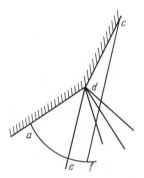

FIGURE 6.11

---

* For the equations of motion of a string, a homogeneous system for the discontinuities of the multipliers is in fact derived in Gyunter's book.[55,p.300]

Of course, if the flow has discontinuity lines of the gas-dynamic parameters z^i (shock waves), then the multipliers are also discontinuous on these lines. To obtain relations between the limiting values of the multipliers on the two sides of the discontinuity, it is necessary to take the relationship between the variations of the limiting values of the basic gas-dynamic parameters into account. Introduction of discontinuities of the multipliers enabled Kraiko and Shipilin to solve a number of interesting problems on optimal supersonic profiles with corners.

To conclude this section, we shall present some results relating to the Legendre condition.

Following Ref. 443, we consider the supersonic gas flow past planar and axisymmetric bodies, assuming that an attached shock wave is formed in the flow, as shown in Fig. 6.12. Here, ac is the body contour, which is to be determined; d is a point of inflection of the contour; ab is the attached shock wave; cb is a characteristic of the second family; and dm and dn are characteristics of the first family bounding the rarefaction wave $dmbnd$. The undisturbed flow is assumed to be uniform and parallel to the x axis.

The flow of the gas in the region of influence abc is described by the equations [see (6.86)–(6.89)]

$$\partial y^\nu p/\partial \psi - \partial u/\partial y = 0, \qquad (\partial/\partial\psi)(u/v) + (\partial/\partial y)(1/y^\nu \rho v) = 0$$

$$\tag{6.118}$$

$$(u^2 + v^2)/2 + \gamma p/(\gamma - 1)\rho = \text{const}, \qquad p/\rho^\gamma = \varphi^{\gamma-1}(\psi)$$

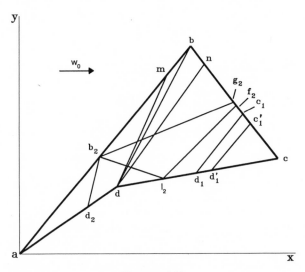

FIGURE 6.12

which are written in terms of the independent variables y and ψ, the latter being the stream function. In (6.118), ν takes the value $\nu = 0$ in the planar case and $\nu = 1$ in the axisymmetric case; $\varphi(\psi)$ is the entropy function.

The isoperimetric condition imposed on the profile of the body has the form

$$R_0 = \int_{y_a}^{y_c} F[y, x(y), x'(y)] \, dy \tag{6.119}$$

where $x(y)$ is the function describing the profile of the body, $x' = dx/dy$, and F is a more or less arbitrary function.

The wave drag is expressed by the functional [see (6.91)]

$$T = \int_{y_a}^{y_c} y^\nu p \, dy$$

The zero-slip condition is satisfied along ab ($\psi = 0$) [see (6.92)]

$$x' - u/v = 0 \tag{6.120}$$

where the position of the point c is assumed fixed.

We pose the following optimization problem: Find the function $x(y)$ which yields the minimum of the functional T, subject to the conditions (6.118) in the region abc, the known (Hugoniot) relations across the shock wave ab, and conditions (6.119) and (6.120) along ac.

For this problem, the necessary conditions of stationarity may be derived in a straightforward manner, along with the expression for the increment of the functional T. With the slope $x'(y)$ of the profile as the control, we take the variation $\delta x'(y)$ to be nonzero over the interval $d_1 d_1'$ in the neighborhood of the point d_1; the increment $\Delta y = \varepsilon$ corresponds to this interval, which is regarded as a small parameter of the problem. We characterize the quantity $\delta x'(y)$ on $d_1 d_1'$ by an additional small parameter ε_1.

Calculation of the perturbed motion generated by this variation in the slope makes it possible to separate the principal part of the increment of the functional in the parameters ε and ε_1. This part is characterized by the equation

$$\Delta T = \Omega(y_0)\varepsilon(\delta x')^2 + 0(\varepsilon^2 \varepsilon_1^2)$$

In the case when $y_0 \in (y_d, y_c)$ (see Fig. 6.12), $\Omega(y_0)$ is given by

$$\Omega(y_0) = \Gamma \left. \frac{\partial^2 F}{\partial x'^2} \right|_{d_1} + \int_{d_1 c_1} (U_+ A U_+) h d\psi - (U_+ B U_+)_{c_1} (hl)_{c_1} \tag{6.121}$$

Here, Γ is the Lagrange multiplier corresponding to the constraint (6.119); the integral is calculated along the characteristic $d_1 c_1$ of the first family that emanates from the point d_1, and the last term is calculated at the point c_1 of intersection of this characteristic with the characteristic cb. The expressions for the scalars h and l, the vector U_+, and the matrices A and B contain the values of the flow variables and the Lagrange multipliers in the optimal range; they are given in Ref. 443. The appearance of the integral term and the term at the point c_1 is a feature of the problem and is related to the fact that the perturbations of the flow variables propagate along a characteristic strip of the first family, whose base is the interval $d_1 d_1'$ on which the control is varied. These perturbations are of the order of ε_1; therefore, they contribute to the principal part of ΔT. The Legendre condition has the form $\Omega(y_0) \geq 0$.

If the function F has the form [see (6.116)]

$$F = \frac{dG(y, x(y))}{dy} = G_y + G_x x'$$

then the condition $\Omega \geq 0$ is transformed into the inequality obtained earlier by a different method[170,450] for problems which admit transition to the control contour.

If the slope is varied on the interval ad of the profile of the body, then the expression for Ω becomes more complicated than (6.121), due to the presence of terms associated with the propagation of perturbations along the characteristics $d_2 b_2$, $b_2 l_2$, $l_2 f_2$ of the first and second families (see Fig. 6.12) due to reflections from the shock wave and the profile of the body, and also along the entropy characteristic (streamline) $b_2 g_2$. The number of such reflections depends on the position of the point d_2 in the neighborhood of which the profile is varied.

6.6. Necessary Conditions for Optimality in Quasi-linear Hyperbolic Systems with Fixed Principal Part

In many applications, we encounter optimization problems for hyperbolic systems of the form

$$z_{xy}^i = f_i(z^1, \ldots, z^n; z_x^1, \ldots, z_x^n; z_y^1, \ldots, z_y^n; u(x, y); x, y) \quad (6.122)$$

Every system of differential equations with the same principal part

$$a z_{xx}^i + b z_{xy}^i + c z_{yy}^i = f_i(z^1, \ldots, z^n, u(x, y); x, y) \quad (i = 1, \ldots, n)$$

where for all of the equations the coefficients a, b, c are the same functions of position satisfying the condition $4ac - b^2 < 0$, may be reduced to the form (6.122).

Optimization problems for systems of the form (6.122) have been investigated by Egorov.[65,67-69] Here, we shall present some of his results, concentrating our attention on estimates of the remainder term in the expression for the increment of the functional. The approach consists of an application of the general derivation scheme for necessary conditions described in Chapter 2.

We shall assume that the function $u(x, y) = (u_1, \ldots, u_r)$ on the right-hand sides of (6.122) is a control. We define the set of admissible controls A as the class of bounded piecewise continuous functions which take on values in some convex region of r-dimensional Euclidean space.

We shall seek solutions $z^i(x, y)$ in a convex region—the rectangle $G(0 \le x \le a, 0 \le y \le b)$, whose sides are characteristics.

We specify initial data on two adjacent sides of the rectangle:

$$z^i(0, y) = \varphi_i(y), \qquad z^i(x, 0) = \psi_i(x) \qquad (i = 1, \ldots, n)$$

(Goursat conditions). Here, φ_i and ψ_i are continuously differentiable functions satisfying the conjugate conditions $\varphi_i(0) = \psi_i(0)$. The right-hand sides f_i of (6.122) are assumed smooth in x and y and twice continuously differentiable with respect to all of the remaining arguments.

For the following discussion, the concept of smoothness of the solutions z corresponding to different admissible controls u needs to be made more precise. In all cases, the solutions z are assumed to be continuous functions.

If the control $u(x, y)$ is continuous, then the solution $z(x, y)$ together with its derivatives z_x, z_y, z_{xy} is continuous. If the control $u(x, y)$ has a discontinuity along a continuously differentiable curve Γ, then the nature of the smoothness of the solution depends on the type of curve. Namely, if Γ is a characteristic, then the derivative of the solution z along the normal to Γ will have a discontinuity of Γ since the Goursat problem (6.122) then decomposes into two such problems in contiguous regions. If Γ is not a characteristic, then a unique extension of the solution through Γ requires additional conjugate conditions. As before, in similar cases (Chapter 3), we shall assume that the solution has continuous first-order derivatives.

Subject to the Goursat conditions, we then pose the following optimization problem for (6.122): to determine $u \in A$ to minimize the functional

$$I = \sum_{i=1}^{n} c_i z^i(a, b) \tag{6.123}$$

Here, the c_i are given constants, and the numbers a and b determine the size of the basic rectangle G.

Many problems that arise in different applications can be reduced to such a scheme; e.g., the study of phenomena in chemical reactors,[18] processes of the adsorption of gases,[18] etc. The solvability of the Goursat problem in this formulation has been studied extensively (see, e.g., Ref. 289).

With the methods set forth in Chapter 2, many other functionals associated with solutions of the problem (6.122)-(6.123) may be reduced to the form (6.123). Examples of such reductions can be found in Ref. 68.

The derivation of necessary conditions for optimality is based on the construction of the expression for the increment of the functional. In this problem, it is convenient to keep the basic equations in the form (6.122). We use the notation introduced in Chapter 2 for quantities in optimal and admissible control ranges and introduce the Lagrange multipliers $\omega_i(x, y)$, we deduce the identity

$$\iint_G \sum_{i=1}^n \omega_i[(Z_{xy}^i - z_{xy}^i) - f_i(Z, Z_x, Z_y, U, x, y)$$

$$+ f_i(z, z_x, z_y, u, x, y)]\, dx\, dy = 0 \qquad (6.124)$$

The following calculations are made in accordance with the usual scheme and require no explanation.

We have

$$\iint_G \omega_i(Z_{xy}^i - z_{xy}^i)\, dx\, dy$$

$$= \iint_G \omega_i \Delta z_{xy}^i\, dx\, dy$$

$$= \int_0^a [\omega_i \Delta z_x^i]_{y=0}^{y=b}\, dx - \iint_G \omega_{iy} \Delta z_x^i\, dx\, dy$$

$$= \int_0^a [\omega_i \Delta z_x^i]_{y=0}^{y=b}\, dx - \int_0^b [\omega_{iy} \Delta z^i]_{x=0}^{x=a}\, dy + \iint_G \omega_{ixy} \Delta z^i\, dx\, dy$$

$$= \omega_i(a, b)\Delta z^i(a, b) - \omega_i(0, b)\Delta z^i(0, b) - \omega_i(a, 0)\Delta z^i(a, 0)$$

$$+ \omega_i(0, 0)\Delta z^i(0, 0)$$

$$- \int_0^a [\omega_{ix} \Delta z^i]_{y=0}^{y=b}\, dx - \int_0^b [\omega_{iy} \Delta z^i]_{x=0}^{x=a}\, dy + \iint_G \omega_{ixy} \Delta z^i\, dx\, dy$$

$$\sum_{i=1}^{n} \iint_{G} \omega_i \Delta f_i \, dx \, dy$$

$$= \sum_{i=1}^{n} \iint_{G} \omega_i [f_i(Z, Z_x, Z_y, U, x, y) - f_i(z, z_x, z_y, u, x, y)] \, dx \, dy$$

$$= \iint_{G} \left[H(p, u + \Delta u, x, y) - H(p, u, x, y) + \sum_{i=1}^{3n} \frac{\partial H(p, u + \Delta u, x, y)}{\partial p^i} \Delta p^i \right.$$

$$\left. + \frac{1}{2} \sum_{i,k=1}^{3n} \frac{\partial^2 H(p + \theta \Delta p, u + \Delta u, x, y)}{\partial p^i \partial p^k} \Delta p^i \Delta p^k \right] dx \, dy$$

$$0 < \theta < 1$$

Here we have introduced the notation

$$p = (p^1, \dots, p^{3n}) = (z^1, \dots, z^n, z_x^1, \dots, z_x^n, z_y^1, \dots, z_y^n)$$

$$H(p, u, x, y) = \sum_{i=1}^{n} \omega_i f_i(z, z_x, z_y, u, x, y)$$

Integrating by parts, we find

$$\sum_{i=n+1}^{3n} \iint_{G} \frac{\partial H(p, u, x, y)}{\partial p_i} \Delta p^i \, dx \, dy$$

$$= \sum_{i=1}^{n} \int_{0}^{b} [H_{z_x^i} \Delta z^i]_{x=0}^{x=a} \, dy + \sum_{i=1}^{n} \int_{0}^{a} [H_{z_y^i} \Delta z^i]_{y=0}^{y=b} \, dx$$

$$- \sum_{i=1}^{n} \iint_{G} \frac{\partial H_{z_x^i}}{\partial x} \Delta z^i \, dx \, dy - \sum_{i=1}^{n} \iint_{G} \frac{\partial H_{z_y^i}}{\partial y} \Delta z^i \, dx \, dy$$

Collecting our results, we reduce the identity (6.124) to the form

$$\sum_{i=1}^{n} [\omega_i(a, b) \Delta z^i(a, b) - \omega_i(0, b) \Delta z^i(0, b) - \omega_i(a, 0) \Delta z^i(a, 0)$$

$$+ \omega_i(0, 0) \Delta z^i(0, 0)]$$

$$= \sum_{i=1}^{n} \int_0^a [\omega_{ix}\Delta z^i]_{y=0}^{y=b}\, dx + \sum_{i=1}^{n} \int_0^b [\omega_{iy}\Delta z^i]_{x=0}^{x=a}\, dy$$

$$+ \iint_G [H(p, u + \Delta u, x, y) - H(p, u, x, y)]\, dx\, dy$$

$$- \sum_{i=1}^{n} \left[\omega_{ixy} - \frac{\partial H}{\partial z^i} + \frac{\partial}{\partial x}\frac{\partial H}{\partial z_x^i} + \frac{\partial}{\partial y}\frac{\partial H}{\partial z_y^i}\right]\Delta z^i\, dx\, dy$$

$$+ \sum_{i=1}^{n} \int_0^a [H_{z_y^i}\Delta z^i]_{y=0}^{y=b}\, dx + \sum_{i=1}^{n} \int_0^b [H_{z_x^i}\Delta z^i]_{x=0}^{x=a}\, dy$$

$$+ \sum_{i=1}^{3n} \iint_G \left[\frac{\partial H(p, u + \Delta u, x, y)}{\partial p^i} - \frac{\partial H(p, u, x, y)}{\partial p^i}\right]\Delta p^i\, dx\, dy$$

$$+ \frac{1}{2}\sum_{i,k=1}^{3n} \iint_G \frac{\partial^2 H(p + \theta\Delta p, u + \Delta u, x, y)}{\partial p^i\, \partial p^k}\Delta p^i \Delta p^k\, dx\, dy$$

If we determine the functions $\omega_i(x, y)$ by means of the equations

$$\omega_{ixy} = \frac{\partial H}{\partial z^i} - \frac{\partial}{\partial x}\frac{\partial H}{\partial z_x^i} - \frac{\partial}{\partial y}\frac{\partial H}{\partial z_y^i} \tag{6.125}$$

and the boundary conditions

$$\left.\begin{aligned}
\omega_{ix}(x, b) &= -\frac{\partial H(p, u, x, b)}{\partial z_y^i} \\[2mm]
\omega_{iy}(a, y) &= -\frac{\partial H(p, u, a, y)}{\partial z_x^i} \\[2mm]
\omega_i(a, b) &= -c_i
\end{aligned}\right\} \tag{6.126}$$

then, taking into account the equations

$$\Delta z^i(x, 0) = \Delta z^i(y, 0) = 0, \qquad 0 \le x \le a, \qquad 0 \le y \le b$$

we obtain the following expression for the increment of the functional I:

$$\Delta I = -\iint_G [H(p, u + \Delta u, x, y) - H(p, u, x, y)]\, dx\, dy - \eta_1 - \eta_2 \tag{6.127}$$

where

$$\eta_1 = \sum_{i=1}^{3n} \int\int_G \left[\frac{\partial H(p, u + \Delta u, x, y)}{\partial p^i} - \frac{\partial H(p, u, x, y)}{\partial p^i} \right] \Delta p^i \, dx \, dy$$

$$\eta_2 = \frac{1}{2} \sum_{i,k=1}^{3n} \int\int_G \frac{\partial^2 H(p + \theta \Delta p, u + \Delta u, x, y)}{\partial p^i \, \partial p^k} \Delta p^i \Delta p^k \, dx \, dy$$

(6.128)

Equation (6.127) is exact. Our aim now is to show that under certain conditions imposed on the admissible increment Δu of the control, the first integral term on the right-hand side in (6.127) is the principal part of the increment of the functional, so that the sign of ΔI will be determined by the sign of this term. To formulate a precise assertion in line with Egorov,[67-69] we introduce a definition, which allows the formulation of a precise assertion.

We shall say that an admissible control $u(x, y)$ satisfies a maximum condition if

$$H(\omega, p, u, x, y) = \max_{U \in A} H(\omega, p, U, x, y) \qquad (6.129)$$

almost everywhere in the region G.

Theorem (Egorov[67-69]).   If the admissible control $u(x, y)$ is to minimize the functional I, it is necessary that it satisfy the maximum condition.

Before we turn to the proof, let us make some remarks concerning the theorem.

If it is to be used in practice, we need to know the $2n + 1$ functions z^i, ω_i, u. These unknowns are obtained from the $2n + 1$ equations (6.122), (6.125), and (6.129). Equations (6.122) and (6.125) are introduced during the integration of $4n$ arbitrary functions for whose determination we have the $4n$ conditions (6.123) and (6.126). Thus, we generally distinguish isolated solutions of the problem (6.122)–(6.123). If an optimal solution exists, it will be included among these distinguished solutions.

Note that the expression in the integrand in (6.127) plays the same role as the expression in the first integral term in the general formula (2.172). It is an important point that the function $H(p, u + \Delta u, x, y)$ only contains increments of the control functions and does not contain increments of any other quantities which could play the role of parametric variables when the basic equations are expressed in normal form. If the first-order derivatives of z^i are continuous, then one could generally take discontinuous second derivatives z^i_{xx}, z^i_{yy}, as parametric variables setting $z^i_{xx} = \zeta^i$, $z^i_{yy} = \zeta^{i+n}$. However, the variables ζ would then occur linearly in the Hamilton function, and their coefficients would vanish by virtue of the stationarity conditions. As a result, the difference $H(z, \zeta + \Delta\zeta, u + \Delta u, x, y) - H(z, \zeta, u, x, y)$, calculated for the equations in normal form in accordance with the rules of Chapter 2, would coincide with the expression in the integrand in (6.127), which is what we wanted to prove.

Our argument is based on the assumption of smoothness of the z^i; omission of this assumption invalidates the argument.* Nevertheless, the expression (6.127) for the increment of the functional still retains its value. The point is that if the controls are varied in regions of small diameter the terms η_1 and η_2 are small or of higher order than the integral in (6.127), so that this integral is the principal part of the increment ΔI in all cases, and the representation of this increment in the form (6.127) corresponds to the particular features of the investigated problem.

We now turn to the proof of the theorem. We introduce the auxiliary functions $\alpha_i = \Delta z^i_x$, $\beta_i = \Delta z^i_y$ and set

$$\alpha = \sum_{i=1}^n |\alpha_i|, \qquad \beta = \sum_{i=1}^n |\beta_i|, \qquad \Delta z = \sum_{i=1}^n |\Delta z^i|, \qquad \Delta u = \sum_{i=1}^r |\Delta u_i| \quad (6.130)$$

Integrating the obvious equations

$$\Delta z^i_{xy} = \Delta \frac{\partial H}{\partial \omega_i}$$

under the additional conditions

$$\Delta z^i(0, y) = \Delta z^i(x, 0) = 0$$

and using the fact that the functions f_i satisfy a Lipschitz condition, we obtain

$$\left.\begin{array}{l} |\alpha_i| \leq N \displaystyle\int_0^y \sum_{i=1}^n (|\Delta z^i| + |\beta_i| + |\alpha_i|)\, dy + N_1 \int_0^y \sum_{i=1}^r |\Delta u_i|\, dy, \\[4mm] |\Delta z^i| \leq \displaystyle\int_0^x |\alpha_i|\, dx \\[4mm] |\beta_i| \leq N \displaystyle\int_0^x \sum_{i=1}^n (|\Delta z^i| + |\alpha_i| + |\beta_i|)\, dx + N_1 \int_0^x \sum_{i=1}^r |\Delta u_i|\, dx, \\[4mm] |\Delta z^i| \leq \displaystyle\int_0^y |\beta_i|\, dy \end{array}\right\} \quad (6.131)$$

---

* As mentioned above, discontinuity of the derivatives along characteristics is possible.

Here, N and N_1 are constants. With the notation (6.130), it then follows that

$$\alpha(x, y) \leq Nn \int_0^y \alpha(x, \eta) \, d\eta + Nn \int_0^Y [\Delta z(x, \eta) + \beta(x, \eta)] \, d\eta$$

$$+ N_1 n \int_0^b \Delta u(x, \eta) \, d\eta \qquad \Delta z \leq \int_0^x \alpha(\xi, y) \, d\xi$$

$$\beta(x, y) \leq Nn \int_0^x \beta(\xi, y) \, d\xi + Nn \int_0^X [\Delta z(\xi, y) + \alpha(\xi, y)] \, d\xi$$

$$+ N_1 n \int_0^a \Delta u(\xi, y) \, d\xi \qquad \Delta z \leq \int_0^y \beta(x, \eta) \, d\eta$$

where $0 \leq x \leq X$, $0 \leq y \leq Y$; (X, Y) is a fixed point of the region G. Using the well-known lemma,[115,p.19] we arrive at the estimates

$$\alpha(x, y) \leq M \int_0^Y [\Delta z(x, \eta) + \beta(x, \eta)] \, d\eta + M_1 \int_0^b \Delta u(x, \eta) \, d\eta$$

$$\beta(x, y) \leq P \int_0^X [\Delta z(\xi, y) + \alpha(\xi, y)] \, d\xi + P_1 \int_0^a \Delta u(\xi, y) \, d\xi$$

where M, P, M_1, P_1 are positive constants.

When the inequalities for Δz are taken into account, we obtain

$$\alpha(x, y) \leq M_2 \int_0^Y \beta(x, \eta) \, d\eta + M_1 \int_0^b \Delta u(x, \eta) \, d\eta$$

$$\beta(x, y) \leq P_2 \int_0^X \alpha(\xi, y) \, d\xi + P_1 \int_0^a \Delta u(\xi, y) \, d\xi$$

From this we obtain the inequalities

$$\alpha(X, Y) \leq M_3 \int_0^Y \int_0^X \alpha(x, y) \, dx \, dy + M_4 \iint_G \Delta u(x, y) \, dx \, dy$$

$$+ M_1 \int_0^b \Delta u(X, y) \, dy$$

$$\beta(X, Y) \le P_3 \int_0^Y \int_0^X \beta(x, y) \, dx \, dy + P_4 \int\int_G \Delta u(x, y) \, dx \, dy$$

$$+ P_1 \int_0^a \Delta u(x, Y) \, dx$$

The integration of the first inequality with respect to X, from 0 to X, and another application of the lemma yield

$$\int_0^X \alpha(X, Y) \, dX \le M_5 \int\int_G \Delta u(x, y) \, dx \, dy$$

Together with the preceding inequalities, this results in

$$\alpha(X, Y) \le Q_1 \int\int_G \Delta u(x, y) \, dx \, dy + S_1 \int_0^b \Delta u(X, y) \, dy$$

and similarly

$$\beta(X, Y) \le Q_2 \int\int_G \Delta u(x, y) \, dx \, dy + S_2 \int_0^a \Delta u(x, Y) \, dx$$

The above estimates for Δz now show that

$$\Delta z(X, Y) \le Q \int\int_G \Delta u(x, y) \, dx \, dy$$

In terms of the notation (6.130), we finally obtain

$$|\Delta z^i(x, y)| \le Q \int_G \sum_{i=1}^r |\Delta u_i(x, y)| \, dx \, dy$$

$$|\Delta z_x^i(x, y)| \le Q_1 \int\int_G \sum_{i=1}^r |\Delta u_i(x, y)| \, dx \, dy + S_1 \int_0^b \sum_{i=1}^r |\Delta u_i(x, y)| \, dy$$

$$|\Delta z_y^i(x, y)| \le Q_2 \int\int_G \sum_{i=1}^r |\Delta u_i(x, y)| \, dx \, dy + S_2 \int_0^a \sum_{i=1}^r |\Delta u_i(x, y)| \, dx$$

It now is easy to estimate the terms η_1 and η_2 in (6.127). Using the fact that $\partial H/\partial p^i$ satisfies a Lipschitz condition, we find

$$|\eta_1| = T\left[\int\int_G \sum_{i=1}^r |\Delta u_i(x, y)|\, dx\, dy\right]^2$$

$$+ T_1 \int_0^a \left[\int_0^b \sum_{i=1}^r |\Delta u_i(x, y)|\, dy\right]^2 dx$$

$$+ T_2 \int_0^b \left[\int_0^a \sum_{i=1}^r |\Delta u_i(x, y)|\, dx\right]^2 dy$$

The functions $\partial^2 H/\partial p^i \partial p^k$ are bounded in G, and therefore

$$|\eta_2| \leq N_1 \int\int_G \sum_{i,j=1}^{3n} |\Delta p^i \, \Delta p^k|\, dx\, dy \tag{6.132}$$

The latter integral allows for an estimate similar to that for $|\eta_1|$.

If the increment $\Delta u_i(x, y)$ is nonzero in a disk G_ε of radius ε, then

$$|\eta| \leq \varepsilon L \int\int_{G_\varepsilon} \left[\sum_{i=1}^r |\Delta u_i(x, y)|\right]^2 dx\, dy \tag{6.133}$$

The proof of the theorem now causes no further difficulties. Suppose the theorem is false. Then there exists a point (ξ, η) at which the maximum condition (6.129) is not satisfied, i.e., for some control $U^1 \in A$ at the point (ξ, η) the opposite inequality is satisfied:

$$H(\omega(\xi, \eta), z(\xi, \eta), z_x(\xi, \eta), z_y(\xi, \eta); U^1; \xi, \eta)$$

$$\geq H(\omega(\xi, \eta), z(\xi, \eta), z_x(\xi, \eta), z_y(\xi, \eta); u(\xi, \eta); \xi, \eta) \tag{6.134}$$

Since $z(x, y)$ and $\omega(x, y)$ are continuous, and z_x, z_y, and $u(x, y)$ are piecewise continuous, there exists a closed region G^1 containing the point (ξ, η) in which both sides of the inequality (6.134) are continuous and, hence, uniformly continuous. If (ξ, η) is a discontinuity point of the control $u(x, y)$, we set $u(\xi, \eta) = u(\xi + 0, \eta + 0)$, $z_x(\xi, \eta) = z_x(\xi + 0, \eta + 0)$, $z_y(\xi, \eta) = z_y(\xi + 0, \eta + 0)$, and we take the point (ξ, η) as a boundary point of G^1.

It follows from what we have said that there exists $\delta > 0$ such that

$$H(\omega(x, y), z(x, y), z_x(x, y), z_y(x, y); U^1; x, y)$$

$$- H(\omega(x, y), z(x, y), z_x(x, y), z_y(x, y); u(x, y); x, y) > \delta \tag{6.135}$$

for all $(x, y) \in G_\varepsilon \subset \dot{G}^1$, where G_ε is a disk of radius ε contained in G^1. Consider the admissible control

$$U(x, y) = \begin{cases} u(x, y), & (x, y) \notin G_\varepsilon \\ U^1, & (x, y) \in G_\varepsilon \end{cases}$$

Equations (6.127), (6.133), and (6.135) show that

$$\Delta I = -\iint_{G_\varepsilon} [H(p; U^1; x, y) - H(p; u; x, y)] \, dx \, dy - \eta$$

$$\leq -\left[\iint_{G_\varepsilon} \delta \, dx \, dy - |\eta| \right] \leq -\iint_{G_\varepsilon} \left[\delta - \varepsilon L \left(\sum_{i=1}^{r} |\Delta u_i(x, y)| \right)^2 \right] dx \, dy$$

The increments Δu_i are bounded, and therefore the number ε can be made sufficiently small to assure that the expression in the square brackets is positive. However, then $\Delta I < 0$, which contradicts the optimality of the control $u(x, y)$. It is readily noted that the decisive element of the proof is the estimate (6.133), which shows that the term $\eta = \eta_1 + \eta_2$ in (6.127) is a quantity of higher order in ε for variations within the disk G_ε, whereas the integral in (6.127) is of first order for such variations.

If the functions f_i have the form

$$f_i = \sum_{k=1}^{n} [c_{ik}(x, y)z_x^i + d_{ik}(x, y)z_y^i + e_{ik}(x, y)z^i] + g_i(u) \tag{6.136}$$

then the maximum condition (6.129) is also sufficient for optimality of the control u. The proof follows directly if one notes that the terms η_1 and η_2 in (6.128) vanish identically for the functions (6.136).

In Refs. 68 and 69, Egorov obtained necessary conditions for optimality of controlled systems whose processes are described by (6.122) subject to additional conditions which differ from those of (6.123). In particular, in Ref. 69 he studied cases where relations between the values of z^i and its derivatives are specified along the boundaries of the basic rectangle G. These relations also contained boundary controls. The corresponding system of necessary conditions was obtained in Ref. 69 subject to certain additional assumptions; a more general case was considered by Kuz'mina.[85]

A detailed study of a number of optimization problems for a second-order hyperbolic equation allowing for possible discontinuities of the Lagrange multipliers was made by Petukhov and Troitskii.[122] In Ref. 123, they also applied the general method to the problem of optimal loading of longitudinally vibrating rod.

7
Parabolic and Other Evolution
Optimization Problems

Many applications involve the optimal control of processes governed by parabolic partial differential equations. Above all, there are various problems involving the propagation of temperature or concentration fields. There are the practically important problems of the melting or solidification of bodies, crystal pulling directly from the melt (e.g., Stepanov's method[159]), and other problems with moving boundaries. All provide nontrivial examples of controlled distributed parameter systems.

Several aspects of the formulation of this class of optimization problems are illustrated in the first sections of this chapter. More general results involving extensions of the maximum principle are given in the following sections.

7.1. Optimal Heating of Bodies: Initial and Boundary Value Control

We consider the problem of heating a layer of unit thickness by changing the temperature of the surrounding medium. The temperature distribution in the layer is described by the heat conduction equation

$$z_t - z_{xx} = 0 \tag{7.1}$$

with boundary conditions

$$z_x(t, 0) = 0, \qquad z_x(t, 1) = h[u(t) - z(t,)] \tag{7.2}$$

and initial condition

$$z(0, x) = 0 \tag{7.3}$$

The constant $h > 0$ is the coefficient of heat exchange. The function $u(t)$, i.e., the temperature of the medium that adjoins the layer along the line $x = 1$, will be regarded as a measurable control whose modulus does not exceed unity.

The optimization problem consists of choosing this function in such a way as to minimize the functional

$$I[u] = \int_0^1 [z(T, x) - z_0(x)]^2 \, dx \qquad (7.4)$$

which expresses the mean square deviation of the temperature at the time $t = T$ from the given function $z_0(x) \in L_2(0, 1)$.

This problem has been studied by many authors; the most important results are due to Butkovskii[18] and Egorov,[74] whose investigations provide the basic material for the present section.

It is obvious that zero is the greatest lower bound of the values of the functional $I[u]$. If function $z_0(x)$ is given and there exists an admissible control $u_0(t)$ such that the solution $z(t, x)$ which it generates coincides with the function $z_0(x)$ at the time $t = T$, then the lower bound is attained: $I[u_0] = 0$. Of course, not all functions $z_0(x) \in L_2(0, 1)$ have this property. Egorov showed that, using admissible controls, one can either "hit" zero exactly or a small neighborhood of zero; later, Gal'chuk showed that any constant of modulus less than unity is attainable[42]; a constant with absolute value unity cannot be attained by means of any admissible control for finite T.

Egorov's result is obtained by the following argument. There exists an infinitely differentiable function $f(t)$ which is identically zero for $t \leq 0$ and $t \geq 1$ and such that

$$f(t) \geq 0, \qquad f(t) \neq 0, \qquad |f^n(t)| \leq A^n \Gamma(3n/2) \qquad (7.5)$$

It is readily verified that the function

$$z_*(t, x) = \gamma \int_0^t f(\xi) \, d\xi + \gamma \sum_{n-1}^{\infty} \frac{f^{(n-1)}(t) x^{2n}}{(2n)!} \qquad (7.6)$$

is a solution of the problem (7.1)–(7.4) if

$$u(t) = \gamma \int_0^t f\left(\frac{\xi}{T}\right) d\xi + \gamma \sum_{n=1}^{\infty} \frac{d^{n-1}}{dt^{n-1}} f\left(\frac{t}{T}\right) \left[\frac{1}{h(2n-1)!} + \frac{1}{(2n)!} \right] \qquad (7.7)$$

By an appropriate choice of the constant γ, one can achieve $|u(t)| < 1$. obviously,

$$z_*(T, x) = \gamma \int_0^T f\left(\frac{\xi}{T}\right) d\xi = \text{const} = u(T) \qquad (7.8)$$

and, therefore, $|z_*(T, x)| < 1$.

Suppose we are given a function $z_0(x)$ (attainable or not). The existence of an optimal control $u(t)$ can be proved comparatively easily[74] by considering a sequence of infinitely differentiable admissible controls $u_n(t)$ each of which vanishes in a corresponding neighborhood of the point $t = 0$ and satisfies $I[u_n] \to \inf I[u]$, where the lower bound is taken over all admissible controls. From the sequence $u_n(t), n = 1, 2, \ldots$, one can choose a subsequence $u_{n_k}(t)$ which converges to $u(t), |u(t)| \leq 1$, weakly in $L_2(0, T)$. The corresponding solutions $z_{n_k}(t, x)$ converge to the function $z(t, x)$, which, as one can show, satisfies (7.1)-(7.3) if the function $u(t)$ is taken as control.

Uniqueness of an optimal control is much harder to prove. First, it is clear that if there exists a control $u_0(t)$ such that the temperature at all points of the layer is zero $[z(T, x) = 0]$ at time T, then the control

$$u_1(t) = u(t) + \gamma u_0(t) \tag{7.9}$$

will lead to the same values of the functional as $u(t)$. Let $u(t)$ be an optimal control; suppose that we have $|u(t)| < 1 - \delta, \delta > 0$ for $t \in [t_1, t_2], 0 \leq t_1 < t_2 < T$. Then any control (7.9) corresponding to a γ sufficiently small to assure $|u_1(t)| \leq 1$ will be optimal. In other words, if the optimal control $u(t)$ is to be unique, the equation $|u(t)| = 1$ must be satisfied almost everywhere on $[0, T]$.

A control $u_0(t)$ having these properties does indeed exist. It is given by

$$u_0(t) = \sum_{n=0}^{\infty} \frac{d^n}{dt^n} f\left(\frac{t - t_1}{t_2 - t_1}\right) \left[\frac{1}{h(2n - 1)!} + \frac{1}{(2n)!}\right] \tag{7.10}$$

where $f(t)$ is the function defined above [see (7.5)]. The corresponding solution

$$z_0(t, x) = \sum_{n=0}^{\infty} \frac{d^n}{dt^n} f\left(\frac{t - t_1}{t_2 - t_1}\right) \frac{x^{2n}}{(2n)!} \tag{7.11}$$

vanishes for $t = T$.

If the control $u(t)$ is such that the corresponding solution $z(t, x)$ vanishes for $u(t)$, then the function $u(t)$ has a countable number of sign changes on $[0, T]$, and the point $t = T$ is an accumulation point for the switching times.* This is to be expected since a zero temperature can be reached only through alternate intervals of cooling and heating.

---

* It is assumed that $u(t)$ is not identically zero in any interval adjoining the point $t = T$. Otherwise, the previous statements concerning the point $t = T$ apply to the left endpoint of such an interval.

The proof is simple. If λ_k are the positive roots of the equation $\lambda \tan \lambda = h$, then the condition $z(T, x) = 0$ clearly is equivalent to the infinite system of equations (moment problem)

$$\int_0^T u_0(t) \, e^{\lambda_k^2(t-T)} \, dt = 0 \qquad (k = 1, 2, \ldots) \tag{7.12}$$

[the system $\{e^{\lambda_k^2(t-T)}\}$ is not complete on any interval since $\lambda_n = n\pi + h/n\pi + o(1/n)$ and the series $\Sigma_{n=1}^\infty 1/\lambda_n^2$ converges; by virtue of this, problem (7.12) can have a nontrivial solution].

If there were to exist $\rho > 0$ such that the function $u_0(t)$ did not change sign for $t \in (T - \rho, T]$, then the equation

$$-\int_0^{T-\rho} u_0(t) \, e^{\lambda_k^2(t-T)} \, dt = \int_{T-\rho}^T u_0(t) \, e^{\lambda_k^2(t-T)} \, dt$$

should hold for all k. The modulus of the left-hand side is bounded above by the expression $(1/\lambda_k^2)e^{-\rho\lambda_k^2}$, and the modulus of the right-hand side is not less than

$$e^{-\rho\lambda_k^2} \int_{T-\rho}^T |u(t)| \, dt$$

and equality cannot be realized for sufficiently large k. This result applies, in particular, to the control $u_0(t)$.

However, if the given optimal control $u(t)$ satisfies the condition $|u(t)| = 1$, then there exist other optimal controls, which do not satisfy this condition.

Indeed, if we set

$$\tilde{u}(t) = \begin{cases} \text{sign } u_0(t) & u_0(t) \neq 0 \\ 1 & u_0(t) = 0 \end{cases}$$

and take $z_0(x)$ in (7.4) equal to $\tilde{z}(T, x)$, where $\tilde{z}(t, x)$ is a solution of the problem (7.1)–(7.3) with control $\tilde{u}(t)$, then the control $\tilde{u}(t) - \gamma u_0(t), 0 < \gamma < 1$, will be optimal along with the control $\tilde{u}(t)$. The absolute value of this new optimal control is not equal to unity.

Generally, if the optimal control $u(t)$ is such that $|u(t)| = 1$, then it will be nonunique if and only if there exists a control $\tilde{u}(t) \neq 0$ such that $u(t)\tilde{u}(t) \geq 0$ and the solution $\tilde{z}(t, x)$ corresponding to $\tilde{u}(t)$ vanishes for $t = T$. At the same time, it is obvious that if $u(t)$ is optimal, then so is the control $u(t) - \gamma\tilde{u}(t)$ for $\gamma > 0$.

Conversely, suppose $u_1(t)$ and $u_2(t)$ are optimal controls, i.e., $I[u_1] = I[u_2]$; then the corresponding solutions $z_1(t, x)$ and $z_2(t, x)$ are equal at $t = T$: $z_1(T, x) = z_2(T, x)$. Assume not. Then the equation

$$I\left[\frac{u_1 + u_2}{2}\right] = \tfrac{1}{2}I[u_1] + \tfrac{1}{2}I[u_2] - \tfrac{1}{4}\int_0^1 [z_1(T, x) - z_2(T, x)]^2 \, dx$$

implies

$$I\left[\frac{u_1 + u_2}{2}\right] < I[u_1] = I[u_2]$$

contradicting the optimality of u_1 and u_2. Therefore, the control $u_3(t) = (1/2)[u_1(t) - u_2(t)]$ generates a solution $z(t, x)$ that is equal to zero for $t = T$. If $|u_1(t)| = 1$, then the control $u_3(t)$ can be taken as $\tilde{u}(t)$ since $u_1(t)u_3(t) \geq 0$.

It is now clear that if the optimal control satisfies $|u(t)| \equiv 1$ and is piecewise constant, it is unique; for otherwise there would exist a function $\tilde{u}(t)$ with the properties listed above. By virtue of the inequality $u(t)\tilde{u}(t) \geq 0$ this function would have only a finite number of sign changes, and therefore could not satisfy (7.12) for sufficiently large k.

Butkovskii[18] showed that the absolute value of the optimal control $u(t)$ is equal to unity for $z_0(x) \equiv$ const and inf $I[u] > 0$.* This control is unique; for if $u_1(t)$ and $u_2(t)$ are two such optimal controls, then the control $(1/2)[u_1(t) + u_2(t)]$ is optimal. However, if $|u_1(t)| = |u_2(t)| = (1/2)|u_1(t) + u_2(t)| = 1$, then $u_1(t) = u_2(t)$. What we have said holds whenever the condition $|u(t)| = 1$ is necessary for optimality of the control $u(t)$. In other words, $u(t)$ fails to be unique only in those cases where this

---

* An integral representation of the solution is given by

$$z(t, x) = \int_0^t K(x, t, \tau)u(\tau) \, d\tau$$

where $K(x, t, \tau)$ is the Green's function for the system. Based on this integral representation the corresponding variation of the functional (7.4) is given by

$$\Delta I = 2\int_0^T [U(\tau) - u(\tau)]|z(T, x) - z_0(x)|K(x, T, \tau) \, \text{sign} \, [z(T, x) - z_0(x)] \, dx \, d\tau$$

For a minimum of $I[u]$, the integrand must be nonnegative.

If inf $I[u] > 0$, then the equation $z(T, x) \equiv z_0(x)$ is impossible for any admissible control; with $K(x, T, \tau) \neq 0$, the necessary result follows.

control takes on values inside the interval $[-1, 1]$ (i.e., the control is singular according to Rozonoér's definition[144]). As Friedman showed in Ref. 162 (see also Ref. 252), this last situation can occur only subject to inf $I(u) = 0$.

If we minimize the functional

$$I_\gamma[u] = \int_0^1 [z(T, x) - z_0(x)]^2 \, dx + \gamma \int_0^T u^2(t) \, dt, \qquad \gamma > 0 \quad (7.13)$$

instead of the functional (7.4), then an optimal control is unique. Assuming otherwise, we would obtain the equality

$$I_\gamma[u_1] = I_\gamma[u_2] = \inf_{|u| \le 1} I_\gamma[u]$$

for two controls $u_1(t) \ne u_2(t)$. At the same time, it would follow from

$$I_\gamma[u_1] + I_\gamma[u_2] = 2I_\gamma\left[\frac{u_1 + u_2}{2}\right] + \frac{1}{2} \int_0^1 [z_1(T, x) - z_2(T, x)]^2 \, dx$$

$$+ \frac{\gamma}{2} \int_0^T [u_1(t) - u_2(t)]^2 \, dt$$

that

$$I_\gamma\left[\frac{u_1 + u_2}{2}\right] < I_\gamma[u_1]$$

contradicting the optimality of $u_1(t)$.

We now assume that the control occurs in the initial conditions of the problem. Suppose we consider (7.1) in the region $0 \le x \le 1, 0 \le t \le T$ subject to the boundary conditions

$$z_x(t, 0) = \beta[z(t, 0) - f_0(t)], \qquad z_x(t, 1) = \alpha[f_1(t) - z(t, 1)] \qquad \alpha, \beta > 0$$

and with initial condition $z(0, x) = u(x)$. The problem consists of choosing a measurable function $u(x)$ which does not exceed unity in absolute value and is such that the functional (7.4) attains a minimum. In contrast to the case where the control occurs in the boundary condition, the solution of this problem is always unique.[74] Indeed, as before, we find that if $u_1(x)$ and $u_2(x)$ are two different controls, and $z_1(t, x)$ and $z_2(t, x)$ are the corresponding solutions, then the function $z(t, x) = z_1(t, x) - z_2(t, x)$ satisfies (7.1) and the boundary conditions

$$z_x(t, 0) = \beta z(t, 0), \qquad z_x(t, 1) = -az(t, 1), \qquad z(T, x) = 0$$

In accordance with Ref. 294, these conditions imply $z(t, x) \equiv 0$ for $t \leq T$ so that $z(0, x) = u_1(x) - u_2(x) = 0$.

From a practical point of view, there is a great interest in problems where a given temperature distribution must be realized with a specified accuracy in the shortest possible time in a given layer.

Suppose, for example, that for a given function $z_0(x)$ and constant δ we want to determine a control $|u(t)| \leq 1$ for which the inequality

$$\int_0^1 [z(T, x) - z_0(x)]^2 \, dx \leq \delta^2 \tag{7.14}$$

is satisfied in the shortest possible time T. The existence of an optimal control is proved in virtually the same manner as above.

Note that condition (7.14) must be satisfied as an equality at $t = T$; otherwise, this time would not be optimal in view of the continuity of the function

$$f(t) = \int_0^1 [z(t, x) - z_0(x)]^2 \, dx$$

The set (7.14) is a ball in $L_2(0, 1)$. On the other hand, the set of functions $z(T, x)$ generated by the admissible controls $u(t)$ is obviously convex; from what has been shown, it follows that the set has a point in common with the ball (7.14). There can only be one such point; for supposing otherwise, we should have the equation

$$\|z_1(T, x) - z_0(x)\|_{L_2(0,1)} = \delta, \qquad \|z_2(T, x) - z_0(x)\|_{L_2(0,1)} = \delta$$

and therefore

$$\|\tfrac{1}{2}[z_1(T, x) + z_2(T, x)] - z_0(x)\|_{L_2(0,1)} < \delta$$

Thus, using the control $(1/2)[u_1(t) + u_2(t)]$ we could satisfy the inequality (7.14) in a time less than T.

By virtue of the convexity of the set of attainable $z(T, x)$, there exists a plane separating this set and the ball (7.14) in the space $L_2(0, 1)$ and containing the unique point they have in common.[17] It is readily seen that this requirement is equivalent to the inequality

$$\int_0^T (u - U)G(\xi) \, d\xi \geq 0$$

where $G(\xi)$ is some analytic function and u and U are, respectively, an optimal and an admissible control. The optimal control is given by

$$u(t) = \text{sign } G(t)$$

i.e., $|u(t)| = 1$ almost everywhere. This property of the optimal control then guarantees its uniqueness (the proof is analogous to the proof given in the paragraph prior to the last footnote).

7.2. The Case When the Solution at a Finite Time is Given

In problem (7.1)–(7.4) it was noted that the optimal control can be singular only if $\inf I[u] = 0$. This condition is equivalent to the solution $z(t, x)$ "hitting" the given function $z_0(x)$ exactly at the time T. In general, optimal control was not uniquely determined. This circumstance shows that for functions $z_0(x)$ for which exact "hitting" is possible when $|u(t)| \leq 1$ the problem (7.1)–(7.4) is not strictly correct in the sense that it admits a certain freedom of control. Clearly, this is due to the fact that the problem is no longer an optimization problem for the functions $z_0(x)$ of this class; rather, it has been transformed into a problem for determining any admissible control that realizes exact hitting.* At the same time, it is obvious that the problem may again be transformed into an optimization problem by imposing some optimization requirement in addition to the condition of exact "hitting." In this context, the requirement that the time T be minimized has the greatest practical interest.

We arrive at the following time optimization problem. Determine a measurable control $u(t)$ whose absolute value does not exceed unity and for which the solution of the problem (7.1)–(7.3) satisfies the condition

$$z(T, x) = z_0(x)$$

in the shortest time T. Of course, it is assumed that the function $z_0(x)$ belongs to the class of attainable functions for $T < \infty$.

For the case $z_0(x) \equiv \text{const} = v, |v| < 1$, Gal'chuk[42] has shown that this problem has a solution. We give his proof. Let $\{T\}$ and $\{u_T(t)\}$ be the sets of times and corresponding controls for which the problem is solvable. We set $T^* = \inf\{T\}$. If $T^* \in \{T\}$, then the corresponding control $u_{T^*}(t)$ is

---

* Rozonoer calls such problems degenerate (Ref. 144, Part II, p. 1443).

optimal. Suppose $T^* \notin \{T\}$. The control $u_T(t)$ satisfies the system of equations [see (7.12)]

$$\int_0^T u_T(t) \, e^{\lambda_k^2(t-T)} \, dt = \frac{v}{\lambda_k^2} \qquad (k = 1, 2, \dots)$$

or

$$\int_0^T \chi_T(t) u_T(t) \, e^{\lambda_k^2(t-T)} \, dt = \frac{v}{\lambda_k^2} \qquad (k = 1, 2, \dots)$$

Here, $\chi_T(t)$ is the characteristic function of the interval $[0, T]$. Since $T^* = \inf\{T\}$, there exists a sequence $I_n \to T^*$ and a corresponding sequence of controls $\chi_{T_n}(t) u_{T_n}(t)$ such that

$$\int_0^T \chi_{T_n}(t) u_{T_n}(t) \, e^{\lambda_k^2(t-T)} \, dt = \frac{v}{\lambda_k^2}$$

The set $|u(t)| \le 1$ is weakly compact in $L_2(0, T)$, and therefore there exists a weak limit of the sequence of functions $\chi_{T_n}(t) u_{T_n}(t) \, e^{-\lambda_k^2 T_n}$, i.e.,

$$\lim_{n \to \infty} \int_0^T \chi_{T_n}(t) u_{T_n}(t) \, e^{\lambda_k^2(t-T_n)} \, dt = \int_0^T \chi_{T^*}(t) u_{T^*}(t) \, e^{\lambda_k^2(t-T^*)} \, dt$$

$$= \int_0^{T^*} u_{T^*}(t) \, e^{\lambda_k^2(t-T^*)} \, dt = \frac{v}{\lambda_k^2}$$

The control $u_{T^*}(t)$ is an optimal control.

The case when the attained function $z_0(x) \ne \text{const}$ is much more complicated. The following theorem holds.

Theorem 7.1 (Egorov[74]). Let $z_0(x)$ be a function which is attained at the time $t = t_1$ with the control $u_1(t)$, and

$$\left| \int_0^1 z_0(x) \cos \lambda_n x \, dx \right| < C e^{-c_0 \lambda_n^{1 + 2\varepsilon_0}}$$

for certain positive $C, c_0, \varepsilon_0 < 1, n = 1, 2, \ldots$. Then there exists a control $u(t)$ for which the solution of the problem coincides with $z_0(x)$ at time T, and the time T is a minimum. This control is unique, and $|u(t)| \equiv 1$ almost everywhere in $[0, T]$.

We shall not give the proof of the theorem, but merely sketch the most important points. As in the problem subject to the constraint (7.14), uniqueness is ensured by the condition $|u(t)| \equiv 1$ almost everywhere. In the mentioned problem, this condition was derived from the construction of a support plane to a convex set in a function space. The existence of such a plane is guaranteed if the convex set has an interior point.[17]

In the problem with the constraint (7.14), the separating plane exists: the ball (7.14) clearly has interior points. However, if one poses the condition of exact "hitting," then this approach cannot be used unless it is first shown that the set of functions $z(T, x)$ at $t = T$, generated by admissible controls, has interior points. In other words, it is necessary to prove that the set of such $z(T, x)$ contains a certain ball.

The requirement that the set of attainable functions have an interior point is essential; if it is not satisfied, then assertions based on the construction of separating planes, and in particular the analogue of Pontryagin's maximum principle, may be false. Egorov[75] has found an example which illustrates such a possibility.

In the space l_2 of sequences $x = (x_1, \ldots, x_n \ldots)$ with norm $\|x\| = (\Sigma_{n=1}^{\infty} x_n^2)^{1/2}$, consider the equation $dx/dt = u(t)$. On the solution of this equation we impose the conditions $x(a) = (0, \ldots, 0, \ldots)$, $x(b) = (1, 1/2, \ldots, 1/n, \ldots)$, and we take the set of admissible controls to be given by $A = \{u : u = (u_1, \ldots, u_n, \ldots), |u_n| \leq 1/n + 1/n^2\}$. We pose the problem of minimizing the transition time $b - a$ from the state $x(a)$ to the state $x(b)$.

An optimal control is $\bar{u}(t) = (1, 1/2, \ldots, 1/n, \ldots)$. Indeed, for such a control the time required for the transition is equal to unity. On the other hand, the minimum transition time of the coordinate x_n from $x_n(a) = 0$ to $x_n(b) = 1/n$ is $n/(n + 1)$; therefore, it is impossible for $x(a)$ to transit to $x(b)$ in less than unit time. If Pontryagin's maximum principle held, then there would exist a nonzero constant vector $\psi = (\psi_0, \psi_1, \ldots, \psi_n, \ldots) \in l_2$ such that the function would attain a maximum for $u_n = \bar{u}_n = 1/n$. At the same time, if $\psi_k \neq 0$, then a maximum of H with respect to $u \in A$ is obviously attained for

$$u_k = \left(\frac{1}{k} + \frac{1}{k^2}\right) \operatorname{sign} \psi_k$$

In the space l_2, the set of attainable $x(b)$ does not have an interior point. However, it may have one in other spaces; e.g., in the space with the norm $\|x\| = \sup_n |nx_n|$. If the set of admissible $x(b)$ in the space B has an interior point, then the maximum principle holds.

The proof is simple. Referring to the theorem on the separability of convex sets,[17] one can guarantee the existence on a nontrivial linear functional $\xi \in B^*$ such that $(\xi, x(b)) \geq (\xi, x)$, where x is any element of the attainable set. From this it follows that $(\xi) \int_a^b [\bar{u}(t) - u(t)] \, dt \geq 0$ for any $u(t) \in A$, so that $(\xi, \bar{u}(t) - u(t)) \geq 0$ almost everywhere on the interval $[a, b]$, which is what we wanted to prove.

We now turn to the theorem formulated above. Consider the set A of sequences $a = (a_1, \ldots, a_n, \ldots)$, where

$$a_n = \int_0^T u(\xi) \, e^{\lambda_n^2(\xi - T)} \, d\xi, \qquad |u(t)| \leq 1, \qquad a \in B$$

As the space B we take the Banach space of vectors $b = (b_1, \ldots, b_n, \ldots)$ such that $b_n \, e^{c_0 \lambda_n^{1+\varepsilon}} \to 0$. We introduce a norm in B by the equation

$$\|b\|_B = \sup_n |b_n \, e^{c_0 \lambda_n^{1+\varepsilon_0}}|$$

The set A is isomorphic to the set of attainable states $z(T, x)$. As shown in Ref. 74, the set A contains the ball $\|b\|_B \leq \rho$; i.e., for every element b in this ball one can find an admissible control $u(t)$ such that

$$b_n = \int_0^T u(\xi) \, e^{\lambda_n^2(\xi - T)} \, d\xi$$

Furthermore, it is proved in Ref. 74 that the element a^* corresponding to the optimal control $u^*(t)$ belongs to the boundary of A. By the theorem on the separability of convex sets, we now prove that there exists a nontrivial linear functional $f \in B^*$ such that

$$(f, a^*) \geq (f, a), \qquad \forall a \in A$$

or, equivalently, that

$$\int_0^T [u^*(\xi) - u(\xi)] F(\xi) \, d\xi \geq 0$$

The function $F(\xi)$ is analytic on the interval $0 \leq t < T$; it follows that $u^*(t) = \operatorname{sign} F(t)$ almost everywhere, i.e., $|u^*(t)| \equiv 1$.

7.3. Optimal Relay Controls

As was pointed out in Sections 7.1 and 7.2, optimal controls frequently belong to the boundary ∂A of the set A of admissible controls. If, for example, this set is defined by the inequality $u_{min} \leq u(t) \leq u_{max}$, then an optimal control can take only the limiting values u_{min} and u_{max}. A two-position control of this kind is known as a *relay* control.

In this section, we shall find some conditions subject to which optimal controls belong to the boundary ∂A of the set A for a certain class of linear optimization problems in a Banach space.* These results are due to Friedman.[251-253]

We consider a parabolic optimization problem with control on the right-hand side of a linear equation. We suppose that in the region $S_T = G \times [0, T]$ the parabolic operator

$$Lz = \frac{\partial z}{\partial t} - A(x, t, D)z \equiv \frac{\partial z}{\partial t} - \sum_{|\alpha| < 2r} a_\alpha(x, t)D^\alpha z$$

with $r > 0, D^\alpha = D_1^{\alpha_1} \ldots D_m^{\alpha_m}, D_j = \partial/\partial x_j, |\alpha| = \alpha_1 + \cdots + \alpha_m$, is defined. We assume that the coefficients $a_\alpha(x, t)$ and the boundary Γ of the region G are sufficiently smooth.

We consider the problem

$$\left.\begin{array}{ll} Lz = u(x, t), & (x, t) \in S_{t_0} \\[2mm] z(0, x) = v(x), & x \in \Gamma \\[2mm] B_j(t, x, D)z = w_j(x, t), & (x, t) \in \Gamma \times [0, t_0) \\[2mm] 1 \leq j \leq r \end{array}\right\} \qquad (7.15)$$

The operators B_j are assumed to be such that the problem (7.15) has a unique and sufficiently smooth solution. With regard to the operator L, it is assumed to possess the following uniqueness property: let L^* be the adjoint operator, let $T > 0$, and suppose $v(x, t)$ is any sufficiently smooth function satisfying: (a) the equation $L^*v = 0$ in S_T, (b) the condition $B_j v = 0$ in $\Gamma \times [0, T]$, (c) the condition $v(t, x) = 0$ on $\Gamma \times \Delta$, where Δ is some set of positive measure on $(0, t)$, then $v(T, x) = 0$. We have already used this uniqueness property "in the opposite direction" in Section 7.1 in connection

---

* In the Western literature, the inclusion $u \in \partial A$ is known as the bang-bang principle.

with Ref. 294. In Ref. 252, it is shown that if L and B_j are analytic operators, and Γ is an analytic boundary, then the property holds. The functions u, v, and w are assumed to be controls whose values belong to closed convex sets A_u, A_v, and A_w of the corresponding spaces. We are given a set P of final states $z(x, t)$ of the system, P containing interior points. It can be assumed that P is a convex body with sufficiently smooth boundary in the corresponding function space. The problem consists of steering the point $z(x, t)$ to the set P in minimum time T.

We consider the case when the functions v and w are fixed, and the control $u(x, t)$ is to be determined. Under the imposed conditions a Green's function for problem (7.15) is given by $K(x, t, \xi, \tau)$. It may be used to represent the solution $z(x, t)$ in the form

$$z(x, t) = \beta_0(x, t) + \int_0^t \int_G K(x, t, \xi, \tau) u(\xi, \tau) \, d\xi \, d\tau \qquad (7.16)$$

where β_0 is a known function.

Existence of an optimal control may be proved in a straightforward manner; we omit the proof. Let us consider uniqueness. Suppose that the admissible controls u are such that the norm $\|u(\cdot, \tau)\|_{L_q(G)}$ is bounded as a function of τ for some $1 < q \leq \infty$. Young's inequality shows that the integral (7.16) belongs to $L_s(G)$, where $1 \leq s \leq q$. One can also show that for $q > m/2r$ this integral belongs to $L_\infty(G)$.

Bearing this in mind, consider the set

$$\Omega_T : \left\{ v \in L_s(G); v(x) = \beta_0(x, T) + \int_0^T \int_G K(x, t, \xi, \tau) u(\xi, \tau) \, d\xi \, d\tau \right\}$$

of elements $v(x) = z(x, t)$ attainable at the time $t = T$ by means of admissible controls. This set is convex in $L_s(G)$. Let $u^*(x, t)$ be an optimal control and let $z^*(x, t)$ be the corresponding solution. Clearly, $z^*(x, T) \in \Omega_T \cap P$. The element $z^*(x, T)$ must belong to the boundary of the set P; for otherwise, by continuity, the element $z^*(x, T - \varepsilon)$ would belong to int P, contradicting the optimality of the time T.

The set P contains interior points; consequently, a support hyperplane $\Pi(z)$ exists. We denote the half-spaces separated by this hyperplane by $\Pi_-(z)$ and $\Pi_+(z)$; suppose $p \in \Pi_+(z)$. We show that $\Omega_T \in \Pi_-(z)$; otherwise, there would exist a point $v \in \Omega_T$ lying within Π_+. Because of the conditions imposed on P, one could then find a point $\bar{z} \in P$ belonging to the interval (z, v). However, one simultaneously has $\bar{z} \in \Omega_T$ (convexity), which contradicts the minimality of T.

Let $\gamma(x)$ be the function in $L_{s'}(G)(1/s + 1/s' = 1)$ which determines the hyperplane $\Pi(z)$. Then one obviously has

$$\int_G v(x)\gamma(x)\,dx \le \int_G z^*(x, T)\gamma(x)\,dx, \qquad \forall v \in \Omega_T$$

Using the representation (7.16), this inequality may be rewritten as

$$\int_0^T \int_G k_0(\xi, \tau)u(\xi, \tau)\,d\xi\,d\tau \le \int_0^T \int_G k_0(\xi, \tau)u^*(\xi, \tau)\,d\xi\,d\tau, \qquad \forall u \in A_u$$

$$(7.17)$$

where

$$k_0(\xi, \tau) = \int_G K(x, T, \xi, \tau)\gamma(x)\,dx$$

We assume that there exists a set Δ of positive measure such that $u^*(x, t)$ for $t \in \Delta$ does not belong to the boundary of the set A_u. From (7.17) we readily conclude that this is possible only under the condition that $k_0(\xi, \tau) \equiv 0$ on $G \times \Delta$. Then, however, by virtue of the "uniqueness in the reverse direction" for the operator L, we conclude that $\gamma(x) = 0$ almost everywhere on $[0, T)$, which is impossible.

Similar theorems also apply to more general equations in Banach spaces. The following is an example of such a theorem.

Consider the equation

$$\frac{dz}{dt} + Az = u(t) \tag{7.18}$$

in which the dependent variable $z(t)$ and the control $u(t)$ are elements of the Banach space B, and A is a t-independent linear (unbounded) operator satisfying the following conditions:

(a) The domain $D(A)$ of definition of A is dense in B and the operator A is closed;
(b) $(\lambda I - A)^{-1}$ exists for Re $\lambda \ge 0$ and

$$\|(\lambda I - A)^{-1}\| \le \frac{c}{1 + |\lambda|}, \qquad \text{Re } \lambda \ge 0, \qquad c = \text{const}$$

We are interested in solutions of (7.18) satisfying the initial condition

$$z(x, 0) = z_0(x) \tag{7.19}$$

and the condition

$$z(x, T) = z_1(x) \tag{7.20}$$

at the time $t = T$. Of course, it is not possible to "hit" (7.20) exactly for every choice of the control $u(t)$. The problem consists of choosing $u \in A_u$, where A_u is a convex set, in such a way as to minimize the time of "hitting." We assume, of course, that the set is sufficiently "spacious" for the hitting of a given function $z_1(x)$ to be realizable in a time $T < \infty$.

Theorem 7.2 (Friedman[252]).   Suppose that under the formulated conditions the operator A generates a strongly continuous semigroup, and A_u is a convex neighborhood of the origin. If an optimal control exists, it belongs to the boundary ∂A_u of the set A_u for almost all t.

Other theorems of this type have been given by Friedman,[253] Balakrishnan,[182-186] and Fattorini.[242-247]

7.4. Necessary Conditions for Optimality in Quasi-linear Systems of Parabolic Type

Consider a region G of m-dimensional Euclidean space $(x_1, \ldots, x_m)$ with sufficiently smooth boundary Γ and define the elliptic operator

$$L_i z = \sum_{p=1}^{n} \sum_{j,k=1}^{m} a_{jk}^{ip} \frac{\partial^2 z^p}{\partial x_j \, \partial x_k}$$

whose coefficients $a_{jk}^{ip}(x, t)$ depend continuously on the parameter t and are twice continuously differentiable with respect to $x = (x_1, \ldots, x_m)$. The operator L_i may then be used to form the system

$$L_{it} z = z_t^i - L_i z = f_i(t, x, z, z_x, u), \qquad 0 \le t \le T, \qquad x \in G \tag{7.21}$$

here $z = (z^1, \ldots, z^n)$, and the functions $f = (f_1, \ldots, f_n)$ are continuous with respect to t and twice continuously differentiable with respect to the remaining arguments. In the following, the function $u(t, x) = (u_1(t, x), \ldots, u_p(t, x))$ will be interpreted as a volume control with values in the convex region A_u of p-dimensional Euclidean space.

For the continuous functions $z(t, x)$ we specify initial and boundary conditions of the form

$$z(0, x) = a(x), \qquad\qquad x \in G$$

$$P_i(t, x)z = \varphi_i(t, x, z, v), \qquad x \in \Gamma, \qquad 0 \le t \le T$$

(7.22)

The functions $a(x) = (a_1(x), \ldots, a_n(x))$ are continuous; the functions φ_i satisfy the same conditions as the f_i; $v(t, x) = (v_1(t, x), \ldots, v_p(t, x))$ is a boundary control whose values belong to the convex region A_v of p-dimensional Euclidean space. The operators P_i act in accordance with the formulas

$$P_i z = \sum_{p=1}^{m} \left(a_i^{ip} \frac{\partial z^p}{\partial l_{ip}} + b_{ip} z^p \right)$$

(7.23)

Here, a_i^{ip} and b_{ip} are continuous functions of the variables (t, x), and l_{ip} are arbitrary directions making an acute angle with the outer normal to Γ.

We shall also assume that the controls $u(t, x)$ and $v(t, x)$ are piecewise continuous and that possible discontinuity surfaces are smooth. Apart from these properties, all of the functions occurring in the formulation of the problem are assumed to satisfy additional conditions that guarantee the existence and uniqueness of a solution of the problem (7.21)-(7.22) for all admissible controls (see, e.g., Ref. 118).

We now pose the optimization problem. We introduce the functional

$$I = \sum_{i=1}^{n} \left[\int_G \alpha_i(x)z^i(T, x) \, dx + \int_0^T \int_G \beta_i(t, x)z^i(t, x) \, dt \, dx \right.$$

$$\left. + \int_0^T \int_\Gamma \gamma_i(t, x)z^i(t, x) \, dt \, dx \right]$$

(7.24)

Here, $\alpha_i, \beta_i, \gamma_i$ are given continuous functions. The problem consists of the determination of controls $u \in A_u, v \in A_v$ such that the functional (7.24) takes on its minimum value.

The functional (7.24) is not particularly general; it is chosen only for simplicity of the arguments. The method used can be extended to more complicated cases when the functional is nonlinear.

The problem we have formulated was posed and studied by Egorov.[68,69] In this section, we shall present some of his results.

With an eye on the derivation of an expression for the increment of the functional, we introduce the Lagrange multipliers $\omega_i(t, x)$ $(i = 1, \ldots, n)$ along with the functions

$$H(t, x, z, z_x, \omega, u) = \sum_{i=1}^{n} \omega_i f_i(t, x, z, z_x, u)$$

(7.25)

$$h(t, x, z, \omega, v) = \sum_{i=1}^{n} \omega_i \varphi_i(t, x, z, v)$$

We proceed from the identity

$$\int_0^T \int_G \left[\sum_{i=1}^{n} \omega_i L_{it} z - H(t, x, z, z_x, \omega, u) \right] dt\, dx$$

$$+ \int_0^T \int_\Gamma \left[\sum_{i=1}^{n} \omega_i P_i z - h(t, x, z, \omega, v) \right] dt\, dx$$

for the optimal state (z, u, v) and the analogous identity for the admissible state (Z, U, V). The use of these identities and of the notation $\Delta z = Z - z$, etc., yield

$$\int_0^T \int_G \left[\sum_{i=1}^{n} \omega_i L_{it} \Delta z - H(t, x, Z, Z_x, \omega, U) + H(t, x, z, z_x, \omega, u) \right] dt\, dx$$

$$+ \int_0^T \int_\Gamma \left[\sum_{i=1}^{n} \omega_i P_i \Delta z - h(t, x, Z, \omega, V) + h(t, x, z, \omega, v) \right] dt\, dx = 0$$

(7.26)

An integration by parts results in

$$\int_0^T \int_G \omega_i L_{it} \Delta z\, dt\, dx = \int_0^T \int_G \omega_i (\Delta z_t^i - L_{it} \Delta z)\, dt\, dx$$

$$= \int_G \omega_i \Delta z^i |_{t=0}^{t=T}\, dx - \int_0^T \int_G \Delta z^i M_{it} \omega\, dt\, dx$$

$$- \int_0^T \int_\Gamma (\omega_i P_i \Delta z - \Delta z^i Q_i \omega)\, dt\, dx$$

(7.27)

Here, M_{it} and Q_i are operators which act in accordance with

$$M_{it}\omega = \omega_{it} + M_i\omega$$

$$= \omega_{it} + \sum_{p=1}^{n} \sum_{j,k=1}^{m} \frac{\partial}{\partial x_j}\left(a_{jk}^{pi}\frac{\partial \omega_p}{\partial x_k}\right) - \sum_{j,k=1}^{m} \frac{\partial}{\partial x_j}\left(\frac{\partial a_{jk}^{pi}}{\partial x_k}\omega_p\right) \qquad (i=1,\ldots,n)$$

$$\tag{7.28}$$

$$Q_i\omega = \sum_{p=1}^{n}\left[a_\lambda^{pi}\frac{\partial \omega_p}{\partial \lambda_{ip}} + d_{ip}\omega_p\right] \tag{7.29}$$

The functions a_λ^{pi} and d_{ip} and the directions λ_{ip} are related to a_l^{ip}, b_{ip}, and l_{ip} in obvious ways. With (7.27) the identity (7.26) may be written in the form

$$\int_G \omega_i \Delta z^i\big|_{t=0}^{t=T}\,dx - \int_0^T \int_G \Delta z^i M_{it}\omega\,dt\,dx + \int_0^T \int_\Gamma \Delta z^i Q_i\omega\,dt\,dx$$

$$= \int_0^T \int_G [H(t,x,Z,Z_x,\omega,U) - H(t,x,z,z_x,\omega,u)]\,dt\,dx$$

$$+ \int_0^T \int_\Gamma [h(t,x,Z,\omega,V) - h(t,x,z,\omega,v)]\,dt\,dx \tag{7.30}$$

A solution of the adjoint boundary-value problem

$$M_{it}\omega = -\frac{\partial H(t,x,z,z_x,\omega.\,u)}{\partial z^i} + \sum_{k=1}^{m} \frac{\partial}{\partial x_k}\left(\frac{\partial H(t,x,z,z_x,\omega,u)}{\partial z_{x_k}^i}\right)$$

$$+\beta_i(t,x), \qquad 0 \le t \le T, \qquad x \in G$$

$$\omega_i(T,x) = -a_i(x), \qquad x \in G \tag{7.31}$$

$$Q_i\omega = \frac{\partial h(t,x,z,\omega,v)}{\partial z^i} + \sum_{k=1}^{m} \frac{\partial H(t,x,z,z_x,\omega,u)}{\partial z_{x_k}^i}X_k(x)$$

$$-\gamma_i(t,x), \qquad x \in \Gamma, \qquad 0 \le t \le T$$

is given by the function $\omega_i(t,x)$. Here, $X_k(x)$ are the direction cosines of the outer normal to Γ. The use of these formulas along with (7.24) to eliminate $M_{it}\omega$ and $Q_i\omega$ from the identity (7.30) yields the following expression for the increment of the functional I:

$$\Delta I = -\int_0^T \int_G [H(t, x, Z, Z_x, \omega, U) - H(t, x, z, z_x, \omega, u)] \, dt \, dx$$

$$+ \sum_{i=1}^n \int_0^T \int_G \left[\frac{\partial H(t, x, z, z_x, \omega, u)}{\partial z^i} \Delta z^i \right.$$

$$+ \sum_{k=1}^m \frac{\partial H(t, x, z, z_x, \omega, u)}{\partial z_{x_k}^i} \Delta z_{x_k}^i \right] dt \, dx$$

$$- \int_0^T \int_\Gamma [h(t, x, Z, \omega, V) - h(t, x, z, \omega, v)] \, dt \, dx$$

$$+ \sum_{i=1}^n \int_0^T \int_\Gamma \frac{\partial h(t, x, z, \omega, v)}{\partial z^i} \Delta z^i \, dt \, dx$$

With the use of Taylor's formula, we then reduce the equation to the form

$$\Delta I = -\int_0^T \int_G [H(t, x, z, z_x, \omega, U) - H(t, x, z, z_x, \omega, u)] \, dt \, dx$$

$$- \int_0^T \int_\Gamma [h(t, x, z, \omega, V) - h(t, x, z, \omega, v)] \, dt \, dx - \eta_1 - \eta_2 \qquad (7.32)$$

$$\eta_1 = \sum_{i=1}^n \int_0^T \int_G \left[\frac{\partial H(t, x, z, z_x, \omega, U)}{\partial z^i} - \frac{\partial H(t, x, z, z_x, \omega, u)}{\partial z^i} \right] \Delta z^i \, dt \, dx$$

$$+ \sum_{i=1}^n \sum_{k=1}^m \int_0^T \int_G \left[\frac{\partial H(t, x, z, z_x, \omega, U)}{\partial z_{x_k}^i} - \frac{\partial H(t, x, z, z_x, \omega, u)}{\partial z_{x_k}^i} \right] \Delta z_{x_k}^i \, dt \, dx$$

$$+ \sum_{i=1}^n \int_0^T \int_\Gamma \left[\frac{\partial h(t, x, z, \omega, V)}{\partial z^i} - \frac{\partial h(t, x, z, \omega, v)}{\partial z^i} \right] \Delta z^i \, dt \, dx$$

$$\eta_2 = \frac{1}{2} \sum_{i,k=1}^{n+mn} \int_0^T \int_G \frac{\partial^2 H(t, x, g + \theta_1 \Delta g, \omega, U)}{\partial g^i \, \partial g^k} \Delta g^i \, \Delta g^k \, dt \, dx$$

$$+ \frac{1}{2} \sum_{i,k=1}^n \int_0^T \int_G \frac{\partial^2 h(t, x, g + \theta_2 \Delta g, \omega, V)}{\partial g^i \, \partial g^k} \Delta g^i \, \Delta g^k \, dt \, dx$$

$$0 < \theta_{1/2} < 1$$

[The notation $g = (g^1, \ldots, g^{n+nm}) = (z^1, \ldots, z^n, z_{x_1}^1, \ldots, z_{x_m}^n)$ is introduced for brevity.] Equation (7.32) is an exact expression for the increment of the functional I. We show that under local strong variations Δu and Δv of the controls the integral terms in this formula form the principal part of ΔI. This will serve as a proof of the following theorem.

Theorem 7.3 (Egorov[69]).   If the admissible controls $u(t, x)$ and $v(t, x)$ are to minimize the functional I, it is necessary that the following maximum conditions be satisfied:

$$H(t, x, z, z_x, \omega, u) = \max_{U \in A_u} H(t, x, z, z_x, \omega, U)$$

$$h(t, x, z, \omega, v) = \max_{V \in A_v} h(t, x, z, \omega, V)$$

$$(7.33)$$

almost everywhere in $G \times [0, T)$ and on the boundary $\Gamma \times [0, T)$.

In order to obtain estimates of the remainder terms η_1 and η_2, we note that the increments Δz^i form a solution of the boundary-value problem

$$\left.\begin{array}{l}
L_{it}\Delta z^i = \Delta \dfrac{\partial H(t, x, z, z_x, \omega, u)}{\partial \omega_i}, \qquad 0 \leq t \leq T, \qquad x \in G \\[3mm]
\Delta z^i(0, x) = 0, \qquad x \in G \\[3mm]
P_i \Delta z = \Delta \dfrac{\partial h(t, x, z, \omega, v)}{\partial \omega_i}, \qquad 0 \leq t \leq T, \qquad x \in \Gamma
\end{array}\right\} \quad (7.34)$$

where

$$\Delta \frac{\partial H}{\partial \omega_i} = \frac{\partial H(t, x, Z, Z_x, \omega, U)}{\partial \omega_i} - \frac{\partial H(t, x, z, z_x, \omega, u)}{\partial \omega_i}$$

$$(7.35)$$

$$\Delta \frac{\partial h}{\partial \omega_i} = \frac{\partial h(t, x, Z, \omega, V)}{\partial \omega_i} - \frac{\partial h(t, x, z, \omega, v)}{\partial \omega_i}$$

The problem (7.34) is equivalent to the following system of integrodifferential equations:

$$\Delta z(t, x) = \int_0^t \int_G K_{11}(t, x, \tau, \xi) \Delta \frac{\partial H}{\partial \omega} \, d\tau \, d\xi$$

$$+ \int_0^t \int_\Gamma K_{12}(t, x, \tau, \xi) \psi(\tau, \xi) \, d\tau \, d\xi, \qquad x \in G \qquad (7.36)$$

$$\psi(t, \xi) = -\Delta \frac{\partial h}{\partial \omega} + \int_0^t \int_G K_{21}(t, \xi, \tau, \zeta) \Delta \frac{\partial H}{\partial \omega} \, d\tau \, d\zeta$$

$$+ \int_0^t \int_\Gamma K_{22}(t, \xi, \tau, \zeta) \psi(\tau, \zeta) \, d\tau \, d\zeta, \qquad \xi \in \Gamma \qquad (7.37)$$

where $\Delta \partial H/\partial \omega = (\Delta \partial H/\partial \omega_1, \ldots, \Delta \partial H/\partial \omega_n)$, $\Delta \partial h/\partial \omega = (\Delta \partial h/\partial \omega_1, \ldots, \Delta \partial h/\partial \omega_n)$, and where the K_{ij} are matrices related to Green's matrix in a known manner. The iteration of (7.37) yields

$$\psi(t, \xi) = -\Delta \frac{\partial h}{\partial z} + \int_0^t \int_G K_n(t, \xi, \tau, \eta) \Delta \frac{\partial H}{\partial \omega} \, d\tau \, d\eta$$

$$+ \int_0^t \int_\Gamma K^n(t, \xi, \tau, \eta) \psi(\tau, \eta) \, d\tau \, d\eta$$

$$- \int_0^t \int_\Gamma \sum_{i=0}^{n-1} K^i(t, \xi, \tau, \eta) \Delta \frac{\partial h}{\partial \omega} \, d\tau \, d\eta \qquad (7.38)$$

Here

$$K_n(t, \xi, \tau, \eta) = K_{n-1}(t, \xi, \tau, \eta) + \int_\tau^t \int_\Gamma K_{n-1}(t, \xi, \alpha, \beta) K_0(\alpha, \beta, \tau, \eta) \, d\alpha \, d\beta$$

$$K^n(t, \xi, \tau, \eta) = \int_\tau^t \int_\Gamma K^{n-1}(t, \xi, \alpha, \beta) K^0(\alpha, \beta, \tau, \eta) \, d\alpha \, d\beta$$

$$K_0 = K_{21}, \qquad K^0 = K_{22}, \qquad n = 1, 2, \ldots$$

Taking n sufficiently large and using the well-known estimates for the Green's matrix and its derivatives, we can achieve a bounded kernel K^n. Choose and fix such an n; (7.38) may then be used to deduce the inequality

$$w(t) \le P \int_0^t w(\tau) \, d\tau + \int_0^t \int_G Q_n(t, \tau, \eta) \left| \Delta \frac{\partial H}{\partial \omega} \right| \, d\tau \, d\eta$$

$$+ \int_0^t \int_\Gamma R_n(t, \tau, \eta) \left| \Delta \frac{\partial h}{\partial \omega} \right| \, d\tau \, d\eta + \int_\Gamma \left| \Delta \frac{\partial h}{\partial z} \right| \, d\eta \qquad (7.39)$$

where P is a positive constant,

$$w(t) = \int_\Gamma |\psi(t, \xi)| d\xi, \qquad Q_n = \int_\Gamma |K_n(t, \xi, \tau, \eta)| d\xi$$

$$R_n = \int_\Gamma \sum_{i=0}^{n-1} |K^i(t, \xi, \tau, \eta)| d\xi$$

We introduce the notation

$$\left.\begin{aligned}
w_k(t) &= \int_0^t w_{k-1}(\tau) \, d\tau, & w_0(t) &= w(t) \\[2ex]
Q_{nk} &= \int_\tau^t Q_{n(k-1)}(t, \tau, \eta) d\tau, & Q_{n0} &= Q_n \\[2ex]
R_{nk} &= \int_0^t R_{n(k-1)}(t, \tau, \eta) \, d\tau, & R_{n1} &= R_n(t, \tau, \eta) + 1 \\[2ex]
& k = 2, 3, \ldots &&
\end{aligned}\right\} \quad (7.40)$$

and integrate the inequality (7.39) k times successively; we obtain

$$w_k(t) \le P \int_0^t w_k(\tau) \, d\tau + \int_0^t \int_G Q_{nk}(t, \tau, \eta) \left| \Delta \frac{\partial H}{\partial \omega} \right| d\tau \, d\eta$$

$$+ \int_0^t \int_\Gamma R_{nk}(t, \tau, \eta) \left| \Delta \frac{\partial h}{\partial \omega} \right| d\tau \, d\eta \qquad (7.41)$$

We now choose k sufficiently large to assure the boundedness of the functions Q_{nk} and R_{nk} for $0 \le \tau \le t \le T, x \in G$, and we set

$$Q(t) = \int_0^t \int_G \left| \Delta \frac{\partial H}{\partial \omega} \right| \max_{0 \le \theta \le t} Q_{nk}(\theta, \tau, \eta) \, d\tau \, d\eta$$

$$R(t) = \int_0^t \int_\Gamma \left| \Delta \frac{\partial h}{\partial \omega} \right| \max_{0 \le \theta \le t} R_{nk}(\theta, \tau, \eta) \, d\tau \, d\eta$$

Under these conditions, (7.41) may be written in the form

$$w_k(\theta) \le P \int_0^\theta w_k(\theta) \, d\theta + Q(t) + R(t)$$

Then, the use of the lemma on p. 19 of Ref. 115 yields

$$w_k(\theta) \leq A[Q(t) + R(t)], \qquad 0 \leq \theta \leq t \leq T, \qquad A = \text{const} > 0$$

and, in particular,

$$w_k(t) \leq A[Q(t) + R(t)]$$

From (7.39)–(7.41) it is apparent that

$$w_k(t) \leq \int_0^t d\tau \left[\int_G M_1(t, \tau, \eta) \left| \Delta \frac{\partial H}{\partial \omega} \right| d\eta + \int_\Gamma N_1(t, \tau, \eta) \left| \Delta \frac{\partial h}{\partial \omega} \right| d\eta \right]$$

(7.42)

where M_1 and N_1 are functions of the type Q_n and R_n. Since the number n is sufficiently large, (7.38) in conjunction with (7.42) may be used to show that

$$|\psi(t, \xi)| \leq \int_0^T d\tau \left[\int_G M_2(t, \tau, \xi, \eta) \left| \Delta \frac{\partial H}{\partial \omega} \right| d\eta \right.$$

$$\left. + \int_\Gamma N_2(t, \tau, \xi, \eta) \left| \Delta \frac{\partial h}{\partial \omega} \right| d\eta \right]$$

where M_2 and N_2 are functions of the Green's function type.

The differentiation of (7.36) with respect to $x_1, \ldots, x_m$ (which is valid under the assumptions made about the functions f and φ) results in the analogous inequalities

$$|\Delta g_i(t, x)| \leq \int_0^T \int_G M_3(t, x, \tau, \eta) \sum_{s=1}^p |\Delta u_s(\tau, \eta)| \, d\tau \, d\eta$$

$$+ \int_0^T \int_\Gamma N_3(t, x, \tau, \eta) \sum_{j=1}^\rho |\Delta v_j(\tau, \eta)| \, d\tau \, d\eta$$

where

$$g = (z^l, \ldots, z^n, z^l_{x_1}, \ldots, z^n_{x_m})$$

This suffices to provide an estimate for the terms η_1 and η_2 in (7.32). We obtain

$$
\begin{aligned}
|\eta_1| \le B_1 & \int_0^T \int_G \sum_{j=1}^p |\Delta u_j(t, x)| \left\{ \int_0^T \int_G M_{11}(t, x, \tau, \eta) \sum_{j=1}^p |\Delta u_j(\tau, \eta)| \, d\tau \, d\eta \right. \\
& \left. + \int_0^T \int_\Gamma N_{11}(t, x, \tau, \eta) \sum_{k=1}^p |\Delta v_k| \, d\tau \, d\eta \right\} dt \, dx \\
+ B_2 & \int_0^T \int_\Gamma \sum_{i=1}^p |\Delta v_j(t, x)| \left\{ \int_0^T \int_\Gamma M_{11}(t, x, \tau, \eta) \sum_{j=1}^p |\Delta u_j| \, d\tau \, d\eta \right. \\
& \left. + \int_0^T \int_\Gamma N_{11}(t, x, \tau, \eta) \sum_{k=1}^p |\Delta v_k| \, d\tau \, d\eta \right\} dt \, dx
\end{aligned}
$$

where

$$
B_i = \text{const} > 0, \qquad M_{11} = M_3 + M_4, \qquad N_{11} = N_3 + N_4
$$

In accordance with the previously stipulated condition, the derivatives $\partial^2 H / \partial g^i \, \partial g^k$ and $\partial^2 h / \partial g^i \, \partial g^k$ are bounded, and it thus follows that

$$
\begin{aligned}
|\eta_2| \le B_3 & \int_0^T \int_G \left[\int_0^T \int_G M_{11}(t, x, \tau, \eta) \sum_{k=1}^p |\Delta u_k| \, d\tau \, d\eta \right. \\
& \left. + \int_0^T \int_\Gamma N_{11}(t, x, \tau, \eta) \sum_{j=1}^p |\Delta v| \, d\tau \, d\eta \right]^2 dt \, dx \\
+ B_4 & \int_0^T \int_\Gamma \left[\int_0^T \int_G M_{11}(t, x, \tau, \eta) \sum_{k=1}^p |\Delta u_k| \, d\tau \, d\eta \right. \\
& \left. + \int_0^T \int_\Gamma N_{11}(t, x, \tau, \eta) \sum_{i=1}^p |\Delta v_j| \, d\tau \, d\eta \right]^2 dt \, dx
\end{aligned}
$$

The remainder term $\eta = \eta_1 + \eta_2$ can now be estimated in accordance with

$$
\begin{aligned}
|\eta| \le & \int_0^T \int_G \left\{ \int_0^T \int_G P(t, x, \tau, \eta) \sum_{k=1}^p |\Delta u_k| \, d\tau \, d\eta \right. \\
& \left. + \int_0^T \int_\Gamma Q(t, x, \tau, \eta) \sum_{j=1}^p |\Delta v_j| \, d\tau \, d\eta \right\}^2 dt \, dx \\
& + \int_0^T \int_\Gamma \left\{ \int_0^T \int_G P(t, x, \tau, \eta) \sum_{k=1}^p |\Delta u_k| \, d\tau \, d\eta \right. \\
& \left. + \int_0^T \int_\Gamma Q(t, x, \tau, \eta) \sum_{j=1}^p |\Delta v_j| \, d\tau \, d\eta \right\}^2 dt \, dx \qquad (7.43)
\end{aligned}
$$

where the functions P and Q are of the same type as M_{11} and N_{11}.

The arguments which follow are standard; they coincide almost literally with those given at the end of the proof of the analogous theorem in Section 6.6

7.5. The Penalty Method

In all of the previous exposition, constraints in optimization problems were taken into account by means of Lagrange multipliers. Courant proposed a different method for taking constraints into account—the so-called *penalty method*. For problems of optimal control, Courant's ideas were developed by Balakrishnan,[182-186] De Julio,[221,222] Jones and McCormick,[271] and Sasai.[361] The advantage of this approach is that it enables one to prove the existence of solutions of a large class of optimization problems, and, being constructive, the proof simultaneously provides a method for obtaining an approximate solution.

In writing the present section, I have used De Julio's paper,[221] which considers linear evolution problems containing control functions in both the equation and the boundary conditions. The results have a bearing on the investigations of Sections 7.1-7.3.

We consider a controlled system described by equations of the form

$$\frac{dz}{dt} = Az(t) + Bu(t), \qquad z(0) = 0 \qquad (7.44)$$

Here, A is an unbounded linear operator which maps the domain $D(A)$, which is dense in the Hilbert space H_1, onto this space. The symbol $u(t)$ denotes a control that belongs to the Hilbert space H_2 for almost every t; this space is mapped onto H_1 by the linear bounded operator B.

We shall assume that a solution of problem (7.44) is an element of the space $L_2(T; H_1)$ of functions $f(t)$ which are measurable with respect to t and take values in H_1. We introduce a norm in $L_2(T; H_1)$ by

$$\|f\|_1 = \left(\int_0^T \|f(t)\|_{H_1}^2 \, dt \right)^{1/2}$$

Here, $\|f(t)\|_{H_1}$ is the ordinary notation for the norm in H_1. The space $L_2(T; H_2)$ and its norm are similarly defined.

At the end of the section, we shall consider a system described by the equations

$$\frac{dz}{dt} = Az(t), \qquad z(0) = 0, \qquad cz(t) = u(t) \qquad (7.45)$$

All that we have said above remains in force with regard to the operator A; c denotes a linear operator which maps the domain $D(c) \supset D(A)$ into the space H_2. As before, a solution of the problem (7.45) is regarded as an element of $L_2(T; H_1)$. In both problems, it is assumed that $u \in A_u$ where A_u is a closed convex set in $L_2(T; H_2)$.

We now turn to problem (7.44). We define the operator

$$S = \frac{\partial}{\partial t} - A$$

its domain of definition $D(S)$ is the set of all continuously differentiable functions $f(t)$ with values in $D(A)$ for almost every $t \in [0, T]$, and $f(0) = 0$; the elements $Sf \in L_2(T, H_1)$ form a dense set in this space. We assume that the operator S admits closure, and we denote the minimal closed extension of this operator by $\bar{S}$.

We define a generalized solution of problem (7.44) as an element $z \in D(\bar{S})$ satisfying the identity

$$(z, S^*\varphi) = (Bu, \varphi) \tag{7.46}$$

for all $\varphi \in D(S^*)$. Here, S^* is the adjoint operator; the usual notation is adopted for the scalar product. The domain $D(S)$ is dense in $L_2(T; H_1)$, and the operator S^* always exists in addition, in accordance with the well-known property of operators in Hilbert spaces, $S^{**} = \bar{S}$ and $D(S^*)$ is dense in $L_2(T; H_2)$. For this reason, (7.46) is equivalent to the equation

$$\bar{S}z = Bu \tag{7.47}$$

We define the real functional $I[z; u]$ on the product space $L_2(T; H_1) \times L_2(T; H_2)$ and we assume that this functional has the following properties:

(a) $I[z; u] \geq 0$ for all t, x;
(b) I is weakly lower semicontinuous;
(c) $I[z_n; u_n] \to \infty$ if $\lim_{n \to \infty} (\|z_n\|_1 + \|u_n\|_2) \to \infty$.

We consider the following optimization problem: find an element $u_0 \in A_u$ and element $z_0 \in D(\bar{S})$ corresponding to it in the sense of problem (7.47) such that

$$I[z_0; u_0] = j_0 = \inf_{\substack{u \in A_u \\ z \in D}} I[z; u]$$

subject to the constraints (7.47).

The penalty method used below to prove the existence of a solution of this problem is based on the fact that one can formulate an auxiliary variational problem without constraints (the so-called ε problem). Applied to problem (7.44), one has the following ε problem. We choose $\varepsilon > 0$ and construct the functional

$$I_\varepsilon[z; u] = I[z; u] + \frac{1}{\varepsilon} \|\bar{S}z - Bu\|_1^2$$

We are to determine elements $u_\varepsilon \in A_u$ and $z_\varepsilon \in D(\bar{S})$ such that

$$I_\varepsilon[z_\varepsilon; u_\varepsilon] = j_\varepsilon = \inf_{\substack{u \in A_u \\ z \in D(\bar{S})}} I_\varepsilon[z; u]$$

In this problem, the choice of the elements u and z is no longer restricted by the condition (7.47).

We show that subject to the imposed conditions the ε problem has a solution.

Suppose the sequence of elements $u_n \in A_u, z_n \in D$ is such that the corresponding numerical sequence $\{I_\varepsilon[z_n, u_n]\}$ tends to j_ε:

$$\lim_{n \to \infty} I_\varepsilon[z_n; u_n] = j_\varepsilon$$

Since it decreases, the sequence $\{I_\varepsilon[z_n, u_n]\}$ must be bounded. As a consequence of property (a) the sequence $\{I[z_n, u_n]\}$ also is, and this, by virtue of (c), means that the norms $\|z_n\|_1$ and $\|u_n\|_2$ of the elements are uniformly bounded:

$$\|z_n\|_1 \le C_1, \qquad \|u_n\|_2 \le C_2$$

Since B is a bounded operator, and the norms $\|u_n\|_2$ are bounded, it follows from what we have said that the norms of the elements Sz_n are bounded:

$$\|Sz_n\|_1 \le C_3$$

By virtue of the weak compactness of bounded sets in a Hilbert space, there exist elements $u_\varepsilon, z_\varepsilon, y_\varepsilon$ which are weak limits of the sequences $\{u_n\}, \{z_n\}, \{Sz_n\}$, respectively; furthermore, since A_u is closed and convex, this set is weakly closed and contains the element u_ε.

Let φ be any function in $D(S^*)$. Using the definition of weak convergence, we find

$$(y_\varepsilon, \varphi) = \lim_{n \to \infty} (Sz_n, \varphi) = \lim_{n \to \infty} (z_n, S^*\varphi) = (z_\varepsilon, S^*\varphi)$$

Since the set $D(S^*)$ is dense in $L_2(T; H_1)$ and $S^{**} = \bar{S}$, it follows that

$$z_\varepsilon \in D(\bar{S}) \quad \text{and} \quad \bar{S}z_\varepsilon = y_\varepsilon$$

Finally, using the weak lower semicontinuity of the functional I and the same property of the norm, we obtain

$$j_\varepsilon = \lim_{n \to \infty} I_\varepsilon[z_n; u_n] = \lim_{n \to \infty} I[z_n; u_n] + \lim_{n \to \infty} \frac{1}{\varepsilon} \|Sz_n - Bu_n\|_1^2$$

$$\geq I[z_\varepsilon; u_\varepsilon] + \frac{1}{\varepsilon} \|\bar{S}z_\varepsilon - Bu_\varepsilon\|_1^2 = I_\varepsilon[z_\varepsilon; u_\varepsilon]$$

In conjunction with the preceding discussion, this means that $j_\varepsilon = I_\varepsilon[z_\varepsilon; u_\varepsilon]$. This proves the theorem.

By choosing ε sufficiently small, we can make the solution of the ε problem arbitrarily close to the solution of the original problem. More precisely, we have the following therem.

Theorem 7.4 (Existence).   Let $\{\varepsilon_n\}$ be a sequence of positive numbers tending to zero. Then there exists a subsequence (which we also denote by $\{\varepsilon_n\}$) such that the weak limits z_0 and u_0 of the sequences $\{z_{\varepsilon_n}\}$ and $\{u_{\varepsilon_n}\}$ belong, respectively, to the sets $D(\bar{S})$ and A_u and satisfy the limit equations

$$\lim_{n \to \infty} I_{\varepsilon_n}[z_{\varepsilon_n}; u_{\varepsilon_n}] = I[z_0, u_0] = j_0 = \inf_{\substack{u \in A_u \\ z \in D}} I[z; u] \tag{7.48}$$

where the elements u and z are related by

$$\bar{S}z = Bu \tag{7.49}$$

To prove this, we note that j_0 can be regarded as a lower bound of the functional I_ε under the additional condition (7.47). The imposition of the constraint can only increase the lower bound, and therefore

$$j_0 \geq j_{\varepsilon_n} = I_{\varepsilon_n}[z_{\varepsilon_n}; u_{\varepsilon_n}]$$

As before, we establish the existence of weak limits u_0, z_0, y_0 of the sequences $\{u_{\varepsilon_n}\}, \{z_{\varepsilon_n}\}, \{y_{\varepsilon_n}\}$, at the same time

$$u_0 \in A_u, \qquad z_0 \in D(\bar{S}), \qquad \bar{S}z_0 = y_0$$

We can also assert that there exists an n-independent constant K such that

$$\frac{1}{\varepsilon_n} \|\bar{S}z_{\varepsilon_n} - Bu_{\varepsilon_n}\|_1^2 \leq K^2$$

or

$$\|\bar{S}z_{\varepsilon_n} - Bu_{\varepsilon_n}\|_1 \leq K\sqrt{\varepsilon_n}$$

It follows that

$$\lim_{n \to \infty} \|\bar{S}z_{\varepsilon_n} - Bu_{\varepsilon_n}\|_1 = 0 \tag{7.50}$$

The elements z_0 and u_0 are related by (7.47). To prove this, we consider the inequality

$$0 \leq \|(\bar{S}z_{\varepsilon_n} - Bu_{\varepsilon_n}) - (\bar{S}z_0 - Bu_0)\|_1^2$$

$$= \|\bar{S}z_{\varepsilon_n} - Bu_{\varepsilon_n}\|_1^2 + \|\bar{S}z_0 - Bu_0\|_1^2 - 2(\bar{S}z_{\varepsilon_n} - Bu_{\varepsilon_n}, \bar{S}z_0 - Bu_0)$$

Passing to the limit as $n = \infty$ and taking (7.50) into account along with the definition of weak convergence, we obtain

$$0 \leq \|\bar{S}z_0 - Bu_0\|_1^2$$

whence

$$\bar{S}z_0 = Bu_0 \tag{7.51}$$

To prove (7.48) and (7.49), we need to use the weak lower semicontinuity of the functional I; we have

$$j_0 \geq \overline{\lim_{n \to \infty}} I_{\varepsilon_n}[z_{\varepsilon_n}; u_{\varepsilon_n}] = \underline{\lim_{n \to \infty}} I_{\varepsilon_n}[z_{\varepsilon_n}; u_{\varepsilon_n}] \geq \lim_{n \to \infty} I[z_{\varepsilon_n}; u_{\varepsilon_n}]$$

$$+ \underline{\lim_{n \to \infty}} \frac{1}{\varepsilon_n} \|\bar{S}z_{\varepsilon_n} - Bu_{\varepsilon_n}\|_1^2 \geq I[z_0; u_0] \tag{7.52}$$

The sign $>$ contradicts the definition of the greatest lower bound; it follows that equality holds (7.52), and (7.48) is proved. Assertion (7.49) follows when (7.51) is taken into account.

We now consider problem (7.45), which contains a control in the boundary condition. The operator S is defined as before. In the set of elements $D(S)$ we introduce the norm

$$\|f\|_V = \left(\int_0^T \|f(t)\|_{H_1}^2 \, dt + \int_0^T \|Sf(t)\|_{H_1}^2 \, dt \right)^{1/2}$$

The closure V of the set $D(S)$ in this norm is a Hilbert space, algebraically equivalent to $D(\bar{S})$. We define the operator C by

$$(Cg)(t) = cg(t)$$

at the same time

$$D(C) = \{g \in L_2(T; H_1): g(t) \in D(c)$$

$$\text{for almost all } t \in [0; T]; \, Cg \in L_2(T; H_2)\}$$

It is assumed that $D(C) \supset D(\bar{S})$ on the set $D(\bar{S})$ the operator C is continuous as an operator from V to $L_2(T; H_2)$.

We define a generalized solution of problem (7.45) as an element $z \in D(\bar{S})$; for given $u \in A_u$ it satisfies the conditions

$$\bar{S}z = 0, \qquad Cz = u \tag{7.53}$$

We introduce the functional $I[z; u]$ on the product space $L_2(T; H_1) \times L_2(T; H_2)$ and endow it with the properties (a) and (b) as before; however, replace property (c) by the requirement (d) $\lim_{n \to \infty} I[z_n; u_n] = \infty$ if $\lim_{n \to \infty} \|z_n\|_1 = \infty$ for all $u_n \in A_u$.

The optimization problem consists of the following: to find an element u_0 in the closed convex set $A_u \in L_2(T; H_2)$ and a corresponding [in the sense of problem (7.53)] element $z_0 \in D(\bar{S})$ such that

$$I[z_0; u_0] = j_0 = \inf_{\substack{u \in A_u \\ z \in D}} I[z; u]$$

subject to the constraints (7.53). The corresponding ε problem is: to determine elements $u_\varepsilon \in A_u$, $z_\varepsilon \in D(\bar{S})$ such that

$$I_\varepsilon[z_\varepsilon; u_\varepsilon] = j_\varepsilon = \inf_{\substack{u \in A_u \\ z \in D}} I_\varepsilon[z; u]$$

subject to the condition $Cz = u$. Here we have introduced the notation

$$I_\varepsilon[z_\varepsilon; u_\varepsilon] = I[z; u] + \frac{1}{\varepsilon} \|\bar{S}z\|_1^2 \qquad (7.54)$$

The proofs of the existence theorems are similar to those of the analogous theorems for problem (7.47). It thus is of interest to briefly consider the differences. In the ε problem, the elements $z_n \in D, u_n \in A_u$ of the weakly converging minimizing sequences are related by the linear equation $Cz_n = u_n$, which by virtue of the continuity of C as an operator from V to $L_2(T; H_2)$ survives the weak passage to the limit. With this in mind, the proof may be completed as above. Similar remarks apply to the proof of the existence of a solution of the basic problem.

The penalty method differs advantageously from other methods in that constraints need not be taken into account explicitly. This is a valuable advantage for practical calculations which are made for the ε problem in accordance with the basic idea of the method. The theorems just given show that for sufficiently small ε this process enables one to determine a solution of the basic optimization problem with known accuracy.

The algorithmic realization of this idea has been studied by Balakrishnan[186] and De Julio.[222]

7.6. Optimization Problems with Moving Boundaries

In many problems of the theory of heat conduction, the shape of the basic region of variation of the independent variables is not specified in advance but is determined from additional conditions which occur in the problem formulation. A classical example is the Stefan problem, which describes the propagation of a melting (solidification) front whose motion is due to a phase transition. A number of problems of diffusion, earth mechanics, etc. are treated by Degtyarev.[61]

A process develops in the region $l_1(t) \le x \le l_2(t), 0 \le t \le T$; it is described by the following system of parabolic equations;

$$z_t^i - z_{xx}^i = F_i(x, t, z) \qquad (i = 1, \ldots, n) \qquad (7.55)$$

Here, $z = (z^1(x, t), \ldots, z^n(x, t))$ are the dependent variables which describe the state of the system, T is the time of the process, and $F_i(x, t, z)$ are given functions of their arguments, continuous with respect to x and t and twice continuously differentiable with respect to the remaining arguments.

We take initial conditions in the form

$$z^i(x, 0) = a_i(x) \qquad (i = 1, \ldots, n) \qquad (7.56)$$

where $a_i(x)$ are given continuous functions.

The boundary conditions can be chosen in the form

$$z^i(l_j(t), t) = b_{ij}(t) \qquad (i = 1, \ldots, n; j = 1, 2) \tag{7.57}$$

or in the form

$$\left.\frac{\partial z^i(x, t)}{\partial x}\right|_{x=l_j(t)} + \gamma_{ij}(t)z^i(l_j(t), t) = b_{ij} \qquad (i = 1, \ldots, n; j = 1, 2) \tag{7.58}$$

Here, $\gamma_{ij}(t)$ and $b_{ij}(t)$ are given continuous functions.

The functions $l_j(t)$, which characterize the motion of the boundaries, satisfy the system of equations

$$\frac{dl_j(t)}{dt} = f_j(l_1(t), l_2(t), u(t), t) \qquad (j = 1, 2) \tag{7.59}$$

with initial conditions

$$l_1(0) = c_1 \qquad (j = 1, 2) \tag{7.60}$$

Here, $u(t) = (u_1(t), \ldots, u_p(t))$ is a control vector, c_j are given positive constants, and f_j are functions which are continuous with respect to t and twice continuously differentiable with respect to the remaining parameters.

We shall take the admissible controls $u(t)$ to be piecewise continuous functions taking on values in some region A of p-dimensional Euclidean space.

It is assumed that problem (7.55)–(7.60) has a unique solution for a given admissible control, and that this solution changes little as a result of small deviations from the law of motion of the boundaries.

We pose the following optimization problem: among admissible controls $u(t) \in A_u$ to determine a control that minimizes (maximizes) the functional

$$I = \int_0^T \int_{l_1(t)}^{l_2(t)} G_1(t, x, z) \, dt \, dx + \int_{l_1(T)}^{l_2(T)} G_2[x, z(x, T)] \, dx$$

$$+ \int_0^T G_3[t, \varphi(l(t), t), l(t), u(t)] \, dt + G_4[l(T)] \tag{7.61}$$

Here, $G_j (j = 1, \ldots, 4)$ are given functions which are continuous with respect to t and x and twice continuously differentiable with respect to the remaining arguments.

Following the general scheme of Chapter 2, we introduce the Lagrange multipliers $\psi(t, x) = (\psi_1(t, x), \ldots, \psi_n(t, x))$ and $\lambda_1(t), \lambda_2(t)$ as solutions of the boundary-value problem

$$\frac{\partial \psi_i}{\partial t} + \frac{\partial^2 \psi_i}{\partial x^2} = -\frac{\partial H}{\partial z^i}$$

$$\frac{d\lambda_j}{dt} = -\frac{\partial h}{\partial l_j(t)} - (-1)^j \left[G_1 + \sum_{i=1}^n f_i \psi_i \frac{\partial z^i}{\partial x} + \psi_i \frac{\partial^2 z^i}{\partial x^2} - \frac{\partial \psi_i}{\partial x} \frac{\partial \varphi_i}{\partial x} \right]_{x=l_j(t)}$$

$$(j = 1, 2)$$

$$\psi_i(x, T) = -\frac{\partial G_2}{\partial z^i(x, T)} \qquad (i = 1, \ldots, n)$$

$$\lambda_j(T) = -\frac{\partial G_4}{\partial l_j(T)} - (-1)^j G_2 |_{x=l_j(T)} \qquad (j = 1, 2)$$

The boundary conditions for $\psi_i(x, t)$ are specified in the form

$$\psi_i(l_i(t), t) = 0$$

or

$$(f_j - \gamma_{ij}) \psi_i(l_i(t), t) - \frac{\partial \psi_i(x, t)}{\partial x} \bigg|_{x=l_j(t)} = (-1)^j \frac{\partial h}{\partial \varphi_i(l_j(t), t)}$$

$$(i = 1, \ldots, n; j = 1, 2)$$

Here, we have introduced the notation

$$H = \sum_{i=1}^n \psi_i(x, t) F_i(x, t, \varphi(x, t)) - G_1$$

$$h = \sum_{j=1}^2 \lambda_j f_j(l(t), u(t), t) - G_3$$

In Ref. 61, Degtyarev proved the following theorem.

Theorem 7.5. If the control $u(t)$ is to minimize (maximize) the functional I, it is necessary that it maximize (respectively, minimize) the function h:

$$h(t, l(t), z(l(t), t), u(t)) = \sup_{u \in A_u} h(t, l(t), z(l(t), t), u)$$

almost everywhere.

The proof is based on the incremental expression

$$\Delta I = -\int_0^T [h(t, l(t), z(l(t), t), u(t) + \Delta u(t))$$

$$-h(t, l(t), z(l(t), t), u(t))] \, dt + o(\varepsilon) \qquad (7.62)$$

which holds if $\Delta u(t)$ is nonzero on a set of measure ε. Equation (7.62) may readily be obtained by the repeatedly tested methods of the preceding sections.

In problems with moving boundaries, the control sometimes enters in a different manner. Budak and Gol'dman studied a problem for an equation of the type (7.55) with $F = 0$ and the additional conditions

$$y(t) < x < l, \qquad 0 < t \le T$$

$$z(t, y(t)) = f(t)u(t), \qquad 0 < t \le T$$

$$z(t, l) = c, \qquad 0 < t \le T$$

$$z(0, x) = a(x), \qquad l_0 \le x \le l$$

$$\gamma(y(t), t)y'(t) = -z_x(y(t), t) + \varphi(y(t), t), \qquad 0 < t \le T$$

$$y(0) = l_0$$

with $\gamma(x, t) \ge \gamma_{\min} > 0$, and with $\psi(x, t)$, $a(x)$, and $\varphi(x, t)$ as known smooth functions. The role of control in this problem is played by the function $u(t)$, a continuously increasing function on $[0, T]$ which satisfies the conditions $u(0) = u_{\min} \geq 0, u(T) = u_{\max}, 0 < L_2 \leq [u(t'') - u(t')]/(t'' - t') \leq L_1$ for all $t', t'' \in [0, T]$; L_1 and L_2 are constants. The problem consists of choosing a u in the given set which minimizes the functional

$$I[u] = \int_0^T z(y(t), t)y(t) \, dy(t)$$

In Ref. 16, Budak and Gol'dman proved the existence of an optimal control. The proof is based on the compactness of the set $\{u_k\}$ of admissible controls and on estimates for the derivatives $y'_k(t)$ that make it possible to prove convergence of the corresponding sequences $\{y_k(x)\}$.

8
Bellman's Method in Variational Problems with Partial Derivatives

The part played by Bellman's method of dynamic programming in the theory of optimal control is well known. Together with Pontryagin's maximum principle, this method is an important tool for solving problems of optimal control and it is especially well suited for the use of computers.

From the point of view of the calculus of variations, Bellman's method bears the same relation to the maximum principle as the Hamilton–Jacobi method does to Euler's method. One can say that Bellman's method is the Hamilton–Jacobi method for variational problems with differential constraints in the presence of restrictions on the controls.

The method has been widely developed by many people, initially by Bellman himself; it exists in both discrete and continuous variants, the latter applying, principally, to systems described by ordinary differential equations.* The derivation of Bellman's equation for a typical problem of this kind was given in Section 6.2.

The arguments on which this derivation is based also remain valid in problems involving partial differential equations. However, there then is an important difference between static problems and problems containing time. Whereas Bellman's optimality principle can be transferred almost literally to many dynamic problems involving partial differential equations, problems of elliptic type are trickier because of their nonevolutionary nature. Nevertheless, under certain conditions, the Hamilton–Jacobi method can be extended to this class of variational problems as well.

Investigations of various generalizations of the Hamilton–Jacobi method to problems involving partial differential equations go back to the pioneering studies of Volterra,[377,378] Fréchet,[250] and Lévy[92]; many results can be found in Prange's dissertation.[340]

In the present exposition, we begin by describing the basic arguments as they apply to variational problems of the simpler kind and we then

---

* For an account of Bellman's method, see Refs. 11, 12, 229, and Rozonóer's paper.[144]

indicate ways in which the method can be extended to problems with constraints specified by partial differential equations.

8.1. Canonical Equations for the Simple Variational Problem with Many Independent Variables: Volterra Form and Hadamard–Lévy Form

We consider the problem of extremizing the functional

$$I = \iint_G L(x, y, z, z_x, z_y) \, dx \, dy \tag{8.1}$$

where the comparison functions $z(x, y)$ are made to satisfy the requirement

$$z = f(t) \qquad \text{on the boundary } \Gamma \text{ of region } G \tag{8.2}$$

We introduce the conditions

$$\frac{\partial z}{\partial x} - z_x = 0$$

$$\frac{\partial z}{\partial y} - z_y = 0 \tag{8.3}$$

and we shall regard Eqs. (8.1)–(8.3) as a problem with three unknown functions z, z_x, z_y. With the use of Lagrange multipliers p, q, and ρ, we may form the following functional for this problem:

$$\iint_G \left[L + p\left(\frac{\partial z}{\partial x} - z_x\right) + q\left(\frac{\partial z}{\partial y} - z_y\right) \right] dx \, dy - \int_\Gamma \rho(t)[z - f(t)] \, dt \tag{8.4}$$

The Euler equations and the natural boundary conditions have the form

$$L_{z_x} - p = 0, \qquad L_{z_y} - q = 0, \qquad L_z - \frac{\partial p}{\partial x} - \frac{\partial q}{\partial y} = 0 \tag{8.5}$$

$$py_t - qx_t - \rho = 0 \qquad \text{on } \Gamma \tag{8.6}$$

Adding one or more of the relations (8.3), (8.5), and (8.6) as subsidiary conditions, we obtain various modifications of the posed variational problem.[86] If the first two equations in (8.5) are taken as subsidiary conditions, the functional (8.4) may be written in the form

$$\iint_G \left(p\frac{\partial z}{\partial x} + q\frac{\partial z}{\partial y} - H \right) dx \, dy - \int_\Gamma \rho(t)[z - f(t)] \, dt \tag{8.7}$$

where

$$H = H(x, y, z, p, q) = pz_x + qz_y - L(x, y, z, z_x, z_y) \qquad (8.8)$$

and the functions z_x and z_y can be expressed in terms of p and q by means of the equations

$$L_{z_x} = p, \qquad L_{z_y} = q$$

With (8.7) regarded as a functional of z, p, q, the corresponding Euler equations become

$$\frac{\partial z}{\partial x} - \frac{\partial H}{\partial p} = 0, \qquad \frac{\partial z}{\partial y} - \frac{\partial H}{\partial q} = 0, \qquad \frac{\partial p}{\partial x} + \frac{\partial q}{\partial y} + \frac{\partial H}{\partial z} = 0 \qquad (8.9)$$

the last of which coincides with the last equation of (8.5), in view of the obvious equality [see (8.8)]

$$\frac{\partial L}{\partial z} = -\frac{\partial H}{\partial z}$$

Equations (8.9) form a system of canonical equations of the variational problem (8.1)–(8.2) in Volterra form. Here the functions p and q (the Lagrange multipliers) are analogous to the canonical momenta in the equations of analytical mechanics.*

Another way of expressing the canonical equations is due to Hadamard and Lévy. The simplest variant of this form is obtained by choosing the second equation of (8.5) and the first equation of (8.3) as subsidiary conditions. In this way, we arrive at the relations[340]

$$\frac{\partial z}{\partial y} = K_q, \qquad \frac{\partial q}{\partial y} = -K_z + \frac{\partial}{\partial x} \frac{\partial K}{\partial z_x}$$

or, using the notation for the functional derivative,

$$\frac{\partial z}{\partial y} = \frac{\delta \mathcal{H}}{\delta q}, \qquad \frac{\partial q}{\partial y} = -\frac{\delta \mathcal{H}}{\delta z} \qquad (8.10)$$

---

* Obviously, the Lagrange multipliers ξ_{ij} in (2.47) which also have the Volterra form, play the same role.

To obtain these equations, we introduce the notation

$$qz_y - L(x, y, z, z_x, z_y) = K(x, y, z, z_x, q) \qquad (8.11)$$

i.e., we take a Legendre transformation with respect to the argument z_y. The double integral (8.4) takes on the form

$$\iint_G \left[q\frac{\partial z}{\partial y} - K(x, y, z, z_x, q) \right] dx\, dy \qquad (8.12)$$

Equations (8.10) are the Euler equations for the functional (8.12) of the two functions z and q or, equivalently, the Hamilton equations for the functional (8.1). The functionals $\mathscr{L}$ and $\mathscr{K}$ are defined by

$$\mathscr{L}[y, z, z_y] = \int_{x_0}^{x_1} L(x, y, z, z_x, z_y)\, dx \qquad (8.13)$$

$$\mathscr{K}[y, z, q] = \int_{x_0}^{x_1} K(x, y, z, z_x, q)\, dx \qquad (8.14)$$

Obviously,

$$\mathscr{K} = \int_{x_0}^{x_1} qz_y\, dx - \mathscr{L} \qquad (8.15)$$

In (8.10), the variable y plays a distinguished role, usually that of the time. In cases when a distinguished variable of this kind does not exist, a different form of the canonical equations is more convenient. To obtain it, we introduce a system of orthogonal curvilinear coordinates (α, β) in the closed region, G, in such a way that the boundary Γ is the coordinate line $\alpha = 0$.

In these new coordinates, the functional (8.1) takes on the form

$$I = \iint_G \tilde{f}(z_\alpha, z_\beta, z, \alpha, \beta) H_\alpha H_\beta\, d\alpha\, d\beta$$

where H_α and H_β are the Lamé coefficients. We transform the Euler equation

$$\frac{1}{H_\alpha} \frac{\partial}{\partial \alpha}(\tilde{f}_{z_\alpha} H_\alpha) + \frac{1}{H_\beta} \frac{\partial}{\partial \beta}(\tilde{f}_{z_\beta} H_\beta) + \frac{1}{H_\alpha H_\beta} \frac{\partial H_\beta}{\partial \alpha} \tilde{f}_{z_\alpha} H_\alpha$$

$$+ \frac{1}{H_\alpha H_\beta} \frac{\partial H_\alpha}{\partial \beta} \tilde{f}_{z_\beta} H_\beta - \tilde{f}_z = 0$$

in terms of the parameters n and t whose differentials

$$dn = H_\alpha \, d\alpha, \qquad dt = H_\beta \, d\beta$$

are, respectively, the differentials of the arcs of the coordinate lines α and β.

We introduce the radii of curvature ρ_n and ρ_t of the lines $\alpha = \text{const}$ and $\beta = \text{const}$:

$$\frac{1}{\rho_n} = \frac{1}{H_\alpha H_\beta} \frac{\partial H_\beta}{\partial \alpha} = \frac{\partial \ln H_\beta}{\partial n}$$

$$\frac{1}{\rho_t} = \frac{1}{H_\alpha H_\beta} \frac{\partial H_\alpha}{\partial \beta} = \frac{\partial \ln H_\alpha}{\partial t}$$

The Euler equation may now be written in the form

$$\frac{\partial \tilde{f}_{z_n}}{\partial n} + \frac{\partial \tilde{f}_{z_t}}{\partial t} + \frac{1}{\rho_n} \tilde{f}_{z_n} + \frac{1}{\rho_t} \tilde{f}_{z_t} - \tilde{f}_z = 0$$

or

$$\delta \tilde{f}_{z_n} = -\frac{\partial \tilde{f}_{z_t}}{\partial t} \delta n - \tilde{f}_{z_t} \frac{1}{\rho_t} \delta n - \left(\frac{1}{\rho_n} \tilde{f}_{z_n} - \tilde{f}_z \right) \delta n$$

Now, it is readily seen that

$$\frac{d}{dt} \delta n = \frac{\delta n}{\rho_t}$$

Taking this into account, the Euler equation may be written in the form

$$\delta \tilde{f}_{z_n} = -\frac{d}{dt} (\tilde{f}_{z_t} \delta n) - \left(\frac{1}{\rho_n} \tilde{f}_{z_n} - \tilde{f}_z \right) \delta n \tag{8.16}$$

We now introduce a new dependent variable:

$$\xi = \frac{\partial \tilde{f}}{\partial z_n} \tag{8.17}$$

and construct the function (Hamiltonian)

$$K(\alpha, \beta, z, z_t, \xi) = \xi z_n - \tilde{f} \qquad (8.18)$$

We have

$$\frac{\partial K}{\partial \xi} = z_n, \qquad \frac{\partial K}{\partial z} = -\frac{\partial \tilde{f}}{\partial z}, \qquad \frac{\partial K}{\partial z_t} = -\frac{\partial \tilde{f}}{\partial z_t}$$

The Euler equation (8.16) is equivalent to the system

$$\delta z = \frac{\partial K}{\partial \xi} \delta n$$

$$\delta \xi = \frac{d}{dt}(K_{z_t} \delta n) - \left(\frac{1}{\rho_n}\xi + K_z\right)\delta n \qquad (8.19)$$

We have arrived at the canonical equations in the Hadamard–Lévy form. These equations determine the increments of the canonical variables z and ξ on the transition from the curve C to a curve C' separated from C by a distance δn along the normal. Formally, (8.19) can be obtained by deriving the Euler equations for the functional

$$I = \int \delta n \int [\xi z_n - K(\alpha, \beta, z, z_t, \xi)] \, dt$$

with respect to the two functional arguments z and ξ:

$$z_n - \frac{\partial K}{\partial \xi} = 0$$

$$\frac{d}{dt}(K_{z_t} \delta n)\, dt - \frac{\delta}{\delta n}(\xi \, dt)\, \delta n - \frac{\partial K}{\partial z}\, \delta n \, dt = 0$$

Bearing in mind that

$$\frac{\delta \, dt}{\delta n} = \frac{dt}{\rho_n}$$

we arrive at the system (8.19).

8.2. The Hamilton–Jacobi Equation for the Simple Variational Problem with Partial Derivatives

We shall base our arguments on the functional (8.1). We take a rectangular coordinate system x, y, z and imagine an extremal surface $z = z(x, y)$.

Let Σ, the contour of this surface, be the space curve defined by its projection $(x(t), y(t))$ onto the x, y plane, and the coordinate $z(t)$.

We consider the extremal value of the functional I as a function of Σ; in other words, we shall study the dependence of the values I^* taken by the functional I on Σ $(\Gamma; z(t))$ on the solutions of the Euler equation

$$\frac{\partial L_{z_x}}{\partial x} + \frac{\partial L_{z_y}}{\partial y} - \frac{\partial L}{\partial z} = 0 \tag{8.20}$$

Variation of an element of the curve Σ can be characterized by the components $\delta n(t)$ and $\Delta z(t)$ along the normal $\mathbf{n}$ to the planar curve γ and along the z coordinate; denoting the corresponding functional derivatives by I_n^* and I_z^*, we write

$$\delta I^* = \int_\Gamma (I_n^* \, \delta n + I_z^* \, \Delta z) \, dt \tag{8.21}$$

On the other hand, by virtue of (8.20)

$$\delta I^* = \int_\Gamma L \, \delta n \, dt + \int_\Gamma (L_{z_x} y_t - L_{z_y} x_t) \, \delta z \, dt$$

or, taking into account the expression for the total variation Δz on the moving boundary (see Section 2.2) $\Delta z = \delta z + z_n \delta n$

$$\delta I^* = \int_\Gamma \{(L - (L_{z_x} y_t - L_{z_y} x_t) z_n] \, \delta n + (L_{z_x} y_t - L_{z_y} x_t) \, \Delta z\} \, dt$$

$$= \int_\Gamma [(L - L_{z_n} z_n) \, \delta n + L_{z_n} \, \Delta z] \, dt$$

A comparison with (8.21) leads to the equations

$$\frac{\delta I^*}{\delta n} = I_n^* = L - L_{z_n} z_n$$

$$\frac{\delta I^*}{\delta z} = I_z^* = L_{z_n} \tag{8.22}$$

which hold along Γ.

We use the orthogonal curvilinear coordinates α and β introduced in Section 8.1 and the corresponding parameters n and t ($dn = H_\alpha \, d\alpha$, $dt = H_\beta \, d\beta$); suppose the constant value $\alpha = \alpha_0$ corresponds to the line Γ.

By means of the formula

$$\xi = L_{z_n} \tag{8.23}$$

which defines the "canonical momentum" ξ, we eliminate z_n from the expression

$$L_{z_n} z_n - L(\alpha_0, \beta, z, z_t, z_n) \tag{8.24}$$

What is left is the Hamiltonian K [see (8.18)]:

$$K(\alpha_0, \beta, z, z_t, \xi) = \xi z_n - L \tag{8.25}$$

Equations (8.22) can now be written in the form

$$I_n^* = -K(\alpha_0, \beta, z, z_t, \xi)$$
$$I_n^* = \xi \tag{8.26}$$

Eliminating ξ, we obtain the equation

$$I_n^* = K(I_z^*, z_t, z, \alpha_0, \beta)$$

Since the value of α_0 is arbitrary, we may assume that the functional $I^*[\Gamma; z(t)]$ satisfies the following equation in partial functional derivatives:

$$I_n^* = K(I_z^*, z_t, z, \alpha, \beta) \tag{8.27}$$

This is the Hamilton–Jacobi equation of the variational problem we are considering.

We now give an example. Consider the Dirichlet integral for

$$L = (\text{grad } z)^2 = z_n^2 + z_t^2$$

as defined by (8.1). Obviously,

$$\xi = 2z_n$$

$$K = \xi z_n - L = \frac{\xi^2}{4} - z_t^2$$

The Hamilton–Jacobi equation has the form

$$I_n^* = \tfrac{1}{4}(I_z^*)^2 - z_t^2$$

In applications, the case when one of the independent variables is the time is important. Let y be such a variable. In the coordinates x, y the basic region is a rectangle $(x_0 \leq x \leq x_1, \; y_0 \leq y \leq y_1)$; we shall be interested in the dependence of the extremal value of the functional I on the parameter y_1. Calculating the increment in accordance with (8.21), we must assume a variation $\delta n = \delta y_1$ that does not depend on x. As a result, we obtain

$$\frac{\partial I^*}{\partial y_1} = I_{y_1}^* = -\int_{x_0}^{x_1} K(\xi, z_x, z, y_1, x)\, dx$$

$$\frac{\delta I^*}{\delta z^*} = I_z^* = \xi$$

(8.28)

or, replacing y_1 by y,

$$\frac{\partial I}{\partial y} + \int_{x_0}^{x_1} K\left(x, y, z, z_x, \frac{\delta I}{\delta z}\right) dx = 0$$

Using the notation (8.12), we may reduce this equation to the form

$$\frac{\partial I}{\partial y} + \mathcal{H}\left[y, z, \frac{\delta I}{\delta z}\right] = 0$$

(8.29)

Here, the symbol $\delta I / \delta z$ denotes the functional derivative

$$\frac{\delta I}{\delta z} = \lim_{\|\delta z\|_{L_1(G)} \to 0} \frac{I[y, z + \delta z] - I[y, z]}{\|\delta z\|_{L_1(G)}}$$

with respect to y; the functional I is an ordinary function. The Hamilton–Jacobi equation (8.29) is generated by the variational problem

$$\int_{y_0}^{y_1} \int_{x_0}^{x_1} L(x, y, z, z_x, z_y)\, dx\, dy = \min$$

(8.30)

Its importance in the investigation of this problem can be demonstrated by means of the concept of self-adjoint and matching (or mutually consistent) boundary conditions.[45,94]

To this end, we introduce the functional $S[z; y]$ and consider the variational problem

$$\int_{y_0}^{y_1} \int_{x_0}^{x_1} L(x, y, z, z_x, z_y) \, dx \, dy - S[z; y_1] = \min$$

We assume that the function $z(x, y_0)$ is known; the natural boundary condition on $y = y_1$ has the form

$$\frac{\delta \mathcal{L}}{\delta z_y} - \frac{\delta S}{\delta z} = 0$$

or, in accordance with the second equation of (8.5),

$$q(x, y_1, z, z_x, z_y) = \frac{\delta S}{\delta z}\bigg|_{y=y_1} \tag{8.31}$$

This condition specifies the quantity $z_y(x, y_1)$, at every point x of the straight line $y = y_1$; note that it is proportional to the corresponding direction cosine of the normal to the integral surface $z = z(x, y)$. It is convenient to write (8.31) as

$$z_y(x, y_1) = \Psi[z; y_1] \tag{8.32}$$

where ψ is a functional of z and a function of x and y_1.

Definition 8.1. The boundary conditions (8.32) associated with the functional (8.30) are *self-adjoint on* $y = y_1$ if and only if there exists a functional $S[z; y]$, depending on y as on a parameter, such that

$$\frac{\delta \mathcal{L}[y, z, \Psi[z; y_1]]}{\delta z_y} = \frac{\delta S}{\delta z} \quad \text{for } y = y_1 \tag{8.33}$$

If the boundary conditions (8.31) are to be self-adjoint on the straight line $y = y_1$, it is necessary and sufficient that the following condition be satisfied:

$$\frac{\delta q(x, y_1, z(x, y_1), z_x(x, y_1), \Psi[z; y_1])}{\delta z(x_2, y_1)}\bigg|_{x=x_1}$$

$$= \frac{\delta q(x, y_1, z(x, y_1), z_x(x, y_1), \Psi[z; y_1])}{\delta z(x_1, y_1)}\bigg|_{x=x_2} \tag{8.34}$$

We first prove necessity. Self-adjoint boundary conditions are determined by (8.31). With $x = x_1$ and a variation with respect to $z(x_2, y_1)$ and, conversely, $x = x_2$ and a variation with respect to $z(x_1, y_1)$, we obtain (8.34) since both expressions are equal to

$$\frac{\delta^2 S}{\delta z(x_1, y_1)\, \delta z(x_2, y_1)}$$

Sufficiency follows from the fact that if the conditions (8.32) involve (8.34), then there exists a functional $S[z; y]$ whose functional derivative on $y = y_1$ is equal to q. The variation of the functional

$$\int_{y_0}^{y_1} \int_{x_0}^{x_1} L\, dx\, dy - S[z; y_1]$$

when equated to zero, yields condition (8.33), thus proving self-adjointness.

Definition 8.2. The boundary condition

$$z_y(x) = \Psi_1[z] \tag{8.35}$$

specified on $y = y_1$, and the boundary condition

$$z_y(x) = \Psi_2[z] \tag{8.36}$$

specified on $y = y_2$, are called *matched* (consistent) *with one another* if every solution of the system (8.5) which satisfies condition (8.35) on $y = y_1$ satisfies condition (8.36) on $y = y_2$ and vice versa.

Definition 8.3. Suppose that, for all $y(a \le y \le b)$, the following boundary conditions are given:

$$z_y(x, y) = \Psi[z; y]$$

These boundary conditions form a *field of the functional* (8.30) if: (i) for every value of y, the conditions are self-adjoint; (ii) for any two y_1 and y_2 in $[a, b]$, the conditions are consistent with one another.

Suppose the self-adjointness of the boundary conditions has been established; what additional requirements must we impose on these conditions to match them with each other? In other words, in what cases do self-adjoint boundary conditions form a field of the functional (8.30)? This question is answered by the following theorem:

Theorem 8.1. If the boundary conditions

$$\frac{\delta \mathcal{L}}{\delta z_y} = \frac{\delta S}{\delta z} \tag{8.37}$$

are to be matched with one another, it is necessary and sufficient that the functional $S[z; y]$ satisfy the Hamilton–Jacobi equation

$$\frac{\partial S}{\partial y} + \int_{x_0}^{x_1} K\left(x, y, z, z_x, \frac{\delta S}{\delta z}\right) dx = 0$$

or

$$\frac{\partial S}{\partial y} + \mathcal{H}\left[y, z, \frac{\delta S}{\delta z}\right] = 0 \tag{8.38}$$

We now show that, subject to the conditions of the theorem, any manifold in the space (x, y, z) along which (8.37) is satisfied is an integral surface of the Euler equations (8.5). From (8.37) it follows that

$$q(x, y, z, z_x, \Psi[z; y]) = \frac{\delta S}{\delta z}$$

therefore, (8.38) can also be written in the form

$$\frac{\partial S}{\partial y} = -\mathcal{H}[y, z, q]$$

Varying this equation with respect to z, we obtain

$$\frac{\partial}{\partial y} \frac{\delta S}{\delta z} = -\frac{\delta \mathcal{H}[y, z, \Psi[z; y]]}{\delta z}$$

or

$$\frac{\partial q}{\partial y} = -\frac{\delta \mathcal{H}[y, z, \Psi[z; y]]}{\delta z} \tag{8.39}$$

and the assertion of the theorem follows. For, let us replace q by

$$\frac{\delta \mathscr{L}[y, z, z_y]}{\delta z_y}\Bigg|_{z_y = \Psi[z; y]}$$

and the functional $\mathscr{K}$ by its expression (8.15). Carrying out the operations indicated in (8.39) and taking the equation $\delta \mathscr{L}/\delta z_y = \partial L/\partial z_y$ into account, we obtain

$$\frac{\partial}{\partial y} \frac{\delta \mathscr{L}}{\delta z_y} + \int_{x_0}^{x_1} \frac{\delta^2 \mathscr{L}[y, z, \Psi[z; y]]}{\delta z_y(x, y) \, \delta z(\xi, y)} \frac{\partial \Psi[z(\xi, y); y]}{\partial y} d\xi$$

$$= \frac{\delta \mathscr{L}}{\delta z} + \int_{x_0}^{x_1} \frac{\delta \mathscr{L}[y, z, \Psi[z; y]]}{\delta z_y(\xi, y)} \frac{\delta \Psi[z(\xi, y); y]}{\delta z(x, y)} d\xi$$

$$- \int_{x_0}^{x_1} \frac{\delta \Psi[z(\xi, y); y]}{\delta z(x, y)} \frac{\delta \mathscr{L}[y, z, \Psi[z; y]]}{\delta z_y(\xi, y)} d\xi$$

$$- \int_{x_0}^{x_1} \Psi[z(\xi, y); y] \frac{\delta^2 \mathscr{L}[y, z, \Psi[z; y]]}{\delta z_y(\xi, y) \delta z(x, y)} d\xi \tag{8.40}$$

The condition of self-adjointness, (8.37), yields

$$\frac{\delta^2 \mathscr{L}}{\delta z_y(\xi, y) \delta z(x, y)} = \frac{\delta^2 \mathscr{L}}{\delta z_y(x, y) \delta z(\xi, y)}$$

We may thus rewrite (8.40) as

$$\frac{\delta \mathscr{L}}{\delta z} = \frac{\partial}{\partial y} \frac{\delta \mathscr{L}}{\delta z_y} + \int_{x_0}^{x_1} \Psi[z(\xi, y); y] \frac{\delta^2 \mathscr{L}[z, \Psi[z; y]]}{\delta z(\xi, y) \delta z_y(x, y)} d\xi$$

$$+ \int_{x_0}^{x_1} \frac{\delta^2 \mathscr{L}[z, \Psi[z; y]]}{\delta z_y(x, y) \delta z_y(\xi, y)} \frac{\partial \Psi[z(\xi, y); y]}{\partial y} d\xi \tag{8.41}$$

Since

$$\frac{\delta^2 \mathscr{L}}{\delta z(\xi, y) \, \delta z_y(x, y)} = \frac{\delta}{\delta z(\xi, y)}\Bigg|_{\Psi = \text{const}} \frac{\delta \mathscr{L}}{\delta z_y(x, y)}$$

$$+ \int_{x_0}^{x_1} \frac{\delta^2 \mathscr{L}}{\delta z_y(x, y) \delta z_y(\zeta, y)} \cdot \frac{\partial \Psi[z(\zeta, y); y]}{\delta z(\xi, y)} d\zeta$$

(8.41) can be represented in the form

$$\frac{\delta\mathcal{L}}{\delta z} = \frac{\partial}{\partial y}\frac{\delta\mathcal{L}}{\delta z_y} + \int_{x_0}^{x_1}\frac{\delta}{\delta z(\xi, y)}\bigg|_{\Psi=\text{const}}\frac{\delta\mathcal{L}}{\delta z_y(x, y)}\Psi[z(\xi, y), y]\,d\xi$$

$$+ \int_{x_0}^{x_1}\frac{\delta^2\mathcal{L}}{\delta z_y(x, y)\,\delta z_y(\xi, y)}\bigg[\frac{\partial\Psi[z(\xi, y); y]}{\partial y}$$

$$+ \int_{x_0}^{x_1}\Psi[z(\zeta, y), y]\frac{\delta\Psi[z(\xi, y); y]}{\delta z(\zeta, y)}\,d\zeta\bigg]\,d\xi \qquad (8.42)$$

Taking into account the condition $z_y = \Psi[z(x, y); y]$ and the relation

$$z_{yy} = \frac{\partial\Psi[z(x, y); y]}{\partial y} + \int_{x_0}^{x_1}\Psi[z(\zeta, y); y]\frac{\delta\Psi[z(x, y), y]}{\delta z(\zeta, y)}\,d\zeta$$

which follows from it, (8.42) may be restated in the form

$$\frac{\delta\mathcal{L}}{\delta z} = \frac{\partial}{\partial y}\frac{\delta\mathcal{L}}{\delta z_y} + \int_{x_0}^{x_1}\frac{\delta}{\delta z(\xi, y)}\bigg|_{\Psi=\text{const}}\frac{\delta\mathcal{L}}{\delta z_y(x, y)}z_y(\xi, y)\,d\xi$$

$$+ \int_{x_0}^{x_1}\frac{\delta^2\mathcal{L}}{\delta z_y(x, y)\delta z_y(\xi, y)}z_{yy}(\xi, y)\,d\xi$$

or

$$\frac{\delta\mathcal{L}}{\delta z} - \frac{d}{dy}\frac{\delta\mathcal{L}}{\delta z_y} = 0$$

We have obtained the Euler equation of the original variational problem; it is equivalent to the canonical equations (8.5). This proves the sufficiency of condition (8.38); its necessity may be confirmed by reversing the previous steps.

8.3. The Hamilton–Jacobi Method for the Simple Variational Problem with Partial Derivatives

The deep analogy existing between one-dimensional and multi-dimensional variational problems is also manifested in the fact that Jacobi's theorem, and with it, the Hamilton–Jacobi method of integration of the canonical system admit a natural generalization to variational problems for multiple integrals.[92,94,340,378]

It is well known that Jacobi's theorem in its classical form establishes a way of obtaining a general solution of the canonical system from a known complete integral of the Hamilton–Jacobi equation.

Regarding the system with infinitely many degrees of freedom whose behavior is described by the functional (8.30) as the limiting case of a holonomic system with a finite number (n) of degrees of freedom (for which the classical Jacobi theorem is valid), it is natural to pose this question: what is a complete integral of the classical Hamilton–Jacobi equation in the limit as $n \to \infty$?

The limiting form of this equation has already been obtained in the form of (8.38). The generalization of the concept of a complete integral to a large class of first-order variational equations is due to Lévy[92]; the Hamilton–Jacobi equation belongs to this class.

According to Lévy, a complete integral S of (8.38) has two functional arguments: the unknown function $z(x, y)$ and the parametric function $\alpha(x)$; in addition, it depends on the variable y as on a parameter. If the parametric function coincides with the initial function $z(x, y_0)$, the complete integral is called a principal functional.* It should be noted that (8.38) has a set of complete integrals like the classical equation; the question of their mutual relationship will not be considered here. In particular, a complete integral does not always have a parametric function $\alpha(x)$ occurring explicitly in its expression as functional argument; this function may be represented, for example, by an infinite set of constants. The parametric function appeared in the complete integral as a result of the passage to the limit from the system of constants that occur in the complete integral of the Hamilton–Jacobi equation of the finite dimensional approximating problem. It may happen that some of these constants remain "isolated" constants in the limit; the possibilities which arise are best considered in examples.

The Hamilton–Jacobi method of integrating the canonical system is based on the validity of the following generalization of Jacobi's theorem.

Theorem 8.2.[92,94] Let $S[z, \alpha; y]$ be a complete integral of the Hamilton–Jacobi equation (8.38) and $\beta(x)$ be an arbitrary function. Then the functional $z = z[x, y; \alpha, \beta]$ determined by the equation†

$$\frac{\delta S}{\delta \alpha} = \beta \tag{8.43}$$

---

* By analogy with principal Hamilton function.

† The remark made above concerning the parametric function applies to this equation.

and the functional

$$q = \frac{\delta S}{\delta z} \tag{8.44}$$

form the general solution of the canonical system.

The proof follows from the fact that

$$\frac{d}{dy} \frac{\delta S}{\delta \alpha} = 0$$

holds on every integral surface. Indeed,

$$\frac{d}{dy} \frac{\delta S}{\delta \alpha} = \frac{\partial}{\partial y} \frac{\delta S}{\delta \alpha} + \int_{x_0}^{x_1} \frac{\delta^2 S}{\delta z(\xi, y) \delta \alpha} z_y(\xi, y) \, d\xi$$

We substitute $S = S[z, \alpha; y]$ in (8.38) and vary the result with respect to α; we obtain

$$\frac{\partial}{\partial y} \frac{\delta S}{\delta \alpha} = -\int_{x_0}^{x_1} \frac{\delta \mathcal{H}}{\delta q(\xi, y)} \frac{\delta^2 S}{\delta z(\xi, y) \delta \alpha} \, d\xi$$

Substitution in the preceding equation yields

$$\frac{d}{dy} \frac{\delta S}{\delta \alpha} = \int_{x_0}^{x_1} \frac{\delta^2 S}{\delta z(\xi, y) \delta \alpha} \left[z_y(\xi, y) - \frac{\delta \mathcal{H}}{\delta q(\xi, y)} \right] d\xi$$

On the integral surface $z_y = \delta \mathcal{H} / \delta q$, whence

$$\frac{d}{dy} \frac{\delta S}{\delta \alpha} = 0$$

which is what we wanted to prove.

If $S[z, \alpha; y]$ is a complete integral, then (8.43) can be solved for z, yielding a general solution which depends on the two arbitrary functions α and β. Equation (8.44) then defines the canonically conjugate momentum q. The functions α and β are determined by the initial conditions. For functions K of certain type, one can give methods for constructing the complete integral (8.38) (separation of variables); in particular, there exist generalizations of the well-known Liouville and Stäckel theorems.[95]

Jacobi's theorem, which now has been proved for (8.38), can also be extended to an equation of the type (8.27) and is convenient to use in cases when the variables x and y are on completely equal footings (e.g., in problems of elliptic type).

We consider the closed curve Σ_0 in the space (x, y, z) defined by its projection $\Gamma_0(x_0(t), y_0(t))$ onto the (x, y) plane and the coordinate $z_0(t)$; suppose, further, that $\Sigma(\Gamma, z(t))$ is another closed space curve such that the projections of Γ_0 and Γ do not have common points.

The extremal value taken by the functional (8.1) on the admissible surfaces $z = z(x, y)$ passing through the curves Σ_0 and Σ is a functional of these curves; this functional satisfies (8.27) with respect to the coordinates $(x(t), y(t), z(t))$ and $(x_0(t), y_0(t), z_0(t))$ [more precisely, with respect to the coordinates α, β, z (α_0, β_0, z_0)]. Therefore, the functional satisfies Lévy's definition[92] of a complete integral (it is a principal functional); knowledge of it makes it possible to obtain an integral of the Euler equations directly.[340] For this, it is sufficient to set

$$\frac{\delta I^*}{\delta z} = \xi, \qquad \frac{\delta I^*}{\delta z_0} = -\xi_0(t)$$

and solve the resulting equations for z and ξ:

$$z(\alpha, \beta) = f(z_0(t), \xi_0(t), \alpha)$$

$$\xi(\alpha, \beta) = g(z_0(t), \xi_0(t), \alpha)$$

The functions $z_0(t)$ and $\xi_0(t)$ can be found from the condition that the integral surface $z(\alpha, \beta)$ passes through the curves Σ_0 and Σ.

8.4. The Principle of Optimality

Bellman's method of solving optimal control problems is based on the principle of optimality—a fundamental assertion characterizing the optimal behavior of systems. Consider a controlled distributed parameter system; suppose its state is described by a vector $z(z^1, \ldots, z^n)$ which is defined and continuous in the closed region $G \cup \Gamma$ of the space $(x_1, \ldots, x_m)$. The vector z satisfies a system of equations of the type in Section 2.1 with corresponding boundary conditions; the equations contain the control functions u.

Suppose that the aim of the control is to minimize a certain additive functional $I[z; u]$, and suppose that there exists a control u^* that minimizes this functional. Consider the optimal integral surface $z(x)$ corresponding to this control. In the open region G, we distinguish a closed connected part δG and in the space z, x the cylindrical surface with base δG and generators parallel to the z axes until they intersect the integral surface $z(x)$. At the intersection, we obtain a space curve whose projection onto the space $(x_1, \ldots, x_m)$ is the manifold $\delta\Gamma$ bounding δG.

We now consider the original optimization problem; this time, however, for the region $G/\delta G$ bounded by the manifold $\Gamma \cup \delta\Gamma$; we take the values of the functions z on the new contour $\Gamma \cup \delta\Gamma$ equal to the values which they assume* at the corresponding points of the original optimal integral surface $z(x)$.

The principle of optimality asserts that under these conditions the optimal control u^{**} in the modified problem does not differ from the optimal control in the original problem (of course, we are considering the control in the region $G/\delta G$ common to both problems). In other words, a control that is optimal for an entire region is also optimal for any part of it if the values of the basic variables z on the new boundary are equal to the optimal values of these variables for the original problems at the same points.

The reservation about the values of the basic variables on the new boundary of the region is extremely important; in examples below it will be shown that the condition manifests itself differently in static and dynamic problems.

8.5. Bellman's Equation for Evolution Control Problems

We consider an optimization problem for the evolution equation

$$
\left.
\begin{aligned}
z_t = f(z, u, x, t) &\equiv z_{xx} + B(x, t)u(x, t) \\[6pt]
t \geq t_0, &\quad \alpha \geq x \geq 0 \\[6pt]
z(t_0, x) &= z_0(x) \\[6pt]
z(t, 0) = f_1(t), &\quad z(t, a) = f_2(t) \\[6pt]
I = \int_0^a z^2(T, x)\, dx &= \min_{u \in A}
\end{aligned}
\right\}
\qquad (8.45)
$$

where A is a given set and z a continuous function.

---

* On some parts of both the new and the old boundary manifolds the values of z need not be specified (see below).

Note that the operator $f(z, u, x, t)$ depends on u as an ordinary function. The region G on the plane (t, x) is the half-strip $t > t_0$, $a \geq x \geq 0$. Suppose $u^*(t, x)$ is an optimal control; we denote the value of the optimal function $z(t, x)$ for $t = t_1 > t_0$ by $z_1(x)$ and consider the problem (8.45) once more; this time, however, for the half-strip $t > t_1$, $a \geq x \geq 0$. The principle of optimality asserts that the optimal control $u^{**}(t, x)$ in the new problem coincides with the optimal control $u^*(t, x)$ in the original problem (of course, within the half-strip $t \geq t_1$, $a \geq x \geq 0$). It then follows, in particular, that the optimal value of the functional I is independent of the choice of initial time on the optimal trajectory.

The solutions of problems containing the time satisfy the principle of causality. This principle here is manifested in that the values of a solution at the time t can depend only on the values of the controls at times before t and cannot depend on the values of the controls at subsequent times. Applied to the present problem, this means that if $u = u^*(t, x)$ for $0 \leq t \leq t_1$, then the solution z for this interval of time does not depend on the behavior of $u(t, x)$ for $t > t_1$; in particular, the value of $z(t_1, x) = z_1(x)$ does not depend on this behavior. For this problem the principle of optimality is completely obvious: if it were not true and the value of the functional on the control $u^{**}(t, x)$ were less than on the control $u^*(t, x)$, then the control $u^*(t, x)$ could be improved by replacing it by the control

$$v(t, x) = \begin{cases} u^*(t, x), & t_0 \leq t \leq t_1 \\ u^{**}(t, x), & t_1 < t \leq T \end{cases}$$

Under the influence of the new control, the system would first be translated into the state $z(t_1, x) = z_1(x)$ by the control $u^*(t, x)$ (completely independent of the subsequent behavior of the control) and then into the final state. It is apparent from this discussion that the condition concerning the basic variables on the new boundary $(t = t_1)$ of the region as specified in the principle of optimality is automatically satisfied in this example.

This result is characteristic of examples containing the time; it is due to the fact that the Cauchy problem is generally well posed in this context. Under these conditions, the principle of optimality enables one to obtain a functional equation satisfied by the optimal value of the functional (Bellman's equation). One can see that this is none other than the Hamilton–Jacobi equation for the optimal control problem.

We derive Bellman's equation for the example (8.45). Let $I_T[z_0; t_0]$ be the value of the functional attained with the control $u^*(t, x)$ for the initial state $z_0(x)$, t_0. Changing the initial state, we obtain a functional $I_T[z; t]$ of the initial function $z(x)$, and on t as on a parameter. Consider the state

$z(x, t) = z'(x)$, $t' > t_0$, on the optimal surface. The optimal value of the functional does not depend on the choice of initial state on the optimal surface, and therefore

$$I_T[z_0; t_0] = I_T[z'; t']$$ (8.46)

Consider the optimal control $u^*(t, x)$ for $t' \geq t \geq t_0$. The state $z'(x)$ into which this control carries the system at the time t' must minimize the function $I_T[z'; t']$ compared with all other values that could be obtained on states z', t' generated by all possible admissible controls $u(t, x) \in A$ on the interval $[t_0, t']$. At the same time, it must be borne in mind that $z(t', x) = z'(x)$ depends on $u(t, x)$, as well as on $z_0(x)$, t_0.

The property of the control $u^*(t, x)$ for $t \in [t_0, t']$ just described may be expressed by the equation

$$I_T[z_0; t_0] = \min_{\substack{u(t,x) \in A, \\ t_0 \leq t \leq t'}} I_T[z'; t']$$ (8.47)

when (8.46) is taken into consideration.

We first consider the case when the control u does not depend on x: $u = u(t)$. Setting $t' = t_0 + \tau$, where τ is sufficiently small, we obtain

$$z'(x) = z(t', x) = z(t_0 + \tau, x) = z_0(x) + \tau f(z_0, u(t_0); x, t_0) + o(\tau) \quad (8.48)$$

Note that an admissible control at the time t_0 here serves as the argument $u(t_0)$ of the function f.

If the functional $I[z; t]$ is continuously differentiable with respect to z and t, then (8.48) may be used to obtain

$$I_T[z'; t'] = I_T[z_0; t_0] + \tau \int_0^a \frac{\delta I[z_0; t_0]}{\delta z_0} f(z_0, u(t_0), x, t_0) \, dx$$

$$+ \tau \frac{\partial I_T[z_0; t_0]}{\partial t_0} + o(\tau)$$ (8.49)

On the right-hand side of this equation, only the function f and the terms $o(\tau)$ depend on the admissible control u. Thus, the substitution of the expression for $I_T[z'; t']$ in (8.47) yields

$$\frac{\partial I_T[z_0; t_0]}{\partial t_0} = -\min_{u(t_0) \in A} \left[\int_0^a \frac{\delta I[z_0; t_0]}{\delta z_0} f(z_0, u(t_0); x, t_0) \, dx + \frac{o(\tau)}{\tau} \right]$$

In the limit $\tau \to 0$ this becomes

$$\frac{\partial I_T[z_0; t_0]}{\partial t_0} = - \min_{u(t_0)\in A} \int_0^a \frac{\delta I[z_0; t_0]}{\delta z_0} f(z_0, u(t_0); x, t_0) \, dx \qquad (8.50)$$

Since z_0 and t_0 are arbitrary, the subscript zero may be omitted and we regard (8.50) as a variational equation for $I_T[z; t]$; the determination of the functional I_T [see (8.45)] provides the initial condition for the solution of this equation:

$$I_T[z(T, x); T] = \int_0^a z^2(T, x) \, dx \qquad (8.51)$$

We set

$$-\frac{\delta I[z; t]}{\delta z} = q \qquad (8.52)$$

and introduce the function

$$\bar{K}(t, x, z, z_x, q, u) = qf \qquad (8.53)$$

along with the functional

$$\mathcal{K}[t, z, q, u] = \int_0^a qf \, dx \qquad (8.54)$$

In this notation, (8.50) takes on the form*

$$\frac{\partial I_T[z; t]}{\partial t} = \max_{u(t)\in A} \int_0^a \bar{K}\left(t, x, z, z_x, \frac{\delta I_T}{\delta z}, u\right) dx \qquad (8.55)$$

or

$$\frac{\partial I_T[z; t]}{\partial t} = \max_{u(t)\in A} \mathcal{K}\left[t, z, \frac{\delta I_T}{\delta z}, u\right] \qquad (8.56)$$

Equation (8.55) [or (8.56)] is Bellman's equation for the considered example; we see that the functional $\max_{u\in A} \mathcal{K}[t, z, q, u]$ plays a role analogous to the functional $\mathcal{K}$ in the Hamilton–Jacobi equation (8.38) of the variational problem of the simplest kind.

---

* We use the obvious identity $-\min f = \max(-f)$.

If the solution $I_T[z; t]$ of Bellman's equation is sufficiently smooth, it is easy to show that the functions z and q satisfy the system (8.8), where we set

$$\mathcal{K} = \max_{u \in A} \bar{\mathcal{K}}[t, z, q, u] = \max_{u \in A} \int_0^a qf\, dx \qquad (8.57)$$

We recall that the arguments z and t of the functional $I_T[z; t]$ are the initial value $z(x)$ of the function z and the initial time t. From the derivation it is apparent that Bellman's equation (8.56) expresses the circumstance that variation of the control $u(t)$ can only increase the functional $I_T[z', t']$ compared with the value which it had on the state z', t' if the control $u(t)$ was optimal for the interval $[t_0, t']$. In other words, Bellman's equation expresses the nonnegativity of the increment of the functional I under a variation of the control $u(t)$ in that interval.

The derivation is not essentially changed if we assume that the control u depends on t and x. If the control is varied as before in the strip $t_0 + \tau \geq t \geq t_0$, then (8.48) remains valid with $u(t_0)$ replaced by $u(t_0, x)$. Equation (8.50) also remains valid; now, however, the minimum of the right-hand side can also be calculated with respect to admissible functions $u(t, x)$ that differ from the optimal function within a small range Δx of variation of the variable x. The usual argument shows that if the integral is to be minimized, it is necessary and sufficient that the integrand attain a minimum; this leads to Bellman's equation

$$\frac{\partial I_T[z; t]}{\partial t} = \max_{u(t,x) \in A} \bar{K}\left(t, x, z, z_x, \frac{\delta I_T}{\delta z}, u\right) \qquad (8.58)$$

which replaces (8.56). As before, the initial condition is given by (8.51).

The Cauchy problem for (8.45) remains well posed if the initial data are specified, not on the interval $t = 0$, $0 \leq x \leq a$, but on any other curve l bounding the basic half-strip below (Fig. 8.1).

When the initial conditions are specified in this manner, the principle of optimality can be expressed by the equation [see (8.47)]

$$I_T[z_m; t_m] = \min_{\substack{u(t,x) \in A \\ t_m \leq t \leq t_l}} I_T[z_l; t_l] \qquad (8.59)$$

where $z_l = z(t_l, x)$ and $t_l = t_l(x)$ are the values of z_l and t_l on the curve l. Whereas t enters the functional I_T in (8.47) as a parameter, it occurs in (8.59) together with z as a functional argument.

To derive Bellman's equation, we shall assume that the curve l is close to m; the position of a point in the neighborhood of the curve m can be

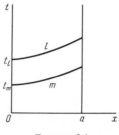

FIGURE 8.1

conveniently specified by means of curvilinear coordinates α and β such that the curve m is the coordinate line $\alpha = 0$ and the curve l is the line $\alpha = \delta\alpha = \text{const}$. The main stages in the derivation are no different from the preceding ones; now, however, it is necessary to express the derivative z_α in terms of the derivatives of z with respect to β. Doing this, we arrive at an equation analogous to (8.58); here, it is important that the control u, as before, be linear in the function $\bar{K}$. It is perfectly clear that the Bellman equation obtained in this way will be (8.58) transformed to the new coordinates.

One can show that the condition for a maximum of $\bar{K}$ with respect to u is not changed as a result of the transition from the old to the new coordinates. This circumstance found its reflection in the preceding chapters, in which it was shown that Weierstrass's condition does not depend on the orientation of the variation strip for examples of the type. Indeed, the increment of the control $u(\alpha, \beta)$ within a small range $\Delta\beta$ of variation of β for $\delta n = H_\alpha \, \delta\alpha \to 0$ is equivalent to a variation of the control in a strip whose orientation is specified by the coordinate lines $\alpha = \text{const}$. For the example considered, what we have said above means that the choice of the coordinates α and β does not affect the condition for a maximum of $\bar{K}$ with respect to the control u.

In the following section, we shall derive Bellman's equation for the problem of Chapter 3; in this example, the shape of a small domain of local variation has a strong influence on the maximizing condition. The strongest condition is obtained for the special choice of coordinates corresponding to the critical orientation of the strip under local variation.

8.6. Bellman's Equation for Elliptic Optimization Problems

Problem 8.1. Suppose the controlled system in the region G with boundary $\Gamma = \Gamma_1 \cup \Gamma_2$ is described by the equation

$$\Delta z = h(z, u, x, y) \tag{8.60}$$

and boundary condition

$$z|_\Gamma = f(t) \tag{8.61}$$

where $f(t)$ is a continuous function. The solution z is assumed to be continuously differentiable in G.

The control $u(x, y)$ belongs to the closed convex set R; the problem is to determine an element $u^*(x, y)$ in R which minimizes the functional

$$F = \int_{\Gamma_2} \frac{\partial z}{\partial n}\, dt \qquad (n \text{ denotes the outer normal}) \tag{8.62}$$

Let $u^*(x, y)$ be an optimal control, and let $z^*(x, y)$ be the corresponding continuously differentiable solution on the problem. Let $z_\gamma^*(t)$ denote the value of this solution of the arc γ, whose ends α and β are on the arc AcB (Fig. 8.2) and which divides G into two parts: G_1 and G_2.

We then pose the following optimization problem.

Problem 8.2. Suppose we are given (8.60), the boundary condition (8.61) on the arc $A\alpha \cup {}_\beta B \cup \Gamma_2$, and the condition $z = z_\gamma^*(t)$ on γ for the region G_2; to find a control $u^{**}(x, y) \in R$ which minimizes the functional (8.62).

In accordance with the principle of optimality, we must have

$$u^{**}(x, y) = u^*(x, y), \qquad (x, y) \in G_2$$

The proof of this assertion in the elliptic case differs from the one given in the preceding section for an evolution problem.

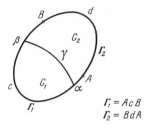

FIGURE 8.2

The point is that in problems containing the time, the value of z on the new boundary of the region [the function $z_1(x)$ of the preceding section is the analogue of $z^*_\gamma(t)$ in the present example] was determined exclusively by the values of the control $u^*(t, x)$ at the preceding instants of time $(t_0 \le t \le t_1)$ and was completely independent of the subsequent behavior of the control. This made it possible to vary the control for $t > t_1$ (the analogue of the region G_2) without disturbing the values of $z = z_1(x)$ on the new boundary of the region.

In the elliptic case, it is impossible to change the control without changing the solutions everywhere in the region or without violating the natural conditions of its continuity.

Suppose that the principle of optimality is invalid and that the control $u^{**}(x, y)$ in Problem 8.2 gives the functional a smaller value than the control $u^*(x, y)$ in the same problem. Returning to Problem 8.1, we then consider the control

$$v(x, y) = \begin{cases} u^*(x, y), & (x, y) \in G_1 \\ u^{**}(x, y), & (x, y) \in G_2 \end{cases}$$

under the additional condition that the function z assumes the previous value $z^*_\gamma(t)$ on γ.

This control is an improving control for Problem 8.1, which now actually splits up into two independent problems for the regions G_1 and G_2; the limiting values of the normal derivatives $\partial z/\partial n$ on different sides of the curve γ are, in general, different, contradicting the requirement of continuous differentiability of the solution. Obviously, nothing essential in the argument is changed if we replace the control $u^*(x, y)$ in G_1 by any other admissible control $u_1(x, y)$. However, one can attempt to choose this control in such a way as to achieve continuity of the normal derivatives of the solution z on γ. If this can be done,* then the control

$$v_1(x, y) = \begin{cases} u_1(x, y), & (x, y) \in G_1 \\ u^{**}(x, y), & (x, y) \in G_2 \end{cases}$$

will be optimal for Problem 8.1, and its solution will be continuously differentiable in the open region G. This contradiction shows that the principle of optimality is valid.

---

* The possibility of doing this is determined, in particular, by the restrictions on the admissible controls.

Comparing this argument with the analogous derivation for the evolution problem, we cannot fail to notice a significant difference. It is the following. If we assume that the principle of optimality is invalid in the evolution problem, then we can always construct a control $v(x, y)$ which gives a better result, as we did in the previous section. In other words, the proof of the validity of the principle of optimality for evolution problems has an unconditional nature.

For problems of elliptic type, the situation is different: if we give up the principle of optimality, we cannot always construct a control which gives a better result for the original problem. Therefore, for problems of such type the principle of optimality may not hold.

If the principle of optimality is valid, then Bellman's equation can be constructed by means of arguments analogous to those given in the preceding section.

As an example, we derive Hadamard's formula for the variation of the Green's function of the Laplace operator by means of dynamic programming.[190]

Let G be a bounded connected region of the (x, y) plane with smooth boundary Γ. We define a function $z(x, y) \in C_2(G)$ and consider the boundary-value problem

$$\Delta z = v, \qquad z|_\Gamma = w \tag{8.63}$$

This problem has a unique solution $(dS = dx\,dy)$

$$z(P) = \int_G g(P, Q) v(Q)\, dS_Q + \int_\Gamma \frac{\partial g(P, Q)}{\partial n_Q} w(Q)\, dt_Q \tag{8.64}$$

where the Green's function $g(P, Q)$ satisfies the conditions

$$\int_{C_P} \Delta g(P, Q)\, dS_Q = \int_{\gamma_P} \frac{\partial g(P, Q)}{\partial n_Q}\, dt_Q = 1$$

Here, C_P is any disk lying entirely in G with center at P; γ_P is the boundary of C_P with outer normal n. The integrals in (8.64) are solutions of the problems

$$\Delta z^1 = v, \qquad z^1|_\Gamma = 0 \tag{8.65}$$

and

$$\Delta z^2 = 0, \qquad z^2|_\Gamma = w \tag{8.66}$$

respectively. The functions z^1 and z^2 are orthogonal in the sense that

$$\int_G (\nabla z^1, \nabla z^2) \, dS = \int_\Gamma z^1 \frac{\partial z^2}{\partial n} \, dt - \int_G z^1 \Delta z^2 \, dS = 0 \qquad (8.67)$$

We set

$$f(v, w) = \min_{\substack{z \\ (z|_\Gamma = w)}} \int_G [2zv + |\nabla z|^2] \, dS$$

It is easy to see that

$$f(v, 0) = \min_{\substack{z \\ (z|_\Gamma = 0)}} \int_G [2zv + |\nabla z|^2] \, dS = \min_{\substack{z \\ (z|_\Gamma = 0)}} \int_G (2v - \Delta z)z \, dS$$

$$= \int_G \int_G g(P, Q)v(P)v(Q) \, dS_P \, dS_Q \qquad (8.68)$$

On the other hand, if z^2 is a solution of the problem (8.66), then (8.67) may be used to deduce

$$f(v, w) = \min_{\substack{z \\ (z|_\Gamma = 0)}} \int_G [2(z + z^2)v + |\nabla z + \nabla z^2|^2] \, dS$$

$$= \min_{\substack{z \\ (z|_\Gamma = 0)}} \int_G [2zv + |\nabla z|^2] \, dS + \int_G [2z^2 v + |\nabla z^2|^2] \, dS$$

$$= f(v, 0) + 2 \int_G z^2 v \, dS + \int_G |\nabla z^2|^2 \, dS$$

Replacing w by εw, we obtain

$$f(v, \varepsilon w) = f(v, 0) + 2\varepsilon \int_G z^2 v \, dS + \varepsilon^2 \int_G |\nabla z^2|^2 \, dS$$

from which it follows that

$$f(v, \varepsilon w) - f(v, 0) = 2\varepsilon \int_G \int_\Gamma g_{n_Q}(P, Q)v(P)w(Q) \, dS_P \, dt_Q + o(\varepsilon) \qquad (8.69)$$

We now consider the curve Γ^* obtained from Γ by displacing the latter along the inner normal through a small amount $\varepsilon\varphi(t) > 0$. Suppose the curve Γ^* bounds a region G^* that lies entirely within G. If $z|_\Gamma = 0$, then $z|_{\Gamma^*} = -z_n\delta n + o(\varepsilon)$ where $\mathbf{n}$ is the direction of the outer normal to Γ, and $\delta n = \varepsilon\varphi$.

We set

$$f(\varepsilon,\cdot v, w) = \min_{\substack{z \\ (z|_{\Gamma^*}=w)}} \int_{G^*} [2zv + |\nabla z|^2]\, dS$$

If $z|_\Gamma = 0$, then $|\nabla z|_\Gamma^2 = (z_n)_\Gamma^2$, and by the principle of optimality

$$f(0, v, 0) = \min_{\substack{z \\ (z|_\Gamma=0)}} \int_G [2zv + |\nabla z|^2]\, dS$$

$$= \min_{\substack{z \\ (z|_\Gamma=0)}} \left\{ \int_{G^*} [2zv + |\nabla z|^2]\, dS + \int_{G/G^*} [2zv + |\nabla z|^2]\, dS \right\}$$

$$= \min_{\substack{z \\ (z|_{\Gamma^*}=-z_n\delta n)}} \left\{ \int_{G^*} [2zv + |\nabla z|^2]\, dS + \int_\Gamma (z_n)^2\delta n\, dt + o(\varepsilon) \right\}$$

$$= \min_{z_n} \left[f(\varepsilon, v, -z_n\delta n) + \int_\Gamma (z_n)^2\delta n\, dt + o(\varepsilon) \right] \tag{8.70}$$

We set $\delta f(v, w) = f(\varepsilon, v, w) - f(0, v, w)$; obviously, $\delta f(v, -z_n\delta n) = \delta f(v, 0) + o(\varepsilon)$. Using this and taking (8.69) into account, we can represent (8.70) in the form

$$\min_{z_n} \left[\delta f(v, 0) - 2 \int_G \int_\Gamma g_{n_Q}(P, Q)v(P)z_n(Q)\delta n_Q\, dS_P\, dt_Q + \int_\Gamma (z_n)^2\delta n\, dt \right]$$

$$= o(\varepsilon) \tag{8.71}$$

The Euler equation of the problem (8.71) has the form

$$z_n(Q) = \int_G g_{n_Q}(P, Q)v(P)\, dS_P$$

Its substitution in (8.71) yields

$$\delta f(v,0) = \int_\Gamma \int_G \int_G g_{n_M}(P,M)g_{n_M}(Q,M)v(P)v(Q)\delta n_M\, dS_P\, dS_Q\, dt_M + o(\varepsilon)$$

$$(8.72)$$

We now denote the Green's function of G^* by $g(\varepsilon, P, Q)$ and set $\delta g(P, Q) = g(\varepsilon, P, Q) - g(0, P, Q)$. Equation (8.68), when applied to G^*, results in

$$f(\varepsilon, v, 0) = \int_{G^*} \int_{G^*} g(\varepsilon, P, Q)v(P)v(Q)\, dS_P\, dS_Q$$

Remembering that the Green's function vanishes on the boundary of the region and that it has a bounded normal derivative there, we can write

$$\delta f(v, 0) = \int_G \int_G \delta g(P, Q)v(P)v(Q)\, dS_P\, dS_Q + o(\varepsilon)$$

Comparing this with (8.72) and remembering that v is arbitrary, we deduce

$$\delta g(P, Q) = \int_\Gamma g_{n_M}(P, M)g_{n_M}(Q, M)\delta n_M\, dt_M + o(\varepsilon)$$

We have obtained Hadamard's formula for the variation of the Green's function. All of the arguments and our result remain true for any finite number of independent variables; the restriction $\delta n > 0$ can be lifted readily.

To conclude this section, we derive Bellman's equation for the optimization problem of Chapter 3.

We consider the case of scalar resistivity and write the basic equations in terms of the local coordinates n, t associated with the variation strip [see (2.120)].

We have

$$z_n^1 = \rho z_t^2 - \frac{1}{c}VBx_t, \qquad z_n^2 = -\frac{1}{\rho}z_t^1 + \frac{1}{c\rho}VBy_t$$

We introduce the Lagrange multipliers λ and μ and form the function L (Lagrangian):

$$L = \lambda\left(z_n^1 - \rho z_t^2 + \frac{1}{c}VBx_t\right) + \mu\left(z_n^2 + \frac{1}{\rho}z_t^1 - \frac{1}{c\rho}VBy_t\right)$$

and the function K (Hamiltonian):

$$K = \sum_{i=1}^{2} z_n^i L_{z_n^i} - L = \lambda\left(\rho z_t^2 - \frac{1}{c} VBx_t\right) + \mu\left(-\frac{1}{\rho} z_t^1 + \frac{1}{c\rho} VBy_t\right)$$

The canonical equations in Hadamard's form are

$$\frac{\partial \lambda}{\partial n} = -\frac{\partial}{\partial t}\left(\frac{\mu}{\rho}\right) - \frac{1}{\rho_t}\frac{\mu}{\rho} - \frac{\lambda}{\rho_n}, \qquad \frac{\partial \mu}{\partial n} = \frac{\partial(\lambda\rho)}{\partial t} + \frac{\lambda\rho}{\rho_t} - \frac{\mu}{\rho_n}$$

[we recall that ρ_n and ρ_t are the radii of curvature of the lines $\alpha = $ const and $\beta = $ const (see Section 8.1)]. We satisfy these equations by the substitution

$$\lambda = -\omega_{1t} = -\frac{1}{\rho}\omega_{2n}, \qquad \mu = -\omega_{2t} = \rho\omega_{1n}$$

where ω_1 and ω_2 are the functions introduced in Section 2.3.

To derive Bellman's equation, we introduce the functional

$$I^*[z^1(t), z^2(t)] = \min_{\rho} I$$

here, $z^1(t)$ and $z^2(t)$ are the values of the optimal functions z^1 and z^2 on any closed smooth curve belonging to the basic region G (including the boundary).

Eliminating λ and μ from the relations

$$I_{z^1}^* = L_{z_n^1} = \lambda$$

$$I_{z^2}^* = L_{z_n^2} = \mu$$

$$I_n^* = \max_{\rho} K = \max_{\rho}\left[\lambda\left(\rho z_t^2 - \frac{1}{c} VBx_t\right) + \mu\left(-\frac{1}{\rho} z_t^1 + \frac{1}{c\rho} VBy_t\right)\right]$$

we arrive at Bellman's equation

$$I_n^* = \max_{\rho}\left[I_{z^1}^*\left(\rho z_t^2 - \frac{1}{c} VBx_t\right) + I_{z^2}^*\left(-\frac{1}{\rho} z_t^1 + \frac{1}{c\rho} VBy_t\right)\right]$$

This equation in terms of functional derivatives contains the operation max, as it must, in accordance with the general results of Chapter 2.

8.7. Linear Systems: Quadratic Criterion

In many practically important cases, the behavior of controlled systems is described by linear equations, and the functional to be minimized is quadratic in the dependent variables and the controls. If at the same time there are no restrictions on the control, it is frequently possible to construct an exact solution of the problem. An example of such a system was considered by Kim and Erzberger,[274] who used Bellman's method to construct a solution. Following their treatment, we consider an optimization problem for an evolutionary system in the region $0 \le t \le T$, $(x, y) \in G$

$$z_t^1 = z^2$$

$$(8.73)$$

$$z_t^2 = \Delta z^1 + az^1 + bz^2, \qquad \Delta = \frac{\partial^2}{\partial x^2} + \frac{\partial^2}{\partial y^2}$$

subject to the initial and boundary conditions

$$z(0, x, y) = \{z^1(0, x, y), z^2(0, x, y)\} = \{z_0^1(x, y), z_0^2(x, y)\} = z_0(x, y) \quad (8.74)$$

$$(z^1 + \alpha z_n^1)|_\Gamma = u(t, s), \qquad \alpha > 0 \qquad (8.75)$$

In (8.75), Γ is the (fixed) boundary of G with outer normal n, and s is the arc length of Γ. The function u is a boundary control.

We consider the following optimization problem: to determine a function $u(t, s)$ in the class of piecewise continuous functions of two variables so as to minimize the quadratic functional

$$\int_0^T F(z, u, t) \, dt + f[z; T] \qquad (8.76)$$

where $z = (z^1, z^2)$ and

$$
\left.
\begin{aligned}
F(z, u, t) &= \tfrac{1}{2} \int_G \int_G (z^1(t, P), z^2(t, P)) \\
&\quad \times \begin{Bmatrix} Q_1(P, P') & 0 \\ 0 & Q_2(P, P') \end{Bmatrix} \begin{pmatrix} z^1(t, P') \\ z^2(t, P') \end{pmatrix} dS \, dS' \\
&\quad + \tfrac{1}{2}\gamma_1 \int_\Gamma u^2(t, s) \, ds \\
f[z; T] &= \tfrac{1}{2}\gamma_2 \int_G \int_G (z^1(T, P), z^2(T, P)) \\
&\quad \times \begin{Bmatrix} Q_1(P, P') & 0 \\ 0 & Q_2(P, P') \end{Bmatrix} \begin{pmatrix} z^1(T, P') \\ z^2(T, P') \end{pmatrix} dS \, dS' \\
dS &= dx_P \, dy_P, \qquad dS' = dx_{P'} \, dy_{P'}
\end{aligned}
\right\} \quad (8.77)
$$

The kernels $Q_i(P, P')$ $(i = 1, 2)$ in (8.76) are assumed to be positive, i.e., for any $\psi \in L_2(G)$

$$\int_G \int_G \psi(P)Q_i(P, P')\psi(P') \, dS \, dS' \geq 0 \qquad (i = 1, 2)$$

The constants γ_1 and γ_2 are also assumed to be positive.

As usual, we consider the minimum I of the functional (8.76) as a functional of the initial state of the system: $I = I[z_0; 0]$. For arbitrary $t \in [0, T]$ the principle of optimality yields

$$I[z; t] = \min_{\substack{u(t') \\ t < t' \leq T}} \left\{ \int_t^T F \, dt' + f[z; T] \right\} \tag{8.78}$$

At time T

$$I[z; T] = \tfrac{1}{2}\gamma_2 \int_G \int_G \{z^1(T, P)Q_1(P, P')z^1(T, P')$$

$$+ z^2(T, P)Q_2(P, P')z^2(T, P')\} \, dS \, dS' \tag{8.79}$$

From (8.78) we obtain the equation

$$I[z(t + \Delta t); t + \Delta t] - I[z(t); t] = -\min_{\substack{u(t', s) \\ t < t' \leq t + \Delta t}} \int_t^{t + \Delta t} F \, dt'$$

from which we arrive at Bellman's equation [see (8.50)]

$$-\frac{\partial I[z; t]}{\partial t} = \min_{u(t, s)} \left\{ F(z, u, t) + \int_G \left[\frac{\delta I[z; t]}{\delta z^1} z^2(t, P) \right. \right.$$

$$\left. \left. + \frac{\delta I[z; t]}{\delta z^2}(\Delta z^1(t, P) + az^1(t, P) + bz^2(t, P)) \right] dS \right\} \tag{8.80}$$

by the usual argument. This equation is to be solved subject to the initial condition (8.79).

We seek the functional $I[z; t]$ in the form

$$I[z; t] = \frac{1}{2} \int_G \int_G (z^1(t, P), z^2(t, P))$$

$$\times \left\{ \begin{matrix} R_1(t, P, P') & R_3(t, P, P') \\ R_3(t, P, P') & R_2(t, P, P') \end{matrix} \right\} \left(\begin{matrix} z^1(t, P') \\ z^2(t, P') \end{matrix} \right) dS\, dS' \quad (8.81)$$

where R_1, R_2, R_3 are twice continuously differentiable functions of their arguments which are symmetric about the points P and P'. Since I cannot take on negative values, the kernels R_1 and R_2 must be positive for all $t \in [0, T]$.

The substitution of (8.81) in (8.80) leads to functional equations for the kernels R_1, R_2, R_3. We now derive these equations.

The functional derivatives (8.81) are given by the equations

$$\frac{\delta I[z; t]}{\delta z^1} = \frac{1}{2} \left[\int_G R_1 z^1(t, P')\, dS' + \int_G R_3 z^2(t, P')\, dS' \right.$$

$$\left. + \int_G R_1 z^1(t, P)\, dS + \int_G R_3 z^2(t, P)\, dS \right] \quad (8.82)$$

$$\frac{\delta I[z; t]}{\delta z^2} = \frac{1}{2} \left[\int_G R_3 z^1(t, P')\, dS' + \int_G R_2 z^2(t, P')\, dS' \right.$$

$$\left. + \int_G R_2 z^2(t, P)\, dS + \int_G R_3 z^1(t, P)\, dS \right] \quad (8.83)$$

The corresponding terms on the right-hand side of (8.80) have the form

$$\int_G \frac{\delta I[z; t]}{\delta z^1} z^2(t, P)\, dS = \frac{1}{2} \int_G \int_G [z^2(t, P) R_1 z^1(t, P') + z^2(t, P) R_3 z^2(t, P')$$

$$+ z^1(t, P) R_1 z^2(t, P') + z^2(t, P) R_3 z^2(t, P')]\, dS\, dS'$$

$$(8.84)$$

$$\int_G \frac{\delta I[z; t]}{\delta z^2} \Delta z^1(t, P)\, dS = \frac{1}{2} \int_G \int_G [\Delta z^1(t, P) R_3 z^1(t, P')$$

$$+ \Delta z^1(t, P) R_2 z^2(t, P') + z^2(t, P) R_2 \Delta z^1(t, P')$$

$$+ z^1(t, P) R_3 \Delta z^1(t, P')]\, dS\, dS' \quad (8.85)$$

etc. We transform the terms on the right-hand side of the last equation by integration by parts subject to the boundary condition (8.75); the result is

$$\int_G \int_G \Delta z^1(t, P) R_3 z^1(t, P') \, dS \, dS'$$

$$= \int_G \int_G z^1(t, P) \, \Delta_P R_3(t, P, P') z^1(t, P') \, dS \, dS'$$

$$+ \int_G \int_\Gamma \left[z^1(t, P) \frac{\partial R_3(t, P, P')}{\partial n_P} - \frac{\partial z^1(t, P)}{\partial n} R_3 \right] z^1(t, P') \, dS' \, ds_P$$

$$= \int_G \int_G z^1(t, P) \, \Delta_P R_3(t, P, P') z^i(t, P') \, dS \, dS'$$

$$+ \frac{1}{\alpha} \int_G \left[\int_\Gamma R_3(t, P, P') u(t, s_P) \, ds_P \right.$$

$$\left. - \int_\Gamma z^1(t, P) G_3(t, P, P') \, ds_P \right] z^1(t, P') \, dS' \tag{8.86}$$

Here, we have

$$G_3(t, P, P') = \left[R_3(t, P, P') + \alpha \frac{\partial R_3(t, P, P')}{\partial n_P} \right] \Bigg|_{P \in \Gamma} \tag{8.87}$$

and Δ_P denotes the Laplace operator with respect to the coordinates of the point P.

The remaining terms in (8.85) are transformed similarly. Substituting (8.77), (8.84), (8.85), etc., in (8.80) we may express the right-hand side explicitly in terms of $u(t, s)$; subsequently, we may then carry out the operation $\min_{u(t,s)}$. Note that if the control $u(t, s)$ is discontinuous on some smooth line situated on the cylindrical surface in the space t, x, y with generator parallel to the t axis and directrix Γ, then the solution $z = (z^1, z^2)$ is continuous within this cylinder right up to its boundary. This can be shown by means of a Green's function representation of the solution.

The right-hand side of (8.80) is minimized by the control

$$u = u^*(t, s_P) = -\frac{1}{\alpha \gamma_1} \int_G [z^2(t, P') R_2(t, P, P')|_{P \in \Gamma}$$

$$+ z^1(t, P') R_3(t, P, P')|_{P \in \Gamma}] \, dS' \tag{8.88}$$

The partial derivative of the functional (8.81) with respect to t is given by

$$\frac{\partial I[z; t]}{\partial t} = \frac{1}{2} \int_G \int_G (z^1(t, P), z^2(t, P)) \begin{Bmatrix} R_{1t}(t, P, P') & R_{3t}(t, P, P') \\ R_{3t}(t, P, P') & R_{2t}(t, P, P') \end{Bmatrix}$$

$$\times \begin{pmatrix} z^1(t, P') \\ z^2(t, P') \end{pmatrix} dS \, dS'$$

Substituting this result, along with (8.77), (8.84)–(8.86), etc., into (8.80) and eliminating $u = u^*(t, s)$ by means of (8.88), we arrive at the relation

$$0 = \int_G \int_G z^1(t, P) \Big[\tfrac{1}{2} R_{1t} + \tfrac{1}{2} Q_1 + a R_3$$

$$- \frac{1}{2\alpha^2 \gamma_1} \int_\Gamma R_3(t, P, P'') R_3(t, P'', P') \, ds_{P''}$$

$$+ \tfrac{1}{2} (\Delta_P + \Delta_{P'}) R_3 \Big] z^1(t, P') \, dS \, dS'$$

$$+ \int_G \int_G z^1(t, P) \Big[R_{3t} + R_1 + a R_2 + b R_3$$

$$- \frac{1}{\alpha^2 \gamma_1} \int_\Gamma R_2(t, P, P'') R_3(t, P'', P') \, ds_{P''}$$

$$+ \tfrac{1}{2} (\Delta_P + \Delta_{P'}) R_2 \Big] z^2(t, P') \, dS \, dS'$$

$$+ \int_G \int_G z^2(t, P) \Big[\tfrac{1}{2} R_{2t} + \tfrac{1}{2} Q_2 + b R_2 - \frac{1}{2\alpha^2 \gamma_1} \int_\Gamma R_2(t, P, P'')$$

$$\times R_2(t, P'', P') \, ds_{P''} + R_3 \Big] z^2(t, P') \, dS \, dS'$$

$$- \frac{1}{2} \int_G \int_\Gamma \frac{\partial z^1(t, s_P)}{\partial n_P} G_3(t, P, P') z^1(t, P') \, ds_P \, dS'$$

$$-\frac{1}{2}\int_G\int_\Gamma \frac{\partial z^1(t, s_{P'})}{\partial n_{P'}} G_3(t, P', P)z^1(t, P)\, ds_{P'}\, dS$$

$$-\frac{1}{2}\int_G\int_\Gamma \frac{\partial z^1(t, s_P)}{\partial n_P} G_2(t, P, P')z^2(t, P')\, ds_P\, dS'$$

$$-\frac{1}{2}\int_G\int_\Gamma \frac{\partial z^1(t, s_{P'})}{\partial n_{P'}} G_2(t, P', P)z^2(t, P)\, ds_{P'}\, dS \qquad (8.89)$$

where

$$G_2(t, P, P') = \left[R_2(t, P, P') + \alpha\frac{\partial R_2(t, P, P')}{\partial n_P} \right]_{P\in\Gamma}$$

Equation (8.89) must be satisfied identically with respect to z^1 and z^2. This requirement leads to a system of integrodifferential equations for the kernels R_1, R_2, R_3:

$$\left.\begin{aligned}
-\frac{\partial R_1}{\partial t} &= Q_1 + 2aR_3 - \frac{1}{a^2\gamma_1}\int_\Gamma R_3(t, P, P'') \\
&\quad \times R_3(t, P'', P')\, ds_{P''} + (\Delta_P + \Delta_{P'})R_3(t, P, P') \\
-\frac{\partial R_3}{\partial t} &= R_1 + aR_2 + bR_3 - \frac{1}{a^2\gamma_1}\int_\Gamma R_2(t, P, P'') \\
&\quad \times R_3(t, P'', P')\, ds_{P''} + \tfrac{1}{2}(\Delta_P + \Delta_{P'})R_2(t, P, P') \\
-\frac{\partial R_2}{\partial t} &= Q_2 + 2bR_2 - \frac{1}{a^2\gamma_1}\int_\Gamma R_2(t, P, P'') \\
&\quad \times R_2(t, P'', P')\, ds_{P''} + 2R_3(t, P, P')
\end{aligned}\right\} \qquad (8.90)$$

Simultaneously, we obtain the boundary conditions

$$G_2 \equiv \left[R_2(t, P, P') + \alpha\frac{\partial R_2(t, P, P')}{\partial n_P} \right]_{P\in\Gamma} = 0$$
$$\qquad\qquad\qquad\qquad\qquad\qquad\qquad\qquad\qquad\qquad (8.91)$$
$$G_3 \equiv \left[R_3(t, P, P') + \alpha\frac{\partial R_3(t, P, P')}{\partial n_P} \right]_{P\in\Gamma} = 0$$

the functions R_1, R_2, R_3 at time T are given by

$$\left.\begin{aligned}
R_1(T, P, P') &= \gamma_2 Q_1(P, P') \\
R_2(T, P, P') &= \gamma_2 Q_2(P, P') \\
R_3(T, P, P') &= 0
\end{aligned}\right\} \tag{8.92}$$

which readily follow from a comparison of (8.81) with (8.76) and (8.77).

The use of (8.91) to replace the functions R_2 and R_3 in the integrands of (8.90) by the normal derivatives $\partial R_2/\partial n$, $\partial R_3/\partial n$ yields the system

$$\left.\begin{aligned}
-\frac{\partial R_1}{\partial t} &= Q_1 + 2aR_3 - \frac{1}{\gamma_1}\int_\Gamma \frac{\partial R_3(t, P, P'')}{\partial n_{P''}} \frac{\partial R_3(t, P'', P')}{\partial n_{P''}} \, ds_{P''} \\
&\quad + (\Delta_P + \Delta_{P'})R_3(t, P, P') \\[2mm]
-\frac{\partial R_3}{\partial t} &= R_1 + aR_2 + bR_3 - \frac{1}{\gamma_1}\int_\Gamma \frac{\partial R_2(t, P, P'')}{\partial n_{P''}} \frac{\partial R_3(t, P'', P')}{\partial n_{P''}} \, ds_{P''} \\
&\quad + \tfrac{1}{2}(\Delta_P + \Delta_{P'})R_2(t, P, P') \\[2mm]
-\frac{\partial R_2}{\partial t} &= Q_2 + 2bR_2 - \frac{1}{\gamma_1}\int_\Gamma \frac{\partial R_2(t, P, P'')}{\partial n_{P''}} \frac{\partial R(t, P'', P')}{\partial n_{P''}} \, ds_{P''} \\
&\quad + 2R_3(t, P, P')
\end{aligned}\right\} \tag{8.93}$$

This system of Riccati equations must be solved subject to the initial conditions (8.92) and the boundary conditions (8.91).

Kim and Erzberger[274] proposed a method for constructing a solution of the system (8.93) in the case when the kernels $Q_i(P, P')$ ($i = 1, 2$) have the form

$$Q_1 = \sum_{i=1}^M q_1^{ii}\psi_i(P)\psi_i(P'), \qquad Q_2 = \sum_{i=1}^M q_2^{ii}\psi_i(P)\psi_i(P') \tag{8.94}$$

where $\psi_i(P)$ are the eigenfunctions of the boundary-value problem

$$\Delta z = \lambda z, \qquad (z + \alpha z_n)|_\Gamma = 0$$

In practice, the values of the solution $z = (z^1, z^2)$ can only be measured at a finite number of points of G. We therefore restrict ourselves to a finite dimensional subspace spanned by the first M eigenfunctions of the problem. The corresponding representation of the solution z has the form

$$z^1(t, P) = \sum_{i=1}^{M} \psi_i(P) z_i^1(t), \qquad z^2(t, P) = \sum_{i=1}^{M} \psi_i(P) z_i^2(t) \qquad (8.95)$$

Setting

$$
\left.
\begin{aligned}
R_1(t, P, P') &= \sum_{i,j=1}^{M} \psi_i(P) \psi_i(P') R_1^{ij}(t) \\[2ex]
R_2(t, P, P') &= \sum_{i,j=1}^{M} \psi_i(P) \psi_i(P') R_2^{ij}(t) \\[2ex]
R_3(t, P, P') &= \sum_{i,j=1}^{M} \psi_i(P) \psi_i(P') R_3^{ij}(t)
\end{aligned}
\right\} \qquad (8.96)
$$

we obtain the following system of Riccati equations for the determination of the coefficients R_k^{ij} ($k = 1, 2, 3$):

$$
\left.
\begin{aligned}
-\frac{dR_1^{ij}}{dt} &= q_1^{ij} \delta_{ij} + 2a R_3^{ij} - \frac{1}{\gamma_1} \sum_{k,l=1}^{M} R_3^{ik} R_3^{lj} \int_{\Gamma} N_k N_l \, ds_{P''} + (\lambda_i + \lambda_j) R_3^{ij} \\[2ex]
-\frac{dR_3^{ij}}{dt} &= R_1^{ij} + a R_2^{ij} + b R_3^{ij} \\[2ex]
& \quad - \frac{1}{\gamma_1} \sum_{k,l=1}^{M} R_2^{ik} R_3^{lj} \int_{\Gamma} N_k N_l \, ds_{P''} + \tfrac{1}{2}(\lambda_i + \lambda_j) R_2^{ij} \\[2ex]
-\frac{dR_2^{ij}}{dt} &= q_2^{ij} \delta_{ij} - \frac{1}{\gamma_1} \sum_{k,l=1}^{M} R_2^{ij} R_2^{lj} \int_{\Gamma} N_k N_l \, ds_{P''} + 2 R_3^{ij} \\[2ex]
& \qquad N_k = \frac{d\psi_k(P'')}{dn_{P''}}, \qquad \delta_{ij} = \begin{cases} 1, & i = j \\ 0, & i \neq j \end{cases}
\end{aligned}
\right\} \qquad (8.97)
$$

The initial conditions

$$R_1^{ij}(T) = \gamma_2 q_1^{ij}, \qquad R_3^{ij}(T) = 0, \qquad R_2^{ij}(T) = \gamma_2 q_2^{ij}$$

$$(i, j = 1, \ldots, M) \tag{8.98}$$

follow from (8.92).

If a solution of the system (8.97) is known, then the optimal control $u^*(t, s)$ can be expressed in terms of the solution z in accordance with (8.88). Using (8.95), we may write

$$u^*(t, s_P) = -\frac{1}{\gamma_1} \sum_{i,j=1}^{M} \left[\frac{\psi_i(P)}{\alpha}\right]_\Gamma [z_i^2(t) R_2^{ji}(t) + z_i^1(t) R_3^{ji}(t)] \tag{8.99}$$

The optimal solution $z^1(t, P)$ can be expressed in terms of the control $u^*(t, s)$ by means of the Green's function $K(t - t', P, P')$ in accordance with the formula[86]

$$z^1(t, P) = \int_G z^1(0, P') \frac{\partial K(t, P, P')}{\partial t} \, dS' + \int_G z^2(0, P') K(t, P, P') \, dS'$$

$$+ \int_0^t dt' \int_\Gamma u^*(t', s_{P'}) \frac{\partial K(t - t', P, P')}{\partial n_{P'}} \, ds_{P'} \tag{8.100}$$

where the initial values z^1 and z^2 are given by

$$z^1(0, P) = \sum_{i=1}^{M} \psi_i(P) z_i^1(0)$$

$$\tag{8.101}$$

$$z^2(0, P) = \frac{\partial z^1(0, P)}{\partial t} = \sum_{i=1}^{M} \psi_i(P) z_i^2(0)$$

The substitution of (8.95), (8.99), and (8.101) in (8.100) and the use of the expansion

$$K(t - t', P, P') = \sum_{i=1}^{\infty} \psi_i(P) \psi_i(P') \frac{\sin n_i(t - t')}{n_i}, \qquad n_i = \sqrt{-\lambda_i}$$

for the Green's function, yield left- and right-hand sides in the form of linear combinations of the functions $\psi_i(P)$; equating the coefficients of these functions, we obtain a system of ordinary differential equations relating the functions $z_i^1(t)$ and $z_i^2(t)$; these equations must be integrated subject to the fact that $z_i^2(t) = dz_i^1(t)/dt$.

APPENDIX A
Calculations and Comments

A.1. Comments Concerning Equation (3.26)

If $\rho = $ const, then Eqs. (3.15) show that $z^1(x, y)$ is a harmonic function. We introduce the analytic function $f(z) = z_x^1 - iz_y^1 = u + iv$, $z = x + iy$. The boundary conditions (3.16) are specified in the form (see Fig. 3.3): $u = 0$ on the electrodes BB' and on the insulators CC and $C'C$, $v = -c^{-1}VB(x)$ on the insulators BC and $B'C'$,

$$IR = \left[G + 2 \int_0^\lambda v(x, \delta) \, dx \right] R = -\rho \int_{-\delta}^\delta v(0, y) \, dy$$

$$G = \frac{V}{c} \int_{-\lambda}^\lambda B(x) \, dx \tag{1}$$

Because of the symmetry of the problem, we can consider the right-hand half of the channel $ACCA$, specifying the condition $u = 0$ along AA.

Consider the elliptic function

$$t = \tau + i\nu = \kappa \, \text{sn} \left(i \frac{K(k)}{\delta} z, k \right) \tag{2}$$

where the constant κ and the modulus k are given by

$$\kappa = \text{dn} \left(\frac{\lambda}{\delta} K(k), k' \right), \qquad \frac{K(k')}{K(k)} = \frac{x_c}{\delta} \tag{3}$$

The function (2) is a conformal mapping of the right-hand half of the channel $ACCA$ (see Fig. 3.3) onto the upper half-plane $\nu > 0$; the corresponding points are shown in Fig. A.1.

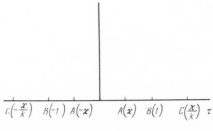

$$c\left(-\tfrac{x}{k}\right) \quad B(-1) \quad A(-x) \qquad A(x) \quad B(1) \qquad C\left(\tfrac{x}{k}\right) \quad \tau$$

FIGURE A.1

The function $f(z)$ becomes the function $f_1(t) = u_1 + iv_1$, and the conditions (1) take on the form (Fig. A.1)

$$u_1 = 0 \quad \text{for } -1 < \tau < 1,\ \nu = 0 \quad \text{and for } |\tau| > \kappa/k,\ \nu > 0$$

$$v_1 = -\frac{1}{c} V B_1(\tau) \quad \text{for } 1 < |\tau| < \kappa/k,\ \nu = 0,\ f_1(\infty) \text{ bounded}$$

$$\frac{R}{\rho}\left[G + \frac{2\delta}{k}\int_\kappa^1 \frac{v_1(-\tau, 0)\, d\tau}{(\tau^2 - \kappa^2)\left(1 - \dfrac{k^2\tau^2}{\kappa^2}\right)} \right] = -\frac{2\delta}{K(k)}\int_0^\kappa \frac{v_1(-\tau, 0)\, d\tau}{(\kappa^2 - \tau^2)\left(1 - \dfrac{k^2\tau^2}{\kappa^2}\right)}$$

$$\tag{4}$$

$$B_1(\tau) = B[x(\tau)]$$

A function $f_1(t)$ satisfying the first three conditions (4) and bounded for $t = \pm\kappa/k$, can be found in accordance with the Keldysh–Sedov formula.[90] We have

$$if_1(t) = \sqrt{\frac{t^2 - \kappa^2/k^2}{t^2 - 1}}\left[\frac{V}{\pi c i}\int_{-\kappa/k}^{-1} \frac{B_1(\rho)}{\rho - t}\sqrt{\frac{\rho^2 - 1}{\rho^2 - \kappa^2/k^2}}\, d\rho \right.$$

$$\left. + \frac{V}{\pi c i}\int_1^{\kappa/k} \frac{B_1(\rho)}{\rho - t}\sqrt{\frac{\rho^2 - 1}{\rho^2 - \kappa^2/k^2}}\, d\rho + \gamma \right]$$

Here, γ is a real constant, and the branch of the square root is chosen in such a way that the root is positive on the ray $(\kappa/k, \infty)$. We set $t = \tau$; for $|\tau| < 1$, we have

$$v_1(\tau, 0) = v_1(-\tau, 0)$$

$$= \sqrt{\frac{\kappa^2/k^2 - \tau^2}{1 - \tau^2}}\left[\frac{2V}{\pi c}\int_1^{\kappa/k} \frac{\rho}{\rho^2 - \tau^2} B_1(\rho)\sqrt{\frac{\rho^2 - 1}{\kappa^2/k^2 - \rho^2}}\, d\rho - \gamma \right]$$

The condition (4) determines the constant γ:

$$\gamma = \frac{kK(k)}{2\kappa K(\kappa)\left(1 + \dfrac{R}{\rho}\alpha^*\right)\delta} \left\{ \frac{R}{\rho}G + \frac{4V\delta}{\pi cK(k)}\frac{\kappa}{k}\right.$$

$$\times \int_1^{\kappa/k} \frac{\rho}{\sqrt{(\rho^2 - 1)(\kappa^2/k^2 - \rho^2)}}$$

$$\times \left[\frac{\rho^2 - 1}{\rho^2}\Pi\left(\frac{\pi}{2}, -\frac{\kappa^2}{\rho^2}, \kappa\right) + \frac{R}{\rho}\Pi\left(\frac{\pi}{2}, \frac{\kappa'^2}{\rho^2 - 1}, \kappa'\right)\right] B_1(\rho)\,d\rho\right\}$$

Here

$$\alpha^* = \frac{K(\kappa')}{K(\kappa)}, \qquad \kappa^2 + \kappa'^2 = 1 \tag{5}$$

$$\Pi\left(\frac{\pi}{2}, h, k\right) = \int_0^{\pi/2} \frac{d\beta}{(1 + h\sin^2\beta)\sqrt{1 - k^2\sin^2\beta}}$$

is the complete elliptic integral of the third kind. Calculating the current I, we arrive at the expression $(\sigma = \rho^{-1})$

$$I = I_{x_c} = \frac{\sigma}{1 + R\sigma\alpha^*}\left\{G + \frac{4V\delta}{\pi cK(k)}\frac{\kappa}{k}\int_1^{\kappa/k}\frac{\rho}{\sqrt{(\rho^2 - 1)(\kappa^2/k^2 - \rho^2)}}\right.$$

$$\times \left[\Pi\left(\frac{\pi}{2}, \frac{\kappa'^2}{\rho^2 - 1}, \kappa'\right) - \alpha^*\frac{\rho^2 - 1}{\rho^2}\Pi\left(\frac{\pi}{2}, -\frac{\kappa^2}{\rho^2}, \kappa\right)\right] B_1(\rho)\,d\rho\right\}$$

We set $B = B_1 = B_0 = $ const; using the well-known representations of the complete elliptic integral of the third kind (Ref. 90, p. 571), we can readily reduce the last formula to the form

$$I_{x_c}[B_0] = \frac{\sigma\varepsilon}{1 + R\sigma\alpha^*}\left[\frac{\lambda}{\delta} - \frac{F\left(\arcsin\dfrac{\kappa'}{k'}, k'\right)}{K(k)} + \alpha^*\right] \tag{6}$$

$$\varepsilon = \frac{2VB_0\delta}{c}$$

where $F(\varphi, k)$ is the incomplete elliptic integral of the second kind. Furthermore,

$$\kappa' = \sqrt{1 - \kappa^2} = \sqrt{1 - \mathrm{dn}^2\left(\frac{\lambda}{\delta} K(k), k'\right)} = k' \, \mathrm{sn}\left(\frac{\lambda}{\delta} K(k), k'\right)$$

$$\arcsin \frac{\kappa'}{k'} = \arcsin \mathrm{sn}\left(\frac{\lambda}{\delta} K(k), k'\right) = \mathrm{am}\left(\frac{\lambda}{\delta} K(k), k'\right)$$

$$F\left(\arcsin \frac{\kappa'}{k'}, k'\right) = \frac{\lambda}{\delta} K(k)$$

Equations (6) can now be reduced to the form of Eq. (3.26). It is easy to show that as the parameter x_c increases from λ to ∞ the function α^* increases monotonically from λ/δ to $\alpha^*_\infty = K(\kappa'_\infty)/K(\kappa_\infty)$, $\kappa^{-1}_\infty = \cosh \pi\lambda/2\delta$.

A.2. Algorithm for the Calculation of a Generalized Solution for the Problem of Section 3.2 (the Case of Scalar Resistivity)

Our point of departure is the inequality $\Delta I \le 0$ or [see Eq. (2.102)]

$$I[U] - I[u] = \iint_G \frac{\Delta u}{u} (\mathbf{J}, \nabla \omega_2) \, dx \, dy$$

$$= \iint_G \frac{\Delta u}{u} (\mathbf{j}, \nabla \omega_2) \, dx \, dy + \iint_G \frac{\Delta u}{u} (\Delta \mathbf{j}, \nabla \omega_2) \, dx \, dy \quad (7)$$

If this inequality is satisfied for every admissible choice of U, then u is an optimal element and $I[U] \le I[u] = I_{\max}$.

Suppose the element u is such that there exists an admissible control U for which the opposite inequality is satisfied, i.e.,

$$I[U] - I[u] = \iint_G \frac{\Delta u}{u} (\mathbf{j}, \nabla \omega_2) \, dx \, dy + \iint_G \frac{\Delta u}{u} (\Delta \mathbf{j}, \nabla \omega_2) \, dx \, dy \ge 0 \quad (8)$$

It is known that the optimal function u can take on only two values: $u_{\min}$ and $u_{\max}$. Considering a nonoptimal function u, we can assume that it takes on the same values but distributed in a nonoptimal manner. Then, for the existence of an element U having the property considered above, it is sufficient that

$$u = u_{\min} \qquad \text{at those points where } (\mathbf{j}, \nabla \omega_2) > 0$$

$$u = u_{\max} \qquad \text{at those points where } (\mathbf{j}, \nabla \omega_2) < 0$$

More precise conditions are obtained by introducing local variations in a strip. If $(\mathbf{n}, \mathbf{t})$ are the unit vectors of the normal and the tangent to the strip, then

$$J_n = j_n, \qquad UJ_t = uj_t$$

$$\frac{\Delta u}{u}(\mathbf{j} + \Delta \mathbf{j}, \nabla \omega_2) = \frac{\Delta u}{u}\left[(\mathbf{j}, \Delta \omega_2) - \frac{\Delta u}{U} j_t \omega_{2t} \right] \tag{9}$$

If $u = u_{\min}$, then the last expression is nonnegative for all directions of the strip if (Fig. A.2)

$$\cos \chi - \frac{U - u_{\min}}{U} \cos^2 \frac{\chi}{2} \geq 0$$

$$\chi \leq \arccos p_1, \qquad p_1 = \frac{U - u_{\min}}{U + u_{\min}}$$

The most stringent condition is obtained by setting $U = u_{\max}$ in the last inequality; we obtain

$$\chi \leq \arccos p, \qquad p = \frac{u_{\max} - u_{\min}}{u_{\max} + u_{\min}}$$

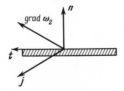

FIGURE A.2

The case $u = u_{\max}$ is treated similarly; as a result, we obtain the following rule of descent*

$$\left.\begin{array}{ll} U = u_{\max} & \text{at points of the set } E_1\colon (\mathbf{j}, \nabla\omega_2) > 0 \\[1em] & 0 \le \chi \le \arccos p \\[1em] U = u_{\min} & \text{at points of the set } E_2\colon (\mathbf{j}, \nabla\omega_2) < 0 \\[1em] & \pi - \arccos p \le \chi \le \pi \end{array}\right\} \tag{10}$$

These inequalities give no information concerning the value of U at the points where $\cos^{-1} p \le \chi \le \pi - \cos^{-1} p$ (in general, such points form a set of nonzero measure, if u is a nonoptimal control). We obtain the corresponding rule by requiring that (9) attain the maximal possible value with respect to U and with respect to the orientation of the strip; using the remark made above, we can assume that U takes on two values: $u_{\max}$ and $u_{\min}$.

Suppose $(\mathbf{j}, \nabla\omega_2) > 0$; if $U = u_{\max}$, then the maximum of (9) is

$$|\mathbf{j}| |\nabla\omega_2| \frac{u_{\max} - u}{u} \left(\cos\chi + \frac{u_{\max} - u}{u_{\max}} \sin^2 \frac{\chi}{2} \right) \tag{11}$$

if $U = u_{\min}$, then the maximum of (9) is

$$|\mathbf{j}| |\nabla\omega_2| \frac{u_{\min} - u}{u} \left(\cos\chi + \frac{u_{\min} - u}{u_{\min}} \sin^2 \frac{\chi}{2} \right) \tag{12}$$

Suppose $(\mathbf{j}, \nabla\omega_2) < 0$; if $U = u_{\max}$, then the maximum of (9) is

$$|\mathbf{j}| |\nabla\omega_2| \frac{u_{\max} - u}{u} \left(\cos\chi + \frac{u_{\max} - u}{u_{\max}} \sin^2 \frac{\chi}{2} \right) \tag{13}$$

if $U = u_{\min}$, then the maximum of (9) is

$$|\mathbf{j}| |\nabla\omega_2| \frac{u_{\min} - u}{u} \left(\cos\chi + \frac{u_{\min} - u}{u_{\min}} \sin^2 \frac{\chi}{2} \right) \tag{14}$$

---

* Note that the direction $\mathbf{t}$, which bisects the angle χ in Fig. A.2, is not the direction of the strip corresponding to the maximal rate of descent (the orientation of the strip in the direction perpendicular to $\mathbf{t}$ corresponds to the maximal rate). The requirement that the expression (9) be nonnegative for all directions of the strip can be fulfilled only if the inequalities (10) are satisfied.

Suppose $u = u_{max}$. If $(\mathbf{j}, \nabla \omega_2) > 0$, then we must compare (11) and (12). For $u = u_{max}$ the first of them is zero, while the second is positive for $\pi/2 > \chi > \cos^{-1} p$ and negative for $\cos^{-1} p > \chi > 0$.

In the first of these cases, the maximal rate of descent corresponds to the choice $U = u_{min}$; in the second, to $U = u_{max}$.

If $(\mathbf{j}, \nabla \omega_2) < 0$ for $u = u_{max}$, then we must compare (13) and (14). The former is zero under the imposed conditions, and the latter is positive; the maximal rate of descent is obtained for $U = u_{min}$.

Similarly, if $u = u_{min}$ then we must take $U = u_{min}$ for $\pi > \chi > \pi - \cos^{-1} p$; for $\pi/2 < \chi < \pi - \cos^{-1} p$, $U = u_{max}$; and for $0 < \chi < \pi/2$, $U = u_{max}$.

Combining our results, we arrive at the following rule for fastest descent:

$$\left.\begin{array}{ll} U = u_{max} & \text{at points of} \\[1ex] \begin{cases} E_1: (\mathbf{j}, \nabla \omega_2) > 0, & 0 < \chi < \arccos p \\ E_1^*: u = u_{min}, & \arccos p < \chi < \pi - \arccos p \end{cases} \\[3ex] U = u_{min} & \text{at points of} \\[1ex] \begin{cases} E_2: (\mathbf{j}, \nabla \omega_2) < 0, & \pi - \arccos p < \chi < \pi \\ E_2^*: u = u_{max}, & \arccos p < \chi < \pi - \arccos p \end{cases} \end{array}\right\} \tag{15}$$

For the difference $I[U] - I[u]$, we obtain the expression

$$I[U] - I[u] = \iint_{E_1} \frac{u_{max} - u}{u} [(\mathbf{j}, \nabla \omega_2) + (\Delta \mathbf{j}, \nabla \omega_2)]\, dx\, dy$$

$$+ \iint_{E_1^*} \frac{u_{max} - u_{min}}{u_{min}} [(\mathbf{j}, \nabla \omega_2) + (\Delta \mathbf{j}, \nabla \omega_2)]\, dx\, dy$$

$$+ \iint_{E_2} \frac{u_{min} - u}{u} [(\mathbf{j}, \nabla \omega_2) + (\Delta \mathbf{j}, \Delta \omega_2)]\, dx\, dy$$

$$+ \iint_{E_2^*} \frac{u_{min} - u_{max}}{u_{max}} [(\mathbf{j}, \Delta \omega_2) + (\Delta \mathbf{j}, \nabla \omega_2)]\, dx\, dy$$

If the variation is performed in a strip, then the integrands of all integrals are nonnegative by virtue of (15); therefore, in the limit $I[U] \to [u]$

we get

$$u \to u_{\max} \qquad \text{by measure on the set } E_1$$

$$u \to u_{\min} \qquad \text{by measure on the set } E_2$$

$$\text{mes } E_1^* \to 0, \qquad \text{mes } E_2^* \to 0$$

A.3. On the Derivation of the Inequality (3.76)

We transform the expressions in the square brackets on the left-hand side of inequality (3.75). The components J_α and J_β of the vector $\mathbf{J}$ can be expressed in terms of the components J_n and J_t in accordance with the expressions (see Fig. 3.13)

$$J_\alpha = J_n \cos\theta - J_t \sin\theta, \qquad J_\beta = J_n \sin\theta + J_t \cos\theta \qquad (16)$$

Replacing J_n and J_t by their expressions (3.73), we obtain

$$J_\alpha = j_\alpha - \frac{\rho_{tt} - P_{tt}}{P_{tt}} j_t \sin\theta - \frac{\rho_{tn} - P_{tn}}{P_{tt}} j_n \sin\theta + O(\varepsilon)$$

$$\qquad\qquad\qquad\qquad\qquad\qquad\qquad\qquad\qquad\qquad\qquad (17)$$

$$J_\beta = j_\beta + \frac{\rho_{tt} - P_{tt}}{P_{tt}} j_t \cos\theta + \frac{\rho_{tn} - P_{tn}}{P_{tt}} j_n \cos\theta + O(\varepsilon)$$

We have [see (3.53) and Fig. 3.13]

$$\rho_{nn} = 1/2[\rho_1 + \rho_2 + (\rho_1 - \rho_2)\cos 2\theta]$$

$$\rho_{tn} = \rho_{nt} = -1/2(\rho_1 - \rho_2)\sin 2\theta$$

$$\rho_{tt} = 1/2[\rho_1 + \rho_2 - (\rho_1 - \rho_2)\cos 2\theta]$$

$$P_{nn} = 1/2[P_1 + P_2 + (P_1 - P_2)\cos 2\psi]$$

$$P_{tn} = P_{nt} = -1/2(P_1 - P_2)\sin 2\psi$$

$$P_{tt} = 1/2[P_1 + P_2 - (P_1 - P_2)\cos 2\psi]$$

Using these relations, we can express the right-hand sides of (17) in terms of the principal values ρ_1, ρ_2 and P_1, P_2 of the tensors ρ and $\mathbb{P}$ in terms of the angles θ and ψ between the normal $\mathbf{n}$ to the variation strip and the axes α and $\mathbf{A}$, respectively. We obtain [henceforth, we omit terms $O(\varepsilon)]$

$$J_\alpha = [P_1 + P_2 - (P_1 - P_2)\cos 2\psi]^{-1}$$

$$\times [m - 2(P_1 - P_2)j_n \sin\psi \sin(\theta - \psi)]$$

$$J_\beta = [P_1 + P_2 - (P_1 - P_2)\cos 2\psi]^{-1}$$

$$\times [n - 2(P_1 - P_2)j_n \cos\psi \sin(\theta - \psi)]$$

(18)

In these equations,

$$m = j_\alpha[P_1 + P_2 - (P_1 - P_2)\cos 2\theta] - 2\sin\theta(\Delta\rho_1 j_\alpha \sin\theta - \Delta\rho_2 j_\beta \cos\theta)$$

$$n = j_\beta[P_1 + P_2 - (P_1 - P_2)\cos 2\theta] + 2\cos\theta(\Delta\rho_1 j_\alpha \sin\theta - \Delta\rho_2 j_\beta \cos\theta)$$

Inequality (3.75) contains the components $P_{\alpha\alpha}$, $P_{\alpha\beta} = P_{\beta\alpha}$, $P_{\beta\beta}$ of the tensor $\mathbb{P}$ in the system of the principal axes α and β of the tensor ρ. These components can be expressed in terms of the principal values P_1 and P_2 of the tensor $\mathbb{P}$ and in terms of the angle $\Delta\gamma = \theta - \psi$ between the principal axes α and $\mathbf{A}$ of the tensors ρ and $\mathbb{P}$ by means of (3.53). In these, the x axis must be replaced by the α axis and the α axis by the $\mathbf{A}$ axis; at the same time, $\gamma = \Delta\gamma$. After some simple calculations, we obtain

$$(P_{\alpha\alpha} - \rho_1)J_\alpha + P_{\alpha\beta}J_\beta = (P_1 - \rho_1)J_\alpha - (P_1 - P_2)$$

$$\times (J_\alpha \sin\Delta\gamma - J_\beta \cos\Delta\gamma)\sin\Delta\gamma$$

$$P_{\beta\alpha}J_\alpha + (P_{\beta\beta} - \rho_2)J_\beta = (P_2 - \rho_2)J_\beta + (P_1 - P_2)$$

$$\times (J_\alpha \cos\Delta\gamma + J_\beta \sin\Delta\gamma)\sin\Delta\gamma$$

The elimination of J_α and J_β by means of (18) leads to the expressions

$$[(P_{\alpha\alpha} - \rho_1)J_\alpha + P_{\alpha\beta}J_\beta][P_1 + P_2 - (P_1 - P_2)\cos 2\psi]$$

$$= (P_1 - \rho_1)m - (P_1 - P_2)f\sin\Delta\gamma$$

$$[P_{\beta\alpha}J_\alpha + (P_{\beta\beta} - \rho_2)J_\beta][P_1 + P_2 - (P_1 - P_2)\cos 2\psi]$$

$$= (P_2 - \rho_2)n + (P_1 - P_2)g\sin\Delta\gamma$$

Here, we have introduced the notation

$$f = [2\rho_1 j_\alpha + (\rho_1 - \rho_2) j_\beta \sin 2\theta]$$

$$\times \sin \Delta\gamma - j_\beta [\rho_1 + \rho_2 - (\rho_1 - \rho_2) \cos 2\theta] \cos \Delta\gamma$$

$$g = [2\rho_2 j_\beta + (\rho_2 - \rho_1) j_\alpha \sin 2\theta]$$

$$\times \sin \Delta\gamma + j_\alpha [\rho_1 + \rho_2 - (\rho_1 - \rho_2) \cos 2\theta] \cos \Delta\gamma$$

The use of these expressions together with (3.77) in the left-hand side of (3.75) yields (3.76).

A.4. Comments on Equation (3.86)

The root $(\tan \gamma)_1$ of Eq. (3.86) is determined by

$$1/2(\rho_1 \zeta^2 \omega_{2x} + \rho_2 \zeta^1 \omega_{2y})^{-1} \{ (\rho_1 + \rho_2)(\omega_{2y}\zeta^2 - \omega_{2x}\zeta^1)$$

$$+ [(\rho_1 + \rho_2)^2 (\omega_{2y}\zeta^2 - \omega_{2x}\zeta^1)^2 + 4(\rho_1^2 + \rho_2^2)\omega_{2x}\omega_{2y}\zeta^1\zeta^2$$

$$+ 4\rho_1\rho_2((\omega_{2y}\zeta^1)^2 + (\omega_{2x}\zeta^2)^2)]^{1/2} \}$$

or, equivalently, by

$$(\tan \gamma)_1 = \frac{-(\rho_1 + \rho_2) \cos (\varphi + \psi) + [4\rho_1\rho_2 + (\rho_1 - \rho_2)^2 \cos^2 (\varphi - \psi)]^{1/2}}{2(\rho_1 \sin \varphi \cos \psi + \rho_2 \cos \varphi \sin \psi)}$$

Here, φ and ψ are the angles formed by the vectors $\mathbf{j}$ and grad ω_2, respectively, with the x axis.

The expression

$$\frac{4}{(\cos^2 \gamma)_1} (|\mathbf{j}||\nabla\omega_2|)^{-1} j_\alpha \omega_{2\alpha} (\rho_1 \sin \varphi \cos \psi + \rho_2 \cos \varphi \sin \psi)^2$$

$$= 4(|\mathbf{j}||\nabla\omega_2|)^{-1}(\rho_1 \sin \varphi \cos \psi + \rho_2 \cos \varphi \sin \psi)^2 [\zeta^1 \omega_{2x}$$

$$+ (\tan^2 \gamma)_1 \zeta^2 \omega_{2y} + (\zeta^2 \omega_{2x} + \zeta^1 \omega_{2y})(\tan \gamma)_1] \tag{19}$$

can be represented in the form

$$\alpha_1 \rho_1^2 + \alpha_2 \rho_2^2 + 2\alpha_3 \rho_1\rho_2 + 2(\alpha_4 \rho_1 + \alpha_5 \rho_2)[4\rho_1\rho_2 + (\rho_1 - \rho_2)^2 \cos^2 (\varphi - \psi)]^{1/2}$$

Simple calculations lead to the following expressions for the coefficients:

$$\alpha_1 = 2\sin^2\varphi\,\cos(\varphi - \psi), \qquad \alpha_2 = 2\sin^2\psi\,\cos(\varphi - \psi)$$

$$\alpha_3 = 4\sin\varphi\,\sin\psi - (\sin^2\varphi + \sin^2\psi)\cos(\varphi - \psi)$$

$$\alpha_4 = \sin^2\varphi, \qquad \alpha_5 = \sin^2\psi$$

The sign of $j_\alpha\omega_{2\alpha}$ coincides with the sign of (19); the latter is nonnegative since the difference

$$[4\rho_1\rho_2 + (\rho_1 - \rho_2)^2\cos^2(\varphi - \psi)](\rho_1\sin^2\varphi + \rho_2\sin^2\psi)^2$$
$$- \{(\rho_1^2\sin^2\varphi + \rho_2^2\sin^2\psi)\cos(\varphi - \psi)$$
$$+ \rho_1\rho_2[4\sin\varphi\,\sin\psi - (\sin^2\varphi + \sin^2\psi)\cos(\varphi - \psi)]\}^2$$

is clearly equal to

$$4\rho_1\rho_2[\rho_1\sin^2\varphi - \rho_2\sin^2\psi - (\rho_1 - \rho_2)\sin\varphi\,\sin\psi\,\cos(\varphi - \psi)]^2$$

For $j_\beta\omega_{2\beta}$, the proof is similar.

A.5. On the Derivation of Equations (3.142) and (3.143)

Before variation, the current line $L(f)$ passes through an arbitrary point P of the basic region; after variation, the line $L_1(F)$ passes through this point (Fig. A.3). These lines are assumed to be close to one another in the space C_1.

We specify the position of any point Q in the neighborhood of the line $L(f)$ by coordinates s and ζ, where s is the arc length of $L(f)$ from the point N' to the point M, the base of the perpendicular dropped from Q onto the line $L(f)$; ζ is the projection of the vector $\overline{MQ}$ onto the normal $\mathbf{n}$ to the curve $L(f)$ at the point M. The directions of the unit vectors $\mathbf{t}$ and $\mathbf{n}$ of the tangent and normal to the curve $L(f)$ are given by

$$\mathbf{t}(x_s, y_s), \qquad \mathbf{n}(y_s, -x_s)$$

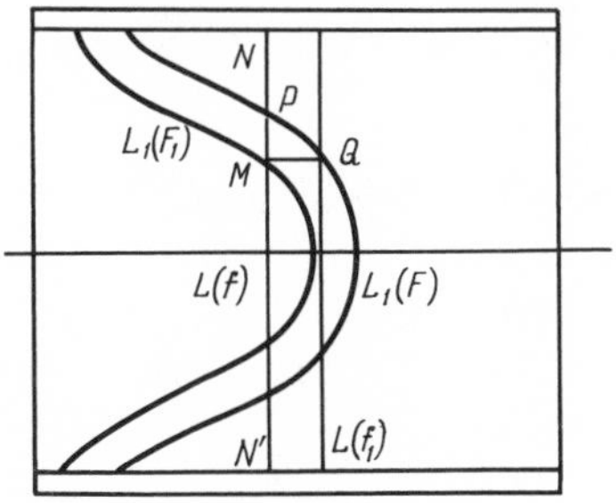

FIGURE A.3

The radius vector $\mathbf{R}$ of the point Q is related to the radius vector $\mathbf{r}$ of the point M by

$$\mathbf{R} = \mathbf{r} + \mathbf{n}\zeta$$

Using the Frenet formula

$$\frac{d\mathbf{n}}{ds} = \frac{\mathbf{t}}{\rho}$$

where ρ is the radius of curvature of the curve $L(f)$ at the point M, we obtain the connection between the differentials:

$$d\mathbf{R} = \mathbf{t}\left(1 + \frac{\zeta}{\rho}\right) ds + \mathbf{n}\, d\zeta$$

from which follow the expressions for the Lamé coefficients:

$$h_s = 1 + \frac{\zeta}{\rho}, \qquad h_\zeta = 1$$

If the point Q is displaced along the curve $L_1(F)$, then, along this curve,

$$(\text{grad } F, d\mathbf{R}) = 0 \quad \text{or} \quad F_s\, ds + F_\zeta\, d\zeta = 0$$

whence

$$d\zeta = -(F_s / F_\zeta)\, ds$$

along the same curve. Simultaneously,

$$d\mathbf{R} = \left[\left(1 + \frac{\zeta}{\rho}\right)\mathbf{t} - (F_s / F_\zeta)\mathbf{n}\right] ds = \mathbf{T}\, dS \tag{20}$$

here, $\mathbf{T}$ denotes the unit vector tangential to $L_1(F)$, and dS is the differential arc length of the same curve.

The curve $L(f_1)$ of the basic (old) family (in Fig. A.3, the curves of the old family are vertical straight lines) passes through the point Q; the variations

$$\varphi_P = F - f, \qquad \varphi_Q = F - f_1$$

$$+ \frac{1}{6} B_{xxx} \int_{-\delta}^{\delta} \zeta^3 \, dy - \frac{3}{8} \frac{B_x^2}{B\delta} \left(\int_{-\delta}^{\delta} \zeta \, dy \right)^2$$

$$- \frac{3}{8} \frac{B_x B_{xx}}{B\delta} \left(\int_{-\delta}^{\delta} \zeta \, dy \right) \left(\int_{-\delta}^{\delta} \zeta^2 \, dy \right)$$

$$+ \frac{5}{32} \frac{B_x^3}{(B\delta)^2} \left(\int_{-\delta}^{\delta} \zeta \, dy \right)^3 \Bigg] + o(\zeta^3) \tag{30}$$

For the function $f = f_0(x)$ defined by (3.189), a corresponding expression for ζ is given by [see (23)]:

$$\zeta = \rho_{\min}(2c\delta/VB_0)^{1/2} e^{\gamma x/2}\{(\varphi_p - \varphi)$$

$$- \rho_{\min}(2c\delta/VB_0)^{1/2} e^{\gamma x/2} (\varphi_p - \varphi)[\varphi_\zeta - (1/4)\gamma(\varphi_p - \varphi)]$$

$$- \rho_{\min}^2(2c\delta/VB_0) e^{\gamma x}(\varphi_p - \varphi)[-\varphi_\zeta^2 + (1/2)\varphi_{\zeta\zeta}(\varphi_p - \varphi)$$

$$+ (3/4)\gamma\varphi_\zeta(\varphi_p - \varphi) - (\gamma^2/12)(\varphi_p - \varphi)^2]\}$$

The substitution of this result in (30) yields

$$- \Delta\varphi_p - \frac{\gamma_0^2}{4}\left(\varphi_p - \frac{1}{2\delta}\int_{-\delta}^{\delta} \varphi \, dy \right)$$

$$= \frac{2\gamma_0 k + k^2}{4}\left(\varphi_p - \frac{1}{2\delta}\int_{-\delta}^{\delta} \varphi \, dy \right)$$

$$- \frac{\gamma_0^2}{8\delta}\rho_{\min}\sqrt{\frac{2c\delta}{VB_0}} e^{\gamma_0 x/2}\left\{ \int_{-\delta}^{\delta} \varphi_\zeta(\varphi_p - \varphi) \, dy \right.$$

$$+ \frac{\gamma_0}{4}\int_{-\delta}^{\delta}(\varphi_p - \varphi)^2 \, dy - \frac{3\gamma_0}{8\delta}\left[\int_{-\delta}^{\delta}(\varphi_p - \varphi) \, dy \right]^2 \right\}$$

$$- \frac{c\gamma_0^2\rho_{\min}^2}{4VB_0} e^{\gamma_0 x}\left\{ -\int_{-\delta}^{\delta} \varphi_\zeta^2(\varphi_p - \varphi) \, dy + \frac{1}{2}\int_{-\delta}^{\delta} \varphi_{\zeta\zeta}(\varphi_p - \varphi)^2 \, dy \right.$$

$$- \frac{\gamma_0}{4}\int_{-\delta}^{\delta} \varphi_\zeta(\varphi_p - \varphi)^2 \, dy + \frac{3\gamma_0}{4\delta}\int_{-\delta}^{\delta}(\varphi_p - \varphi) \, dy$$

$$\times \int_{-\delta}^{\delta} (\varphi_p - \varphi)\varphi_\zeta \, dy + \frac{3}{16} \frac{\gamma_0^2}{\delta} \int_{-\delta}^{\delta} (\varphi_p - \varphi) \, dy$$

$$\times \int_{-\delta}^{\delta} (\varphi_p - \varphi)^2 \, dy - \frac{5}{32} \frac{\gamma_0^2}{\delta^2} \left[\int_{-\delta}^{\delta} (\varphi_p - \varphi) \, dy \right]^3 \bigg\}$$

$$+ \sqrt{\frac{VB_0}{2c\delta}} \frac{e^{-(1/2)\gamma_0 x}}{2\rho_{\min}} \left(1 - \frac{\gamma_0 x}{2}\right) k - \sqrt{\frac{VB_0}{2c\delta}} \frac{e^{-(1/2)\gamma_0 x}}{8\rho_{\min}} x \left(1 + \frac{\gamma_0 x}{2}\right) k^2$$

from which we obtain (3.189) for the coefficients F_{ij}.

The determination of the L_{ij} [see (3.205)] entails calculations in accordance with (3.208); these calculations are very cumbersome and are therefore omitted.

A.7. An Alternate Solution to the Problem of Section 4.4

We present an alternate demonstration of the main result of Section 4.4. It was shown there that the second invariant $I_2(\mathbb{D}^0)$ of an effective tensor $\mathbb{D}^0$ remains constant and equal to that of the initial materials provided that the compounds making up the composite differ from one another only in the components and not in the invariant of $\mathbb{D}$ which must remain the same.

To determine effective characteristics of a medium with microstructure, one has to solve the following boundary-value problems for a cell of periodicity which is assumed to be a square with sides of unit length $(0 \le x \le 1, 0 \le y \le 1)$[421]:

$$\psi_i = \psi_i(x, y), \qquad i = 1, 2$$

$$\nabla \cdot \mathbb{P} \cdot \nabla \psi_1 = 0, \qquad \psi_1(1, y) - \psi_1(0, y) = 1; \qquad \psi_1(x, 1) - \psi_1(x, 0) = 0;$$
$$\tag{31}$$
$$[n \cdot \mathbb{P} \cdot \nabla \psi_1] = 0$$

$$\nabla \cdot \mathbb{P} \cdot \nabla \psi_2 = 0, \qquad \psi_2(1, y) - \psi_2(0, y) = 0; \qquad \psi_2(x, 1) - \psi_2(x, 0) = 1;$$
$$\tag{32}$$
$$[n \cdot \mathbb{P} \cdot \nabla \psi_2] = 0$$

Here, $\mathbb{P}$ is the conductivity tensor of some nonhomogeneous materials filling the cell, and $[\,\cdot\,]$ denotes a difference of the values of the argument at congruent points on opposite sides of a square. Note the $[\mathbb{P}] = 0$.

The second invariant $I_2(\mathbb{D}^0)$ of the effective tensor is determined by the formula

$$I_2(\mathbb{D}^0) = AC - B^2 \tag{33}$$

where

$$A = \iint_S \nabla \psi_1 \cdot \mathbb{P} \cdot \nabla \psi_1 \, dx \, dy, \qquad B = \iint_S \nabla \psi_1 \cdot \mathbb{P} \cdot \nabla \psi_2 \, dx \, dy$$

$$C = \iint_S \nabla \psi_2 \cdot \mathbb{P} \cdot \nabla \psi_2 \, dx \, dy$$

(34)

and the integrals are evaluated over the area S of the cell.

To prove the proposed constancy of $I_2(\mathbb{D}^0)$, it is necessary that its first variation, calculated subject to the conditions (31), (32), vanish identically. This variation is provided (a) by terms initiated by rotation of the principal axes of $\mathbb{P}$, and (b) by terms arising from separate alterations of the eigenvalues α and β of $\mathbb{P}$ with their product $\alpha\beta$ being kept constant.

The calculation of the first variation is carried out in the usual manner (see Chapter 2). Denote the eigenvectors which enter the dyadic representation of $\mathbb{P}$ by $\mathbf{a}$ and $\mathbf{b}$; i.e.,

$$\mathbb{P} = \alpha \mathbf{aa} + \beta \mathbf{bb}$$

Then the necessary conditions for constancy of $I_2(\mathbb{D}^0)$ will take on the form

$$C\psi_{1a}\psi_{1b} + A\psi_{2a}\psi_{2b} - B(\psi_{1a}\psi_{2b} + \psi_{1b}\psi_{2a}) = 0 \quad (35)$$

$$C(\alpha\psi_{1a}^2 - \beta\psi_{1b}^2) + A(\alpha\psi_{2a}^2 - \beta\psi_{2b}^2) - 2B(\alpha\psi_{1a}\psi_{2a} - \beta\psi_{1b}\psi_{2b}) = 0 \quad (36)$$

Here, $\psi_{1a}, \ldots$ denote the projections of $\nabla \psi_1$, $\nabla \psi_2$ in the directions of $\mathbf{a}$ and $\mathbf{b}$.

It is by no means obvious that the necessary conditions (35), (36) are satisfied *identically* by the solutions ψ_1 and ψ_2 of the boundary-value problems (31) and (32). To demonstrate that this is nevertheless the case, we begin with the first of Eqs. (32) and introduce the "conjugate" function ω by the relationship

$$\mathbb{P} \cdot \nabla \psi_2 = \mathbf{i}\omega_y - \mathbf{j}\omega_x = \mathbb{O} \cdot \nabla \omega \tag{37}$$

where $\mathbb{O} = \mathbf{ij} - \mathbf{ji}$ is a plane rotation tensor for an angle $\pi/2$. In view of the identity (4.19), (37) may be rewritten in the form

$$\alpha\beta \mathbb{O}^T \cdot \nabla \psi_2 = \mathbb{P} \cdot \nabla \omega \tag{38}$$

The left-hand side of this equality is a solenoidal vector which means that ω satisfies the same equation as ψ_1 and ψ_2. As to the boundary conditions, they are readily seen to be

$$[\omega_t] = [\omega_n] = 0$$

where $\mathbf{t}$ and $\mathbf{n}$ are the unit tangent and unit normal vectors on the boundary, respectively. It now is obvious that ω allows the representation

$$\omega = \mu\psi_1 + \nu\psi_2 \tag{39}$$

where μ and ν are constants. To determine these constants, one should use the boundary conditions (32) as deduced from the differences between the values of ψ_2 at the congruent points of the opposite sides of a square. These conditions may also be written in the form

$$\iint_S \psi_{2x}\, dx\, dy = 0, \qquad \iint_S \psi_{2y}\, dx\, dy = 1 \tag{40}$$

The use of (38) and of the identities

$$\iint_S \mathbf{i} \cdot \mathbb{P} \cdot \nabla\psi_1\, dx\, dy = \iint_S \nabla\psi_1 \cdot \mathbb{P} \cdot \nabla\psi_1\, dx\, dy = A$$

$$\iint_S \mathbf{i} \cdot \mathbb{P} \cdot \nabla\psi_2\, dx\, dy = \iint_S \mathbf{j} \cdot \mathbb{P} \cdot \nabla\psi_1\, dx\, dy = \iint_S \nabla\psi_1 \cdot \mathbb{P} \cdot \nabla\psi_2 = B$$

$$\iint_S \mathbf{j} \cdot \mathbb{P} \cdot \nabla\psi_2\, dx\, dy = \iint_S \nabla\psi_2 \cdot \mathbb{P} \cdot \nabla\psi_2\, dx\, dy = C$$

in (40) leads to the following linear system for determination of μ and ν

$$B\mu + C\nu = 0, \qquad A\mu + \beta\nu = -\alpha\beta$$

with solutions [see (33)]

$$\mu = -\frac{\alpha\beta}{I_2(\mathbb{D}^0)} C \quad \text{and} \quad \nu = \frac{\alpha\beta}{I_2(\mathbb{D}^0)} B \tag{41}$$

Combining this with (39) and (38), we arrive at the following relationships connecting the projections of $\nabla \psi_1$ and $\nabla \psi_2$ along the eigenvectors **a**, **b**:

$$\psi_{2a} = -\frac{C\beta}{[I_2^2(\mathbb{D}^0) + B^2\alpha\beta]}[\psi_{1b}I_2(\mathbb{D}^0) - \psi_{1a}B\alpha]$$

$$\psi_{2b} = \frac{C\alpha}{[I_2^2(\mathbb{D}^0) + B^2\alpha\beta]}[\psi_{1a}I_2(\mathbb{D}^0) + \psi_{1b}B\beta]$$

(42)

So far we have been dealing with the function ω which was a "conjugate" to ψ_2 [see (32)]. However, the same approach may be used for (31); in quite the same way we then arrive at the relationships

$$\psi_{1a} = \frac{A\beta}{[I_2^2(\mathbb{D}^0) + B^2\alpha\beta]}[\psi_{2b}I_2(\mathbb{D}^0) + \psi_{2a}B\alpha]$$

$$\psi_{1b} = -\frac{A\alpha}{[I_2^2(\mathbb{D}^0) + B^2\alpha\beta]}[\psi_{2a}I_2(\mathbb{D}^0) - \psi_{2b}B\beta]$$

(43)

We now use (43) to eliminate ψ_{1a}, ψ_{1b} from (42), and require that the determinant of the resulting homogeneous system for ψ_{2a}, ψ_{2b} vanish. This requirement reduces to the relationship

$$AC\alpha\beta = I_2^2(\mathbb{D}^0) + B^2\alpha\beta$$

or

$$I_2(\mathbb{D}^0) = \alpha\beta \qquad (44)$$

which is the desired result. We have thus shown that it is an immediate consequence of (31) and (32); the necessary conditions (35) and (36) are also easily shown to be satisfied identically. Toward this end, it suffices to eliminate ψ_{2a} and ψ_{2b} from the left-hand sides of (35) and (36) with the aid of (42), and subject to (44).

APPENDIX B
Remarks and Guide to the Literature

B.1. Chapter 2

A method for writing partial differential equations in a form analogous to Cauchy's normal form is given in Rashevskii's book.[141] In optimal control problems, this method was used by the author,[96,99,300] Cesari,[208] and Armand.[177] To a large extent, the material of Sections 2.1 and 2.2 is based on the author's papers.[96,99,300]

In his monograph, Rashevskii did not consider optimization problems, so that the control functions u are not emphasized. For optimization problems, it is extremely important to distinguish the parametric variables ζ and the controls u. In this connection, see the remarks on Chapter 3.

A method for replacing inequality constraints by equivalent equality constraints is well known in the calculus of variations (see Gernet[47]).

The necessary conditions of stationarity derived in Section 2.2 of course refer to the optimization problem formulated there. The basic features of this derivation are also retained in other problems, as can be seen from the subsequent exposition. At this stage, it does not appear to be worthwhile to strive for excessive generality of the stationarity conditions.

Canonical equations in the form of Eqs. (2.47) apparently were first introduced by Volterra,[377,378] and were subsequently used in many investigations (see also the comments on Chapter 8).

Some of the examples used to illustrate necessary conditions in Section 2.2 are analyzed, in detail, in subsequent chapters. The dependence of the eigenvalues of an elliptic operator on its coefficients was studied by Pucci.[341-344] Example 2.5 contains problems described by Pólya and Szegö (see also Raibouschinsky[349]).

A considerable amount of the material of Sections 2.3 to 2.6 is new, although the basic concept of variation in a strip and the necessary conditions (2.114) and (2.116) had already been introduced by me earlier.[97,99]

The derivation of Weierstrass's necessary conditions for the general case, given in Section 2.7, is based on the formula (2.172) for the increment as constructed by Rozonoér's method.[144] An important feature in this formula is the separation of the principal terms containing the increments of the parametric variables. The remaining terms (the remainder term) are estimated differently, in accordance with the particular problem. The proof for the nonlinear case, given in Section 2.7, is similar to Cesari's.[208] Similar proofs for a number of other cases are given in the later chapters.

For systems of a very general form, Plotnikov[128,129] has proposed a method for the derivation of necessary optimality conditions, including Weierstrass's necessary conditions. This method is based on the consideration of variations of admissible controls which depend on a finite number of parameters, where the variation is carried out in such a way that the principal parts of the corresponding increments of the dependent variables depend on these parameters linearly. For example, in Eq. (2.107) the role of such a parameter is played by the area $\pi \varepsilon^2 ab$ of the ellipse of variation which characterizes the admissible control U [see Eq. (2.106)]. A similar method of variation can be introduced in virtually all optimization problems.

A very general functional-theoretic approach to derivation of necessary optimality conditions is due to Dubovitskii and Milyutin,[64] an approach which recently has been extended to problems involving partial differential equations. The first results in this direction are due to Moskalenko.[111]

With regard to Weierstrass's necessary conditions and the related generalized maximum principle of Pontryagin, see also the investigations of Gabasov,[40] Baum and Cesari,[188] Brogan,[196-198] Halkin and Neustadt,[266] Lions,[93,291,292] Lions and Stampacchia,[295,296] Neustadt,[324] Oguztöreli,[326,328] Seierstad,[363] Tzafestas,[371,372] Wang,[379] and Klötzler's monograph.[278] Other references are given in the applications in subsequent chapters.

Krotov[84] indicates ways for extending his earlier method for deriving sufficient conditions for a minimum in problems with one independent variable to problems involving partial differential equations. In this connection, see also Stoddart's paper.[366]

Legendre's necessary condition was considered by the author[96] and in connection with hyperbolic problems by Petukhov and Troitskii[122]; Jacobi's necessary condition is studied in detail in an example in Chapter 3; see also Ref. 401.

B.2. Chapter 3

Many investigations have analyzed end effects in MHD channels. Many are cited in the text; a detailed exposition of the problem is given by Vatazhin, Lyubimov, and Regirer[33] in their monograph. The formulation and solution of the basic optimization problem studied in this chapter was given by myself in previous papers,[97,99,100,102,300] which form the basis of the exposition in Sections 3.2 to 3.12. The influence of the distribution of the magnetic induction on the characteristics of an MHD generator (Section 3.1) was investigated by Shercliff[164] and by the author.[101] The qualitative features of the optimal regime and its numerical characteristics given in Section 3.13 were studied by the author and Simkina.[103]

The need to introduce a tensor control is due to the absence of an optimal scalar control for $B(x) \neq$ const. This question is the subject of a discussion in Lions's monograph[93] and Cesari's paper[207]; Cesari proposed a relaxation of the problem by a modification of the equations by Gamkrelidze's method.[43] This idea is related to theorems on the existence of optimal controls in problems involving partial differential equations established by Cesari in a series of papers.[203-207,209,210] Among the conditions that guarantee the existence of such controls according to Cesari, an important role is played by the requirement of convexity of the set of functions $X_i^j(x, z, \zeta, u)$ in Eqs. (2.2) with respect to the set of variables ζ, u. In general, this convexity must be verified allowing for the fact that the parametric variables ζ are not independent of u; of course, if convexity can be proved without allowing for this dependence, Cesari's theorems guarantee the existence of optimal controls. As Cesari proved in Ref. 207, such "unconditional" convexity does not hold for Eqs. (3.15); it is observed, however, in the equations as modified by Gamkrelidze, and the modified regularized problem has an optimal solution.

It should be noted that the verification of "unconditional" convexity of the set of right-hand sides of Eqs. (3.15) is in no way related to the influence of the term $c^{-1}VB(x)$, which can be ignored here. Also, if $B(x) \equiv$ const, the existence of a solution can be proved in a different way (this is done in Section 3.14). The existence of a solution in this case must be related to the fact that although there is no "unconditional" convexity of the set $X_i^j(x, z, \zeta, u)$ with respect to ζ and u, the set is nevertheless convex if one takes into account the connection between ζ and u under the condition $B(x) =$ const (the actual verification of this convexity is difficult).

The introduction of a tensor control is associated with the notion of G-closure of the initial set of controls (see Chapters 4 and 5). This method is suggested by physical considerations discussed in the Introduction. Concerning the connections between G-closures and Gamkrelidze's relaxation procedure, see Section 4.6. In Section 3.14, whose material is new, we prove the existence under some additional assumptions concerning the existence of an optimal solution for the case when one of the principal values of the resistivity tensor ρ is infinitely large.

Various questions of existence and methods of regularization are discussed by Ekeland,[231,234] Furi and Vignoli,[256,257] Ghouila-Houri,[258,259] Olech,[331] Warga,[385] and Zolezzi.[395] See also Lions's paper "The optimal control of distributed systems" [*Russian Mathematical Surveys*, **28**, No. 4, 13 (1973)], which gives an extensive bibliography.

The problem considered in this chapter is an example of an optimization problem for a system described by equations of elliptic type. Such problems, at a general level, have been considered by Vaisbord,[26] Girsanov,[48] Gulenko and Ermol'ev,[54] Egorov,[66] Nenakhov and Gorchakov,[116] Raitum,[135] Ekeland,[234] and Zolezzi.[396]

B.3. Chapter 4

The notation of G-convergence of differential operators was introduced by De Giorgi and Spagnolo and has been discussed in many papers.[448,481,525,526] This notion is associated with the problem of existence of solutions which is traditional for the calculus of variations, as well as with the necessity of suitable extensions of the variational problems.[465,498,499,523,524] The related topics in homogenization were studied in Refs. 403, 405–407, 421–427, 441, 525, and 526.

Weak convergence of energy was considered in Ref. 448 as well as in Refs. 404, 484, 485, and 515, in connection with the notion of "compensated compactness." The estimates of Section 4.2 are essential for subsequent conclusions; they were obtained by the author, Cherkaev, and Fedorov.[476] Sections 4.3 to 4.6 follow the papers of the author and Cherkaev[477,479]; see also Tartar,[514,515] Raitum,[498] Dykhne,[442] and Kozlov.[466] The optimization problem of elastic torsion (Sections 4.7 and 4.8) was studied by Cea and Malanowski[202] as well as by Klosowicz and the author,[277] where the last paper contained some ambiguities which have been eliminated in the present exposition.

In its relaxed form (Sections 4.9 to 4.11), this problem was solved by the author and Cherkaev,[473] by Lavrov, the author, and Cherkaev,[468] and by the author, Cherkaev, and Fedorov.[475] Concerning additional optimization problems in the theory of elastic torsion, see also Refs. 397, 412, 416, 417, 463, and 467.

B.4. Chapter 5

Problems involving optimal plate design were studied by many authors. The presentation of the rigidity tensor $\mathbb{D}$ as used in the text was introduced by Rychlewski[539]; see also Ref. 471. Contradictions within the necessary optimality conditions as outlined in Section 5.3 were described in Ref. 475. Weak convergence of energy and of the second invariant of strain which form the basis for the estimates in Section 5.5 were considered in Ref. 476. Sections 5.6 to 5.8 follow Refs. 476 and 478. Important results on bounds of effective properties of elastic media are due to Hill.[459,460]

The problem of optimal design of plates of variable thickness has evoked a vast bibliography. Analysis of necessary conditions for optimality in the nonregularized version of this problem[408] has led to smooth stationary controls. On the other hand, it has been discovered that the numerical procedures based on these conditions are unstable,[402,434-436,490] and Weierstrass's necessary condition is violated[474,510]; see also the survey in Ref. 452. The relaxed formulations are now available only for the one-dimensional problem of a circular plate of variable thickness.[433,437,492] At the same time, there are optimization problems for plates which (like their torsional prototype[202]) do not require relaxation since the corresponding problem of existence can be solved independently. Examples may be found in Refs. 399 and 440 for bounded measurable controls and in Refs. 469 and 470 for smooth controls.

Various problems in the optimal design of plates equipped with stiffening ribs are discussed in Refs. 505-508 and 512.

There have been many investigations concerning the optimal distribution of external forces applied to elastic systems. Here, we must mention Egorov and Rafatov,[71] Komkov,[279-281] Yavin,[388,389] and Yavin and Sivan.[393] We should also mention the new estimates obtained in the investigation of the torsional rigidity of rods and the frequencies of normal vibrations of membranes and plates in terms of isoperimetric inequalities. Many results in this field, some of which have become classics, are contained in the monograph of Pólya and Szegö.[132] Of the more recent investigations, we mention only those of Banks,[187] McNabb,[313] Payne,[333,334] and Payne and Weinberger.[335]

B.5. Chapter 6

Systems described by first-order partial differential equations have attracted the interest of many authors, principally in connection with problems of the heating of "thin" bodies (see, e.g., Butkovskii[18] and the bibliography given there, and also Sirazetdinov[149]). In these problems, the control occurs in the right-hand side of a linear equation.

Problems of the type considered in Section 6.1, containing the control in the principal part, were also considered by the author.[301] I am indebted to Dr. K. G. Guderley for pointing out that the "shadow" zone is unstable if a discontinuous control is replaced by a slightly different continuous control.

The material in Section 6.3 is based largely on the paper of Majerczyk-Gómulkowa and Mioduchowski,[303] who obtained conditions for a weak maximum of the functional. In the present exposition, we obtain a condition for a strong relative maximum which is the same as that given for a weak maximum in Ref. 303. Important new results for optimization problems in plasticity were obtained by Strang and Kohn.[463,513]

The problems studied in Section 6.4 have been analyzed in detail by Jackson.[270] These studies give an interesting example involving the situation when the necessary conditions depend essentially on the method of variation (the variations of different types described at the end of Section 6.4). Pontryagin's maximum principle here is generalized in a nontrivial way for such problems.

Section 6.5 reviews investigations concerning the optimal shapes of surfaces immersed in a supersonically flowing gas. The large number of investigations into this subject is of course explained by its importance for practical applications. Basically, the exposition follows the paper of Guderley and Armitage[53]; this paper, together with the investigations of Sirazetdinov,[147,148] not only laid the foundations for this new stage in the development of this topic, but also had a stimulating influence on the investigation of the optimization of distributed parameter systems, in general.

Many results on the optimal shapes of bodies around which a flow takes place are contained in Ref. 160, whose Russian translation has an Appendix written by Gonor and Kraiko. Legendre's condition for optimization problems in supersonic gas dynamics was established by Fedorov[443]; see also the paper by Fedorov and the author.[480]

The last section of the chapter is based on material from Egorov's papers[67,68]; Egorov has obtained many results involving optimization problems for equations of hyperbolic type. He has shown the fruitfulness of applying Rozonoér's method[144] to a large class of optimization problems involving partial differential equations. The necessary conditions of optimality constructed in Refs. 67 and 68 have frequently been reproduced by other authors using different methods (see, e.g., Cesari[208]).

Hyperbolic optimization problems have also been studied by Volin and Ostrovskii,[35-37] Petukhov and Troitskii,[122,123] Plotnikov and Sikorskaya,[130] Plotnikov and Sumin,[131] Brezis and Lions,[194] Benker and Erfurt,[191,192] Brogan[196,198] Cirina,[216] Goldwin, Sriram, and Graham,[260] Lions,[93,291,292] Lions and Magenes,[293] and Miranker.[315] See also the papers of Mochi,[317] Pulvirenti,[345,346] Santagati,[359,360] Schmaedeke,[362] Rolewicz,[353] and the investigations of Mangeron and Oguztöreli.[305-308]

B.6. Chapter 7

Problems of the control of evolution processes in a Banach space and, in particular, parabolic problems are among the ones that have been most fully developed. This is due to the fact that problems of this type have a strong similarity to optimization problems for systems described by ordinary differential equations.

In Chapter 7, we have only touched on various questions of interest in connection with the extension of Pontryagin's maximum principle to this class.

The exposition of Sections 7.1 and 7.2 is based largely on the papers of Egorov,[74,75] Butkovskii,[18] and Gal'chuk.[42] Problems of optimal heating are of great practical interest and have been studied extensively. Here, we mention only Aliferov and Egorov,[3,4] Andreev and Butkovskii,[5,6] Andreev and Ogul'nik,[7] Vyrk,[39] Golub,[49,50] Dzhimsheleshvili,[62] Plotnikov,[126] Rapoport,[136] Raspopov,[137-140] Sirazetdinov,[150-153] Axelband,[178-181] Chaudhuri,[214] Duffin and McLain,[230] McCausland,[311,312] Mizel and Seidmann,[316] Sakawa,[358] and Yavin and Rasis.[390]

The possibility of extending the maximum principle has been studied for both linear and nonlinear systems. The first of these cases is considered in Section 7.4, which is based on Friedman's investigations.[251-253] The derivation for the nonlinear case given in Section 7.4 is due to Egorov.[69] It should be noted that the maximum principle was derived for the general nonlinear case by Egorov,[75] as cited above.

Apart from these authors, investigations into the maximum principle for control problems in a Banach space have been undertaken by Aslanov and Domshlak,[9] Akhiev and Akhmedov,[10] Volin and Ostrovskii,[36,37] Vostrova,[38] Degtyarev,[56,57] Degtyarev and Sirazetdinov,[59,60] Kante Kabine,[76,77] Mil'shtein,[106] Plotnikov,[128,129] and Sirazetdinov.[147] See also the papers of Axelband,[181] Brogan,[196-198] Chaudhuri,[213] Cole,[217] Connor,[218] Conti,[219,220] Denn,[225,226] Falb,[238] Fattorini,[242-247] Lattes,[287] Lions,[93,291,292] Lions and Magenes,[293] and Lions and Malgrange.[294] A very general approach to the derivation of necessary conditions in control problems in a Banach space is due to Neustadt.[324] Investigation of functional equations to which some control problems can be reduced by the use of the maximum principle has been made by Oguztöreli.[326,328] He has also established sufficient conditions for the existence of optimal controls for a number of problems of the type considered.[327] Generalizations of Pontryagin's maximum principle to control problems in a Banach space have also been considered by Rogak, Kazarinoff, and Scott-Thomas,[351] Rogak and Kazarinoff,[352] Seierstad,[363] Tzafestas,[371] Uzgiris and D'Souza,[374] Varaya,[375] Wang,[379] Yavin,[387] and Yavin and Sivan.[391]

The penalty method presented in Section 7.5 was introduced into the theory of optimal control of distributed parameter systems by Balakrishnan.[182-186] The exposition of Section 7.5 follows De Julio's paper[221]; he also proposed a numerical algorithm for the calculation of optimal regimes based on the penalty method[222]; see also Sasai.[361]

Problems of the control of processes with moving boundaries have been studied by Degtyarev,[61] Budak and Gol'dman,[16] Vasil'ev,[27] and Fasano.[241]

B.7. Chapter 8

The first three sections of the chapter are devoted to classical problems, which, however, are seldom set forth in textbooks on the calculus of variations. The canonical equations for the simplest problems involving partial derivatives were introduced by Volterra[377] and also by Hadamard and Levy.[92] The Hamilton–Jacobi equation and Jacobi's method in the case of the simplest problem was studied by Volterra[377,378] and by Lévy[92] (see also Prange[340]). The concepts of self-adjoint and matched boundary conditions for the simplest hyperbolic problem were introduced by the author[94] in analogy with the procedure by Gel'fand and Fomin[45] for variational problems with one independent variable. Canonical transformations in connection with Jacobi's method have been studied by Musicki.[322]

The application of the theory of integration in a function space to the solution of variational equations was considered by Donsker and Lions.[228] The optimality principle for evolution problems of the control of infinite-dimensional systems and Bellman's equation for such problems have been studied by many people. We mention Egorov and Krushel',[70] Moskalenko,[112] Bellman and Kalaba,[189] Brogan,[196] Graham,[263] Oguztöreli,[326,329] and Tzafestas and Nightingale.[373] The specialized problems for systems described by "polywave" equations (with operators of the type $\partial^4/\partial t^2\,\partial x^2$) have been studied by Mangeron and Oguztöreli by means of dynamic programming.[305]

General questions concerning the extension of Bellman's method to problems involving partial differential equations have been studied by Szefer[369]; in his paper he presents different forms of Bellman's equations and some applications to optimization problems in the theory of elasticity.

Section 8.7 is based on the work of Kim and Erzberger[235,236,274]; with regard to the infinite-dimensional Riccati equation, see also Falb and Kleinmann.[239]

I know of only one example of the application of the principle of optimality to an elliptic optimization problem—Bellman and Osborn's paper[190] quoted in Section 8.6. This is the simplest problem of this kind; the method of dynamic programming has not yet been applied to problems involving differential constraints in the form of equations of elliptic type.

References

1. ABDIKERIMOV, T. A., *Optimal Processes in Some Discrete Systems with Distributed Parameters* [in Russian], Avtomatika i Telemekhanika, Vol. 25, No. 2, 1964.
2. ABDIKERIMOV, T. A., AND KRASNITSKII, M. S., *On the Theory of the Invariance of Systems of Automatic Control with Distributed Parameters* [in Russian], Proceedings of the 13th Scientific Conference of the Professors and Teaching Members of the Physics and Mathematics Faculty at the Kirghiz University, Sektsia Matematiki, Frunze, 1965.
3. ALIFEROV, V. V., AND EGOROV, A. I., *On Optimal Control of a Heat Transfer Process* [in Russian], Izvestiya Akademii Nauk Kirgizskoi SSR, No. 4, 1970.
4. ALIFEROV, V. V., AND EGOROV, A. I., *On a Problem of Optimal Control of a Linear Object with Distributed Parameters* [in Russian], Applied Mathematics and Programming, No. 4, Akademia Nauk Moldavskoi SSR, Kishinev, 1971.
5. ANDREEV, Y. P., AND BUTKOVSKII, A. G., *Optimal Control of the Heating of Massive Bodies* [in Russian], Izvestiya Akademii Nauk SSR, Tekhnicheskaya Kibernetika, No. 5, 1964.
6. ANDREEV, Y. P., AND BUTKOVSKII, A. G., *A Problem of Optimal Control of the Heating of Massive Bodies* [in Russian], Inzhenerno-Fizicheskii Zhurnal, Vol. 8, No. 1, 1965.
7. ANDREEV, Y. P., AND OGUL'NIK, M. G., *Fastest Heating of Massive Bodies with Allowance for a Constraint* [in Russian], Analysis and Synthesis of Automatic Control Systems, Nauka, Moscow, 1968.
8. ARKIN, V. I., KOLEMAEV, V. A., AND SHIRYAEV, A. N., [in Russian], Trudy Matematicheskogo Instituta imeni Steklova Akademiya Nauk SSSR, Vol. 71, 1964 [Collection of Papers on Probability Theory].
9. ASLANOV, D., AND DOMSHLAK, Y. I., *On Some Variational Problems in a Hilbert Space Associated with the Solution of Differential Equations* [in Russian], Izvestiya Akademii Nauk Azerbaidzhanskoi SSR, Seriya Fiziko-Tekhnicheskikh i Matematicheskikh Nauk, No. 4, 1969.
10. AKHIEV, S. S., AND AKHMEDOV, K. T., *On Optimal Control in Systems with Distributed Parameters and Retarded Argument* [in Russian], Uchenye Zapiski Azerbaidzhanskogo Gosudarstvennogo Universiteta, Seriya Fiziko-Matematicheskikh Nauk, No. 1, 1970.
11. BELLMAN, R. E., *Dynamic Programming*, Princeton University Press, Pr8nceton, New Jersey, 1957.
12. BELLMAN, R. E., GLICKSBERG, I., AND GROSS, O. A., *Some Aspects of the Mathematical Theory of Control Processes*, Rand Corporation, Santa Monica, 1958.
13. BORISOV, V. M., *On the Optimal Shape of Bodies in a Supersonic Gas Flow*, Zhurnal Vychislitel'noi Matematiki i Matematicheskoi Fiziki, Vol. 3, No. 4, 1965.
14. BORISOV, V. M., *Variational Problems of Three-Dimensional Supersonic Flows* [in Russian], Prikladnaya Matematika i Mekhanika, Vol. 29, No. 1, 1965.
15. BORISOV, V. M., AND SHIPILIN, A. V., *On Nozzles of Maximal Thrust with Arbitrary Isoperimetric Conditions* [in Russian], Prikladnaya Matematika i Mekhanika, Vol. 28, No. 1, 1964.

16. BUDAK, B. M., AND GOL'DMAN, N. L., *Optimal Stefan Problem for a Quasilinear Parabolic Equation, Solutions of Stefan Problems* (Russian collection), Moscow, 1970.

17. BOURBAKI, N., *Espaces Vectoriels Topologiques* (Livre V, Fascicule de Resultats), Hermann, Paris, 1955.

18. BUTKOVSKII, A. G., *Theory of Optimal Control of Systems with Distributed Parameters* [in Russian], Nauka, Moscow, 1965.

19. BUTKOVSKII, A. G., *Finite Control and Controllability in Distributed Systems* [in Russian], Doklady Akademii Nauk SSSR, Vol. 191, No. 6, 1970.

20. BUTKOVSKII, A. G., AND POLTAVSKII, L. N., *Finite Control of Systems with Distributed Parameters* [in Russian], Avtomatika i Telemekhanika, Vol. 30, No. 4, 1969.

21. BUTKOVSKII, A. G., AND POLTAVSKII, L. N., *Optimal Control of a Distributed Vibrational System* [in Russian], Avtomatika i Telemekhanika, Vol. 26, No. 11, 1965.

22. BUTKOVSKII, A. G., AND POLTAVSKII, L. N., *Optimal Control of a Two-Dimensional Distributed Vibrational System* [in Russian], Avtomatika i Telemekhanika, Vol. 27, No. 4, 1966.

23. BUTKOVSKII, A. G., AND POLTAVSKII, L. N., *Optimal Control of Wave Processes* [in Russian], Avtomatika i Telemekhanika, Vol. 27, No. 9, 1966.

24. VAINBERG, M. M., *Variational Methods for the Investigation of Nonlinear Operators* [in Russian], Gostekhizdat, Moscow, 1956.

25. VAINBERG, M. M., AND TRENOGIN, V. A., *Theory of Bifurcation of Solutions of Nonlinear Equations* [in Russian], Nauka, Moscow, 1969.

26. VAISBORD, E. M., *A Problem of Optimal Control for Systems with Distributed Parameters* [in Russian], Izvestiya Akademii Nauk SSSR, Tekhnicheskaya Kibernetika, No. 5, 1966.

27. VASIL'EV, F. P., *On the Existence of a Solution of an Optimal Stefan Problem* [in Russian], Computational Methods and Programming, No. 12, Moscow State University, Moscow, 1969.

28. VATAZHIN, A. B., *On the Solution of Some Boundary-Value Problems of Magnetohydrodynamics* [in Russian], Prikladnaya Matematika i Mekhanika, Vol. 25, No. 5, 1961.

29. VATAZHIN, A. B., *Magnetohydrodynamic Flow in a Planar Channel with Finite Electrodes* [in Russian], Izvestiya Akademii Nauk SSSR, Otdelenie Tekhnicheskikh Nauk, No. 1, 1962.

30. VATAZHIN, A. B., AND REGIRER, S. A., *Approximate Calculation of the Current Distribution in the Case of Flow of a Conducting Liquid Along a Channel in a Magnetic Field* [in Russian], Prikladnaya Matematika i Mekhanika, Vol. 26, No. 3, 1962.

31. VATAZHIN, A. B., *Determination of the Joule Dissipation in the Channel of MHD* [in Russian], Prikladnaya Mekhanika i Tekhnicheskaya Fizika, No. 5, 1962.

32. VATAZHIN, A. B., *Some Two-Dimensional Problems of the Distribution of Current in an Electrically Conducting Medium Moving Along a Channel in a Magnetic Field* [in Russian], Prikladnaya Mekhanika i Tekhnicheskaya Fizika, No. 2, 1963.

33. VATAZHIN, A. B., LYUBIMOV, G. A., AND REGIRER, S. A., *Magnetohydrodynamic Flows in Channels* [in Russian], Nauka, Moscow, 1970.

34. VATAZHIN, A. B., AND NEMKOVA, N. G., *Some Two-Dimensional Problems on the Distribution of the Electric Current in the Channel of an MHD Generator with Nonconducting Baffles* [in Russian], Prikladnaya Mekhanika i Tekhnicheskaya Fizika, No. 2, 1964.

35. VOLIN, Y. M., AND OSTROVSKII, G. M., *On an Optimal Problem* [in Russian], Avtomatika i Telemekhanika, Vol. 25, No. 10, 1964.

36. VOLIN, Y. M., AND OSTROVSKII, G. M., *On a Problem of Optimization of a System with Distributed Parameters* [in Russian], Prikladnaya Matematika i Mekhanika, Vol. 29, No. 3, 1965.

37. VOLIN, Y. M., AND OSTROVSKII, G. M., *On the Maximum Principle in a Banach Space* [in Russian], Kibernetika, No. 5, 1969.

38. VOSTROVA, Z. I., *Optimal Processes in Discrete Systems Containing Objects with Distributed Parameters* [in Russian], Avtomatika i Telemekhanika, Vol. 27, No. 5, 1966.

39. VYRK, A. K., *Control of the Heating of a Massive Body with Allowance for Restrictions on the Thermal Load* [in Russian], Avtomatika i Telemekhanika, Vol. 33, No. 5, 1972.

40. GABASOV, R., *On Necessary Conditions of Optimality for Systems Described by Partial Differential Equations* [in Russian], Doklady Akademii Nauk BSSR, Vol. 12, No. 7, 1968.

41. GABASOV, R., AND KIRILLOVA, F. M., *The Present State of the Theory of Optimal Systems* [in Russian], Avtomatika i Telemekhanika, Vol. 33, No. 9, 1972.

42. GAL'CHUK, L. I., *On Some Problems of Optimal Control of Systems Described by Parabolic Equations* [in Russian], Vestnik Moskovskogo Universiteta, Seriya Matematiki i Mekhaniki, No. 3, 1968.

43. GAMKRELIDZE, R. V., *Optimal Sliding States* [in Russian], Doklady Akademii Nauk SSSR, Vol. 143, No. 6, 1962.

44. GASANOV, K. K., *On a Problem of the Calculus of Variations for a Second-Order Partial Differential Equation in the Presence of Constraints* [in Russian], Uchenye Zapiski Azerbaidzhanskogo Gosudarstvennogo Universiteta, Seriya Fiziko-Matematicheskikh Nauk, No. 1, 1970.

45. GEL'FAND, I. M., AND FOMIN, S. V., *Calculus of Variations*, Prentice Hall, Englewood Cliffs, New Jersey, 1963.

46. GEL'FAND, I. M., AND SHILOV, G. E., *Generalized Functions, Properties and Operations*, Vol. 1, Academic Press, New York, 1964.

47. GERNET, N. N., *On a Basic Problem of the Simplest Kind in the Calculus of Variations* [in Russian], St. Petersburg, 1913.

48. GIRSANOV, I. V., *Minimax Problems in the Theory of Diffusion Processes* [in Russian], Doklady Akademii Nauk SSSR, Vol. 136, No. 4, 1961.

49. GOLUB, N. N., *Optimal Control of the Process of Heating of Massive Bodies with Internal Heat Sources* [in Russian], Avtomatika i Telemekhanika, Vol. 28, No. 12, 1967.

50. GOLUB, N. N., *Optimal Control of Linear and Nonlinear Systems with Distributed Parameters* [in Russian], Avtomatika i Telemekhanika, Vol. 30, No. 9, 1969.

51. GOLUZIN, G. M., *Geometrical Theory of Functions of a Complex Variable*, A. M. S. Providence, R.I., 1969.

52. GRINBERG, G. A., *Selected Questions of the Mathematical Theory of Electric and Magnetic Phenomena* [in Russian], Izdatel'stvo Akademii Nauk SSSR, 1948.

53. GUDERLEY, K. G., AND ARMITAGE, J. V., *A General Method for the Determination of Best Supersonic Rocket Nozzles*, paper presented at the Symposium on Extremal Problems in Aerodynamics, Boeing Scientific Research Laboratories, Seattle, Washington, 1962.

54. GULENKO, V. P., AND ERMOL'EV, Y. M., *On Some Problems of the Optimal Control of Equations of Elliptic Type* [in Russian], Theory of Optimal Solutions, Proceedings of Seminar, No. 1, Kiev, 1968.

55. GYUNTER, N. M., *Course in the Calculus of Variations* [in Russian], Gostekhizdat, Moscow, 1941.

56. DEGTYAREV, G. L., *On Optimal Control of Linear Stationary Processes with Distributed Parameters* [in Russian], Trudy Kazanskogo Aviatsionnogo Instituta, No. 92, 1966.

57. DEGTYAREV, G. L., *On Optimization of an Industrial Process Described by Systems of Partial Differential Equations* [in Russian], Proc. Third Conference of Young Scientific Workers at Kazan', Sektsia Mekhaniki i Matematiki, Kazan', 1967.

58. DEGTYAREV, G. L., AND SIRAZETDINOV, T. K., *On a Problem of the Optimal Control of Systems with Distributed Parameters* [in Russian], Izvestiya Akademii Nauk SSSR, Tekhnicheskaya Kibernetika, No. 1, 1967.

59. DEGTYAREV, G. L., AND SIRAZETDINOV, T. K., *On Optimal Control of One-Dimensional Processes with Distributed Parameters* [in Russian], Avtomatika i Telemekhanika, Vol. 28, No. 11, 1967.

60. DEGTYAREV, G. L., AND SIRAZETDINOV, T. K., *Optimal Control of One-Dimensional Processes with Retarded Argument* [in Russian], Avtomatika i Telemekhanika, Vol. 31, No. 1, 1970.

61. DEGTYAREV, G. L., *Optimal Control of Distributed Processes with Moving Boundary* [in Russian], Avtomatika i Telemekhanika, Vol. 33, No. 10, 1972.

62. DZHIMSHELESHVILI, M. O., *Optimal Control of Some Two-Dimensional Thermal Objects with Distributed Parameters* [in Russian], Soobshcheniya Akademii Nauk Gruzinskoi SSR, Vol. 52, No. 1, 1968.

63. ZAVELANI-ROSSI, A., *Minimum-Weight Design for Two-Dimensional Bodies*, Meccanica, Vol. 4, pp. 1–10, 1969.

64. DUBOVITSKII, A. Y., AND MILYUTIN, A. A., *Extremalization Problems in the Presence of Constraints* [in Russian], Zhurnal Vychislitel'noi Matematiki i Matematicheskoi Fiziki, Vol. 5, No. 3, 1965.

65. EGOROV, A. I., *Optimal Control of Processes in Distributed Objects* [in Russian], Prikladnaya Matematika i Mekhanika, Vol. 27, No. 4, 1963.

66. EGOROV, A. I., *On a Variational Problem in the Theory of Equations of Elliptic Type* [in Russian], Sibirskii Matematicheskii Zhurnal, Vol. 5, No. 3, 1964.

67. EGOROV, A. I., *Necessary Conditions of Optimality for Systems with Distributed Parameters* [in Russian], Avtomatika i Telemekhanika, Vol. 25, No. 5, 1964.

68. EGOROV, A. I., *Necessary Conditions of Optimality for Systems with Distributed Parameters* [in Russian], Matematicheskii Sbornik, Vol. 69, No. 3, 1966.

69. EGOROV, A. I., *Optimal Processes in Systems with Distributed Parameters and Some Problems of Invariance Theory* [in Russian], Izvestiya Akademii Nauk SSSR, Seriya Matematicheskaya, Vol. 29, 1965.

70. EGOROV, A. I., AND KRUSHEL', E. G., *Dynamic Programming in the Optimal Control of Distributed Objects* [in Russian], Izvestiya Akademii Nauk Kirgizskoi SSR, No. 6, 1968.

71. EGOROV, A. I., AND RAFATOV, R., *On the Approximate Solution of an Optimal Control Problem* [in Russian], Zhurnal Vychisliteli'noi Matematiki i Matematicheskoi Fiziki, Vol. 12, No. 2, 1972.

72. EGOROV, Y. V., *On Some Problems in the Theory of Optimal Control* [in Russian], Doklady Akademii Nauk SSSR, Vol. 145, No. 4, 1962.

73. EGOROV, Y. V., *Optimal Control in a Banach Space* [in Russian], Doklady Akademii Nauk SSSR, Vol. 150, No. 2, 1963.

74. EGOROV, Y. V., *Some Problems in the Theory of Optimal Control* [in Russian], Zhurnal Vychislitel'noi Matematiki i Matematicheskoi Fiziki, Vol. 3, No. 5, 1963.

75. EGOROV, Y. V., *Necessary Conditions of Optimality in a Banach Space* [in Russian], Matematicheskii Sbornik, Vol. 64 (106), No. 1, 1964.

76. KABINE KANTE, *Necessary and Sufficient Conditions of Optimal Processes with Distributed Parameters and Deviating Argument* [in Russian], Aspirantskii Sbornik Universiteta Druzhby Narodov, Moscow, 1968.

77. KABINE KANTE, *On Optimal Control in the Theory of Nonlinear Partial Differential Equations and with Deviating Argument* [in Russian], Fourth Conference of the Faculty of Physics, Mathematics and Natural Sciences, Mathematics Section, Patrice Lumumba Friendship of Nations University, Moscow, 1968.

78. KOPETS', M. M., *On a Problem of Control in Hilbert Space* [in Russian], Dopovidi Akademii Nauk URSR, No. 8, 1972.

79. KRAIKO, A. N., *Variational Problems of the Gas Dynamics of Equilibrium and Nonequilibrium Flows* [in Russian], Prikladnaya Matematika i Mekhanika, Vol. 28, No. 2, 1964.

80. KRAIKO, A. N., *Solution of Variational Problems of Supersonic Gas Dynamics* [in Russian], Prikladnaya Matematika i Mekhanika, Vol. 30, No. 2, 1966.

81. KRAIKO, A. N., NAUMOVA, I. N., AND SHMYGLEVSKII, Y. D., *Construction of Bodies of Optimal Shape in a Supersonic Flow* [in Russian], Prikladnaya Matematika i Mekhanika, Vol. 28, No. 1, 1964.

82. KRASOVSKII, N. N., *Theory of Optimally Controlled Systems* [in Russian], Mechanics in the Soviet Union During the Last Fifty Years, Vol. 1, Nauka, Moscow, 1968.

83. KRIVOLUTSKII, V. S., *On a Problem with Distributed Parameters* [in Russian], Controlled Systems, No. 1, Nauka, Novosibirsk, 1968.

84. KROTOV, V. F., *Methods of Solution of Variational Problems Based on Sufficient Conditions of an Absolute Minimum, III* [in Russian], Avtomatika i Telemekhanika, Vol. 25, No. 7, 1964.

85. KUZ'MINA, A. L., *On an Optimal Control Problem* [in Russian], Commentationes Mathematicae Universitatis Carolinae, Vol. 7, No. 3, 1966.

86. COURANT, R., AND HILBERT, D., *Methods of Mathematical Physics*, Vols. 1 and 2, Interscience Publishers, New York, 1953, 1962.

87. COURANT, R., AND FRIEDRICHS, K. O., *Supersonic Flow and Shock Waves*, Springer-Verlag, Berlin, 1977.

88. LAVRENT'EV, I. V., *Influence of the Position of Baffles on the Characteristics of an MHD Channel* [in Russian], Magnitnaya Gidrodinamika, No. 4, pp. 89–95, 1967.

89. LAVRENT'EV, I. V., *Influence of the Length of a Nonconducting Baffle on the Characteristics of an MHD Channel* [in Russian], Magnitnaya Gidrodinamika, No. 1, pp. 27–32, 1968.

90. LAVRENT'EV, M. A., AND SHABAT, B. V., *Methods of the Theory of Functions of a Complex Variable* [in Russian], Gostekhteoretizdat, Leningrad, 1951.

91. LADYSZHENSKAIA, O. A., AND URAL'TSEVA, N. N., *Linear and Quasilinear Elliptic Equations*, Academic Press, New York, 1968.

92. LÉVY, P., *Problemes concrets d'analyse fonctionelle*, Gauthier-Villars, Paris, 1951.

93. LIONS, J. L., *Optimal Control Systems Governed by Partial Differential Equations*, Springer-Verlag, Berlin, 1970.

94. LURIE, K. A., *On the Hamilton–Jacobi Method in Variational Problems Involving Partial Differential Equations* [in Russian], Prikladnaya Matematika i Mekhanika, Vol. 27, No. 2, 1963.

95. LURIE, K. A., *The Liouville and Stäckel Theorems for Hyperbolic Variational Problems* [in Russian], Prikladnaya Matematika i Mekhanika, Vol. 27, No. 3, 1963.

96. LURIE, K. A., *The Mayer–Bolza Problem for Multiple Integrals and Optimization of the Behavior of Systems with Distributed Parameters* [in Russian], Prikladnaya Matematika i Mekhanika, Vol. 27, No. 5, 1963.

97. LURIE, K. A., *Optimal Control of the Conductivity of a Liquid Moving Along a Channel in a Magnetic Field* [in Russian], Prikladnaya Matematika i Mekhanika, Vol. 28, No. 2, 1964.

98. LURIE, K. A., *On a Method of Deriving Bellman's Equation* [in Russian], Prikladnaya Matematika i Mekhanika, Vol. 21, No. 1, 1967.

99. LURIE, K. A., *Optimal Problems for Distributed Systems* [in Russian], Optimal Systems, Statistical Methods, Nauka, Moscow, 1967.

100. LURIE, K. A., *On the Optimal Distribution of the Resistivity Tensor of the Working Medium in the Channel of a MHD Generator* [in Russian], Prikladnaya Matematika i Mekhanika, Vol. 34, No. 2, 1970.

101. LURIE, K. A., *Optimal Distribution of the Magnetic Field Along the Channel of a MHD Generator with Electrodes of Finite Length* [in Russian], Zhurnal Tekhnicheskoi Fiziki, Vol. 40, No. 12, 1970.

102. LURIE, K. A., *Optimal Distributions of the Resistivity Tensor of the Working Medium in a MHD Channel (the Case of Nonlocal Variations)* [in Russian], Prikladnaya Matematika i Mekhanika, Vol. 35, No. 2, 1971.

103. LURIE, K. A., AND SIMKINA, T. Y., *Load Characteristics of a MHD Channel in the Case of Optimal Distribution of the Resistivity of the Working Medium* [in Russian], Magnitnaya Gidrodinamika, No. 2, 1971.

104. LIUSTERNIK, L. A., AND SOBOLEV, V. I., *Elements of Functional Analysis*, Gordon and Breach, New York, 1961.

105. MEEROV, M. V., AND LITVAK, B. L., *Optimization of Systems of Multiconstraint Control* [in Russian], Nauka, Moscow, 1972.

106. MIL'SHTEIN, G. N., *On the Analytic Construction of Regulators for Equations of Parabolic and Hyperbolic Type* [in Russian], Differentsial'nye Uravneniya, Vol. 8, No. 4, 1972.

107. MILYUTIN, A. A., *General Schemes for Obtaining Necessary Conditions of an Extremum and Optimal Control Problems* [in Russian], Uspekhi Matematicheskikh Nauk, Vol. 25, No. 5, 1970.

108. MIKHLIN, S. G., *Variational Methods in Mathematical Physics*, Pergamon Press, Elmsford, New York, 1964.

109. MIKHLIN, S. G., *Numerical Realization of Variational Methods* [in Russian], Nauka, Moscow, 1966.

110. MIKHLIN, S. G., *Advanced Course in Mathematical Physics*, North-Holland, Amsterdam, 1971.

111. MOSKALENKO, A. I., *On a Class of Optimal Control Problems* [in Russian], Zhurnal Vychislitel'noi Matematiki i Matematicheskoi Fiziki, Vol. 9, No. 1, 1969.

112. MOSKALENKO, A. I., *Bellman's Equation for Optimal Processes in Some Systems with Distributed Parameters* [in Russian], Trudy Vsesoyuznogo Nauchno-Issledovatel'skogo Instituta Tsellyulozno-Bumazhnoi Promyshlennosti, No. 54, 1969. (Proceedings All-Union Scientific Research Institute of the Cellulose and Paper Industry).

113. NAZARENKO, V. V., *Simulation of Magnetohydrodynamic Flows in a Channel in an Electric Bath* [in Russian], Prikladnaya Mekhanika i Tekhnicheskaya Fizika, No. 5, 1963.

114. NATANSON, I. P., *Theory of Functions of a Real Variable* [in Russian], Gostekhteoretizdat, Leningrad, 1957.

115. NEMYTSKY, V. V., AND STEPANOV, V. V., *Qualitative Theory of Differential Equations*, Princeton University Press, Princeton, New Jersey, 1960.

116. NENAKHOV, E. I., AND GORCHAKOV, V. N., *Necessary Conditions of an Extremum in Problems of Automatic Control of Systems Described by Partial Differential Equations of Elliptic Type* [in Russian], Kibernetika, No. 2, 1972.

117. NIKOL'SKII, A. A., *On Figures of Revolution with Channel Having the Least Wave Resistance in Supersonic Flow* [in Russian], Collection of Theoretical Studies in Aerodynamics, Oborongiz, Moscow, 1957.

118. OLEINIK, O. A., *Boundary Value Problems for Linear Equations of Elliptic and Parabolic Type with Discontinuous Coefficients* [in Russian], Izvestiya Akademii Nauk SSSR, Seria Matematicheskaya, Vol. 25, 1961.

119. OSTROVSKII, G. M., *On Optimization of Complex Chemicotechnical Processes* [in Russian], Ivzestiya Akademii Nauk SSSR, Tekhnicheskaya Kibernetika, No. 5, 1964.

120. OSTROVSKII, G. M., AND SHUSHUNOV, L. N., *On a Method of Quasilinearization for Calculation of Optimal Systems with Distributed Parameters* [in Russian], Methods of Optimization of Systems of Multiconstraint Control, Nauka, Moscow, 1972.

121. PAVLOV, V. G., AND CHEPRASOV, V. P., *Group-Invariant Properties of a Nonlinear Optimally Controlled Process with Distributed Parameters* [in Russian], Prikladnaya Matematika i Mekhanika, Vol. 22, No. 3, 1968.

122. PETUKHOV, L. V., AND TROITSKII, V. A., *Variational Optimization Problems for Equations of Hyperbolic Type* [in Russian], Prikladnaya Matematika i Mekhanika, Vol. 36, No. 4, 1972.

123. PETUKHOV, L. V., AND TROITSKII, V. A., *Some Optimal Problems in the Theory of Longitudinal Vibrations* [in Russian], Prikladnaya Matematika i Mekhanika, Vol. 36, No. 5, 1972.

124. PLOTNIKOV, V. I., *On a Problem of the Control of Stationary Systems with Distributed Parameters* [in Russian], Doklady Akademii Nauk SSSR, Vol. 170, No. 2, 1966.

125. PLOTNIKOV, V. I., *On the Optimal Control of Systems with Distributed Parameters*, Doklady Akademii Nauk SSSR, Vol. 175, No. 6, 1967.

126. PLOTNIKOV, V. I., *On the Convergence of Finite-Dimensional Approximations (in the Problem of the Optimal Heating of an Inhomogeneous Body of Arbitrary Shape)* [in Russian], Zhurnal Vychislitel'noi Matematiki i Matematicheskoi Fiziki, Vol. 8, No. 1, 1968.

127. PLOTNIKOV, V. I., *Existence Theorems for Optimizing Functions for Optimal Systems with Distributed Parameters* [in Russian], Izvestiya Akademii Nauk SSSR, Seriya Matematicheskaya, Vol. 34, No. 3, 1970.

128. PLOTNIKOV, V. I., *Necessary Conditions of Optimality for Controlled Systems of General Form* [in Russian], Doklady Akademii Nauk SSSR, Vol. 199, No. 2, 1971.

129. PLOTNIKOV, V. I., *Necessary and Sufficient Conditions of Optimality and Uniqueness Conditions for Optimizing Functions for Controlled Systems of General Form* [in Russian], Izvestiya Akademii Nauk SSSR, Seriya Matematicheskaya, Vol. 36, No. 3, 1972.

130. PLOTNIKOV, V. I., AND SIKORSKAYA, E. R., *Optimization of a Controlled Object Described by a Nonlinear System of Hyperbolic Equations* [in Russian], Izvestiya Vysshikh Uchebnykh Zavedenii, Radiofizika, Vol. 15, No. 3, 1972.

131. PLOTNIKOV, V. I., AND SUMIN, V. I., *Optimization of Objects with Distributed Parameters Described by Goursat-Darboux Systems* [in Russian], Zhurnal Vychislitel'noi Matematiki i Matematicheskoi Fiziki, Vol. 15, No. 3, 1972.

132. PÓLYA, G., AND SZEGÖ, G., *Isoperimetric Inequalities in Mathematical Physics*, Annals of Mathematical Studies, Vol. 27, Princeton University Press, Princeton, New Jersey, 1951.

133. PONTRYAGIN, L. S., BOLTYANSKII, V. G., GAMKRELIDZE, R. V., AND MISHCHENKO, E. F., *The Mathematical Theory of Optimal Processes*, Interscience Publishers, New York, 1962.

134. RABINOVICH, A. B., *Optimization of Some Linear Systems with Distributed Parameters* [in Russian], Zhurnal Vychislitel'noi Matematiki i Matematicheskoi Fiziki, Vol. 8, No. 1, 1968.

135. RAITUM, U. E., *On Some Extremal Problems Associated with a Linear Elliptic Equation* [in Russian], Latviiskii Matematicheskii Ezhegodnik, Vol. 4, 1968.

136. RAPOPORT, E. Y., *On a Control Problem of the Most Rapid Heating of Massive Bodies* [in Russian], Avtomatika i Telemekhanika, Vol. 4, 1971.

137. RASPOPOV, B. M., *A Time Optimal Problem for Uncoupled Heat and Mass Transfer* [in Russian], Avtomatika i Telemekhanika, Vol. 26, No. 10, 1965.

138. RASPOPOV, B. M., *An Optimal Problem of the Most Rapid Drying of a Plate* [in Russian], Elements of the Theory and Technology of Automatic Control, Ilim Publishers, Frunze, 1966.

139. RASPOPOV, B. M., *Control of Various Transport Processes* [in Russian], Inzhenerno-Fizicheskii Zhurnal, Vol. 12, No. 4, 1967.

140. RASPOPOV, B. M., *Some Problems of Optimal Control of Heat and Mass Transfer During Drying* [in Russian], Avtomatika i Telemekhanika, Vol. 29, No. 2, 1968.

141. RASHEVSKII, P. K., *Geometrical Theory of Partial Differential Equations* [in Russian], Gostekhizdat, Moscow, 1947.

142. REITMAN, M. I., AND SHAPIRO, G. S., *Theory of Optimal Design in Constructional Mechanics and the Theory of Elasticity and Plasticity* [in Russian], Itogi Nauki, Mekhanika (Uprogost' i Plastichnost') [Reviews of Science, Mechanics (Elasticity and Plasticity)], 1964; VINITI, Moscow, 1966.

143. ROZHDESTVENSKII, B. L., AND YANENKO, N. N., *Systems of Quasilinear Equations* [in Russian], Nauka, Moscow, 1968.

144. ROZONOÉR, L. U., *Pontryagin's Maximum Principle in the Theory of Optimal Systems (I-III)* [in Russian], Avtomatika i Telemekhanika, Vol. 20, No. 10, pp. 1320-1334, No. 11, pp. 1441-1458, No. 12, pp. 1561-1578, 1959.

145. SENATOROV, P. K., *Coefficient Stability of Solutions of Ordinary Second-Order Differential Equations and Parabolic Equations on a Plane* [in Russian], Differentsial'nye Uravneniya, Vol. 7, No. 4, 1971.

146. SENATOROV, P. K., *On the Stability of the Dirichlet Problem for the Equation $div\{k(x) \, grad \, u\} - q(x)u = -f(x)$* [in Russian], Doklady Akademii Nauk SSSR, Vol. 208, No. 6, 1973.

147. SIRAZETDINOV, T. K., *Pontryagin's Maximum Principle in the Theory of Linear Optimal Processes with Distributed Parameters* [in Russian], Trudy Kazanskogo Aviatsionnogo Instituta, No. 80, 1963.

148. SIRAZETDINOV, T. K., *Optimal Problems in Gas Dynamics* [in Russian], Izvestiya Vysshikh Uchebnykh Zavedenii, Aviatsionnaya Tekhnika, No. 2, 1963.

149. SIRAZETDINOV, T. K., *On the Analytic Construction of Regulators in Processes with Distributed Parameters* [in Russian], Avtomatika i Telemekhanika, Vol. 25, No. 4, 1964.

150. SIRAZETDINOV, T. K., *On the Analytic Construction of Regulators in Processes with Distributed Parameters* [in Russian], Avtomatika i Telemekhanika, Vol. 26, No. 9, 1965.

151. SIRAZETDINOV, T. K., *Optimal Regulation of the Distribution of the Temperature of a Solid* [in Russian], Optimal Control Systems, Nauka, Moscow, 1967.

152. SIRAZETDINOV, T. K., *On the Analytic Construction of Regulators for Magnetohydrodynamic Processes* [in Russian], (I) Avtomatika i Telemekhanika, Vol. 28, No. 10, 1967; (II) Avtomatika i Telemekhanika, Vol. 28, No. 12, 1967.

153. SIRAZETDINOV, T. K., *On the Analytic Construction of Regulators in Systems with Distributed Parameters* [in Russian], Trudy Universiteta Druzhby Narodov im. P. Lumumby (Proceedings of the Patrice Lumumba Friendship of Nations University), Vol. 27, 1968.

154. SLIN'KO, M. G., FEDOTOV, A. V., AND KUZNETSOV, Y. I., *Determination of Optimal Boundary Control for a Process with Variable Activity of the Catalyzer* [in Russian], Discrete Analysis, No. 13, Nauka, Novosibirsk, 1968.

155. SMYTHE, W. R., *Static and Dynamic Electricity*, McGraw-Hill, New York, 1950.

156. SRETENSKII, L. N., *Theory of the Newtonian Potential* [in Russian], Gostekhizdat, Moscow, 1946.

157. STEPANOV, A. V., *Artificial Anisotropy as a Means for Giving Manufactured Articles Required Mechanical Properties* [in Russian], Collection in Honour of the 70th Birthday of Academician A. F. Ioffe, Izdatel'stvo Akademii Nauk SSSR, 1950.

158. STEPANOV, A. V., *Fundamentals of the Physical Theory of Strength and Plasticity of Crystals* [in Russian], Izvestiya Akademii Nauk SSSR, Seriya Fizicheskaya, Vol. 17, No. 3, 1953.

159. STEPANOV, A. V., *The Future of Metal Processing* [in Russian], Lenizdat, Leningrad, 1963.

160. MIELE, A. (Ed.), *Theory of Optimum Aerodynamic Shapes*, Academic Press, New York, 1969.

161. TIKHONOV, A. N., *On Methods of Regularization of Optimal Control Problems* [in Russian], Doklady Akademii Nauk SSSR, Vol. 162, No. 4, 1965.

162. FRIEDMAN, A., *Partial Differential Equations of Parabolic Type*, Prentice–Hall, Englewood Cliffs, New Jersey, 1964.

163. CHERNOUS'KO, F. L., AND BANICHUK, N. V., *Variational Problems of Mechanics and Control* [in Russian], Nauka, Moscow, 1973.

164. SHERCLIFF, J. A., *Theory of Electromagnetic Flow Measurement*, Cambridge University Press, London, 1963.

165. SHIPILIN, A. V., *Optimal Shapes of Bodies with Attached Shock Waves* [in Russian], Mekhanika Zhidkosti i Gaza, No. 4, 1966.

166. SHMYGLEVSKII, Y. D., *Some Variational Problems of the Gas Dynamics of Axisymmetric Supersonic Flows* [in Russian], Prikladnaya Matematika i Mekhanika, Vol. 21, No. 2, 1957.

167. SHMYGLEVSKII, Y. D., *On Some Properties of Axisymmetric Supersonic Gas Flows* [in Russian], Doklady Akademii Nauk SSSR, Vol. 122, No. 5, 1958.

168. SHMYGLEVSKII, Y. D., *On Supersonic Profiles Having Minimal Resistance* [in Russian], Prikladnaya Matematika i Mekhanika, Vol. 22, No. 2, 1958.

169. SHMYGLEVSKII, Y. D., *On a Class of Figures of Revolution with Minimal Wave Resistance* [in Russian], Prikladnaya Matematika i Mekhanika, Vol. 24, No. 5, 1960.

170. SHMYGLEVSKII, Y. D., *Variational Problems for Supersonic Figures of Revolution and Nozzles* [in Russian], Prikladnaya Matematika i Mekhanika, Vol. 26, No. 1, 1962.

171. SHMYGLEVSKII, Y. D., *Some Variational Problems in Gas Dynamics* [in Russian], Trudy Vychislitel'nogo Tsentra Akademii Nauk SSSR, Moscow, 1963.

172. SHOLOKHOVICH, F. A., *On Controllability in Hilbert Spaces* [in Russian], Differentsial'nye Uravneniya, Vol. 3, No. 3, 1967.

173. SHUMUNOV, L. N., *On the Gradient Method for an Optimal System with Distributed Parameters* [in Russian], Control of Multiconstraint Systems, Nauka, Moscow, 1967.

174. YAGUBOV, M. A., *Optimal Control Problem for a Process with Distributed Parameters* [in Russian], Uchenye Zapiski Azerbaidzhanskogo Gosudarstvennogo Universiteta, Seriya Fiziko-Matematicheskikh Nauk, No. 3, 1972.

175. ALVARADO, F. L., AND MUKUNDAN, R., *An Optimization Problem in Distributed Parameter Systems*, International Journal of Control, Vol. 9, No. 6, 1966.

176. ANGEL, E., *Dynamic Programming and Linear Partial Differential Equations*, Journal of Mathematical Analysis and Applications, Vol. 23, No. 3, 1968.

177. ARMAND, J.-L., *Minimum Mass Design of a Plate-Like Structure for Specified Fundamental Frequency*, AIAA Journal, Vol. 9, No. 9, 1971.

178. AXELBAND, E. I., *Function Space Methods for the Optimum Control of Class of Distributed Parameter Systems*, Proceedings of the Automatic Control Conference, Troy, New York, 1965.

179. AXELBAND, E. I., *An Approximation Technique for the Optimal Control of Linear Distributed Parameter Systems with Bounded Inputs*, IEEE Transactions on Automatic Control, Vol. 11, No. 1, 1966.

180. AXELBAND, E. I., *The Structure of the Optimal Tracking Problem for Distributed Parameter Systems*, IEEE Transactions on Automatic Control, Vol. 13, No. 1, 1968.

181. AXELBAND, E. I., *Optimal Control of Linear Distributed Parameter Systems*, Advances in Control Systems, Edited by C. T. Leondes, Vol. 7, Academic Press, New York, 1969.

182. BALAKRISHNAN, A. V., *Optimal Control Problems in Banach Spaces*, Journal of Society for Industrial and Applied Mathematics, Ser. A3, No. 1, 1965.

183. BALAKRISHNAN, A. V., *Semigroup Theory and Control Theory*, Proceedings of IFIP Congress (1965), Spartan Books, Washington, D.C., 1965.

184. BALAKRISHNAN, A. V., *Linear Systems with Infinite Dimensional State Spaces*, Proceedings of a Symposium on System Theory, Brooklyn Polytechnic Institute, No. 4, 1965.

185. BALAKRISHNAN, A. V., *Foundations of the Statespace Theory of Continuous Systems,* International Journal of Computer and System Sciences, Vol. 1, No. 1, 1967.

186. BALAKRISHNAN, A. V., *Introduction of Optimization Theory in a Hilbert Space,* Lecture Notes in Operations Research and Mathematical Systems, Springer-Verlag, Berlin, 1971.

187. BANKS, D. O., *Some Inequalities for the Fundamental Frequency of a Non-Homogeneous Membrane,* Journal of the Society for Industrial and Applied Mathematics, Vol. 13, No. 3, 1965.

188. BAUM, R. F., AND CESARI, L., *On a Recent Proof of Pontryagin's Necessary Conditions,* SIAM Journal on Control, Vol. 10, No. 1, 1972.

189. BELLMAN, R., AND KALABA, R., *Dynamic Programming Applied to Control Processes Governed by General Functional Equations,* Proceedings of the National Academy of Sciences USA, Vol. 48, 1962.

190. BELLMAN, R., AND OSBORN, H., *Dynamic Programming and the Variations of Green's Functions,* Journal of Mathematics and Mechanics, Vol. 7, No. 1, 1958.

191. BENKER, H., *Numerische Behandlung spezieller optimaler Systeme mit verteilen Parametern,* Wissenschaftliche Zeitschrift, Technische Hochschule Chemie, Leuna-Merseburg, Vol. 12, No. 1, 1970.

192. BENKER, H., AND ERFURT, H., *Numerische Behandlung optimaler Systeme mit verteilen Parametern,* Wissenschaftliche Zeitschrift, Technische Hochschule Chemie, Leuna-Merseburg, Vol. 11, No. 3, 1969.

193. BOUCHER, R. A., AND AMES, D. B., *End Effect Losses in DC Magnetohydrodynamic Generators,* Journal of Applied Physics, Vol. 32, No. 5, 1961.

194. BREZIS, H., AND LIONS, J.-L., *Sur certain problemes unilateraux hyperboliques,* Comptes Rendus Academie des Sciences, Serie A, Vol. 264, 1967.

195. BRIGGS, D. L., AND SHEN, C. N., *Distributed Parameter Optimum Control of a Nuclear Rocket with Thermal Stress Constraints,* Transactions of ASME, Basic Engineering, Series D, Vol. 89, 1967.

196. BROGAN, W. L., *Dynamic Programming and a Distributed Parameter Maximum Principle,* Joint Automatic Control Conference, Preprint Papers, New York, New York, 1967.

197. BROGAN, W. L., *Theory and Application of Optimal Control for Distributed Parameter Systems* [review], *I. Theory,* Automatica, Vol. 4, No. 3, 1967; *II. Computational Results,* Automatica, Vol. 4, No. 3, 1967.

198. BROGAN, W. L., *Optimal Control Theory Applied to Systems Governed by Partial Differential Equations* [review], Advances in Control Systems, Edited by C. T. Leondes, Vol. 6, Academic Press, New York, 1968.

199. CAINAIELLO, E. R., Editor, *Functional Analysis and Optimization,* Academic Press, New York, 1966 [contains a number of papers on optimization of distributed systems].

200. CEA, J., *Les methodes de descente dans la theorie de l'optimization,* Revue Francaise Informatique et Recherche Operationelle, Vol. 2, No. 13, 1968.

201. CEA, J., *Optimization: théorie et algorithmes,* Dunod, Paris, 1971.

202. CEA, J., AND MALANOWSKI, K., *An Example of a Max-Min Problem in Partial Differential Equations,* SIAM Journal on Control, Vol. 8, No. 3, 1970.

203. CESARI, L., *Existence Theorems for Multidimensional Problems of Optimal Control,* Differential Equations and Dynamical Systems, Academic Press, New York, 1967.

204. CESARI, L., *Existence Theorems for Multidimensional Lagrange Problems,* Journal of Optimization Theory and Applications, Vol. 1, 1967.

205. CESARI, L., *Multidimensional Lagrange and Pontryagin Problems,* Mathematical Theory of Control, Academic Press, New York, 1967.

206. CESARI, L., *Sobolev Spaces and Multidimensional Lagrange Problems of Optimization,* Annali di Scuola Normale Superiore di Pisa, Classe di Science, Vol. 22, Fasc. II, 1968.

207. CESARI, L., *Multidimensional Lagrange Problems of Optimization in a Fixed Domain and an Application to a Problem of Magnetohydrodynamics*, Archive for Rational Mechanics and Analysis, Vol. 29, No. 2, 1968.

208. CESARI, L., *Optimization with Partial Differential Equations in a Dieudonne-Rashevsky Form and Conjugate Problems*, Archive for Rational Mechanics and Analysis, Vol. 33, No. 5, 1969.

209. CESARI, L., *Seminormality and Upper Semicontinuity in Optimal Control*, Journal of Optimization Theory and Applications, Vol. 6, No. 2, 1970.

210. CESARI, L., *Existence Theorems for Abstract Multidimensional Control Problems*, Journal of Optimization Theory and Applications, Vol. 6, No. 3, 1970.

211. CESARI, L., LA PALM, J. R., AND NISHIURA, T., *Remarks on Some Existence Theorems for Optimal Controls*, Journal of Optimization Theory and Applications, Vol. 3, No. 6, 1970.

212. CESARI, L., AND COWLES, J. E., *Existence Theorems in Multidimensional Problems of Optimization with Distributed and Boundary Controls*, Archive for Rational Mechanics and Analysis, Vol. 46, No. 5, 1972.

213. CHAUDHURI, A. K., *Concerning Optimum Control of Linear Distributed Parameter Systems*, International Journal of Control, Vol. 2, No. 4, 1965.

214. CHAUDHURI, S. P., *Distributed Optimal Control in a Nuclear Reactor*, International Journal of Control, Vol. 16, No. 5, 1972.

215. CHIN-KUO CHU, AND YEN-PING SHIH, *Low Sensitivity Optimal Control of a Class of Linear Distributed Systems*, International Journal of Control, Vol. 16, No. 2, 1972.

216. CIRINA, M., *Boundary Controllability of Non-Linear Hyperbolic Systems*, SIAM Journal on Control, Vol. 7, No. 2, 1969.

217. COLE, J. K., *Time Optimal Controls of the Equation of Evolution in a Separable and Reflexive Banach Space*, Journal of Optimization Theory and Applications, Vol. 2, No. 3, 1968.

218. CONNOR, M. A., *Optimal Control of Systems Represented by Differential-Integral Equations*, IEEE Transactions on Automatic Control, Vol. 17, No. 1, 1972.

219. CONTI, R., *Contribution to Linear Control Theory*, Journal of Differential Equations, Vol. 1, 1965.

220. CONTI, R., *Time Optimal Solution of a Linear Evolution Equation in Banach Spaces*, Journal of Optimization Theory and Applications, Vol. 2, No. 5, 1968.

221. DE JULIO, S., *On the Optimization of Infinite Dimensional Linear Systems*, Proceedings of the 2nd International Conference on Computational Methods in Optimization Problems, San Remo, 1968.

222. DE JULIO, S., *Numerical Solution of Dynamical Optimization Problems*, SIAM Journal on Control, Vol. 8, No. 2, 1970.

223. DELFOUR, M. C., AND MITTER, S. K., *Controllability and Observability for Infinite Dimensional System*, SIAM Journal on Control, Vol. 10, No. 2, 1972.

224. DEMORE, Q., *Una problema di controllo ottimale nel cambiamento di stato in uno strato*, Annali di Matematica Pura ed Applicata, Seria quarta, Vol. 84, 1973.

225. DENN, M. M., *Optimal Boundary Control for a Nonlinear Distributed System*, International Journal of Control, Vol. 4, No. 2, 1964.

226. DENN, M. M., *Optimal Linear Control of Distributed Systems*, Industrial and Engineering Chemistry Fundamentals, Vol. 7, No. 3, 1968.

227. DIVE, P., *Attraction des ellipsoides homogenes et reciproque d'un theoreme de Newton*, Bulletin Societe Mathematique de France, Vol. 59, 1931.

228. DONSKER, M. D., AND LIONS, J.-L., *Frechet-Volterra Variational Equations*, Boundary Value Problems and Function Space Integrals, Acta Mathematica, Vol. 108, 1962.

229. DREYFUS, S. E., *Dynamic Programming and the Calculus of Variations*, Academic Press, New York, 1965.

230. DUFFIN, R. J., AND MCLAIN, D. K., *Optimum Shape of a Cooling Fin on a Convex Cylinder*, Journal of Mathematics and Mechanics, Vol. 17, No. 8, 1968.

231. EKELAND, I., *Formes equivalentes d'un probleme relaxe*, Comptes Rendus de l'Academie des Sciences, Vol. A271, No. 4, 1970.

232. EKELAND, I., *Relaxation de problemes de controle pour des systemes decrits par des equations aux derivees partielles*, Comptes Rendus de l'Academie des Sciences, Vol. A270, No. 19, 1970.

233. EKELAND, I., *Suites minimisantes et solutions de certaines problemes de controle optimal*, Lecture Notes in Mathematics, Vol. 132, Springer-Verlag, Berlin, 1972.

234. EKELAND, I., *Sur le controle optimal de systemes gouvernes par des equations elliptiques*, Journal of Functional Analysis, Vol. 9, No. 1, 1972.

235. ERZBERGER, H., AND KIM, M., *Optimum Boundary Control of Distributed Parameter Systems*, Information and Control, Vol. 9, No. 3, 1966.

236. ERZBERGER, H., AND KIM, M., *Optimum Distributed Parameter Systems with Distributed Control*, Proceedings of IEEE, Vol. 54, No. 4, 1966.

237. FAHMY, M., *Optimal Control of Distributed Parameter Systems*, Proceedings of IEEE, Vol. 115, No. 4, 1968.

238. FALB, P., *Infinite Dimensional Control Problems. On the Closure of the Set of Attainable States for Linear Systems*, Journal of Mathematical Analysis and Applications, Vol. 9, No. 1, 1964.

239. FALB, P. L., AND KLEINMANN, D. L., *Remarks on the Infinite Dimensional Riccati Equation*, IEEE Transactions on Automatic Control, Vol. 11, No. 3, 1966.

240. FANSELAU, R. W., *Comments on Exhaust Nozzle Contour for Maximum Thrust*, Journal of American Rocket Society (J. ARS), Vol. 29, No. 6, 1959.

241. FASANO, A., *Un esempio di controllo ottimale in un problems del tipo di Stefan*, Bolletino di Unione Matematica Italiano, Vol. 4, No. 6, 1971.

242. FATTORINI, H. O., *Time-Optimal Control of Solution of Operational Differential Equations*, SIAM Journal on Control, Series A, Vol. 2, No. 1, 1964.

243. FATTORINI, H. O., *Control in Finite Time of Differential Equations in Banach Space*, Communications on Pure and Applied Mathematics, Vol. 19, 1966.

244. FATTORINI, H. O., *On Complete Controllability of Linear Systems*, Journal of Differential Equations, Vol. 3, 1967.

245. FATTORINI, H. O., *Boundary Control Systems*, SIAM Journal on Control, Vol. 6, No. 3, 1968.

246. FATTORINI, H. O., *A Remark on the Bang-Bang Principle for Linear Control Systems in Infinite-Dimensional Space*, SIAM Journal on Control, Vol. 6, No. 1, 1968.

247. FATTORINI, H. O., *An Observation on a Paper of A. Friedmann*, Journal of Mathematical Analysis and Applications, Vol. 22, No. 2, 1968.

248. FATTORINI, H. O., AND RUSSELL, D. L., *Exact Controllability Theorems for Linear Parabolic Equations in One Space Dimension*, Archive for Rational Mechanics and Analysis, Vol. 43, No. 4, 1971.

249. FRAIDENRAICH, N., MCGRATH, I. A., MEDIN, S. A., AND THRING, M. W., *A Theoretical Analysis of the Thermodynamics of the Straited Magnetohydrodynamic System*, British Journal of Applied Physics, Vol. 15, No. 1, 1964.

250. FRÉCHET, M., *Sur une extension de la methode de Jacobi-Hamilton*, Annali di Matematica, Vol. 3, No. 11, 1905.

251. FRIEDMAN, A., *Optimal Control for Parabolic Equations*, Journal of Mathematical Analysis and Applications, Vol. 18, No. 3, 1967.

252. FRIEDMAN, A., *Optimal Control in Banach Spaces*, Journal of Mathematical Analysis and Applications, Vol. 19, No. 1, 1967.

253. FRIEDMAN, A., *Optimal Control in Banach Space with Fixed End Points*, Journal of Mathematical Analysis and Applications, Vol. 24, No. 1, 1968.

254. FRIEDMAN, A., *Differential Games of Pursuit in Banach Space*, Journal of Mathematical Analysis and Applications, Vol. 25, No. 1, 1969.

255. FUHRMANN, P. A., *On Weak and Strong Reachability and Controllability of Infinite-Dimensional Linear Systems*, Journal of Optimization Theory and Applications, Vol. 9, No. 2, 1972.

256. FURI, M., AND VIGNOLI, A., *On the Regularization of a Nonlinear Ill-Posed Problem in Banach Spaces*, Journal of Optimization Theory and Applications, Vol. 4, No. 3, 1969.

257. FURI, M., AND VIGNOLI, A., *A Characterization of Well-Posed Minimum Problems in a Complete Metric Space*, Journal of Optimization Theory and Applications, Vol. 5, No. 6, 1970.

258. GHOUILA-HOURI, A., *Sur la generalization de la notion de commande d'un systeme guidable*, Revue d'Informatique et de Recherche Operationelle, No. 4, 1967.

259. GHOUILA-HOURI, A., *Problemes d'existence en theorie de la commande*, Identification, Optimalisation et Stabilite-Systemes Automatiques, Paris, 1967.

260. GOLDWIN, R. M., SRIRAM, K. P., AND GRAHAM, M. H., *Time-Optimal Control of a Linear Hyperbolic System*, International Journal of Control, Vol. 12, No. 4, 1970.

261. GOODSON, R. E., AND KHATRI, H. C., *Optimal Control of Systems with Distributed Parameters*, Transactions of ASME, Journal of Basic Engineering, Series D, Vol. 88, No. 2, 1966.

262. GRAHAM, J. W., *Time-Optimal Control of a Class of Linear-Distributed Parameter Systems with Non-Linear Boundary Conditions*, International Journal of Control, Vol. 12, No. 2, 1970.

263. GRAHAM, J. W., *A Hamilton–Jacobi Approach to the Optimal Control of Distributed Parameter Systems*, International Journal of Control, Vol. 12, No. 3, 1970.

264. GRÖTZSCH, H., *Über einige Extremalprobleme der Konformen Abbildung*, Berichte Mathematisch Physikalische Klasse der Sächsischen Akademie der Wissenschaften, Vol. 80, 1928.

265. GUDERLEY, C. G., AND HANTSCH, E., *Beste Formen für achsensymmetrische Überschallschubdüsen*, Zeitschrift für Flugwissenschaften, Vol. 3, No. 9, Sept. 1955.

266. HALKIN, H., AND NEUSTADT, L. W., *General Necessary Conditions for Optimization Problems*, Proceedings of the National Academy of Sciences USA, Vol. 56, 1966.

267. HERGET, C. J., *On the Controllability of Distributed Parameter Systems*, International Journal of Control, Vol. 11, No. 5, 1970.

268. HOLDER, E., *Über eine potentialtheoretische Eigenschaft der Ellipse*, Mathematische Zeitschrift, Vol. 35, 1932.

269. HURWITZ, H., JR., KILB, R. W., AND SUTTON, G. W., *Influence of Tensor Conductivity on Current Distribution in a MHD Generator*, Journal of Applied Physics, Vol. 32, No. 2, 1961.

270. JACKSON, R., *Optimization Problems in a Class of Systems Described by Hyperbolic Partial Differential Equations, Part I, Variational Theory*, International Journal of Control, Vol. 4, No. 2, 1966, *Part II, A Maximum Principle*, Vol. 4, No. 6, 1966.

271. JONES, A. P., AND MCCORMICK, G. P., *A Generalization of the Method of Balakrishnan: Inequality Constraints and Initial Conditions*, SIAM Journal on Control, Vol. 8, No. 2, 1970.

272. KATZ, S., *A General Maximum Principle for End-Point Control Problems*, Journal of Electronics and Control, Vol. 16, No. 2, 1964.

273. KIM, M., *Successive Approximation Method in Optimum Distributed-Parameter Problems,* Journal of Optimization Theory and Applications, Vol. 4, No. 1, 1969, err. corr. Vol. 5, No. 1, 1970.

274. KIM, M., AND ERZBERGER, H., *On the Design of Optimum Distributed Parameter System with Boundary Control Function,* IEEE Transactions on Automatic Control, Vol. 12, No. 1, 1967.

275. KIM, M., AND GAJWANI, S. H., *Variational Approach to Optimum Distributed Parameter Systems,* Transactions on Automatic Control, Vol. 13, No. 2, 1968.

276. KLOSOWICZ, B., *Sur la nonhomogeneite d'une barre tordue,* Bulletin de l'Academie Polonaise des Sciences, Serie des Sciences Techniques, Vol. 18, 1970.

277. KLOSOWICZ, B., AND LURIE, K. A., *On the Optimal Non-Homogeneity of a Torsional Elastic Bar,* Archivum Mechaniki Stosowanej, Vol. 24, No. 2, 1971.

278. KLÖTZLER, R., *Mehrdimensionale Variationsrechnung,* Birkhauser Verlag, Basel, 1970.

279. KOMKOV, V., *The Optimal Control of a Transverse Vibration of a Beam,* SIAM Journal on Control, Vol. 6, No. 3, 1968.

280. KOMKOV, V., *Optimal Control of Vibrating Thin Plates,* SIAM Journal on Control, Vol. 8, No. 2, 1970.

281. KOMKOV, V., *Optimal Control Theory for the Damping of Vibrations of Simple Elastic Systems,* Lecture Notes in Mathematics, Springer-Verlag, Berlin, 1972.

282. KOPPEL, L. B., AND YEN PING SHIN, *Optimal Control of a Class of Distributed Parameter Systems with Distributed Controls,* Industrial Chemical Fundamentals, Vol. 7, No. 3, 1968.

283. KUO, M. C. Y., *Differential Games on Function Space,* Journal of the Franklin Institute, Vol. 289, No. 6, 1970.

284. KUO, M. C. Y., *Duality in Distributed Parameter Systems,* Journal of Engineering Mathematics, Vol. 6, No. 3, 1972.

285. KUSIC, G. L., JR., *Finite Differences to Implement the Solution for Optimal Control of Distributed Parameter Systems,* IEEE Transactions on Automatic Control, Vol. 14, No. 4, 1969.

286. LANDY, R. J., *Distributed Parameter Differential Games, I. A Dynamic Programming Approach, II. A Variational Calculus Approach,* International Journal of Control, Vol. 14, No. 3, 1971.

287. LATTES, R., *Control of Boundary Conditions for Non-Stationary Problems,* Proceedings of the Conference on Mathematical Theory of Control, University of South California, Los Angeles, 1967.

288. LEE, M. E., AND SHEN, D. W. C., *Optimal Control of a Class of Distributed Parameter Systems Using Gradient Methods,* Proceedings of IEEE, Vol. 116, No. 7, 1969.

289. LEES, M., *The Goursat Problem,* Journal of the Society for Industrial and Applied Mathematics, Vol. 8, No. 3, 1960.

290. LEMAIRE, B., *Jeux dans les equations aux derivees partielles,* Lecture Notes in Mathematics, Vol. 132, Springer-Verlag, Berlin, 1970.

291. LIONS, J.-L., *Sur le controle optimal de systemes decrits par des equations aux derivees partielles lineares, I–III,* Comptes Rendus de l'Academie des Sciences, Serie A, Vol. 263, 1966.

292. LIONS, J.-L., *On Some Non-Linear Partial Differential Equations Related to Optimal Control Theory,* Proceedings of Symposium on Non-Linear Functional Analysis, Chicago, 1968.

293. LIONS, J.-L., AND MAGENES, E., *Controle optimal et espaces du type de Gevrey, I, II,* Atti Accademia Nazionale dei Lincei, Classe di Scienze Fisiche, Matematiche e Naturali Rendiconti, Vol. 44, 1968.

294. LIONS, J.-L., AND MALGRANGE, B., *Sur l'unicite retrograde dans les problemes mixtes paraboliques*, Mathematica Scandinavica, Vol. 8, No. 2, 1960.

295. LIONS, J.-L., AND STAMPACCHIA, G., *Variational Inequalities*, Communications on Pure and Applied Mathematics, Vol. 20, 1967.

296. LIONS, J.-L., AND STAMPACCHIA, G., *Inequations variationnelles non coercivites*, Comptes Rendus de l'Academie des Sciences, Serie A, Vol. 261, 1965.

297. LIONS, J.-L., AND TEMAN, R., *Une methode d'eclatement des operateurs et des constraintes en calcul des variations*, Comptes Rendus de l'Academie des Sciences, Serie A, Vol. 263, 1966.

298. TOI LIU-PAN, *A Minimax Problem for Distributed Parameter System*, International Journal of Control, Vol. 13, No. 5, 1971.

299. LUKES, D. L., AND RUSSELL, D. L., *The Quadratic Criterion for Distributed Systems*, SIAM Journal of Control, Vol. 7, No. 1, 1969.

300. LURIE, K. A., *The Mayer–Bolza Problem for Multiple Integrals: Some Optimum Problems for Elliptic Differential Equations Arising in Magnetohydrodynamics*, Topics in Optimization, Edited by G. Leitmann, Academic Press, New York, 1967.

301. LURIE, K. A., *Discontinuities in Hyperbolic Optimum Problems*, Journal of Optimization Theory and Applications, Vol. 7, No. 5, 1971.

302. LYNN, L. L., AND ZAHRADNIK, R. L., *The Use of Orthogonal Polynomials in the Near-Optimal Control of Distributed Systems by Trajectory Approximations*, International Journal of Control, Vol. 12, No. 6, 1970.

303. MAJERCZYK-GÓMULKOWA, J., AND MIODUCHOWSKI, A., *Optymalna niejednorodnosc plastyczna skrecanego preta*, Instytut Podstawowych Problemow Techniki Polskiej Akademii Nauk, Preprint, Warsaw, 1969.

304. MALANOWSKI, K., *O zastosowaniu pewnego algorytmu programowania wypuklego do problemow sterowania optymalnego w przestrzeni Hilberta*, Archiwum Automatyki i Telemechaniki, Vol. 15, No. 3, 1970.

305. MANGERON, D., AND OGUZTÖRELI, M. N., *Programmazione dinamica di una classe di equazoni polivibranti*, Rendiconti Accademia Lombardo, Ser. 8, Vol. 102, 1968.

306. MANGERON, D., AND OGUZTÖRELI, M. N., *Problemi al contorno spettanti ai sistemi di controllo con parametri distribuiti*, Rendiconti Accademia Nazionale Lincei, Ser. 8, Vol. 44, 1968.

307. MANGERON, D., AND OGUZTÖRELI, M. N., *Problemi di controllo con parametri distribuiti per sistemi polivibranti*, Rendiconti Accademia Nazionale Lincei, Ser. 8, Vol. 44, 1968.

308. MANGERON, D., AND OGUZTÖRELI, M. N., *Problemi di controllo con parametri distribuiti. Teoria generale ed applicazioni allo studio dei sistemi polivibranti*, Rendiconti Accademia Scienze Fisiche e Matematiche Napoli, Ser. 4, Vol. 35, 1968.

309. MATSUMOTO, J., AND ITO, K., *Feedback Control of Distributed Parameter Systems with Spatially Concentrated Controls*, International Journal of Control, Vol. 12, No. 3, 1970.

310. MCCAMY, R. C., MIZEL, V. J., AND SEIDMANN, T. I., *Approximate Boundary Controllability for the Heat Equation, I.* Journal of Mathematical Analysis and Applications, Vol. 28, No. 3, 1969.

311. MCCAUSLAND, I., *On Optimum Control of Temperature Distribution in a Solid*, Journal of Electronics and Control, Vol. 14, No. 6, 1963.

312. MCCAUSLAND, I., *Time Optimal Control of a Linear Diffusion System*, Proceedings of IEEE, Vol. 112, 1965.

313. MCNABB, A., *Partial Steiner Symmetrization and Some Conduction Problems*, Journal of Mathematical Analysis and Applications, Vol. 17, No. 2, 1967.

314. MCSHANE, E. J., *Optimal Controls, Relaxed and Ordinary, Mathematical Theory of Control*, Edited by A. V. Balakrishnan and L. W. Neustadt, Academic Press, New York, 1967.

315. MIRANKER, W. L., *Approximate Controllability for Distributed Linear Systems*, Journal of Mathematical Analysis and Applications, Vol. 28, No. 2, 1969.

316. MIZEL, V. J., AND SEIDMANN, T. I., *Observation and Prediction for the Heat Equation*, Journal of Mathematical Analysis and Applications, Vol. 28, No. 2, 1969.

317. MOCHI, G., *Su un problema di controllo ottimo connesso ad una equazione alle derivati paraziali di tipo iperbolico*, Bollettino d'Unione Matematica Italiana, Vol. 21, No. 1, 1966.

318. MOSZYNSKI, J. R., *End Losses in MHD Channels of Variable Cross-Section and Their Reduction*, Electricity from MHD, Vol. 2, Vienna, 1966.

319. MROZ, Z., *Optimal Design of Structures of Composite Materials*, International Journals of Solids and Structures, Vol. 6, No. 7, 1970.

320. MURAT, F., *Un contre-example pour le probleme du control dans coefficients*, Comptes Rendus de l'Academie des Sciences, Serie A, Vol. 273, No. 16, 1971.

321. MURAT, F., *Theoremes de non-existence pour des problemes de controle dans les coefficients*, Comptes Rendus de l'Academie des Sciences, Serie A, Vol. 274, No. 5, 1972.

322. MUSICKI, D., *Canonical Transformations and the Hamilton–Jacobi Method in the Field Theory*, Publications of Institute of Mathematics, Vol. 2, No. 16, 1962.

323. NAGAHISA, Y., AND SAKAWA, Y., *Nonlinear Programming in Banach Spaces*, Journal of Optimization Theory and Applications, Vol. 4, No. 3, 1969.

324. NEUSTADT, L. W., *An Abstract Variational Theory with Applications to a Broad Class of Optimization Problems*, SIAM Journal on Control, (I) Vol. 4, No. 3, 1966, (II) Vol. 5, No. 1, 1967.

325. NIORDSON, F. I., AND PEDERSEN, P., *A Review of Optimal Structural Design*, Report of the Danish Center for Applied Mathematics and Mechanics (DCAMM), No. 31, Sept. 1972.

326. OGUZTÖRELI, M. N., *Una classe di equazioni funzionali nella teoria dei controlli ottimi*, Atti Accademia Nazionale Lincei Rendiconti, Classe Scienze Fisiche Matematiche e Naturali, Ser. 8, Vol. 44, No. 5, 1968.

327. OGUZTÖRELI, M. N., *On Sufficient Conditions for the Existence of Optimal Policies in Distributed Parameter Control Systems*, Atti Accademia Nazionale Lincei Rendiconti, Classe Scienze Fisiche Matematiche e Naturali, Vol. 46, No. 6, 1969.

328. OGUZTÖRELI, M. N., *Construction of Optimal Controls for a Distributed Parameter Control System*, Atti Accademia Nazionale Lincei Rendiconti, Classe Scienze Fisiche Matematiche e Naturali, Vol. 47, 1969.

329. OGUZTÖRELI, M. N., *Optimization in Distributed Parameter Control Systems. A Dynamic Programming Approach*, Revue Roumaine des Sciences Techniques, Mechaniques, Appliquees, Vol. 14, Nos. 1-3, 1969.

330. OGUZTÖRELI, M. N., AND MANGERON, D., *Problemi ottimali spettanti ai sistemi di controllo con parametri distribuiti*, Atti Accademia Nazionale Lincei Rendiconti, Classe Scienze Fisiche Matematiche e Naturali, Ser. 8, Vol. 45, Nos. 1-2, 3-4, 5, 1968.

331. OLECH, C., *Existence Theorems for Optimal Control Problems Involving Multiple Integrals*, Journal of Differential Equations, Vol. 6, No. 3, 1969.

332. OLHOFF, N., *Optimal Design of Vibrating Rectangular Plates*, Report of the Danish Center for Applied Mathematics and Mechanics (DCAMM), No. 37, 1972.

333. PAYNE, L. E., *Isoperimetric Inequalities in the Torsion Problem for Multiply Connected Regions*, Studies on Mathematical Analysis and Related Topics, Stanford, California, 1962.

334. PAYNE, L. E., *Isoperimetric Inequalities and Their Applications* [review], SIAM Review, Vol. 9, No. 3, 1967.

335. PAYNE, L. E., AND WEINBERGER, H. F., *Some Isoperimetric Inequalities for Membrane Frequencies and Torsional Rigidity*, Journal of Mathematical Analysis and Applications, Vol. 2, 1961.

336. PIERRE, D. A., *Minimum Mean Square Error Design of Distributed Parameter Control System*, Proceedings of the 1965 Joint Automatic Control Conference, Troy, New York, 1965.

337. PORTER, W. A., *Extensions of the Minimum Effort Control Problem*, Journal of Mathematical Analysis and Applications, Vol. 13, 1966.

338. PRAGER, W., *Optimization of Structural Design* [review], Journal of Optimization Theory and Applications, Vol. 6, No. 1, 1970.

339. PRAGER, W., AND TAYLOR, J. E., *Problems of Optimal Structural Design*, Journal of Applied Mechanics, Series E, Vol. 35, No. 1, 1968.

340. PRANGE, G., *Die Hamilton–Jacobische Theorie für Doppelintegrale*, Dissertation, Göttingen, 1915.

341. PUCCI, C., *Un problema variazionale per i coefficienti di equazioni differentiali di tipo ellitico*, Annali Scuola Normale Superiore di Pisa, Vol. 3, No. 16, 1962.

342. PUCCI, C., *Su le equazioni ellitiche estremanti*, Rendiconti del Seminario Matematico e Fisico di Milano, Vol. 35, 1965.

343. PUCCI, C., *Operatori ellitichi estremanti*, Annali di Matematica Pura ed Applicata, Vol. 72, 1966.

344. PUCCI, C., *Maximum and Minimum First Eigenvalues for a Class of Elliptic Operators*, Proceedings of the American Mathematical Society, Vol. 17, 1966.

345. PULVIRENTI, G., *Su alcuni problemi di teoria dei controlli relativi ad una equazione alle derivati parziali di tipo iperbolico*, Matematiche, Vol. 23, No. 2, 1968.

346. PULVIRENTI, G., *Existence Theorems for an Optimal Control Problem Relative to a Linear Hyperbolic Partial Differential Equation*, Journal of Optimization Theory and Applications, Vol. 7, No. 2, 1971.

347. PULVIRENTI, G., AND SANTAGATI, G., *Controlli lineari nel problema di Dirichlet per una equazione di tipo iperbolico*, Matematiche, Vol. 23, No. 2, 1963.

348. PULVIRENTI, G., AND SANTAGATI, G., *Nuovi contributi alla teoria dei controlli per sistemi materiali retti la una equazione alle derivati parziali di tipo iperbolico*, Matematiche, Vol. 24, No. 2, 1969.

349. RIABOUSCHINSKY, D., *Sur une probléme de variation*, Comptes Rendus de l'Academie des Sciences, Vol. 185, 1927.

350. ROCKAFELLAR, R. T., *Conjugate Convex Functions in Optimal Control and the Calculus of Variations*, Journal of Mathematical Analysis and Applications, Vol. 32, No. 1, 1970.

351. ROGAK, E. D., KAZARINOFF, N. D., AND SCOTT-THOMAS, J. F., *Sufficient Conditions for Bang-Bang Control in Hilbert Space*, Journal of Optimization Theory and Applications, Vol. 5, No. 1, 1970.

352. ROGAK, E. D., AND KAZARINOFF, N. D., *Remarks on Bang-Bang Control in Hilbert Space*, Journal of Optimization Theory and Applications, Vol. 10, No. 4, 1972.

353. ROLEWICZ, S., *On Controllability of Systems of Strings*, Studia Mathematica, Vol. 36, No. 2, 1970.

354. RUSSELL, D. L., *Optimal Regulation of Linear Symmetric Hyperbolic Systems with Finite Dimensional Controls*, SIAM Journal on Control, Vol. 4, No. 2, 1966.

355. RUSSELL, D. L., *On Boundary Value Controllability of Linear Symmetric Hyperbolic Systems*, Mathematical Theory of Control, Edited by A. V. Balakrishnan and L. W. Neustadt, Academic Press, New York, 1967.

356. RUSSELL, D. L., *Linear Stabilization of the Linear Oscillator in Hilbert Space*, Journal of Mathematical Analysis and Applications, Vol. 25, No. 3, 1969.

357. RUSSELL, D. L., *Boundary Value Control Theory of the Higher Dimensional Wave Equation*, SIAM Journal on Control, (I) Vol. 9, No. 1, 1971; (II) Vol. 9, No. 3, 1971.

358. SAKAWA, Y., *Solution of an Optimal Control Problem in a Distributed Parameter System*, IEEE Transactions on Automatic Control, Vol. 9, No. 4, 1964.

359. SANTAGATI, G., *Alcuni problemi di controllo ottimo relativi ad una equazione alle derivate parziali di tipo iperbolico*, Matematiche, Vol. 23, No. 2, 1968.

360. SANTAGATI, G., *On an Optimal Control Problem for a Hyperbolic Partial Differential Equation*, Journal of Optimization Theory and Applications, Vol. 7, No. 1, 1971.

361. SASAI, H., *A Note on the Penalty Method for Distributed Parameter Control Problems*, SIAM Journal on Control, Vol. 10, No. 4, 1972.

362. SCHMAEDEKE, W., *Mathematical Theory of Optimal Control for Semi-Linear Hyperbolic System in Two Independent Variables*, SIAM Journal on Control, Vol. 5, No. 1, 1967.

363. SEIERSTAD, A., *A Pontryagin Maximum Principle in Banach Space*, IEEE Transactions on Automatic Control, Vol. 13, No. 3, 1968.

364. SEIERSTAD, A., *A Local Attainability Property for Control Systems Defined by Nonlinear Differential Equations in a Banach Space*, Journal of Differential Equations, Vol. 8, No. 3, 1970.

365. SHEN, C. N., AND LIU, T. C., *Distributed Parameter Type of Control for a Bilinear System with References to Control of a Nuclear Rocket*, Proceedings of the Third Congress of International Federation of Automatic Control, London, June 1966.

366. STODDART, A. W. J., *Estimation of Optimality for Multi-Dimensional Control Systems*, Journal of Optimization Theory and Applications, Vol. 3, No. 6, 1969.

367. SUTTON, G. W., *Design Considerations of a Magnetohydrodynamic Electrical Power Generator*, Vistas in Astronautics, New York, Vol. 3, 1960.

368. SUTTON, G. W., AND CARLSON, A. W., *End Effects in Inviscid Flow in MHD Channel*, Journal of Fluid Mechanics, Vol. II, p. I, 1961.

369. SZEFER, G., *Deformable Material Continuum as a Control System with Spatially Distributed Parameters*, Archivum Mechaniki Stosowanej, Vol. 26, No. 6, 1971.

370. TSUJIOKA, K., *Remarks on Controllability of Second Order Evolution Equations in Hilbert Space*, SIAM Journal on Control, Vol. 8, No. 6, 1971.

371. TZAFESTAS, S. G., *A Minimum Principle in Hilbert Space*, International Journal of Control, Vol. 11, No. 6, 1970.

372. TZAFESTAS, S. G., *Optimal Distributed Parameter Control in Function Space*, Journal of the Franklin Institute, Vol. 295, No. 4, 1973.

373. TZAFESTAS, S. G., AND NIGHTINGALE, J. M., *Differential Dynamic Programming Approach to Optimal Nonlinear Distributed Parameter Control Systems*, Proceedings of IEEE, Vol. 116, No. 6, 1969.

374. UZGIRIS, S. C., AND D'SOUZA, A. F., *Optimal Control of Distributed Parameter Systems with Nonlinear Boundary Conditions*, Proceedings of 1966 Joint Automatic Control Conference, Seattle, Washington, 1966.

375. VARAYA, P. P., *Nonlinear Programming in a Banach Space*, SIAM Journal on Applied Mathematics, Vol. 15, No. 2, 1967.

376. VIDYASAGAR, M., *On the Controllability of Infinite-Dimensional Linear Systems*, Journal of Optimization Theory and Applications, Vol. 6, No. 2, 1970.

377. VOLTERRA, V., *Lecons sur les fonctions des lignes*, Collection Borel, Gauthier-Villars, Paris, 1913.

378. VOLTERRA, V., *Sopra una extensione della teoria Jacobi–Hamilton del calcolo delle variazioni*, Accademia Lincei Rendiconti, 1890.

379. WANG, P. K. C., *Control of Distributed Parameter Systems*, Advances in Control Systems, Edited by C. T. Leondes, Vol. 1, Academic Press, New York, 1964.

380. WANG, P. K. C., *On the Feedback Control of Distributed Parameter Systems*, International Journal of Control, Vol. 3, No. 3, 1966.

381. WANG, P. K. C., *Control of a Distributed Parameter System with a Free Boundary*, International Journal of Control, Vol. 5, No. 4, 1967.

382. WANG, P. K. C., *Theory of Stability and Control for Distributed Parameter Systems* [a bibliography], International Journal of Control, Vol. 7, No. 2, 1968.

383. WANG, P. K. C., *Optimal Control of a Class of Linear Symmetric Hyperbolic Systems with Application to Plasma Confinement*, Journal of Mathematical Analysis and Applications, Vol. 28, No. 9, 1969.

384. WANG, P. K. C., AND TUNG, F., *Optimum Control of Distributed Parameter Systems*, Transactions of ASME, D86, No. 1, 1964.

385. WARGA, J., *Relaxed Controls for Functional Equations*, Journal of Functional Analysis, Vol. 5, No. 1, 1970.

386. WEINBERGER, H. F., *An Effectless Cutting of a Vibrating Membrane*, Pacific Journal of Mathematics, Vol. 13, No. 4, 1963.

387. YAVIN, Y., *Conditions upon Optimal Control in Class of Nonlinear Distributed Parameter Systems*, IEEE Transactions on Automatic Control, Vol. 14, No. 1, 1969.

388. YAVIN, Y., *Optimal Control of the Transverse Vibration of a Beam with a Bound on the Potential Energy*, Journal of Optimization Theory and Applications, Vol. 5, No. 5, 1970.

389. YAVIN, Y., *On the Optimal Control of Two Classes of Nonlinear Distributed Parameter Systems*, International Journal of Control, Vol. 16, No. 6, 1972.

390. YAVIN, Y., AND RASIS, Y., *The Bounded Energy Optimal Control for a Class of Heat Conduction Systems*, International Journal of Control, Vol. 11, No. 1, 1970.

391. YAVIN, Y., AND SIVAN, R., *The Optimal Control of a Distributed Parameter System*, IEEE Transactions on Automatic Control, Vol. 12, No. 6, 1967.

392. YAVIN, Y., AND SIVAN, R., *Properties of Optimal Control Functions for a Class of Distributed Parameter Systems*, IEEE Transactions on Automatic Control, Vol. 13, No. 1, 1968.

393. YAVIN, Y., AND SIVAN, R., *The Bounded Energy Optimal Control for a Class of Distributed Parameter Systems*, International Journal of Control, Vol. 8, No. 5, 1968.

394. YEH, H.-H., AND TOU, J. T., *Optimum Control of a Class of Distributed Parameter Systems*, IEEE Transactions on Automatic Control, Vol. 12, No. 1, 1967.

395. ZOLEZZI, T., *Teoremi d'esistenza per problemi di controllo ottimo retti dei equazioni ellitiche o paraboliche*, Rendiconti del Seminario Matematico Universita Padova, Vol. 44, 1970.

396. ZOLEZZI, T., *Necessary Conditions for Optimal Controls of Elliptic or Parabolic Problems*, SIAM Journal on Control, Vol. 10, No. 4, 1972.

397. ANNIN, B. D., *Optimal Design of Non-Homogeneous Anisotropic Elastic Bodies*, Proceedings of III National Congress on Theoretical and Applied Mechanics, Varna, Bulgaria, 1977.

398. ARISTOV, M. V., AND TROITSKII, V. A., *Minimal Weight Elastic Annular Plate* [in Russian], Izvestiya Akademii Nauk SSSR, Mekhanika Tverdogo Tela, No. 3, 1975.

399. ARMAND, J.-L., *Application of the Theory of Optimal Control of Distributed-Parameter Systems to Structural Optimization*, Ph.D. Thesis, Department of Aeronautics and Astronautics, Stanford University, 1971.

400. ARMAND, J.-L., *Numerical Solutions in Optimization of Structural Elements*, Paper presented at the 1st International Conference on Computational Methods in Nonlinear Mechanics, Austin, Texas, 1974.

401. ARMAND, J.-L., LURIE, K. A., AND CHERKAEV, A. V., *On Solution of Optimization Problems for Eigenvalues Arising in Structural Design* [In Russian], Izvestiya Akademii Nauk SSSR, Mekhanika Tverdogo Tela, No. 5, 1978.

402. ARMAND, J.-L., AND LODIER, B., *Optimal Design of Bending Elements*, International Journal of Numerical Methods in Engineering, Vol. 13, pp. 373–384, 1978.

403. BABUSKA, I., *Homogenization Approach in Engineering*, Colloquium IRIA, Dec. 1975.

404. BALL, J. M., *Convexity Conditions and Existence Theorems in Nonlinear Elasticity*, Archive for Rational Mechanics and Analysis, Vol. 63, 1977.

405. BAKHVALOV, N. S., *Averaged Characteristics of Bodies with Periodic Structure* [in Russian], Doklady Akademii Nauk SSSR, Vol. 218, No. 5, 1974.

406. BAKHVALOV, N. S., *Averaging of Partial Differential Equations with Rapidly Oscillating Coefficients* [in Russian], Doklady Akademii Nauk SSSR, Vol. 221, No. 3, 1975.

407. BAKHVALOV, N. S., *Averaging of Nonlinear Partial Differential Equations with Rapidly Oscillating Coefficients* [in Russian], Doklady Akademii Nauk SSSR, Vol. 225, No. 2, 1975.

408. BANICHUK, N. V., KARTVELISHVILI, V. M., AND MIRONOV, A. A., *Numerical Solution of Two-Dimensional Optimization Problems for Elastic Plates* [in Russian], Izvestiya Akademii Nauk SSSR, Mekhanika Tverdogo Tela, No. 1, 1977.

409. BANICHUK, N. V., KARTVELISHVILI, V. M., AND MIRONOV, A. A., *Optimization Problems with Local Performance Criteria in the Theory of Plate Bending* [in Russian], Izvestiya Akademii Nauk SSSR, Mekhanika Tverdogo Tela, No. 1, 1978.

410. BANICHUK, N. V., AND MIRONOV, A. A., *Optimization Problems for Plates Oscillating in an Ideal Fluid* [in Russian], Prikladnaya Matematika i Mekhanika, Vol. 40, No. 3, 1976.

411. BANICHUK, N. V., AND MIRONOV, A. A., *Optimization of Vibration Frequencies of an Elastic Plate in an Ideal Fluid* [in Russian], Izvestiya Akademii Nauk SSSR, Mekhanika Tverdogo Tela, No. 5, 1975.

412. BANICHUK, N. V., *On Certain Variational Problem with Unknown Boundary and Determination of Optimal Shape of Elastic Bodies* [in Russian], Prikladnaya Matematika i Mekhanika, Vol. 39, No. 6, 1975.

413. BANICHUK, N. V., AND MIRONOV, A. A., *Optimization of Vibration Frequencies of an Elastic Plate in an Ideal Fluid* [in Russian], Prikladnaya Matematika i Mekhanika, Vol. 39, No. 5, 1975.

414. BANICHUK, N. V., *Minimax Approach to Structural Optimization Problems*, Journal of Optimization Theory and Applications, Vol. 20, No. 1, 1976.

415. BANICHUK, N. V., *Optimizing of Hole Shape in Plates Working in Bending* [in Russian], Izvestiya Akademii Nauk SSSR, Mekhanika Tverdogo Tela, No. 3, 1977.

416. BANICHUK, N. V., *On Certain Extremal Problem for a Distributed-Parameter System and Determination of Optimal Material Properties of Elastic Medium* [in Russian], Doklady Akademii Nauk SSSR, Vol. 242, No. 5, 1978.

417. BANICHUK, N. V., *On the Optimal Anisotropy of Elastic Bars in Torsion* [in Russian], Izvestiya Akademii Nauk SSSR, Mekhanika Tverdogo Tela, No. 4, 1978.

418. BANICHUK, N. V., *Optimization of Characteristics of Anisotropy of Deformable Media in Plane Problems of the Theory of Elasticity* [in Russian], Izvestiya Akademii Nauk SSSR, Mekhanika Tverdogo Tela, No. 1, 1979.

419. BANICHUK, N. V., *Shape Optimization of Elastic Bodies* [in Russian], Nauka, Moscow, 1980.

420. BERT, C. W., *Optimal Design of a Composite Material Plate to Maximize Its Fundamental Frequency*, Journal of Sound and Vibration, Vol. 30, No. 2, 1977.

421. BENSOUSSAN, A., LIONS, J.-L., AND PAPANICOLAOU, G., *Asymptotic Analysis for Periodic Structures*, North-Holland, Amsterdam, 1978.

422. BERDICHEVSKII, A. L., *On Effective Heat Conductivity of Media with Periodically Disposed Inclusions* [in Russian], Doklady Akademii Nauk SSSR, Vol. 247, No. 6, 1979.

423. BERDICHEVSKII, A. L., *On Averaged Description of a Fluid Containing Gas Bubbles* [in Russian], Izvestiya Akademii Nauk SSSR, Mekhanika Zhidkosti i Gaza, No. 6, 1980.

424. BERDICHEVSKII, A. L., AND BERDICHEVSKII, V. L., *Demonstration of the Hypothesis of Polya* [in Russian], Doklady Akademii Nauk SSSR, Vol. 224, No. 2, 1975.

425. BERDICHEVSKII, A. L., AND BERDICHEVSKII, V. L., *Flow of Ideal Fluid Around a Periodic System of Bodies* [in Russian], Izvestiya Akademii Nauk SSSR, Mekhanika Zhidkosti i Gaza, No. 6, 1978.

426. BERDICHEVSKII, V. L., *Space Averaging of Periodic Structures* [in Russian], Doklady Akademii Nauk SSSR, Vol. 222, No. 3, 1975.

427. BERDICHEVSKII, V. L., *On Averaging of Periodic Structures* [in Russian], Prikladnaya Matematika i Mekhanika, Vol. 41, No. 6, 1977.

428. BERKE, L., AND VENKAYYA, V. B., *Review of Optimality Criteria Approaches to Structural Optimization*, Structural Optimization Symposium, AMD, Vol. 7, Edited by L. A. Schmit, Jr., ASME, New York, 1974.

429. BRACH, R. M., *On Optimal Design of Vibrating Structures*, Journal of Optimization Theory and Applications, Vol. 11, No. 6, 1973.

430. BRATUS', A. S., AND KARTVELISHVILI, V. M., *Perturbation Method with Respect to Optimization Problems Concerning Instability, Vibration Frequencies, Bending and Strength of Elastic Plates With Variable Thickness* [in Russian], Preprint No. 180, Institute of Problems in Mechanics, Academy of Sciences of the USSR, Moscow, 1981.

431. CEA, J., *Identification de domaines*, Lecture Notes in Computer Science, Volume 3, Springer-Verlag, Berlin, 1973.

432. CHENAIS, D., *On the Existence of a Solution in a Domain Identification Problem*, Journal of Mathematical Analysis and Applications, Vol. 52, No. 2, 1975.

433. CHENG, K.-T., *On Some New Optimal Design Formulations for Plates*, DCAMM Report No. 189, The Danish Center for Applied Mathematics and Mechanics, 1980.

434. CHENG, K.-T., *On Non-Smoothness in Optimal Design of Solid Elastic Plates*, International Journal of Solids and Structures, Vol. 17, pp. 795–810, 1981.

435. CHENG, K.-T., *Optimal Design of Solid Elastic Plates*, Ph.D. Thesis, Department of Solid Mechanics, Technical University of Denmark, 1980.

436. CHENG, K.-T., AND OLHOFF, N., *An Investigation Concerning Optimal Design of Solid Elastic Plates*, International Journal of Solids and Structures, Vol. 17, pp. 305–323, 1981.

437. CHENG, K.-T., AND OLHOFF, N., *Regularized Formulation for Optimal Design of Axisymmetric Plates*, DCAMM Report No. 203, The Danish Center for Applied Mathematics and Mechanics, 1981.

438. CHERKAEV, A. V., *On the Statement of Optimal Design Problems for Freely Oscillating Bodies* [in Russian], Prikladnaya Matematika i Mekhanika, Vol. 42, No. 1, 1978.

439. DERVIEUX, A., AND PALLMERIO, B., *Une formule de Hadamard dans les problemes d'optimal design*, Lecture Notes in Computer Science, Vol. 40, Springer-Verlag, Berlin, 1976.

440. DIDENKO, N. I., *Optimal Distribution of Bending Rigidity of Simply Supported Elastic Plate* [in Russian], Izvestiya Akademii Nauk SSSR, Mekhanika Tverdogo Tela, No. 1, 1981.

441. DUVAUT, G., *Analyse fonctionelle et mécanique des milieux continus. Application a l'etude des materiaux composites elastiques a structure periodique-homogenization*, Proceedings of the 14th IUTAM Congress, Delft, 1976; North-Holland, Amsterdam, 1977.

442. DYKHNE, A. M., *Conductivity of a Two-Dimensional System* [in Russian], Zhurnal Eksperimental'noi i Teoreticheskoi Fiziki, Vol. 59, No. 1 (7), 1970.

443. FEDOROV, A. V., *Legendre Condition in Optimal Problems of Supersonic Gasdynamics* [in Russian], Prikladnaya Matematika i Mekhanika, Vol. 39, No. 6, 1975.

444. FEDOROV, A. V., AND CHERKAEV, A. V., *Search of an Optimal Orientation of the Axes of Elastic Symmetry in an Orthotropic Plate* [in Russian], Izvestiya Akademii Nauk SSSR, Mekhanika Tverdogo Tela, No. 3, 1983.

445. FELTON, L. P., *Structural Index Methods in Optimal Design*, Structural Optimization Symposium, AMD, Vol. 7, edited by L. A. Schmit, Jr., ASME, New York, 1974.

446. FENG, T. T., HAUG, E. J., AND ARORA, J. S., *Optimization of Distributed Parameter Structures under Dynamic Loads*, Intertechnics Corporation Report, NTIS ADA 025368, June 1976.

447. FILIPPOV, A. F., *On Certain Questions in the Theory of Optimal Control* [in Russian], Vestnik Moskovskogo Universiteta, Matematika i Astronomia, No. 2, 1959.

448. DE GIORGI, E., AND SPAGNOLO, S., *Sulla convergenza degli integrali della energia per operatori ellittici del secondo ordine*, Bollettino Unione Mathematica Italiana, Vol. 8, pp. 391-411, 1973.

449. GRINEV, V. B., AND FILIPPOV, A. P., *Optimal Circular Plates* [in Russian], Izvestiya Akademii Nauk SSSR, Mekhanika Tverdogo Tela, No. 1, 1977.

450. GUDERLEY, K. G., TABAK, D., BREITER, M. S., AND BHUTANI, O. P., *Continuous and Discontinuous Solutions for Optimum Thrust Nozzles of Given Length*, Journal of Optimization Theory and Applications, Vol. 12, No. 6, 1973.

451. GURVICH, E. L., *On Isoperimetric Problems for Domains with Partly Known Boundaries*, Journal of Optimization Theory and Applications, Vol. 20, No. 1, 1976.

452. HAFTKA, R. T., AND PRASAD, B., *Optimum Structural Design with Plate Bending Elements—A Survey*, AIAA Journal, Vol. 19, No. 4, April 1981.

453. HAUG, E. J., PAN, K. C., AND STREETER, T. D., *A Computational Method for Optimal Structural Design, I: Piecewise Uniform Structures*, International Journal for Numerical Methods in Engineering, Vol. 5, pp. 171-184, 1972.

454. HAUG, E. J., PAN, K. C., AND STREETER, T. D., *A Computational Method for Optimal Structural Design, II: Continuous Problems*, International Journal for Numerical Methods in Engineering, Vol. 9, pp. 649-667, 1975.

455. HAUG, E. J., JR., ARORA, J. S., AND MATSUI, K., *A Steepest Descent Method for Optimization of Mechanical Systems*, Journal of Optimization Theory and Applications, Vol. 119, No. 3, 1976.

456. HAUG, E. J., AND ARORA, J. S., *Applied Optimal Design*, Wiley, New York, 1979.

457. HAUG, E. J., JR., AND FENG, T. T., *Optimal Design of Dynamically Loaded Continuous Structures*, International Journal for Numerical Methods in Engineering, Vol. 12, pp. 299-317, 1978.

458. HAUG, E. J., JR., AND KOMKOV, V., *Sensitivity Analysis in Distributed-Parameter Mechanical System Optimization*, Journal of Optimization Theory and Applications, Vol. 23, No. 3, 1977.

459. HILL, R., *Elastic Properties of Reinforced Solids; Some Theoretical Principles*, Journal of the Mechanics and Physics of Solids, Vol. 11, No. 5, 1963.

460. HILL, R., *Theory of Mechanical Properties of Fibre-Strengthened Materials. I. Elastic Behaviour*, Journal of the Mechanics and Physics of Solids, Vol. 12, No. 4, 1964.

461. HILL, R., *Theory of Mechanical Properties of Fibre-Strengthened Materials. II. Inelastic Behaviour*, Journal of the Mechanics and Physics of Solids, Vol. 12, No. 4, 1964.

462. KLOSOWICZ, B., AND LURIE, K. A., *On the Optimal Distribution of Elastic Moduli of a Non-Homogeneous Body*, Journal of Optimization Theory and Applications, Vol. 12, No. 1, 1973.

463. KOHN, R., AND STRANG, G., *Optimal Design for Torsional Rigidity*, Paper presented at the Conference on Mixed and Hybrid Finite Element Methods, Atlanta, 1981.

464. KOMKOV, V., *Formulation of Pontryagin's Principle in a Problem of Structural Mechanics*, International Journal of Control, Vol. 17, No. 3, 1973.

465. KOMKOV, V., *Control Theory, Variational Principles and Optimal Design of Elastic Systems*, in Proceedings of International Conference on Control, Norman, Oklahoma, March 1977, Wiley, New York, 1978.

466. KOZLOV, S. M., *Averaging of Random Operators* [in Russian], Mathematicheskii Sbornik, Vol. 109 (151), No. 2 (6), 1979.

467. KURSHIN, L. M., *On the Problem of Determination of a Shape of a Bar of Maximal Torsional Rigidity* [in Russian], Doklady Akademii Nauk SSSR, Vol. 223, No. 3, 1975.

468. LAVROV, N. A., LURIE, K. A., AND CHERKAEV, A. V., *Non-Homogeneous Bar of Extremal Rigidity in Torsion* [in Russian], Izvestiya Akademii Nauk SSSR, Mekhanika Tverdogo Tela, No. 6, 1980.

469. LITVINOV, V. G., *Optimal Control of Coefficients in Elliptic Systems* [in Russian], Preprint No. 79, 4 of the Institute of Mathematics, Academy of Sciences of the Ukrainian SSR, Kiev, 1979.

470. LITVINOV, V. G., *The Optimization Problem for the Eigenfrequency of a Plate of Variable Thickness* [in Russian], Zhurnal Vychislitel'noi Matematiki i Matematicheskoi Fiziki, Vol. 19, No. 4, 1979.

471. LURIE, K. A., *Some Problems of Optimal Bending and Tension of Elastic Plates* [in Russian], Izvestiya Akademii Nauk SSSR, Mekhanika Tverdogo Tela, No. 6, 1979.

472. LURIE, K. A., *Optimal Design of Elastic Bodies and the Problem of Regularization*, Technische Hochschule Leipzig, Wissenschaftliche Zeitschrift, Jahrgang 4/1980, No. 6, pp. 339–347.

473. LURIE, K. A., AND CHERKAEV, A. V., *Non-Homogeneous Bar of Extremal Torsional Rigidity* [in Russian], Nonlinear Problems in Structural Mechanics, Structural Optimization, Kiev, 1978.

474. LURIE, K. A., AND CHERKAEV, A. V., *Prager Theorem Application to Optimal Design of Thin Plates* [in Russian], Izvestiya Akademii Nauk SSSR, Mekhanika Tverdogo Tela, No. 6, 1976.

475. LURIE, K. A., CHERKAEV, A. V., AND FEDOROV, A. V., *Regularization of Optimal Design Problems for Bars and Plates and Elimination of Contradictions Within the Necessary Conditions of Optimality, I, II*, Journal of Optimization Theory and Applications, Vol. 37, No. 4, 1982.

476. LURIE, K. A., CHERKAEV, A. V., AND FEDOROV, A. V., *On the Existence of Solutions of Certain Optimal Design Problems for Bars and Plates*, Journal of Optimization Theory and Applications, Vol. 42, No. 2, 1984.

477. LURIE, K. A., AND CHERKAEV, A. V., *G-Closure of a Set of Anisotropically Conducting Media in the Two-Dimensional Case*, Journal of Optimization Theory and Applications, Vol. 42, No. 2, 1984; corrig. Vol. 53, p. 319, 1987.

478. LURIE, K. A., AND CHERKAEV, A. V., *G-Closure of Some Particular Sets of Admissible Material Characteristics for the Problem of Bending of Thin Plates*, Journal of Optimization Theory and Applications, Vol. 42, No. 2, 1984; corrig. Vol. 53, p. 319, 1987.

479. LURIE, K. A., AND CHERKAEV, A. V., *G-Closure of a Set of Anisotropic Conductors in the Two-Dimensional Case* [in Russian], Doklady Akademii Nauk SSSR, Vol. 259, No. 2, 1981.

480. LURIE, K. A., AND FEDOROV, A. V., *Weierstrass Condition in Optimal Problems of Supersonic Gas Dynamics of Weakly Inhomogeneous Flows* [in Russian], Izvestiya Akademii Nauk Armyanskoi SSR, Mekhanika, Vol. 29, No. 6, 1976, addendum in Vol. 30, No. 6, 1977.

481. MARINO, A., AND SPAGNOLO, S., *Un tipo di approssimazione dell operatore* $\sum_i \sum_j \partial/\partial x_i a_{ij}(x) \partial/\partial x_j$ *con operatori* $\sum_i \partial/\partial x_i a(x) \partial/\partial x_i$, Annali Scuola Normale Superiore di Pisa, Vol. 23, pp. 657–673, 1969.

482. MCCONNELL, W. H., *On the Approximation of Elliptic Operators with Discontinuous Coefficients*, Annali Scuola Normale Superiore di Pisa, Serie IV, Vol. 3, pp. 121–137, 1976.

483. MEYERS, N. G., *An L_p-estimate for the Gradient of Solutions of Second Order Elliptic Divergence Equations,* Annali Scuola Normale Superiore di Pisa, Vol. 17, No. 3, pp. 191–206, 1963.

484. MURAT, F., *H-convergence.* Seminaire d'analyse fonctionelle et numerique de l'Université d'Alger, ronéotype, 34 pp., 1977–78.

485. MURAT, F., *Compacité par compensation,* Annali Scuola Normale Superiore di Pisa, Scienze Fisiche Matematiche (IV), 5, 1978.

486. MURAT, F., AND SIMON, J., *Quelques résultats sur le controle par un domain geométrique,* Rapport No. 74003, Laboratoire d'Analyse Numérique, Université Paris VI, 1974.

487. MURAT, F., AND SIMON, J., *Etude de problèmes d'optimal design,* Lecture Notes in Computer Science, Vol. 40, Springer-Verlag, Berlin, 1976.

488. NIORDSON, F., AND OLHOFF, N., *Variational Methods in Optimization of Structures,* DCAMM Report No. 161, The Danish Center for Applied Mathematics and Mechanics, 1979.

489. OLHOFF, N., *Optimal Design of Vibrating Circular Plates,* International Journal of Solids and Structures, Vol. 6, No. 1, 1970.

490. OLHOFF, N., *On Singularities, Local Optima and Formation of Stiffeners in Optimal Design of Plates,* IUTAM Symposium on Optimization in Structural Design, Edited by A. Sawczuk, Z. Mroz, Warsaw, 1973, Springer-Verlag, Berlin, 1975.

491. OHOFF, N., *Optimal Design with Respect to Structural Eigenvalues,* Paper presented at the XVth International Congress on Theoretical and Applied Mechanics, Toronto, 1980.

492. OLHOFF, N., LURIE, K. A., CHERKAEV, A. V., AND FEDOROV, A. V., *Sliding Regimes and Anisotropy in Optimal Design of Vibrating Axisymmetric Plates,* International Journal of Solids and Structures, Vol. 17, No. 10, 1981.

493. PIRONNEAU, O., *Optimization de structure. Application a la mecanique des fluids,* Lecture Notes in Economics and Mathematical Systems, Vol. 107, Springer-Verlag, Berlin, 1974.

494. PIRONNEAU, O., *On Optimum Profiles in Stokes Flow,* Journal of Fluid Mechanics, Vol. 59, pp. 117–128, 1973.

495. PIRONNEAU, O., *On Optimum Design in Fluid Mechanics,* Journal of Fluid Mechanics, Vol. 64, pp. 97–111, 1974.

496. PRAGER, W., *Optimality Criteria in Structural Design,* Proceedings of the National Academy of Sciences USA, Vol. 61, No. 3, 1968.

497. PYATIGORSKII, R. I., AND SEYRANIAN, A. P., *On a Dynamical Problem of Optimal Design* [in Russian], Uchenye Zapiski TsAGI, Vol. 2, No. 6, 1971.

498. RAITUM, U. E., *Optimal Control Problems for Elliptic Equations* [in Russian], Zinatne, Riga, 1989.

499. RAITUM, U. E., *On Optimal Control Problems for Linear Elliptic Equations* [in Russian], Doklady Akademii Nauk SSSR, Vol. 244, No. 4, 1979.

500. RAO, S. S., *Structural Optimization Under Shock and Vibration Environment,* Shock and Vibration Digest, Vol. 11, Feb. 1979.

501. REITMAN, M. I., AND SHAPIRO, G. S., *Optimal Design of Deformable Solid Bodies,* Mechanics of Deformable Solid Body [in Russian], Advances of Science and Technology, All-Union Institute of Scientific and Technological Information, Vol. 12, Moscow, 1978.

502. ROUSSELET, B., *Problems inverses de valeur propres,* Lecture Notes in Computer Science, Vol. 40, Springer-Verlag, Berlin, 1976.

503. ROZVANY, G., *Optimum Design of Flexural Systems,* Pergamon Press, Elmsford, New York, 1976.

504. ROZVANY, G. I. N., OLHOFF, N., CHENG, K.-T., AND TAYLOR, J., *On the Solid Plate Paradox in Structural Optimization,* DCAMM Report No. 212, The Danish Center for Applied Mathematics and Mechanics, 1981.

505. SAMSONOV, A. M., *Optimal Position of a Thin Elastic Rib on an Elastic Plate* [in Russian], Izvestiya Akademii Nauk SSSR, Mekhanika Tverdogo Tela, No. 1, 1978.

506. SAMSONOV, A. M., *On the Optimal Discrete Distribution of Rigidities of a Rib on a Circular Plate* [in Russian], Prikladnaya Mekhanika, Vol. XIV, No. 11, 1978.

507. SAMSONOV, A. M., *The Weierstrass Condition in a Dynamical Optimization Problem for a Ribbed Plate* [in Russian], Izvestiya Akademii Nauk SSSR, Mekhanika Tverdogo Tela, No. 3, 1979.

508. SAMSONOV, A. M., *Necessary Conditions of Optimality of Distribution of Rigidities of a Rib on an Elastic Plate* [in Russian], Izvestiya Akademii Nauk SSSR, Mekhanika Tverdogo Tela, No. 1, 1980.

509. SEYRANIAN, A. P., *On the Criterion of Optimality for Choosing the Thickness of Plates under Bending Loads* [in Russian], Izvestiya Akademii Nauk SSSR, Mekhanika Tverdogo Tela, No. 5, 1973.

510. SEYRANIAN, A. P., *Investigation of an Extremum in an Optimal Problem of Vibrations of a Circular Plate* [in Russian], Izvestiya Akademii Nauk SSSR, Mekhanika Tverdogo Tela, No. 6, 1978.

511. SEYRANIAN, A. P., *Homogeneous Functions and Structural Optimization Problems*, International Journal of Solids and Structures, Vol. 15, pp. 749-759, 1979.

512. SIMITSES, G. J., *Optimal vs the Stiffened Circular Plate*, AIAA Journal, Vol. 11, No. 10, 1973.

513. STRANG, G., AND KOHN, R., *Optimal Design of Cylinders in Shear*, Paper presented at MAFELAP Conference, Brunel University, 1981.

514. TARTAR, L., *Problemes de controle des coefficients dans des equations aux derivees partielles*, Lecture Notes in Economics and Mathematical Systems, Vol. 107, No. 420, pp. 420-426 Springer-Verlag, Berlin, 1975.

515. TARTAR, L., *Compensated Compactness and Applications to Partial Differential Equations*, Lectures held at the Heriot-Watt University, 1978.

516. TROESCH, B. A., *Upper Bounds for the Fundamental Eigenvalue for a Domain of Unknown Shape*, Journal of Optimization Theory and Applications, Vol. 12, No. 5, 1973.

517. TROESCH, B. A., *Elliptical Membranes with Smallest Second Eigenvalue*, Mathematics of Computation, Vol. 27, pp. 767-772, 1973.

518. TROESCH, B. A., AND TROESCH, H. R., *Eigenfrequencies of an Elliptic Membrane*, Mathematics of Computation, Vol. 27, pp. 755-765, 1973.

519. VENKAYYA, V. B., *Structural Optimization: A Review and Some Recommendations*, International Journal for Numerical Methods in Engineering, Vol. 13, pp. 203-228, 1978.

520. VIGDERGAUZ, S. B., *On a Certain Case of an Inverse Problem of Plane Theory of Elasticity* [in Russian], Prikladnaya Matematika i Mekhanika, Vol. 41, No. 5, 1977.

521. VIGDERGAUZ, S. B., *Optimality Conditions in Axisymmetric Problems of Elasticity Theory* [in Russian], Prikladnaya Matematika i Mekhanika, Vol. 46, No. 2, 1983.

522. VIGDERGAUZ, S. B., *Inverse Problem of Three-Dimensional Theory of Elasticity*, Izvestiya Akademii Nauk SSSR, Mekhanika Tverdogo Tela, No. 2, 1983.

523. WARGA, J., *Optimal Control of Differential and Functional Equations*, Academic Press, New York, 1972.

524. YOUNG, L. S., *Lectures on the Calculus of Variations and Optimal Control Theory*, Saunders, Philadelphia, 1969.

525. ZHIKOV, V. V., KOZLOV, S. M., OLEINIK, O. A., AND KHA THIENG NGOAN, *Homogenization and G-Convergence of Differential Operators* [in Russian], Uspekhi Matematicheskikh Nauk, Vol. 34, No. 5, 1979.

526. ZHIKOV, V. V., KOZLOV, S. M., AND OLEINIK, O. A., *On G-Convergence of Parabolic Operators* [in Russian], Uspekhi Matematicheskikh Nauk, Vol. 36, No. 1, 1981.

527. Zubov, V. I., *On the Optimal Supersonic Profile with Prescribed Characteristic of Thickness* [in Russian], Izvestiya Akademii Nauk SSSR, Mekhanika Zhidkosti i Gaza, No. 1, 1976.

528. Armand, J.-L., Lurie, K. A., and Cherkaev, A. V., *Optimal Control Theory and Structural Design, New Directions in Optimum Structural Design*, Edited by E. Atrek, R. H. Gallagher, K. M. Ragsdell, O. C. Zienkiewicz, Wiley, New York, 1984.

529. Bendsoe, M., *Optimization of Plates*, Mathematical Institute, The Technical University of Denmark, 1983.

530. Berdichevskii, V. L., *Variational Principles of Continuum Mechanics* [in Russian], Nauka, Moscow, 1983.

531. Bratus', A. S., *Asymptotic Solutions in Problems of Optimal Control of Coefficients in Elliptic Systems* [in Russian], Doklady Akademii Nauk SSSR, Vol. 259, No. 5, 1981.

532. Gibianskii, L. V., and Cherkaev, A. V., *Design of Composite Plates of Extremal Rigidity* [in Russian], Preprint No. 914, A. F. Ioffe Physico-Technical Institute, Academy of Sciences of the USSR, Leningrad, 1984.

533. Haug, E. J., and Cea, J., Editors, *Optimization of Distributed Parameter Structures*, Vols. 1 and 2, Proceedings of NATO ASI Meeting, Iowa City, 1980, Noorhoff, 1981.

534. Kohn, R. V., and Strang, G., *Structural Design Optimization Homogenization and Relaxation of Variational Problems, Macroscopic Properties of Disordered Media*, Edited by R. Burridge, G. Papanicolaou, S. Childress, Lecture Notes in Physics, No. 154, Springer-Verlag, Berlin, 1982.

535. Kohn, R. V., and Strang, G., *Optimal Design and Relaxation of Variational Problems*, Communications on Pure and Applied Mathematics, Vol. 39, pp. 113, 139, 353, 1986.

536. Kohn, R. V., and Vogelius, M., *A New Model for Thin Plates with Rapidly Varying Thickness*, International Journal of Solids and Structures, Vol. 20, pp. 333–350, 1984.

537. Niordson, F., *Optimal Design of Elastic Plates with a Constraint on the Slope of the Thickness Function*, DCAMM Report No. 225, The Danish Center for Applied Mathematics and Mechanics, 1981.

538. Olhoff, N., and Taylor, J., *On Structural Optimization*, Journal of Applied Mechanics, Vol. 50, pp. 1139–1151, 1983.

539. Rychlewski, J., *On the Hooke's Law* [in Russian], Prikladnaya Matematika i Mekhanika, Vol. 48, No. 3, 1984.

540. Schulgasser, K., *Relationship Between Single-Crystal and Polycrystal Electrical Conductivity*, Journal of Applied Physics, Vol. 47, No. 5, 1976.

541. Troitskii, V. A., and Petukhov, L. V., *Shape Optimization of Elastic Bodies* [in Russian], Nauka, Moscow, 1982.

Index